Technisches Wörterbuch

für

Telegraphie und Post.

Deutsch-englisch und englisch-deutsch.

Von

F. Hennicke.

Berlin.

Verlag von Julius Springer.

1889.

ISBN-13: 978-3-642-98584-3   e-ISBN-13: 978-3-642-99399-2
DOI: 10.1007/978-3-642-99399-2

Softcover reprint of the hardcover 1st edition 1889

Vorrede.

Bei der Bearbeitung des vorliegenden Wörterbuches hat den Verfasser die Absicht geleitet, Denjenigen, welche genöthigt sind, aus dem Deutschen in's Englische zu übersetzen, sowie den Lesern englischer Veröffentlichungen über das Telegraphen= und Postwesen ein Hülfsbuch in die Hand zu geben, welches sämmtliche in jenen beiden Verkehrszweigen gebräuchlichen technischen Ausdrücke enthält. Das Buch erhebt keinen Anspruch darauf, andere englische Wörter= bücher überflüssig zu machen: es will dieselben in Bezug auf Post und Telegraphie, welche in den gewöhnlichen Wörterbüchern ganz vernachlässigt und selbst in den technischen Wörterbüchern recht stiefmütterlich bedacht sind, nur ergänzen.

Das Material ist den in englischer Sprache erschienenen Werken über Telegraphen=Technik und Telegraphen=Bau sowie den englischen (einschl. kolonialen) und amerikanischen Dienstanweisungen für Post und Telegraphie entnommen.

Besondere Sorgfalt ist in der deutsch=englischen Abtheilung darauf gerichtet worden, durch Einflechtung von Redensarten und selbst ganzen Sätzen gewisse Stichworte im sprachlichen Zusammen= hange erscheinen zu lassen.

Der Rechtschreibung ist des Amerikaners Webster „Complete Dictionnary of the English Language" zu Grunde gelegt. Das System, welches derselbe gegenüber der vielfach schwankenden englischen Rechtschreibung aufgestellt hat, scheint mir das natürlichste und konsequenteste zu sein.

Die im Englischen besonders zahlreichen Abkürzungen sind in einem besonderen Anhange am Schlusse des Werkes untergebracht.

Denjenigen, welche mich bei der nicht mühelosen Arbeit bereitwilligst mit Rath und That unterstützt haben, sage ich hiermit meinen verbindlichsten Dank, insbesondere den Herren Robert Franck und Chas. L. Kiewert aus Milwaukee.

Berlin, im Oktober 1888.

Der Verfasser.

Inhaltsverzeichniß.

	Seite
Erklärung der Zeichen	VII
Abkürzungen	VII
Münzen, Maaße und Gewichte in Großbritannien und den Vereinigten Staaten von Amerika	IX
Das metrische System und englisches Maaß	XI
Erster Theil: Deutsch-Englisch	1
Zweiter Theil: Englisch-Deutsch	175
Anhang enthaltend die im Englischen am häufigsten vorkommenden Abkürzungen	299

Erklärung der Zeichen.

= (Gleichheitszeichen) gleich.
~ (Wiederholungszeichen) vertritt das zu Anfang des Artikels stehende fettgedruckte Wort, bez. die fettgedruckte Form.
+ bedeutet, daß in den Ländern englischer Zunge der durch das deutsche Wort ausgedrückte Begriff fehlt, der englische Ausdruck daher nur eine mehr oder weniger freie Uebersetzung ist.

Abkürzungen.

a	Eigenschaftswort; adjective.
adv	Umstandswort; adverb.
Am	in den Vereinigten Staaten von Amerika gebräuchlich; Americanism.
Arithm.	Arithmetik; arithmetics.
Chem.	Chemie; chemistry.
Eisenb.	Eisenbahn; railroad.
Elektriz.	Elektrizität; electricity.
etw.	etwas; something.
F	Fernsprechwesen; telephony.
f	weibliches Hauptwort; feminine.
Geom.	Geometrie; geometry.
HA	Hughes-Apparat; Hughes' type printing instrument.
J—d	Jemand; somebody.
J—m	Jemandem; somebody, to somebody.
J—s	Jemandes; somebody's.
m	männliches Hauptwort; masculine.
MA	Morse-Apparat; Morse apparatus.
M & HA	Morse- und Hughes-Apparat; Morse & Hughes apparatus.
Masch.	Maschinenwesen; machinery.
Magn.	Magnetismus; magnetism.
Math.	Mathematik; mathematics.
Mech.	Mechanik; mechanics.
MT	Militär-Telegraphie; military telegraphy.

VIII   Abkürzungen. — Münzen, Maaße und Gewichte.

n	sächliches Hauptwort; neuter.
Opt.	Optik; optics.
P	postalischer Ausdruck; expression relating to the Post Office Service.
Phys.	Physik; physics.
pl	Mehrzahl; plural.
f.	siehe; vide, see.
f. a. T.	siehe auch den andern Theil; vide second part (of the dictionnary).
s	Hauptwort; substantive.
smb	somebody.
smt	something.
t.	telegraphisch; telegraph- . . .; telegraphic.
t.	telegraph- . . .; telegraphic.
T.	Telegraphen= . . .; Telegraphie= . . . u. s. w.
T	telegraph, telegrapher, telegraphy etc.
T	in der Telegraphie gebräuchlicher Ausdruck; expression relating to telegraph business.
Techn.	Technik, Gewerbekunde; technics, technology.
v	Zeitwort; verb.

Die übrigen Abkürzungen sind ohne Weiteres verständlich.

Münzen, Maaße und Gewichte in Großbritannien und den Vereinigten Staaten von Amerika.

I. Geld.

1. Großbritannien.

4 farthings, *qr* . . .	=	1 penny, *a*.
12 pence	=	1 shilling, *s*.
20 shillings	=	1 pound, £.
A sovereign	=	20 shillings.
A guinea	=	21 -
A crown	=	5 -
A groat	=	4 pence.

2. Vereinigte Staaten.

10 mills	=	1 cent, *c*.
10 cents	=	1 dime, *d*.
10 dimes	=	1 dollar, $
10 dollars	=	1 eagle, *e*.

II. Maaße.

Großbritannien und Vereinigte Staaten.

1. Längenmaaß; Long Measure.

3 barleycorns . . . . . =	1 inch
12 inches . . . . =	1 foot
3 feet . . . . =	1 yard
5½ yards . . . . =	1 rod, perch, or pole
40 rods or perches . . . . =	1 furlong
8 furlongs . . . . . . =	1 mile
6 feet . . . . =	1 fathom
4 inches . . . . =	1 hand
3 miles . . . . =	1 league
60 nautical or geographical miles . . . . =	1 degree
69½ statute miles . . . . =	1 degree
9 inches . . . . =	1 span
18 inches . . . . =	1 cubit.

2. Geviertmaaß; Square Measure.

144 inches . . . . . . =	1 sq. foot
9 sq. feet . . . . . . =	1 sq. yard
30¼ sq. yards, or	
272¼ sq. feet . . . . . . =	1 sq. rod, perch, or pole
40 sq. rods . . . . . . =	1 rood
4 roods, or	
160 sq. rods . . . . . . =	1 acre
640 acres . . . . =	1 sq. mile.

Kubisches Maaß; Cubic Measure.

1728 cu. inches . . . . . . =	1 cu. foot
27 cu. feet . . . . . . =	1 cu. yard
40 feet of round, or	
50 feet of hewn timber . . =	1 ton, or load
42 cu. feet . . . . . . =	1 ton of shipping
16 cu. feet . . . . . . =	1 foot of wood, or a cord foot
8 cord feet, or	
128 cubic feet . . . . . . =	1 cord.

4. Entfernungsmaaß; Measuring Distances.

7,92 inches . . . . =	1 link
25 links . . . . =	1 pole
100 links . . . . =	1 chain
10 chains . . . . =	1 furlong
8 furlongs . . . . =	1 mile.

5. Flüssigkeitsmaaß; Liquid Measure.

4 gills	=	1 pint
2 pints	=	1 quart
4 quarts	=	1 gallon
9 gallons	=	1 firkin
2 firkins	=	18 gal. = 1 kilderkin
2 kilderkins	=	36 gal. = 1 barrel
1½ barrel	=	54 gal. = 1 hogshead (Bier)
1⅓ hogshead	=	72 gal. = 1 puncheon (Bier)
1½ puncheon	=	108 gal. = 1 butt
42 gallons	=	1 tierce
1½ tierce, or 63 gal.	=	1 hogshead (Wein)
1⅓ hogshead, or 84 gal.	=	1 puncheon (Wein)
1½ puncheon, or 126 gal.	=	1 pipe
2 pipes	=	1 tun
231 cubic inches	=	1 gallon
10 gallons	=	1 anker
18 gallons	=	1 rundlet
31½ gallons	=	1 barrel

6. Hohlmaaß; Dry Measure.

2 pints	=	1 quart, *qt*.
4 quarts	=	1 gallon, *gal*.
2 gallons	=	1 peck, *pk*.
4 pecks	=	1 bushel, *bu*.
36 bushels	=	1 chaldron, *ch*.
4 bushels in England	=	1 coomb
2 coombs	=	1 quarter
5 quarters	=	1 wey
2 weys	=	1 last.

III. Gewicht; Weight.

Großbritannien und Vereinigte Staaten.

1. Handelsgewicht; Avoir du poids Weight.

16 drams	=	1 ounce, *oz*.
16 oz.	=	1 pound, *lb*.
28 lbs.	=	1 quarter, *qr*.
4 qrs.	=	1 hundred, *cwt*.
20 cwt.	=	1 ton
175 troy pounds	=	144 pounds avoir du poids

Das metrische System und englisches Maaß.

```
1 pound troy . . . . . . . .  = 5760 grains
1 lb. avoir . . . . . . . . . = 7000 grains
```

2. Goldgewicht; Troy Weight.

```
24 grains, gr. . . . . . . .  = 1 pennyweight, dwt.
20 dwt. . . . . . . . . . . . = 1 ounce, oz.
12 oz. . . . . . . . . . . .  = 1 pound, lb.
```

(Gold, Silber und Edelsteine werden mit diesem Gewicht gewogen.)

3. Apothekergewicht; Apothecaries' Weight.

```
20 grains . . . . . . . . . . = 1 scruple
3 scruples . . . . . . . . .  = 1 dram
8 drams . . . . . . . . . . . = 1 ounce
12 ounces . . . . . . . . . . = 1 pound.
```

Das metrische System und englisches Maaß.

```
1 Millimeter .  =      0,03937 in.
1 Zentimeter .  =      0,39371 -
1 Dezimeter  .  =      3,93710 -
1 Meter . .  =        39,37100 -
1 Dekameter .  =      32,80916 feet
1 Hektometer .  =    328,09167 -
1 Kilometer .  =    1093,63890 yds = 0,62137 mile, or 3280 ft. 10 in.
1 Myriameter  =   10936,38900 yds, or 6,2137 miles.
```

Die englische Meile (statute mile) = 1760 yards = 1609,315 Meter.
Die englische geographische Meile,
  Seemeile . . . . = 2028 yards = 1855 Meter.
Die deutsche geographische Meile . = 4 Seemeilen = 7420,438 Meter.
Die metrische Meile, deutsche Reichs=
  meile . . . . = 4,66 engl. Meilen = 7500 Meter.

```
1 Milligramm . . . . . . =    0,0154 grains
1 Zentigramm . . . . . . =    0,1543 -
1 Dezigramm  . . . . . . =    1,5434 -
1 Gramm  . . . . . . . . =   15,4340 -
1 Dekagramm od. 5,64 drams
   avoir du poids . . . . =  154,3420 -
1 Hektogramm . . . . . . =    3,2154 oz. troy
```

Das metrische System und englisches Maaß.

1 Kilogramm	=	2 lbs. 8 oz. 3 dwt. 2 grs. troy, ob. 2 lbs. 3 oz. 4,652 drams avoir du poids.
1 Myriagramm	=	26,795 lbs. troy, ob. 22,0485 lbs. avoir du poids.
1 Quintal	=	1 cwt. 3 qrs. 25 lbs.

An inch	=	2,54 Zentimeter
A foot	=	0,3048 Meter
A yard	=	0,9144 "
A rod	=	5,029 "
A mile	=	1,6093 Kilometer
A square inch	=	6,452 Quadratzentimeter
A square foot	=	0,0929 Quadratmeter
A square yard	=	0,8361 Quadratmeter
A square rod	=	25,29 "
An acre	=	0,4047 Hektar
A square mile	=	259 "
A cubic inch	=	16,39 Kubikzentimeter
A cubic foot	=	0,02832 Kubikmeter
A cubic yard	=	0,7646 "
A cord	=	3,624 Stere
A liquid quart	=	0,9465 Liter
A gallon	=	3,786 "
A dry quart	=	1,101 "
A peck	=	8,811 "
A bushel	=	35,24 "
An ounce *Avoir du poids* . . .	=	28,35 Gramm
A pound *Avoir du poids* . . .	=	0,4536 Kilogramm
A ton	=	0,9072 Tonnaar
A grain *Troy*	=	0,0648 Gramm
An ounce *Troy*	=	31,104 "
A pound *Troy*	=	0,3732 Kilogramm.

Erster Theil:

Deutsch-englisch.

A.

abändern v to alter (the tract of a t. line); to correct (figures); to change (occupations); to modify (rates); to amend (laws).

Abänderung f alteration, change, correction, modification, amendment; eine ~ beantragen to move an amendment; die ~ einer Adresse beantragen to ask for (to demand) the alteration (correction) of an address; ~s-Antrag application for alteration (of an address).

abästen v to prune, to clear a tree of its branches, to lop the branches of a tree.

abbauen v to pull down, to take down, to dismantle, to remove (a t. line).

abberufen v to recall.

Abberufung f recall, recalling.

abbestellen v to countermand, to revoke (the order for a newspaper).

Abbestellung f countermand, revocation (of an order), counterordering (of an extra mail coach).

abbinden v (Briefe zu einem Bunde vereinigen) to tie up letters, to bind together; (losbinden) to untie, to unbind, to detach.

abblättern v to scale, to exfoliate; der Zinküberzug darf nicht ~ the coating of zinc must not spring off.

abbrechen v (Techn.) to break off, to break up, to pull down; to dismantle (an old t. line); ein Telegramm während der Beförderung ~ to discontinue forwarding a telegram.

Abbruch m breaking off, breaking up, dismantling; auf den ~ verkaufen to sell the materials (of an old t. line); ~ thun to injure.

Abdruck, Abzug m impression, copy, print; (eines Siegels) impress; (Gepräge) stamp; (eines Buchs) copy print; (Stempelabdruck) impression; erhabener ~ embossed stamp; deutlicher Stempel-~ clear impression of the stamp.

abdrucken v to impress, to stamp, to print, to imprint; den Stempel ~ to impress the stamp; wieder ~ to reprint.

abfahren v (abreisen) to depart, to leave, to start, to set out (for a journey); (in See gehen) to set sail, to put to sea; (abkarren) to cart, to wheel, to carry away; die Erde ~ to cart the ground (earth).

Abfahrt f departure; (Ausfahrt v. Schiffen) port clearing; (Eisenb.) setting off, starting; ~punkt m point of departure; ~zeit f time of departure.

Abfall m (Neigung f) fall, falling off, declivity, slope; ~ am Gewicht short weight, deficiency; Abfälle pl refuse, parings; Papier-~ waste paper; Bindfaden-~ waste twine.

abfassen v to draw up (in writing); to write, to compose, to pen down, to word, to draft (instructions); to catch (a thief).

Abfassung f composing, drawing up (of a document), penning, wording, tenor; catching (of a thief).

abfaulen v to rot off.

abfeilen v to file, to file off.

abfertigen v to despatch the mail; (die Post mit Reisenden) to despatch the mail coach; (das Publikum am Schalter) to at-

1*

tend to the public at the counter (office window).

Abfertigung *f* despatch (of a mail); attending (to the public at the counter). **Abfertigungs** . . . ~gebühr (+) fee for despatching a horse express; ~frist *f* despatching time; time allowed (fixed) for despatching the mail (the special messenger, the horse express etc.); ~spind *n* despatching cupboard; ~tisch *m* despatching table; ~übersicht *f* mail list of arrivals and departures, schedule.

abfinden *v* sich, to settle with *smb*, to come to an agreement (to terms) with *smb*, to satisfy *smb*.

Abfindung *f* settlement, agreement; ~ssumme *f* indemnity, sum of acquittance.

abflachen, abebenen *v* to level, to make even.

abfließen, ablaufen *v* to flow off, to run down, (the current flows to the ground).

Abfluß *m* flowing off, discharge; ~graben *m* drain, draining ditch, water course; ~öffnung, ~mündung *f* (Masch.) overflow shoot; ~rohr *n* waste pipe; ~ventil *n* delivery valve, escape valve.

Abgabe *f* (Bestellung) delivery; (Steuer) tax, rate, royalty; (Zoll) custom, duty; (Gebühr) fee; directe ~n assessed taxes; ~nfreiheit *f* immunity from taxes.

abgaben . . . ~frei *a* free from taxes, duty free; ~pflichtig ratable, liable (subject) to taxation.

Abgang *m* departure (of the mail coach, of a vessel); starting (of the mail); (an Gewicht) diminution; (Verlust) deficiency, loss; ~sbuch *n* register of departures; ~spostanstalt *f* despatching office; ~sprüfung *f* bestehen to go through the course of study at a gymnasium (college), to take a degree; ~sverzeichniß, Packet-~sverzeichniß *n* parcels bill; ~szeit *f* time of departure (starting); ~szettel, Brief-~zettel *m* bag list; Packet-~szettel *m* parcels bill.

abgängig *a* waste, used; ~ gewordene Stangen used up poles.

abgeben *v* (Briefe an Korrespondenten) to hand over, to deliver; (Briefe zur Post geben) to post (letters), to drop a letter at the post office; Telegramme ~ (befördern) to forward, to transmit telegrams; „abzugeben bei . . ." care of . . .

abgehen *v* (Post, Zug) to start, to depart, to leave; (Schiff) to sail; ~der Strom *m* (*T*) outgoing current.

abgekehrt *a*, die den herrschenden Winden ~e Seite (*T*) side where the prevailing winds would tend to blow the poles off the roadway (rails).

abgleichen *v* to make even, to level, to adjust; ~, zwei Ebenen in eine bringen to flush a surface with another.

abhandenkommen *v* to go astray; abhanden gekommen lost.

Abhandenkommen *n* loss.

Abhang *m* slope, declivity, descent; ~ eines Hügels slope of a hill; ~ eines Dammes side slope, inner side of a dam; ein steiler ~ a hang.

abhängig *a* (abfallend) sloping; ~er Boden *m* shelving ground; ~e Fläche *f* inclined plane, fall.

abhaspeln *v* to reel, to reel off, to unwind.

abhelfen *v* to remedy, to redress (a grievance); to remove (difficulties), to redress (wrongs).

abholen *v* (Briefe v. d. Post) to call for (to fetch) one's correspondence from the post office; to obtain mail matter on application at the post office; to hold a private box; „nicht abgeholt" (bei postlagernden Sendungen) not called for.

Abholer *m* (abholender Korrespondent) private box holder.

Abholung *f* window delivery; ~serklärung *f* application to have one's letters etc. delivered at the office window; ~sfahrt s. Bahnhofsfahrt; ~sgebührenkarte (+) docket showing the postage due on town letters to be delivered at the window.

Abhülfe *f* remedy, redress; ~ schaffen = abhelfen.

abkehren v ſ. abgekehrt.
abklingeln v (nach beendeter telephoniſcher Unterredung) to ring off.
abkürzen v to abbreviate (a word); to shorten, to lessen (the distance); abgekürzte Adreſſe abbreviated (arbitrary) address.
Abkürzung f abbreviation, abbreviatur.
Ablaß, Waſſerablaß m outlet sluice.
ablaſſen v to despatch, to transmit, to forward, to send (letters, telegrams); eine (poſt-)dienſtl. Meldung ~ to send an official report, to report; eine (telegr.-)dienſtl. Meldung ~ to send a service message.
Ablaſſung f (TP) despatch, transmission.
Ablauf m expiration (of a term, of a bill), maturity (of a bill); ~röhre f (Maſch.) waste pipe, escape; ~tag m day of expiration.
ablaufen v (von der Zeit) to pass away, to expire; (fällig werden) to become due; (von den Schienen ~, entgleiſen) to run off (the rails); (von Uhren, Federn, Gewicht des HA u. ſ. w.) to run down; (vom Waſſer) to flow off.
ablegen v (Rechnung) to account for, to give an account; (einen Eid) to take an oath; (Zeugniß) to bear evidence (witness); (Prüfung) to pass an examination.
ableiten v to turn off;'(einen Fluß) to turn (a river) off his course; einen elektr. Strom ~ to deviate an electric current.
Ableiter m conductor.
Ableitung f turning off; ~ des Stromes zur Erde (in Folge ungenügender Iſolirung) escape; ~sgraben, ~skanal m drain, ditch; ~sröhre f waste pipe, conduit pipe, drain.
ablenken to turn off, to divert, to avert, to deviate, to deflect; (the magnetic needle is deflected or deviated); ~de Kraft f deflective force.
Ablenkung f turning off; (der Magnetnadel) deflection of the magnetic needle; ~swinkel m angle of deviation.

abliefern v to deliver, to hand smt to smb; an die fremde Verwaltung ~ to hand smt to the foreign office.
Ablieferung f delivery; nach erfolgter ~ when delivered; ~sſchein m certificate of delivery, receipt; ~szeit time of delivery.
ablohnen v (Arbeiter) to pay off, to discharge (workmen).
ablöſen v to discharge, to pay (a debt); to take off, to detach (postage stamps from envelopes); ſich im Dienſt ~ to relieve one another, to change places with smb; (ſich ~ z. B. Freimarken vom Briefumſchlage) to come off, to drop off, to become detached.
Ablöſung f (Dienſtablöſung) relief, change, substitution; (das Ablöſen, Techn.) loosening, dropping off; ~szeit f (TP) relieving time.
abmachen v to settle, to wind up, to adjust.
Abmachung f settlement.
abmeſſen v to measure off, to survey, to take the gauge of; (nach der Schnur) to lay out; (mit der Waſſerwaage) to level; (mit dem Zirkel) to compass.
Abmeſſung f measuring, measurement, surveying, alignment; Stangen von verſchiedenen ~en poles of different dimensions (lengths).
Abnahme f (einer ankommenden Poſt) checking of an arriving mail; (eines ausländ. Kartenſchluſſes) receiving of a foreign mail; (der Siegel) unsealing; (von Materialien) accepting, receiving materials; (von Rechnungen, Belegen) audit; (einer Telegr; Linie) survey & settling of accounts. (von Telegr. Draht) testing, examining & approving of the line wire; (= Schwinden) decrease (of traffic, of the current), decrement; ~beſcheinigung f receipt; ~verhandlung f protocol, record.
abnehmen v (die Poſt) to check, to receive the mail; (Siegel) to unseal; (eine Rechnung) to audit an account; (Materialien) to accept, to receive materials (stores); (ſchwächer werden) to decrease; eine Telegr.

Linie ~ to survey and settle the accounts; Draht ~ to test, examine & approve the line wire; nicht ~ to reject.

abnutzen *v* to wear off, to wear away, to waste.

Abnutzung *f* waste, wearing, wasting, wear and tear.

Abonnement *n* subscription (to a newspaper); ~sbetrag *m* amount (price) of subscription; ~sbillet (Eisenb.) contract ticket, season ticket; ~sliste *f* list of subscribers; aus der ~sliste streichen to cancel one's name in the list of subscribers; aufgehobenes ~ subscription suspended; Annahme von ~s receiving of subscriptions.

Abonnent *m* subscriber; (auf einen Platz in der Eisenb.) season ticket holder.

abonniren *v* to subscribe to, to take in (newspapers); to rent (a telephone wire); (auf feste Plätze in der Eisenb.) to take a season ticket.

Abpfählbuch *n* field book (containing the sites marked out for the poles of a t. line).

abpfählen *v* to mark out (a t. line) with pegs (staves), to peg out a line.

Abpfählung *f* marking out with pegs (staves), staking, staking out, pegging out.

abplatten *v* to flatten.

abrechnen *v* (abziehen) to deduct; (eine Abrechnung aufstellen) to prepare an account, to make up an account; (eine Rechnung begleichen) to settle, to make up an account, to balance accounts.

Abrechnung *f* (Abzug) deduction; (die Rechnung selbst) account, settlement; gegenseitige ~ reciprocal account; General-~ general account; monatliche ~ monthly account; vierteljährliche ~ quarterly account; halbjährliche, jährliche ~ halfyearly, annual account; eine ~ aufstellen to prepare (to make up) an account; (die Begleichung einer Rechnung) settlement, adjustment of an account, liquidation; auf ~ on account; sbogen *m* delivery docket, balance sheet; ~sbüreau *n* office of accountability; ~stag settling (clearing) day, day of liquidation.

abreiben *v* to rub off, to scrape; den Draht mit Sandstein ~ to scour the wire.

Abreise *f* departure.

abreisen *v* to depart, to leave, to start.

abreißen *v* to tear off.

Abreißfeder *f* retracting spring, antagonistic spring.

abrinden *v* to decorticate, to bark, to peel off the bark.

Abriß *m* draught, plan, sketch, delineation.

abrollen *v* (Draht, Kabel) to pay out, to roll out.

abrunden *v* to round, to round off; einen Bruch auf eine volle Zahl ~ (nach unten) to omit a fraction; (nach oben) to round off a fraction upwards; abgerundete Spitze *f* (Stift *m*) rounded pin.

Abrundung *f* rounding off; zur ~ der Bausumme adding of a percentage on the total amount to provide against unforeseen contingencies.

absägen *v* to saw off.

Absatz *m*, (von Postfreimarken 2c. an das Publikum) sale of postage stamps etc. to the public; ~, Alinea *n* fresh paragraph.

Absatzpostanstalt s. Post.

abschaben *v* to scrape off, (die Zinkplatte des Zink-Kupfer-Elements) to clean (to wash) the zinc plate.

abscheiden *v* to separate; (Chem.) to eliminate, to part, to set free; (absondern) to secrete, to secern; Metalle ~ to refine; sich ~, sich setzen (Chem.) to settle down, to precipitate, to fall down to the bottom.

abschätzen *v* to tax, to estimate, to appraise.

abschicken s. absenden.

Abschied *m* (das Ausscheiden aus dem Dienst) retirement (from service); resignation; seinen ~ beantragen to tender (to send in) one's resignation; den ~ nehmen to retire (to withdraw) from service; den ~ erhalten to superannuate; (unfreiwillig) to be dismissed; to get pensioned; ~sgesuch *n* tender of resignation.

abschiefern *v* to split off, to scale off, to peel off in scales (said of wire).

Abschlag *m* (Abschlagszahlung *f*) instalment, payment by instalments; (Aufgeld) payment on account, beforehand payment.

abschlägig *a* denying, refusing; ~e Antwort *f*, ~er Bescheid *m* refusal, denial.

abschließen *v* (die Post, die Briefkarte) to make up the mail, the letter bill; (eine Rechnung) to settle, to close, to balance an account; (einen Vertrag) to make an agreement, to enter into a contract, to conclude a treaty; (den Dampf) to cut off the steam.

Abschluß *m* (einer Rechnung) balance of an account, settlement; täglicher ~ daily settlement; (eines Kartenschlusses) making up of a mail; (eines Vertrages) conclusion (of a treaty); Kassen-~ settling, balance of account.

Abschmelzdraht *m* fusible wire, fine spiral wire.

abschneiden *v* to cut off; ~, abgleichen, zwei Ebenen in eine bringen to flush a surface with another.

abschnellen *v* to release (said of the armature of the *HA* electromagnet).

Abschnitt *m* (einer Postanweisung, einer Packetadresse) (+) coupon, counterfoil; den ~ abtrennen to detach the coupon; (Geom.) segment, intercept; (in Büchern) section, paragraph; (einer Baustrecke) section; (der Dienstanweisung) section.

abschnüren, mit der Schnur abmessen *v* to line, to line out, to lay out, to mark by a line.

abschrägen *v* to scarf, to slope off; die Stangen sind oben dachartig abgeschrägt the poles are at the top sloped off in a roof-like manner.

abschrauben *v* to screw off, to unscrew.

abschreiben *v* to copy, to take (a) copy; (von der Rechnung) to deduct (from a bill); (eine Abschlagszahlung gutschreiben) to carry a payment to one's credit.

Abschrift *f* copy, duplicate; gerichtliche ~ exemplification; beglaubigte ~ attested (certified) copy; gleichlautende ~ counterpart, duplicate, true copy; für gleichlautende ~ for the conformity of the copy; durch gleichlautende ~ belegen to exemplify; ~ nehmen to take (to draw up) a copy; flüchtige ~ rough copy.

abschriftlich *a*, *adv* copied, copied out, by way of copy, in duplo.

abschroten *v* (ein Faß) to roll down (a barrel).

abschüssig *a* steep, sloping, shelving.

Abschüssigkeit *f* declivity, steepness.

abschwächen *v* to weaken.

absehen *v* von etw. to leave a thing out of consideration.

absenden *v* to send; ~, befördern to forward (a telegram), to remit (money), to issue (a money order); Waaren ~ to address goods; das ~de Amt the forwarding office.

Absender *m* sender, forwarder; (von Postanweisungen oder Werthsendungen) remitter.

Absendung *f* despatch, sending, forwarding; ~sgeschäft *n* despatching business; ~sland *n* country of origin; ~sort *m* eines Telegramms place of origin of a telegram; ~spostanstalt *f* office of despatch.

absetzen *v* (vom Amte) to dismiss; (von der Rechnung) to deduct, to strike off the bill; (niederschlagen) to annul, to make void, to cancel, to nullify; (Porto) to deduct the postage due, to allow; das gezahlte Porto soll auf Verlangen abgesetzt werden the postage brought to account shall be *allowed* on being claimed; Telegramme ~ to send, to transmit telegrams; Postfreimarken 2c. ~ (verkaufen) to sell.

Absetzung *f* (vom Amte) dismissal; (von der Rechnung) deduction; vorläufige ~ vom Amte suspension; ~sbefugniß *f* power of dismissal.

absorbiren v to absorb.
Absorption f absorption.
abspannen v (eine Feder) to relax a spring; (die Leitung T) to stop the (overground) wire at the terminal insulator, to terminate the wire, to shackle off.
Abspannstange f terminal pole.
absperren v to shut off, to stop; (den Dampf, den elektrischen Strom) to cut off (the steam, the electric current).
Abstand m distance; ~ der Zeichen (T) space between the signs; (zwischen zwei Stangen) space intervening between two posts; in gleichem ~ equidistant.
abständig a dead, decayed (said of wood); ~ werden v to decay.
abstatten v einen Bericht to report on, to give account.
abstecken (eine Linie) = abpfählen.
Absteckpfahl = Markirpfählchen.
abstehen v (Techn.) to stand off, to stand far from, to be at a distance; (entsagen) to desist from.
absteifen v to stay, to support by beams and props, to prop; ein Loch ~ to support the sides of a hole by planks.
abstellen v to correct (a fault), to redress (a grievance), to reform (abuses).
Abstellung f redress, correction, reform; (Abschaffung) abolition.
abstempeln v to stamp, to postmark.
Abstempelung f stamping, postmarking.
abstoßen v to repel.
abstoßend a repulsive'; ~e Kraft repulsive power.
Abstoßung f repulsion.
abtelegraphiren v to forward, to transmit a message.
Abtelegraphirung f forwarding, transmission.
abteufen v to sink, to deepen, to dig a shaft.
abtheilen v to divide, to division off; in Grade ~ to graduate.
Abtheilung f (Theil eines Ganzen) division, section; ~singenieur divisional engineer; ~vorstand chief (head) of the division superintendent; (Theil eines Raumes) compartment, partition, part; (unbestimmte Zahl von Gegenständen) lot; (in Büchern, Schriftstücken u. s. w. Kapitel, Titel) head; Arbeiter-~ gang of workmen; ~ in Grade graduation.
Abtrag m payment, contribution (of an amount of money).
abtragen v to pay off (a debt), to deliver (a letter).
Abtragung f delivery (of letters, parcels etc.); acquittal (of a debt).
abtreiben v (vom Kabel) to drift.
Abtrieb m drift; eine gewisse Länge muß für ~ berechnet werden a certain length of cable must be allowed for drift.
abwärts adv downwards; strom~ down the stream.
abwechseln v (mit J—d im Dienst) to alternate with *smb*.
abwechselnd adv alternately, by turns.
Abwechslung f alternation.
abweichen v to deviate (said of the magnetic needle); (verschieden sein) to vary, to differ from; (in Bezug auf Uebereinstimmung der T.-Apparate) not to work synchronously, not to keep time with each other.
Abweichung f (der Nadel) deviation, deflection; (vom Bauplan) departure from the original design; bei der Drahtabnahme ist eine ~ von 0,5 mm gestattet in examining the wire a margin of 0,5 mm is permitted; ~swinkel m angle of deviation (emergence); (Verschiedenheit) discrepancy.
abweisen v (P) to direct, to send (letters etc.).
Abweiser m (Abweise- oder Prellstein) corner post, corner stone, guard stone, curb stone; (Prellpfahl) m fender.
Abweisung f (abweisender Bescheid) rebuff, refusal.
abwickeln v to unroll, to wind off; sich ~ to wind up, (the correspondence has been promptly wound up).

Abzeichen n (Brust-, Armschild) attribute, badge, arm badge.

abziehen v (abrechnen) to deduct.

Abzug m (Abrechnung f) deduction; (vom Lohne) shortening one's wages; (vom Gehalt) deduction from one's salary; nach ~ der Kosten deducting all charges, charges deducted; ~sgraben m channel, drain; gemauerter ~sgraben stone drain; ~srinne f drain; ~sröhre drain pipe.

abzweigen v to branch off (a line, a battery etc.); eine gemeinsame Batterie für mehrere Leitungen ~ to supply a number of lines from one battery.

Abzweigung f branching off (of a line).

Accept m, **Acceptation** f acceptance, accept, acceptation; protection (of a bill); mit ~ versehen to provide with acceptance; zum ~ gesandt sent (being) out for acceptance; ~s-Zeit time (term) of acceptation; ~Geschäfte acceptances.

Acceptant m acceptor; (der Bezogene) drawee.

acceptiren v (einen Wechsel) to accept, to pay, to honor (a draft); nicht ~ to refuse acceptance; acceptirter Wechsel bill accepted; acceptirt werden to meet due honor.

Accise f excise; unter ~-Verschluß under bond; ~-Zettel permit; ~-Beamter excise officer.

Achat m agate; ~hütchen n agate stud.

Achse, Axe f axis; (Wellbaum m) arbor, axle tree, shaft; (am Rade) axle, axle tree; ~ des Typenrades (HA) type wheel shaft; per ~, auf der ~, zur ~ by land carriage.

Achsen ... ~baum m axle tree; ~futter n axle tree bed; ~lager (am Wagen) axle journal; ~lager (an der Maschine) bearing neck of an axle (shaft), gip; ~schraube f axle nut

Acten s. Akten.

Achtung f (Hoch-~) respect, regard, esteem; ~svoll a respectful; (Schluß von Briefen: I am, Sir, very [most] Respectfully Your Obedient Servant ...).

Ader f (Seele f) des Kabels core.

Adreß ... ~amt n receiving office; ~buch n city directory, guide; ~zettel, Beklebezettel m label, ticket.

Adressat m addressee, receiver; (bei Postanweisungen) payee; (bei Wechseln) drawee.

Adresse f (eines Briefes o. dergl.) address, direction (seltener superscription); unrichtige ~ incorrect direction; wegen mangelhafter (unvollständiger) Adresse for defective (incomplete) address; ein Packet mit folgender ~ a parcel directed as follows; „per ~" care of .....; Noth-~ f (+) supplementary despatch note; Packet-~ f (+) despatch note.

adressiren v to address, to direct a letter; (selten) to superscribe.

Agent m agent.

Agentur f (eines Handelsgeschäfts) agency; Post-~ s. unter Post ...

Aich-Apparat m gauging apparatus; ~maaß n gauge, standard.

aichen v to gauge, to size, to admeasure (weights).

Akkord m contract, agreement; Arbeit auf ~ geben, nehmen to give, to take work in job.

Akkumulator, Kraftsammler m accumulator.

Akten f/pl official correspondence; records; reports, deeds, documents, public papers, acts; ~deckel m cover, wrapper; ~heft bundle of papers (deeds etc.); ~stücke documents, the particulars of a case; zu den ~ legen to put by, to subjoin to the records, to file away.

Aktie f share.

Aktien ... ~bank f joint-stock bank; ~coupon coupon; ~gesellschaft joint-stock company; ~inhaber stockholder, shareholder.

Akustik f acoustics, phonics.

akustischer Telegraph m acoustic telegraph.

Alaun m alum; ~erde, Thonerde, Aluminiumoxyd alum earth, aluminium oxide; ~schiefer m alum slate.

alaunartig a alumish, alumy.

alaunhaltig a aluminous.

Alphabet n alphabet; konventionelles ~ für Telegramme 2c. code.

alternirend = abwechselnd.
Alterszulage f additional pay.
Aluminium n aluminium; ~oxyd n, Thonerde f aluminium oxide.
Amalgam n amalgam, amalgama.
amalgamiren, verquicken v to amalgamate, to amalgamize.
Amalgamirung f amalgamation.
Amboß m anvil (auch für Arbeits- od. Telegraphirschiene).
Ammoniak n ammonia, volatile alkali; kohlensaures ~ carbonate of ammonia; salzsaures ~, Salmiak m hydrochlorate (muriate) of ammonia, sal ammoniac.
Ammoniakflüssigkeit f, **Salmiakgeist** m aqueous ammonia, caustic ammonia, ammonia solution.
Ammonium n ammonium; ~chlorid n, Salmiak m chloride of ammonium, sal ammoniac; ~haltig a ammoniacal.
Amt n office, place, situation, post; Post-~, Telegraphen-~ Post Office, Telegraph Office, Telegraph Station; ein ~ bekleiden to hold (to fill) an office; er ist jünger, älter im ~e als ich he is my junior, my senior in office; J—n des ~s entheben to suspend *smb* from office; Ehren~ honorary office.
Amts... ~antritt m entering upon office; ~bedürfnisse n/pl supplies, equipment of an office; ~bestellbezirk m delivery of the office, (im ~bestellbezirk wohnen to dwell within the delivery of the office); ~bibliothek f official library; ~blatt n official gazette; Post-~-blatt n Post Office Circular; ~bruder m colleague, ~einführung f (T) leading-in wires; ~einrichtung f office establishment; ~führung f administration, management; ~geschäft n official duty (business); ~jubiläum n, das 50jährige ~jubiläum feiern to celebrate the fiftieth anniversary of one's entering upon office; ~-kaution f official bond; ~kosten pl office expenses (expenditures); ~lagernd a to be kept until called for; ~pflicht f duty of office; ~schriftenbuch n correspondence register, index of correspondence; ~siegel n official seal; ~stube f office; ~stunden f/pl office hours, hours of attendance, hours of business; ~telegramm n service message, official telegram; ~thätigkeit f performance of official duties; ~unkosten pl administrative expenses; Entschädigung auf ~unkosten allowance for expenses incurred by the administration of an office; ~verbrechen n offence committed in office; ~vergehen n breach of duty, misconduct, misdemeanor in office; ~verrichtungen f/pl official function, duty, J—m die ~verrichtungen untersagen to suspend *smb* from duty; ~verschwiegenheit f official secrecy; ~vorsteher m Chief (head) of an Office; ~wohnung f official lodgings (residence); von ~wegen adv officially, ex officio (ex offo); s. auch Dienst~.

amtlich a official; nicht der ~en Form entsprechend informal; ~e Einschreibung f (Rekommandirung) registration; ~e Meldung (T) service message; ~es Schreiben n letter on official business; ~e Wiederholung f (T) repetition.

anbei adv enclosed (inclosed), annexed, herewith; ~ der Morsestreifen mit der Quittung the Morse slip containing the acknowledgment of receipt is enclosed; das betreffende Telegramm wird abschriftlich ~ übersandt a copy of the telegram in question is herewith transmitted.

anbelangen s. anlangen.
Anbetracht (in) concerning, considering; in ~ der geltend gemachten Gründe considering the reasons set forth.
anbetreffen v to concern; was die diesseitige Verwaltung anbetrifft as far as this administration is concerned.
anbieten v (bei Lieferungen) to tender, to propose.
Anbieter m one who makes a proposal.
Anbietung f (Submission) tender, proposal; ~sverfahren n inviting tenders (proposals); beschränktes ~sver-

fahren inviting tenders only from a limited number of contractors; schriftliches ~verfahren inviting tenders to be transmitted in sealed letters.

anbinden v to tie (the wire to the insulator).

anbringen v to fix, to affix (a stay to a pole).

anbrüchig a decayed, rotten (said of wood).

Ancienntät f seniority in office.

ändern s. abändern.

Änderung s. Abänderung.

Andrang m pressure; Geschäfts-~ pressure of business; (Anhäufung von Telegrammen) accumulation; den ~ am Schalter durch geeignete Maßnahmen beseitigen to take the necessary measures (steps) to prevent pressure at the counter.

an einander adv together; ~fügen to join; ~stoßen (sich reiben) to grate, to rub; (kreischen) to screech.

anerkannt a certified, accepted, found correct (Anerkenntnißformel bei geprüften Rechnungen).

anerkennen v to acknowledge, to admit, to allow; (würdigen) to appreciate; (vor Gericht ~) to recognize; eine Rechnung ~ to certify, to accept (an account); Wechsel ~ to accept, to honor a bill; nicht ~ to disown.

Anerkenntniß n acknowledgment, acceptation; (Würdigung) appreciation.

anfahren v to carry, to cart (materials); (eine Stange mit dem Rade) to run against a pole.

Anfangsbuchstabe m initial letter; die ~n geben (T) to repeat the initial letters.

anfertigen v (Briefkarten u. dgl.) to prepare (the letter bill); to draw up (instructions, a protocol etc.).

anfeuchten v to wet, to moisten, to damp; ein angefeuchteter Brei von Quecksilberoxydul a moistened paste of protosulphate of mercury.

Anforderung f claim, pretence; ~ stellen auf ... to lay claim to ...

Anfrage f enquiry (inquiry), request, letter of enquiry.

anfragen v to enquire.

anfressen v to corrode (the zinc plate becomes much corroded at the level of the liquid).

anfrischen v (Chem.) to reduce.

anfügen, anfugen v (Techn.) to join.

Anfuhr f und **Abfuhr** f carting, cartage.

Angabe f (Erklärung) declaration; nach ~ des Absenders according to the declaration of the sender; (Darlegung) statement, specification; (Techn.) instruction, order; ~ beim Zollamte entry at the custom house; dienstl. ~n im Kopf des Telegramms service instructions; eine ~ machen (vor Gericht) to lodge an information; ~, das Angegebene data pl.

angeben v to state, to mention; (erklären) to declare; (darlegen) to state; (einzeln ~) to specify; (Waaren beim Zollamt) to enter goods at the custom house; (vor Gericht) to denounce, to inform against; wie vorher angegeben as stated above.

angeblich a so-called, pretended, nominal; ~er Werth m nominal value.

Angebot n offer, bidding; (auf eine Submission) tender, proposal; ~e ausschreiben to invite tenders (proposals).

Angelegenheit f concern, business, affair, matter; Minister der Auswärtigen ~en secretary (minister) of foreign affairs.

angemessen a due (in due time); conformable, suitable (language, price); proportionate, deserved (punishment); adequate (reward).

angeschwemmt a alluvial, diluvial, alluvious; ~es Land n alluvium.

angesichts adv in the face of; ~ dieses Befehls at the receipt of this order.

angestrengt a, über Kraft ~ (Mech.) strained.

angreifen v (Mech.) to catch, to hold fast; to touch (said of a file); to attack (said of acids).

Angriff m in ~ nehmen to commence operations; ~spunkt m einer Kraft working point, point of application.

anhaken v (z. Zeichen der erfolgten rechnerischen Prüfung mit einem Anstrich versehen) to tick, to check.

Anhalte ... ~stelle f (Eisenb.) station; **~stift** m detent pin; **~zeit** f (Eisenb.) stoppage.

anhalten v to stop (a telegram in the course of transmission).

Anhaltspunkt m eines Hebels fulcrum of a lever.

Anhang m appendage, appendix, addition; (eines Briefes) postscript; (eines Kontrakts) conditional clause; **~, Fortsatz** m (Techn.) continuation.

anhauen v einen Baum to mark a tree (for felling).

anhäufen v to accumulate, to amass.

Anhäufung f accumulation (of telegrams, of electricity); rush (of traffic), pressure (of business).

anheften v to affix; (mit Stecknadeln) to pin to.

anheimstellen v to commit to, to defer to.

Anilin n aniline; **~farben** f/pl aniline colors; **~farbstoffe** m/pl aniline dyes; **~firniß** m aniline varnish.

Anion n anione.

Anker m (einer T.-Stange) stay made of wire rope, wire stay, stay, tie, truss wire; (am Magnet) armature; **~bolzen** m tie bolt; **~draht** m stay wire; **~feder** f (HA) spring (tending to raise the armature); **~gestell** n (HA) frame; **~haken** m hook (of a tie), tie hook; **~hebel** m armature lever; **~kloß** m stay block, log of wood (to which the ties are anchored); **~pfahl** m picket of timber (f. ~kloß); **~stein** m large stone (f. ~kloß); **~** eines Elektromagneten von weichem Eisen armature of an electromagnet of soft iron; eine Stange mit einem ~ versehen, verankern to truss a pole.

ankleben v to fix, to fasten to (on), to paste on; to adhere, to stick (said of the armature of an electromagnet).

ankohlen v das Stammende der Stangen to char the butt end of the poles.

ankommen v to arrive; **~der Strom** incoming current; das Telegramm ist am Bestimmungsorte angekommen the message has reached its place of destination.

ankündigen v to announce; (öffentlich) to publish, to promulgate; (in öffentl. Blättern) to advertise.

Ankündigung f declaration, announcement; (die ~ eines Telegramms geschieht mit den Worten: „here message"); (~ in einer Zeitung) advertisement.

Ankunft f arrival, coming; **~s-amt** n office of destination; **~sbuch** n book of arrivals; **~sstempel** m stamp of arrival; **~stelegramm** n received telegram; **~szeit** f time of arrival; nicht erfolgte ~ non-arrival.

Anlage f (Techn.) laying out, plan, planning; die fertige ~ (Betriebs-, Telegraphen-~ 2c.) plant, establishment; **~-Kapital** n cost of the first establishment; aus der ~ (Beilage) werden Sie ersehen by the annexed (enclosure) you will see; **~kosten** pl expenses for the first establishment.

anlangen v (ankommen) to arrive, (betreffen) to concern; was diese Sache betrifft as respects this affair, as to, as for that affair.

Anlaß ... ~apparat m starter; **~maschine, Maschine zum Ingangsetzen** starting machine, starting engine; **~ofen** tempering furnace; **~ventil** starting valve.

anlassen v (Hüttenw.) to temper, to anneal; ~, in Gang setzen to start; den Dampf ~ to put the steam on; Stahl ~ to soften steel.

anlaufen v (Chem.) to oxidize, to be discolored, to tarnish; Stahl blau ~ lassen to blue steel; einen Hafen ~ to call at a harbour, to touch a port; Land ~ to make the land.

anlegen v to found (a city, a house); to establish, (a post office, post routes etc.); to lay out, to build, to construct (a t. line); (Bücher) to arrange a set of books; (Geld) to invest money; (ein Konto) to open an account; (= anlanden) to land, to put on shore, to touch at . . . ;

(Siegel) to seal up; (an die Batterie, an den Umschalter u. s. w.) to join (the battery to the instrument), to supply (a line) with a battery, to bring (the cable to the switch).

anleimen v to glue on, to fasten to ... with glue.

anliegen v to fit well, to sit close.

anliegend a (benachbart) adjacent, adjoining, contiguous; (beiliegend, als Anlage) enclosed, annexed.

anlöthen v to solder, to join by means of soldering.

anmachen v festmachen to fasten, to fix; Farben ~ to dilute (to blend) the colors; den Kalk, Mörtel, Gips ~ to mix, to wet, to dilute the lime, mortar, plaster into a paste; ein Feuer ~ to make, to kindle a fire.

anmalen v to paint, to paint over.

annähernd adv approximately; ~ berechnen to calculate approximately.

Annahme f acceptance, receiving (of letters) acceptation, protection (of a bill); (von Beamten) appointment; (Annahmestelle einer Postanstalt) office counter, office window; (v. Telegrammen) taking in of telegrams; die ~ verweigern to refuse acceptance (of a letter); to dishonor (a bill of exchange); „~ verweigert" (P) refused.

Annahme ... ~beamter m counterman (counterwoman), counter clerk, clerk at the counter, officer in attendance at the counter; ~buch n book of receipts, receipt book; (für Einschreibsendungen) book of registration, registered letter receipt book; (für Telegramme) abstract book; ~dienst m duty at the counter; ~schalter m counter, window; ~stelle f (P) receiving station; ~verweigerung f refusal to receive (a letter, parcel); in der ~ daß ... presuming that ....

annehmen v to accept (a bill), to receive (a letter), to take in (a telegram); Arbeiter ~ to hire workmen; ein Gesetz ~ to pass a law; eine Meinung ~ to embrace an opinion; der angenommene (giltige) Münzfuß m the recognized standard.

annietheu v to rivet, to fasten with a rivet.

annulliren v to annul, to make void, to nullify; (ein Telegramm) to cancel a telegram.

Anode f anode (the positive pole of electricity).

anordnen v to dispose, to order.

anpassen v to fit, to adapt, to adjust.

Anpassung f adaption, adaptation.

Anordnung f disposition, arrangement, plan; ~en treffen to dispose.

aupflöcken v to treenail, to peg, to pin, to fasten with pegs (pins).

anquicken, verquicken v to amalgamate.

anrechnen v (in Anrechnung bringen) to put down to one's account; (Dienstjahre ~) to count the years passed in the service; (Einem etw. ~) to ascribe to; (als Fehler) to impute to.

Anrechnung f charge; (der Dienstjahre) counting of the years passed in the service.

anrosten v to begin to rust, to rust to; (festrosten) to become fixed by means of rust.

Anruf m (T) call, station call; ~zeichen n „call" signal.

anrufen v (T) to call; (F) to ring up (a subscriber).

ansammeln v to collect together; (sich ~) to accumulate.

Ansammler m (Apparat zum Aufspeichern von Electricität) accumulator, condenser.

Ansammlung f accumulation; (von Gehaltsabzügen) collection; ~s= apparat s. Ansammler.

Ansatz m (Techn.) adjoint piece, nose, shoulder, prolongation; (Math. Arithm.) formation; (= Anschlag, Etatsansatz m) estimate, estimation; (= Niederschlag m) crust, deposit, sediment; (in einer Rechnung) rate, charge in an account; (von Porto) charging of postage.

Ansatz ... ~preis m taxation; ~rechnung f (Math.) differential calculus; ~stück n joining piece, adjoint piece; ~stück (Verlängerung) lenghtening piece; ~stücke n/pl der Pole des Elektromagneten, Polschuhe m/pl pole pieces.

ansäuern v (Chem.) to acidify, to acidulate.

anschaffen v to procure, to provide with, to purchase.

Anschaffung f acquisition, purveyance, providing for, furnishing with; ~skosten pl first cost, prime cost; ~skosten (Selbstkostenpreis) purchase price, net cost.

anschäften v ein Werkzeug to handle, to helve a tool.

Anschlag m (Bau) valuation, estimation, rate, calculation, (T) contact, blow (of the lever against the adjusting piece), striking (of the lever against the stop pieces), click; ~, Bekanntmachung f placard; ~schraube f (MA) contact screw; ~stift m striking peg, metallic pin; Kosten=~ estimate, statement, account for building costs, bill of cost; ~ssumme amount of costs; in ~ bringen to take into account.

anschlagen v (veranschlagen) to estimate, to value; ~, befestigen to fix, to fasten; (vom Apparat) to make contact, to strike against the contact screw (stop pieces); zu hoch ~ (im Kostenanschlag) to overrate.

anschließen v (passen) to fit; (Anschluß haben bei Eisenb.) to join, to time with another train; eine T.=Linie an eine andere ~ to connect a t. line with another.

Anschluß m (Verbindung) joining; (Anlage) annexure, enclosure; (Eisenb.) junction, joining; (der Eisenb.=Züge) coincidence of the trains; (an einen Vertrag) accession, adhesion to a treaty; in ~ an mein Schreiben in pursuance of my letter; ~ haben to coincide with, to time with another train; ~ verfehlen to miss the connection with the train; ~bahn f junction line, junction railroad; ~post f mail coach connecting with the train; ~punkt m junction point; ~station f junction, junction station; ~zug m junction train.

anschmelzen v to join by casting; (anfangen zu schmelzen) to begin to melt; (haften) to adhere by melting.

anschmieden v to join to . . . by forging, to hammer together.

anschneiden v to begin to cut.

Anschnitt m first cut; (Kerbe f) notch, nick.

anschrauben v to screw on, to fasten with a screw.

anschreiben v Porto ~ to charge postage.

anschuhen v eine Stange to shoe a post, to nail shoes to poles.

anschütten v to fill up.

Anschüttung f filling up; ~, Damm m embankment.

anschweißen v to weld on, to weld together.

Anschweißstelle, Anschweißung f welding point, weld, welding.

Ansehung f in ~ in consideration of, in regard of, as for, as to.

ansegeln v (einen Hafen) to touch at a port, to make to a port, to call at a port.

ansetzen, v **anstellen, vorbereiten, zurichten** to prepare, to fit; (einen Hebel) to apply a lever; (eine Batterie) to gather rust, to rust; (eine Batterie) to set up a battery; (Porto) to charge postage.

Ansetzen n (Chem.) efflorescence, incrustating.

Ansicht f (Techn.) view; (Grundriß m) plan view; (hintere) back view; (perspektivische) view in perspective; (von der Seite) side view; (vordere) front (frontal) view.

Ansichtszeichnung f scenography.

anspannen v (Pferde) to put the horses to . . .; (eine Feder) to bend a spring; (eine Schraube) to screw; (den Draht) s. recken.

ansprechen v (T), der Apparat spricht an the instrument works.

Ansprechen n (T), der Strom bringt den Apparat zum ~ the current actuates the instrument (puts the instrument in motion).

Anspruch m claim, right; (unbegründeter) pretension; ~ auf etwas haben to be entitled to . . ., to have a right to...; ~ auf etwas erheben to claim, to lay claim to..., to raise a claim to...; ~ geltend machen to assert one's claims, to come in for...; sich eines ~s begeben to disclaim; Rück~ recourse, appeal against....

anstählen v verstählen to steel, to provide with a steel point.

Anstalt f (Techn.) establishment, institution.

Ansteigen n des Bodens ascending slope, acclivity.

anstellen v (dauernd) to appoint; (vorübergehend) to employ; eine Untersuchung ~ to make enquiries into, to institute an examination; eine Berechnung ~ to calculate.

Anstellung f (im Dienst) appointment; (vorübergehende) employment; (= Ernennung) nomination; Staats~ government appointment; kommissarische ~ provisional appointment; probeweise ~ probationary appointment; ~surkunde f certificate of appointment.

anstoßend a (benachbart) contiguous, adjacent.

anstreichen v to paint, to color; (mit Firniß) to varnish; (mit Theer) to tar; weiß ~ to whiten, to whitewash.

Anstrich m paint, color, painting, washing, coat; erster ~, Grundanstrich priming, first coat; wasserdichter, wasserabhaltender ~ waterproof paint.

Antheil m share, part, quota; ~ an etwas haben to participate, to share in...; ~ an etwas nehmen to take an interest in...; ~gebühr f percentage, quota; Porto-~ s. Porto.

antheilig a proportionate, contingent.

Antimon n Spießglanz m antimony, stibium.

antiseptisch, fäulnißwidrig a antiseptic; ~e Substanzen f/pl antiseptics.

Antrag m (Anerbieten n) proposal, proposition, offer; (in öffentlicher Versammlung) motion; (Ersuchen n) request; (Gesuch n) application; einen Urlaubs-~ stellen to apply (to make application) for leave of absence; (auf Gebührenerstattung) claim for refund of the charges paid; ~steller m applicant.

antragen v auf, to make a proposition, to move for..., to request, to apply for....

antreiben v to impel; (einen Keil, Nagel, Bolzen u. dgl.) to drive in (a wedge, nail, bolt etc.).

Antrieb m (Mech.) impetus, impulsion, impellent power.

Antwort f reply, answer; vorausbezahlte ~ reply prepaid, reply paid in advance; ~sformular n prepaid reply form, form of authority.

anwachsen v to increase (said of the current).

Anwachsen n increase, increment (of the current).

Anwärter m candidate, expectant; der schon probeweise beschäftigte ~ candidate who has received a probationary appointment.

Anwartschaft f expectancy; ~ auf Anstellung im Civildienst expectancy for an appointment in the civil service.

anweisen v (unterweisen) to instruct; (zurechtweisen) to direct; (hinweisen) to assign to; Geld ~ to assign money; zur Bezahlung auf eine Kasse ~ to give (to write) an order for the payment of a sum; eine Rechnung ~ to order the payment of an account

Anweisung f (Unterweisung) instruction; Dienst-~ official instructions, regulations; (Befehl, Auftrag) order, injunction; Zahlungs-~ check, draft, bill of exchange, order for payment, warrant; Post-~ money order; telegraphische ~ telegraphic money order.

anwenden v to employ, to apply.

Anwendung f use, practice; (eines Gesetzes) application of a law.

anzapfen v to tap; (sich heimlich in eine T.-Leitung einschalten, um Telegramme abzufangen) to tap a t. line.

Anzeige f (Bekanntmachung f) advertisement, notice; (Meldung f) report; (gerichtliche) denunciation; (telegraphische) telegraphic notice, report by telegraph.

anzeigen v (Techn.) to indicate; (in öffentlichen Blättern) to advertise; (den Empfang) to acknowledge the receipt; (berichten, melden) to report; (verklagen) to inform against.

anziehen v (Electr.) to attract (said of a magnet); (Mech.) to draw, to pull, to draw tight, to stretch, to tighten (a spring); (eine Schraube) to screw; (Stricke) to haul (ropes).

Anziehung f attraction; Haarröhrchen~ f capillary attraction; ~kraft f attractive power, force of attraction; (Chem.) combining power.

Apparat m (T) apparatus, instrument, register; ~beamter m operator, instrument clerk, clerk of the working staff; ~farbe f ink, blue (colored) fluid; ~farbekasten m ink reservoir; ~gehäuse n case; ~gestell n frame; ~journal n f. ~tagebuch; ~saal m instrument room; ~schiene f bar of the commutator connecting with the instrument; ~system n telegraphic system, set of apparatus; ~tagebuch n tablet check, docket; ~tisch m table, instrument counter; ~verbindungen f/pl instrument connections; ~werkstatt f workshop of the Telegraph Department; ~zubehör n instrument appliances, fittings; auf ~ nehmen, legen to connect up, to join, to switch in the speaking instrument; Blauschreib~, Blauschreiber m inker, inkwriter, Morse printer; Doppelschreib~ m double style apparatus; Doppelsprech~ m apparatus for simultaneous transmission (in the same direction); Doppelstift~ m double style app.; Empfangs~ m receiver, receiving app.; Gebe~ m sender, transmitter, sending (transmitting) app., manipulator; Linienschreib~, Monostychograph m app. writing in one line; Nadel~ m needle app. (single needle, double needle instrument); Normal~ m standard register; Reliefschreib~ m embosser, embossing register; Schreib~ m recording register; Selbstauslöse~ m self-starter, selfstarting register, app. provided with a self-starting device; Signal~ m signalling app.; Trockenschreib~ m (dass. wie Reliefschreiber); Typendruck~ m type-printing telegraph; Uebertragungs~ m translator, repeater; Wechselstrom~ m current reverser; Zeiger~ m dial telegraph, pointer instrument; (mit Selbstunterbrechung) step-by-step instrument; ~ mit Federaufzug app. with (wheelwork) clockwork driven by a coiled spring; ~ mit Gewichtsaufzug app. with clockwork driven by a weight.

Äquator m, **Gleicher** m equator, line; der magnetische ~ aclinic line, magnetic equator, line of no dip.

Äquivalent n equivalent; (chemisches, mechanisches) chemical, mechanical equivalent; ~gewicht n equivalent weight.

Ar n (Flächenmaaß von 100 qm) are.

Arbeit f work, working; (Arbeitsweise f, Gang m) working, working order; ~ im Akkord, im Gedinge piece work, task work; ~ einstellen to strike; (auf einer Leitung zu arbeiten aufhören) to cease working; ~geber m employer; ~nehmer m employee.

Arbeiter m workman, worker; ~kolonne f set (gang) of workmen, working party.

Arbeits . . . ~kraft f working power; ~lohn m wages, hire, pay, earnings; ~maaß n amount of work; ~schiene f (MA) anvil; ~strecke f (eines Tages) section; ~strom m open circuit; ~stromschaltung f arrangement to work a line on the open circuit plan; ~stunde hour of duty.

Archiv n archives, records pl.

Archivar m archivist, keeper of the archives, recorder.

Area f, **Areal** n, **Flächeninhalt** m, **Flächenraum** m superficial contents of a figure, area, surface.

Arm m (des Fangapparats) (P) crane, catcher crane; Hebel~ arm, branch (of a lever).

Armatur f armature, keeper (of a magnet).

Armband n, **Armbinde** f arm badge.

armiren v to arm, to cap (a magnet); to fit out (a pole).

Arrest f. Beschlag.

Arretirung f (HA) braking mechanism, stopping and starting brake; ~sfeder f, ~shebel m spring lever; ~srad n wheel of the braking mechanism.

Asphalt m, **Judenpech**, **Erdpech** n asphalt, asphaltum, bitumen, jew's pitch, mineral pitch; Belegung f mit Asphalt asphalting; ~decke f asphalt covering, sheet of asphaltum; ~pflaster n asphalt pavement.

asphaltiren v to asphalt.

Assistent m assistant; Post-~ (+) assistant post office clerk; Telegraphen-~ (+) assistant telegraph clerk.

Ast m (Zweig m) bough, branch; (im Holze) knag, knot; ~frei a free of knags (knots); ~holz n branch wood; ~knorren m knob, knag, knurl; ~loch n knot hole.

astatisch a astatic.

Attest n attestation, testimony, certificate; ärztliches ~ medical certificate; obrigkeitliches ~ government certificate.

attestiren v to certify.

Äther m ether.

ätherisch a (Physi.) aërial, etherious, etherial; ~es (flüchtiges) Oel n volatile oil.

atmosphärisch a atmospheric; ~er Druck m atmospheric pressure.

Atom n atom; ~gewicht n atomic weight; ~theorie f atomic theory.

aufarbeiten v to finish; (T) to dispose (of the entire correspondence).

aufbewahren v to keep.

Aufbewahrung f keeping; ~sfrist f period (term) for keeping smt; ~sort m depository, receptacle.

Auf- und Niederbewegung f (Mech. Masch.) up and down motion.

aufbringen v den Draht auf die Isolatoren to place the wire on the insulators, to erect the wire, to run the wire.

aufeinanderfolgend a consecutive; adv consecutively.

Aufdruck m (Werthänderung bei Postmarken) surcharge.

Aufenthalt m stay, sojourn; (Ort) abode; (Wohnung) residence, domicile; ~sort place of residence (abode); ~szeit sojourn, time of sojourn; (Hinderniß) hinderance, detention; (Verzögerung) delay, stop; ohne ~ without delay.

auferlegen v to inflict (a punishment); to impose (taxes).

Auffang . . . ~apparat m (P) f. Fangapparat; (Techn.) apparatus for collecting gases, gas collector; ~stange f collecting rod of a lightning rod.

auffangen v (Gase) to collect (gases); (Briefe, Telegramme abfangen) to intercept (letters, telegrams).

auffordern v to invite; die ferne Station zum Geben ~ to invite the distant station to transmit.

Aufgabe f (Arbeit) task; (Masch. Physi.) problem, question; (eines Amtes) discontinuance of an office; (eines Briefes) posting of a letter f. auch aufgeben; (einer Stellung) resignation; (eines Telegramms) handing in of a telegram; ~amt n office of origin; ~formular n forwarded telegram form; ~gebiet n territory (country, land) of origin; ~nummer f (T) number of message, (P) number of registration; Packet-~nummer f. Packet . . .; ~ort m place of origin; ~stempel m date stamp, stamp indicating the date & place of posting; ~telegramm n original message; ~zeit f (T) time of handing in; (in Engld) code time f. auch a. T. unter code..; ~zettel, ~nummerzettel f. Packet . . .

aufgeben v (verlassen) to discontinue, to abandon (an office); (zur Post einliefern) to post, to mail, to issue (a money order); to hand in, to deposit, to tender (a telegram), to drop a letter (in the letter box); (am Schalter) to hand over the counter; (Gepäck) to have one's

2

luggage registered; (ein Packet ~) to tender (a parcel) for posting; (eine Stelle) to resign an office, to withdraw from office, to give up a place; bei Eisenb.-Stationen aufgegebene Telegramme telegrams originating at railway stations.

Aufgeber *m* sender; (einer Postanweisung) remitter.

aufgehauen *a* (das vordere Ende der Isolator-Stütze) roughed, grooved.

aufgehen *v* (Arithm.) to remain naught; ~ lassen (eine Rechnung) to balance, to strike a balance; (wechselseitige Forderungen) to balance accounts.

aufhalten *v* (ein Telegramm in der Beförderung) to cancel (a telegram during its transmission), to stop, to arrest, to retain.

aufhängen *v* to hang up; eine Magnetnadel ~ to suspend a magnetic needle.

Aufhängung *f* suspension.

aufhaspeln *v* to reel, to reel up, to hoist, to wind up, to raise with a windlass.

aufhäufen *v* to accumulate.

Aufhäufung *f* accumulation.

aufheben *v* (Mech.) to lift, to lift up; (aufbewahren) to keep, to preserve, to retain (Morse slip is to be retained 6 months); (ungiltig machen) to annul (a contract); to cancel, to repeal (laws); to revoke (orders); to abolish, to abrogate (measures); to cancel, to invalidate (a treaty, a contract); to discontinue, to stop, to abandon (a Post Office); to destroy (the magnetism); sich (gegenseitig) ~ to compensate.

Aufhebung *f* cancelling, repeal, abrogation; (einer Strafe) abolition; (einer gerichtl. Klage) non-suit; (einer Postanstalt) discontinuance.

aufhören *v* (zu arbeiten) to cease, to leave off; (zu zahlen) to stop (to suspend) payment.

aufkleben *v* (Freimarken) to affix (stamps); (Gepäckzettel, Aufgabenummern auf Packete ꝛc.) to label, to ticket (a parcel); (den Streifen des (*HA*), das auf ein Stück Papier geschriebene Telegramm auf das Formular) to gum the slip, the piece of paper to the form.

aufkommen *v* für etw. to make compensation for *smt*; bei der Postagentur N. sind 1000 M. an Porto aufgekommen the receipts for postage at the Sub Office N. have amounted to 1000 Marks.

aufkündigen *v* (einen Vertrag) to give notice (warning); (ein Kapital) to recall.

Aufkündigung *f* notice, warning, recalling (of an investment).

auflegen *v* to lade, to load upon.

aufladen *v* to lade, to load upon.

aufliefern *v* s. aufgeben.

Auflieferung *f* s. Aufgabe.

auflösen *v* to dissolve (a salt in water), to solve, to resolve, to analyze.

Auflösung *f* solution, analysis; gesättigte ~ saturated solution.

auflöthen *v* to solder to, to solder upon; (trennen) to unsolder.

aufmauern *v* to build up with bricks (stones), to raise a wall.

aufnageln *v* to nail on, to spike on.

Aufnahme *f* (Situationsplan) site plan, plan of site; (Vermessung) survey, surveying, measure, note; (nach dem Augenmaaß, flüchtige ~; eye sketch, rough sketch; (eines Telegramms) receipt of a telegram, (eines Telegramms nach dem Gehör) reading by sound, acoustic reading; ~apparat *m* s. Apparat; ~blatt *n*, ~formular *n* received telegram form; ~vermerk *m* notice of reception, entering (inserting) the prescribed items at the top of the form; ~vorrichtung *f* (für eine Stange) cylinder (of cast iron or earthenware), socket; ~zeit *f* time (at which the telegram is received).

aufnehmen *v* (durch Vermessung) to survey, to measure; flüchtig ~ to take an eye sketch; (Geld) to raise money; (einen Posten in eine Rechnung) to enter into an account; (ein Telegramm) to receive, to write down a telegram; (ein Telegramm nach dem Gehör) to read a despatch by the sound; (in den Text) to insert in the text; (ein Protokoll) to draw up a protocol (a record, the minutes).

aufpassen v (Techn.) to fit on, to fit to.

aufräumen v die Baustrecke (T) to clear the route.

aufrechtstehend a upright, vertical.

aufrichten v to erect, to raise, to straighten up.

Aufrichtung f einer Stange (T) erection of a post.

aufrollen v to roll, to roll up, to unroll; (spiralförmig) to coil.

Aufsatz m (schriftlicher) composition.

aufschichten v to pile up, to stack, to staple.

aufschieben v to delay, to adjourn (a business); to prorogue (a meeting); to put off (payment); bis zum nächsten Jahre ~ to adjourn till next year.

aufschießen v (rund zusammenlegen) ein Kabel to coil, to coil up a cable.

Aufschlag m (an Gebühren) additional fee (tax, rate).

aufschrauben v (losschrauben) to screw off, to unscrew, to slacken the screw; (daraufschrauben) to screw (the insulators on the stalks).

Aufschrift s. Adresse.

Aufschub m delay; adjournment (of a meeting); respite (of a payment), suspense; die Sache duldet keinen ~ the matter cannot (must not) be delayed; ~ der Zahlungsfrist prolongation (postponement) of payment; um ~ bitten to apply for prolongation (of payment).

aufschütten v Erde to deposit earth, to fill up earth.

Aufschüttung f embankment, embanking.

Aufschwung m development (of the postal business).

Aufseher m overseer, controller, inspector, surveyor, supervising officer.

aufsetzen v (auf ein Ding) to set up, to put up, to put on, to lay on; (eine Schrift) to write, to compose, to draw up in writing; (eine Rechnung) to make up, to cast an account; (einen Vertrag) to make a contract.

Aufsicht f inspection, control, superintendence; ~sbeamter m (P) controlling officer, controller; (T) Chief operator, Chief clerk, (Inspektor [P]) Surveyor; ~spersonal n superintending staff; Ober-~sbeamter m superintendent on duty.

aufspeichern v to store, to store up, to accumulate.

Aufspeicherung f storing up, storage (of electricity).

aufstellen v to raise, to erect, to place, to plant (a pole); to fit up, to fix, to erect (a machine, an engine); to cast, to prepare (an account); to draw up (a list, an inventory).

Aufstellung f (Techn.) putting up, erection, (einer Abrechnung) preparing of an account.

Aufstellungsjahr n (der Stangen T) year of erection.

Auftrag m (des Bodens) embankment, filling up; (geschäftliche Anweisung) order, commission, charge; Post-~ s. Post ...

aufweichen v to soften.

aufwärts adv upwards; strom-~ up (the) stream; berg-~ up hill.

Aufwärtsbewegung f upward motion.

aufwickeln v to wind up, (den Papierstreifen) to roll up the (Morse) slip.

aufwinden v to twist up, to wind up, to pull up, to raise up, to heave.

Aufwurf m (Erde) jetty, dam, embankment.

aufzählen v (Geldstücke) to pay down; (besonders erwähnen) to count, to enumerate.

Aufzählung f enumeration.

aufziehen v (in die Höhe) = aufwinden; das Gewicht (HA) to wind, to rewind the weight; (den MA) to wind up; (eine neue Papier-

2*

rolle *T*) to put on a new roll of paper.

Aufziehvorrichtung *f* raising machine, raising apparatus.

Aufzug *m* (*HA*) pedal; (Hebevorrichtung zur Beförderung von Telegrammen, Briefbeuteln, Packeten u. f. w.) lift, shoot.

Augenbolzen *m* eye bolt.

Augenmaaß *n* measuring by the sight; nach dem ~ reguliren (*T*) to regulate (the wire) by the eye sight.

ausästen *v* to clear a tree of its superfluous branches, to prune, to lop the branches of trees (upon the route devised for a t. line).

ausbaggern *v* to dredge, to clear of mud, to deepen by dredging.

ausbauchen *v* to hollow out.

ausbilden *v* to instruct.

Ausbildung *f* instruction; fachliche ~ technical (professional) instruction.

ausbleiben *v* (fehlen) to fail; der Strom bleibt aus the current fails; die Zeichen bleiben aus the signals miss, the signals fail to appear on the slip; (nicht ankommen) not to arrive, not to come; (nicht anwesend sein) to be absent; (fernbleiben) to stay away, to remain away; mehrere Posten sind ausgeblieben there are several mails due.

Ausbleiben *n* (Nichtankunft) non-arrival, absence; (unbegründetes ~ vom Dienst) malingering; (einer Zahlung) failing in the payment; (vor Gericht) défaillance, non-appearance in court.

ausbohren *v* to bore, to bore up, to bore out.

ausbreiten *v* to spread, to diffuse, to extend; sich ~ to propagate, to spread, to diffuse (the electricity diffuses itself over the surface).

ausdehnen, strecken *v* to stretch, to spread, to enlarge; (von Gasen, Metallen) to dilate, to expand; sich ~ to expand.

Ausdehnung *f* extension, dimension, stretch; ~ des Dampfes expansion of the steam; ~ durch Wärme dilation by heat; lineare ~, Längenausdehnung *f* linear expansion; lineare ~ des Drahtes (in die Länge) elongation; Postpackete dürfen in keiner ~ 3 Fuß überschreiten parcels must not exceed 3 feet in any dimension; ~sgrenze *f* (für Packete) limitation of size.

auseinander ... ~nehmen *v* to take (an apparatus) to pieces; to dismantle (an exhausted battery); to clean (a battery) throughout.

Auseinandersetzung *f* (schriftl.) exposé.

Ausfall *m* (an Gebühren ꝛc., Wegfall) deficiency, deficit; (Ergebniß) result.

ausfertigen *v* (ein Schriftstück) to draw up, to make out.

Ausfertigung *f* drawing up; (= Reinschrift) fair copy; in doppelter ~ in duplicate.

Ausfuhr *f* export, exportation; ~verbot *n* non-exportation; ~zoll *m* export-duty, duty on exportation.

ausführen *v* (Waaren) to export; (vollziehen) to carry out, to execute, to achieve, to realize.

ausführlich *adv* at large, in detail, minutely.

Ausführung *f* execution; ~arbeiten *f/pl* execution of the work, realization; ~sbestimmungen *f/pl*, ~übereinkunft *f* detailed regulations for the execution (of a convention); zur ~ bringen to carry into effect (execution).

ausfüllen *v* to fill, to fill up.

ausfuttern *v* (Techn.) to line.

Ausfutterung *f* (Techn.) lining, casing.

Ausgabe *f* (in Rechnungen ꝛc.) expense, expenditure, cost; (Aushändigung) delivery, distribution; (Auflage) edition; ~buch *n* cash book; ~schalter *m* delivery window; ~stelle *f* distributing office, delivery office; ~stempel *m* stamp of distribution; in ~ stellen to place (to carry) to one's debit, to charge one with; die ~n sind ebenso groß wie die Einnahmen the expenses balance the receipts; Brief-~ f. Brief...; Zeitungs-~ f. Zeitungs...

Ausgang *m* (Herkunft) issue, origin; (Auslaß *m*) egression, outlet;

~shafen *m* port of embarkation; ~spunkt *m* starting point; ~szoll *m* export-duty, customs outwards.

ausgeben *v* (verauslagen) to expend; (aushändigen) to deliver, to distribute; (Fahrscheine) to book (to issue) tickets.

ausgleichen *v* (wechselseitige Forderungen) to balance, to adjust, to settle (accounts), to strike a balance; (vergüten) to compensate; (sich gegenseitig aufheben) to compensate; to neutralize each other.

Ausgleichung *f* balance, compensation, liquidation, adjustment, settlement, clearing (of drafts), neutralization; ~sbetrag *m* balance, amount of balance, sum of acquittance; ~sstrom *m* compensation current.

ausglühen *v* to anneal.

ausgraben *v* to dig, to dig out, to excavate (the earth).

Ausgußröhre *f* outlet pipe, waste, spout.

aushaken *v* to unhook.

aushändigen *v* to hand, to deliver.

Aushändigung *f* delivery.

Aushang *m* (Bekanntmachung zur Benachrichtigung des Publikums) notice; (Postbericht) regulation notice; einen ~ in die Augen fallend anbringen to exhibit a notice in a conspicuous place.

aushängen *v* (eine Bekanntmachung *f*) to post, to post up, to exhibit (a placard, a notice).

ausheben *v* (Räder 2c.) to lift off the detents, to lift out.

aushöhlen *v* to hollow, to make hollow, to cut hollow; rinnenförmig ~, auskehlen to groove, to rebate, to carve, to gutter, to flute.

Aushöhlung *f* hollowing, excavation; (= Auskehlung *f*) channel, flute, groove, hollow.

Aushülfe *f* assistance, succor, support; (Person) auxiliary, help; ~-Apparat *m* auxiliary instrument.

auskehlen *v*, (ausnutzen) to channel, to groove; ausgekehlte Holzleiste *f* grooved batten.

auskunden *v* (eine Linie [*T*]) to reconnoitre, to make a preliminary survey.

Auskundung *f* (einer Linie [*T*]), Rekognoszirung reconnoissance, preliminary survey.

Auskunft *f* information; J—m ~ ertheilen to inform *smb*; ich bitte um ~ I beg to be informed.

ausladen *v* to unload, to unlade, to discharge (a cart, a waggon); to unship (a ship); (sich erweitern Techn.) to gain.

Auslader *m* (der Elektrizität) discharger, discharging rod; (Transportschiff, Leichter) lighter; (Lastträger) lighterman, lighter.

Auslaudung *f* (Masch. Techn.) shoulder, shoulder piece; (Erweiterung *f*) gain; (einer Böschung) projection, sally of a talus; (Abladung *f*) unloading, discharging, discharge, landing; ~shafen *m* port of discharge; ~sstelle place (station) of discharge.

Auslage *f* expense, cost; (Vorschuß) disbursement, advance; die ~n wieder erstatten to reimburse the expenses; zu seinen ~n kommen to have (to get) one's expenses repaid.

Ausland *n* foreign country; im ~ abroad; ~s- (in Zusammensetzungen) foreign; ~sbureau foreign branch (department) of the Post Office; ~spost foreign mail; ~starif foreign tariff; ~staxe foreign rate (tax); ~ssendungen foreign (letters) correspondence.

ausländisch *a* foreign.

auslassen *v* (ein Wort) to omit, to leave out; (eine Schlittenumbrehung [*HA*]) to omit (to let go) a rotation of the sledge.

Auslassung *f* (das Fehlen) omission; (Äußerung) expression, utterance; die ~en zu Protokoll nehmen to take down the depositions (declarations, evidence).

auslaufen *v* (Techn.) to run out, to terminate in; (von Schiffen) to put (out) to sea, to sail, to depart.

Auslaugebottich *m* lixiviating vat, dissolving vat.

auslaugen *v* (die löslichen Theile durch Behandlung mit Flüssigkeiten beseitigen) to lixiviate, to dissolve, to edulcorate, to wash, to wash out.

auslegen v (Draht, Kabel) to pay out, to lay out; (erklären) to explain, to interpret; (Geld ~, anlegen) to disburse, to lay out money.

Auslege... ~karren m (bei der Kabellegung) paying-out cart, travelling van; ~vorrichtung f paying-out apparatus.

Auslegung f (Draht, Kabel) paying out, laying out; (Erklärung) explanation.

ausliefern v to deliver, to hand over.

Auslieferung f delivery; stückweise ~ v. Postpacketen an die Transitverwaltung single delivery (delivery à decouvert) of postal parcels to the intermediary office; ~schein m delivery receipt; Einzel-~ delivery in open transit (in open mails); ~spostanstalt f office of exchange.

auslösen v (Techn. Masch.) to uncouple, to release; (einen Wecker) to liberate (start) an alarm (a bell); (die Sperrklinke) to release the ratchet.

Auslösung f uncoupling, escapement; ~ des Druckwerks (HA) escapement for releasing the printing mechanism; ~shebel m detent lever; ~sdaumen m releasing cam; ~smagnet m releasing magnet.

ausmessen v to measure, to measure out, to survey; nach dem Inhalt ~ to gauge; den Rauminhalt ~ to cube.

Ausmessung f measuring, measurement, survey; (von Ebenen) planimetry; (fester Körper) stereometry; (körperliche ~ nach dem Rauminhalt) cubature.

ausmitteln v to ascertain.

ausnuthen v s. auskehlen.

ausnützen v to utilise (every rotation of the chariot [HA]); voll ~ utilise fully (the skill of able telegraphists).

ausrecken v den Draht to strain, to kill, to stretch, to draw out the wire.

ausrichten v, gerade richten (Techn.) to straighten; (einen Auftrag) to execute an order.

ausrollen s. abrollen.

Ausrufungszeichen n (T) signal of exclamation.

ausrüsten v to equip, to fit out (poles with insulators); to supply (a t. line with instruments).

Ausrüstung f equipment, fitting out, outfit; ~sgegenstände m/pl fitting materials.

ausschachten v einen Graben to deepen a trench.

ausschalten v (T) to cut out, to place (an instrument) out of circuit.

Ausschalter m (T) commutator (for breaking contact), current breaker, cut-out, switch.

Ausschaltung f (T) cutting out (of an apparatus).

ausschaufeln v to scoop, to throw out with a shovel, to shovel out.

ausscheiden v (Chem.) to eliminate, to separate, to part, to set free, to educe; (= ausfällen, niederschlagen) to precipitate; (= gewinnen) to eliminate, to extract (metal from its ores); (aus dem Dienste, Amte) to retire from the service, (from Office).

Ausscheidung, Abscheidung, Auslösung (Chem.) elimination, separation, parting; (= Ausfällung f) precipitation; (ausgeschiedener Bestandtheil einer chemischen Verbindung) educt; (Niederschlag m) precipitate; (Gewinnung von Metallen aus ihren Erzen) extraction of metals from their ores.

Ausschlag m (der Nadel) deflection, deviation, throw (of the needle); reading (of the galvanometer); ~swinkel m angle of deflection; größter ~ (bei Schwingungsbewegungen) amplitude.

ausschlagen v to deflect, to deviate (said of the needle); (= effloresziren [Chem.]) to effloresce, to form excrescences.

ausschließen v to prohibit (the sending by post); von der Beförderung ausgeschlossene Gegenstände prohibited (forbidden) articles.

Ausschluß m prohibition (to send

by post); ~feber *f* (*HA*) insulated spring (having the object to place the received currents to earth).

ausschneiden *v* to cut out, to carve.

Ausschnitt *m* (Techn.) notch, indent; (Geom. Math.) sector; (aus einer Zeitung) cut from a newspaper.

ausschreiben *v* (in Buchstaben) to write out in full; (eine Lieferung) to invite tenders.

Ausschreibung *f* proposal, inviting tenders; (öffentliche) inviting tenders by advertising in a public paper; (beschränkte, mit Ausschluß der Öffentlichkeit) inviting tenders only from a limited number of contractors.

Ausschuß *m* refuse, trash, waste; ~ zur Eröffnung unbestellbarer Postsendungen) Returned Letter Office; (*Am*) Dead Letter Office (D.L.O); ~papier *n* waste paper.

ausschweifen *v* (Techn.) to sweep, to cut out, to scallop; (bogenartig) to channel on edge.

Ausschweifung *f* (Techn.) sweep, slope, curvature.

Außen ... ~fläche *f* face, surface, superficies; ~stücke *n/pl* (*P*) objects (parcels etc.) conveyed loosely, loose articles; ~winkel *m* external angle.

außereuropäisch *a* extra-european.

äußerst *a* utmost, extreme; von ~er Wichtigkeit of utmost importance, of the last consequence.

aussetzen *v* (vom Strome *T*) to intermit.

ausspringen *v* (*HA*) to run out.

ausstatten *v* to equip.

Ausstattung *f* outfit, fitting up, equipment, furnishing of the offices; ~sgegenstände *m/pl* articles of outfit, office furniture, fittings, equipments; Haupt=~sgegenstände principal furniture of an office; Neben=~sgegenstände secondary furniture of an office; Postkurs=~sgegenstände *m/pl* mail equipments; ~sverzeichniß *n* inventory of property.

ausstellen *v* to issue (a money order); to draw, to issue (a bill of exchange); to draw up (a document, deed); (zur Schau) to exhibit; (tadeln) to censure, to blame, to find fault with, to except to (against).

Aussteller *m* remitter, issuer (of a money order); drawer (of a bill of exchange).

Ausstellung *f* issuing (of a money order); drawing (of a bill); drawing up (of a document); exhibition (of articles); (Tadel) censure, fault; eine ~ machen to find fault with *smb* (*smt*); eine ~ erledigen to remove the fault found; ~s=gegenstand *m* article (object) of exhibition.

ausstemmen *v* to hollow with a mortise chisel.

ausstrahlen *v* to radiate, to emit.

Ausstrahlung *f* radiation.

ausstrecken, ausziehen *v* to stretch, to draw out (iron).

ausstreichen *v* to cancel, to strike out, to efface; ausgestrichene Stelle words (line, passage) cancelled (obliterated.).

Ausstreichung *f* obliteration, cancellation, defacement.

ausströmen *v* emanate; (vom elektr. Strom) to flow out; (vom Dampf) to escape; ~ lassen to emit.

Ausströmung *f* flow (of the current); emanation, emission; escape (of the steam).

Austausch *m* exchange; s. auch Auslieferung; stückweiser ~ von Postsendungen exchange in open mails; (der Batterieflüssigkeiten) diffusion.

austauschen *v* to exchange, to interchange, to diffuse.

austaxiren *v* to tax, to tax with postage, to charge postage.

austragen *v* (Rechnungsposten in den Büchern) to cancel; (Briefe ꝛc.) to deliver, to distribute.

austreiben *v* (den Saft der Bäume) to turn out, to separate, to set free (the sap).

austrocknen *v* to dry, to drain; (Holz an der Luft) to season timber.

ausüben *v* to exercise, to practice, to execute, to exert.

Ausübung *f* exercise, execution; in ~ des Dienstes in execution of one's duty; zur ~ bringen to carry into execution.

auswalzen *v* to draw out, to roll out (iron, wire etc.).

auswärtig (ausländisch) *a* foreign; auswärtiges Amt foreign office.

auswechseln *v* to replace (instruments, poles etc.), to exchange (correspondence).

Auswechselung *f* renewal, replacing (of materials); (Austausch) exchange; ~s-Postanstalt office of exchange; die absendende ~s-Postanstalt the despatching office of exchange; die ~s-Postanstalt des Bestimmungslandes the receiving office of exchange; ~ von Kartenschlüssen exchange of closed mails; ~ von Briefen im Einzeltransit exchange of correspondence in open transit.

Ausweis *m* (Nachweis, Feststellung) statement, account; (Legitimation) proof of identity; (Rechtfertigung) justification, defence; zum ~ auffordern to ask *smb* to give satisfactory information concerning...; ~karte *f* pass card; ~schreiben *n* pass; letter of introduction.

ausweisen *v* to show, to prove; sich ~ to give a proof of one's identity, to establish one's identity, to prove one's self to be; (sich rechtfertigen) to render account for, to answer for, to justify one's self, to show cause.

ausstanzen, lochen *f* to perforate, to punch, to punch out (the paper slip).

ausweiten *v* ein Bohrloch to ream, to rime, to broach.

auswerfen *v* Porto to charge postage.

auszahlen *v* to pay, to pay out; (baar) to pay down in ready money.

Auszahlung *f* payment; ~sstelle the paying office; ~s-Postanstalt the paying post office, the post office of payment.

auszeichnen *v* to mark out, to note out (trees to be felled), to note down (postage due).

ausziehen *v* (Metalle) s. ausstrecken.

Aversum s. Bauschsumme.

automatisch *a* automatic, self-acting; ~er Absteller *m* self-acting switch & contact breaker.

Axe *f* s. Achse.

Axt *f* ax, axe, hatchet; (zum Behauen der Bäume) lopping axe; kleine ~ adze, addice; ~blatt *n* blade of an axe; ~helm *m* handle, helve.

B.

baar *a* in cash; ~er Bestand *m* cash in hand, balance of cash; ~er Ertrag *m* net proceeds; ~es Geld *n* cash, ready money; ~ zahlen to pay down; ~e Zahlung *f* payment in cash.

Baar... ~bestand *m* s. baar; ~kasse, Schalterkasse *f* till; ~sendung *f* remittance of cash; ~vorrath *m* cash in hand, amount of cash, reserve balance; ~vorschuß *m* cash advance; ~zahlung *f* payment in cash.

Backen *f/pl* (des Apparats) side pieces; (der Drehbank) bearers, cheeks, shears, sides of the turning lathe; (am Schraubstock) jaws, cheeks, chops of a vice.

Bagger *m*, **Baggermaschine** *f* dredging engine, dredger; ~schaufel *f* drag.

baggern *v* to dredge, to clean, to clear.

Bahn *f* (Fahrbahn) run, track; (Eisenb.) railway, railroad, line; im Bau befindl. Eisenb. railway in progress; im Betriebe befindl. ~ railway in operation; breitspurige ~ broad gauge railway; eingeleisige ~ single railway, single way, single line; ~ frei all right, line clear; die ~ ist nicht frei line not clear, caution; schmalspurige ~ narrow-gauge railway; unterirdische ~ underground railway; zweigeleisige ~ double railway, double way, double line.

Bahn ... ~arbeiter *m* railway worker; ~aufseher, Streckenaufseher *m* overseer, inspector of a road; ~beamte *m* employee of a railway; ~betrieb *m* working; ~breite, ~fläche *f* außerhalb der Schiene, Bankett *n* side space; ~hof *m* station, railway station; ~hofsbrief (+) letter to be delivered at the railway station immediately upon the arrival of the train; ~hofsfahrt *f* (P) sending of a mail cart to the railway station to fetch the arriving mail bags etc.; ~hofsinspektor *m* (~direktor, ~vorstand *m*) station master; ~hofsrestauration *f* refreshment room; ~körper *m* length; ~kreuzung *f* crossing at a right angle; ~linie *f* line, railway line, way; ~meister *m* police inspector, watchmen's inspector; ~meisterwagen *m*, Draisine *f* trolly; ~netz *n* system of railroads; ~oberbau *m* permanent way, superstructure; ~planum *n*, Kronlinie *f* formation level; ~polizei *f* police at the railway station; ~polizei-Reglement *n* (+) administrative regulations to secure the safety and regular working of railways; ~post *f*, ~postamt *n* travelling post office (T.P.O), railway post office; ~postbeamter *m* travelling mail clerk; ~postdienst *m* travelling (railway) mail service; ~postschaffner *m* railway mail-guard; ~postwagen *m* railway mail-carriage; ~profil *n* section of a railway; ~räumer *m*, Kuhfänger *m* cow catcher; ~schiene *f* rail; ~schwelle *f* sleeper; ~strecke *f* (~stück *n*) portion of a line, section; ~telegraph *m* railway telegraph; ~verkehr *m* railway traffic; ~wärter, ~wächter *m* railway guard, watchman of the line; ~wärterhaus *n* watchman's house; ~zug *m* train, railway train; s. auch Eisenb.

Bake *f* (Schwimmzeichen für die Grenze des Fahrwassers) buoy; (Feuerzeichen am Strand) beacon; ~ngeld *n* beaconage.

Balken *m* beam, baulk, balk, girder; (Eisenbalken, eiserner Träger *m*) iron girder; (hohler eiserner Träger *m*) box girder; (an der Waage) beam of the scales.

Ballon *m* balloon.

Band *n* (Techn.) band, tie, ligature; ~draht *m* wire of a middle sort; ~eisen, Reifeisen *n* hoop iron, band iron, small flat iron; ~schiene *f* iron band.

Bank *f* (Handels-Anstalt) bank; ~agent *m* broker, bank broker; ~anweisung *f* cheque, check, bank bill; ~conto *n* bank (banking) account; ~depositum *n* deposit in bank; ~geschäft *n* banking business, banking transactions; ~note bank note; ~noten (Papiergeld im Allgem.) *f/pl* paper currency, paper money; ~noten-Umlauf paper circulation; ~statut (Ordnung) statute (regulations) of a bank; ~valuta (Werth) *pl* bank money.

Bankeisen *n* cramp iron (for being beaten in wood); ~eisen, ~haken *m* bench hook; ~schraube *f*, stehender Schraubstock *m* bench vice, bench screw.

Bankett *n* (an Eisenb.) step, side space, banquet; (an Straßen) banquette, raised footpath.

Barometer *m* u. *n* barometer, weather glass; das ~ steigt, fällt the b. is rising, falling; ~stand *m* barometric height.

Barrière *f* (Eisenb.) barrier, guard, railway gate; ~nwärter *m* barrier waiter, gate keeper.

Base *f* (Chem.) base; (einsäurige) monoacid base; (einwerthige) monovalent base; (metallische) metallic base; ~nbildung *f* basification.

Basis *f* base, basis; (einer Säule) foot, base of a column; (Grundlinie *f* beim Vermessen) base line, datum line.

Batterie *f* battery, pile; Ausgleichungs~ equating battery; elektrische ~ electric b.; Feld-~ field b.; galvanische ~ galvanic b.; Gegen-~ counteracting b.; gemeinschaftl. ~ the same b. to work two or more lines from; Graphit-~ graphite b.; hintereinander geschaltete ~ b. in which the cells are placed one after the other (ob. in single series);

Hülfs=~ auxiliary b.; Linien=(Telegraphir=)~ main (line) b.; Lokal=~ local ob. secondary b.; nebeneinander geschaltete ~ quantity battery ob. doubling the quantity, ob. placing two sets of batteries side by side, ob. batteries connected in multiple arc (in parallel series); Polarisations=~ polarization b.; Sand=~ sand b.; sekundäre ~ secondary b.; thermo=elektrische ~ thermo electric b.; Trog=~ trough b.; Uebertragungs=~ translating b.; Untersuchungs=~ testing b.; Verstärkungs=~ (s. Hülfs=~); ~gestell *n* battery stand, rack, frame; Brett des ~gestells shelf; ~klemme *f* (brass) binding screw, terminal clamp; ~kontakt *m* contact to which the b. is connected; ~platte *f* battery plate; ~pol *m* pole; ~prüfer *m* detector; ~rückstände *m/pl* black mud; ~schaber *m* scraper; ~schaltung *f* connection of the b.; ~schrank *m* cupboard, battery box; ~umschalter *m* commutator, battery switch; ~zimmer battery room, closet; eine ~ ansetzen to set up a battery; eine ~ auseinandernehmen to take a battery to pieces; eine ~ speisen to supply a battery; s. auch Element.

Bau *m* construction, erection of a building; ~bedürfniß=Nachweisung estimate of the costs for the purchase of sites and buildings; ~behörde *f* board of works; ~gerüst *n* scaffold; ~holz *n* store timber, lumber, timber; ~materialien *n/pl* building materials; ~meister *m* architect; im ~ befindliche T. Linie t. line in progress; Telegraphenbau s. Telegraphen....

bauen *v* to build (a house), to construct (a t. line), to make, to construct (a road); (im Tagelohn) to build by the day.

Baum *m* tree; ~isolator, ~träger *m* swinging insulator; ~scheere *f* pruning shears, stock shears.

Bausch und Bogen, in ~ in the lump.

Bausch ... ~kauf, Kauf in ~ und Bogen *m* purchase in the lump (in the bulk); ~summe *f*, Bauschquantum *n* average sum, sum total.

Bause, Pause, Durchzeichnung *f* calking.

bausen, pausen, durchzeichnen *v* to pounce.

Bayonet=Verschluß *m* bayonet catch.

beachten *v* to mind, to take notice of; nicht ~ to disregard.

Beachtung *f* consideration; unter ~ der Vorschriften in consideration of the instructions; zur ~ to be guided (by the instructions given).

Beamtenmangel *m* insufficiency of the staff.

Beamter *m* officer, public functionary, one holding an office; Apparat=~ s. App.; Büreau=~ clerk; Zivil=~ officer in the civil service.

beanspruchen *v* to claim; (Mech.) to strain.

beanstanden *v* to object to, to refuse; beanstandete Sendungen correspondence excluded from conveyance, prohibited mail matter.

beantragen *v* to make application (request), to request, to apply to *smb* for *smt*; to propose.

beaufsichtigen *v* to survey, to superintend, to inspect, to control.

Beaufsichtigung *f* superintendence, inspection, control, supervision (of the telegraph lines).

beauftragen *v* to commission, to charge one with.

Beauftragter *m* agent, deputy; ständiger ~ der Ober=Postdirektion representative of the Post Office Department.

Bedarf *m* want, demand; ~s=Nachweisung *f* demand for supply, estimate of demand, estimate of the necessary materials, requisition.

bedienen *v* einen Apparat to work an apparatus, to have charge of a telegraph instrument; eine vom Landbriefträger bediente Ortschaft a place served by the rural postman.

Bedienung *f*, einem Anwärter wird die Befähigung zur selbständigen ~ des T.=Apparats zugesprochen the candidate is declared to be capable (ob. is found competent) to take sole charge of a telegraph instrument, ob. the candidate is fully competent for the transmission of public messages.

bedingt *a* conditional, qualified; ~e Annahme *f* conditional (qualified, partial) acceptance.

Bedingung *f* condition, clause, stipulation, terms; ~en für Lieferungsdraht (*T*.) specification of the wire; unter der ~ daß on the condition that; unter jeder ~ at all events; ~sweise *a* upon condition, in a qualified sense.

Bedürfniß *n* need, want, requisite, exigency; nach Maßgabe des dienstlichen ~es according to the exigencies of the service; im ~falle if wanted (required); je nach ~ as it may appear necessary.

Befähigung *f* für den T.-Dienst zeigen to display aptitude for the duties of a telegraphist; ~ zur Bedienung des Apparats (*T*.) s. Bedienung.

befestigen *v* to fasten, to fix, to attach; (eine Stange mit Bändern) to tie a pole; (mit Bolzen) to bolt, to fasten with bolts; (mit Klammern) to cramp.

Befestigung *f* fastening, fixing; ~sklemme *f* clamp; ~smittel *n/pl* retaining materials.

befördern *v* to despatch, to forward, to transmit (telegrams); to transport, to convey (stores); (Telegramme mit der Eisenb.) to send telegrams by train; (in ein höheres Amt) to promote; weiter~ (Briefe) to re-direct (letters); (Telegramme) to re-transmit (telegrams).

Beförderung *f* forwarding, conveyance, despatch; (in ein anderes Amt) promotion; von der ~ ausgeschlossene Gegenstände forbidden articles; zur ~ zugelassen admitted for conveyance, allowed to be transmitted; in der ~ begriffen while being transmitted; (Weiterbeförderung) re-direction (of letters); re-transmission (of telegrams).

Beförderungs...~dienst *m* conveyance (of mails), transmission; Beamte im Telegramm-~dienst operators employed in the actual transmission of telegrams; ~gebühr *f* charge for transmission; ~gelegenheit *f*, ~mittel *n* conveyance, means of conveyance; außerordentliche ~mittel *n/pl* in außergewöhnlichen Bedarfsfällen supplementary means of conveyance in extraordinary emergencies; ~vermerk *m* notice of (completed) transmission; ~weg *m* route of transmission, route (line), viâ; ~weise *f* method of transmission; ~zeit *f* time of transmission.

befragen *v* to examine, to interrogate *smb* about *smt*.

befreit *a* von Porto exempt from postage.

Befreiung *f* von Portozahlung, vom Dienste exemption from postage, from duty (service).

Befugniß *f* right, competence, authority; J—m die ~ ertheilen to authorize *smb*; seine ~e überschreiten to overstep (to go beyond) one's competence.

befugt sein *v* to be authorized to, to be entitled to, to have a right to.

Befund *m* (Gutachten) opinion, estimation; nach ~ according to circumstances.

beglaubigen *v* to attest, to confirm, to verify, to certify; beglaubigte Abschrift authenticated (attested) copy; Unterschrift beglaubigt durch ... signature verified by ...

Beglaubigung *f* attestation, authenticity; ~ der Unterschrift verification of the signature; ~sschein *m* certificate; zur ~ dessen in witness whereof; ~sschreiben *n* credentials, letter of credence (accreditation).

Begleit..~adresse s. Packetadresse; ~papiere *n/pl* way bills and other papers relating to the mail carried by a driver or a guard; ~schein *m* (zu zollpflichtigen Stücken) declaration; ~verzeichniß *n* accompanying list, statement as to the contents of a mail, way bill; ~verzeichniß für Packete parcel bill; ~zettel *m* way bill; ~zettel zu Extrapost u. Kurieren (+) way bill of contents.

Begleitung *f* der Eisenbahnzüge

durch Postpersonal railway mail transportation in charge of postal officers.

Begleiter m (einer Post) mail guard; (von Bahnhofstransporten) railway messenger; (von Bahn oder Dampfschiffsposten) route-agent.

begrenzen v to border, to bound, to limit.

Begrenzungs... ~schraube f stop screw; ~stift m stop, pin.

Behälter m reservoir; Wasser-~, Zisterne f tank.

behandeln v (Techn.) to work, to manipulate; to treat (telegrams), to deal (with the mail); (Holz mit Dampf) to treat timber with steam, to steam timber; (ein Thema) to deal with (in this book the technical expressions of P & T are dealt with).

behändigen v to deliver, to hand, to hand over.

Behändigung f delivery; ~sgebühr f (+) fee in addition to postage for the delivery of judicial summons; ~sschein m (+) bill accompanying judicial summons.

Behandlung f der Telegramme treatment of the telegrams; ~ der Post dealing with the mail.

Beharrungs... ~moment, Trägheitsmoment n moment (momentum) of inertia; ~vermögen n, Trägheit f inertia, force of inertia, force of continuance; ~zustand m permanence, resistance.

behauen v to hew; (Bauholz ~) to square, to hew timber.

behobeln v to plane, to plane smooth.

Behörde f authority, office; Staats-~ office of state, government; höhere ~ superior authority.

bei... (praefix), an-~, ~gefügt, ~gehend, ~geschlossen adv herewith, hereby, enclosed, annexed, by the present, under this cover.

beidrücken v das Siegel to set the seal to ...

beifügen v to add, to enclose, to annex, to affix.

beigefügt a enclosed, annexed, affixed.

Beihülfe f allowance, aid, assistance.

Beil n hatchet; großes ~ ax, axe; ~futteral n hatchet case; ~stiel m hatchet helve.

Beilage f einer Zeitung supplement of a newspaper; (eines Berichts) enclosed piece, enclosure.

Beimengung, Beimischung f addition, admixture.

Beispiel n, zum ~ exempli gratia (e. g.); for example (f. e.); for instance (f. i.).

Beißzange, Kneipzange f nippers, cutting nippers, tweezers, pincers.

Beitrag m contribution; (an Geld zu Kosten u. dgl.) share, quota; (zu einem Bericht, einem Buch o. dgl.) materials pl.

beitreiben v (Zahlung) to collect, to get in (money).

Beitreibung f collection; ~ von Porto im Wege der Pfändung (forcible) collection of postage by means of seizure (attachment); ~sbeschluß m (+) decree (by an administrative authority) to collect forcibly fees or any other amount due to the Treasury; executory decree.

beitreten v to give one's adhesion; (einem Vertrage) to accede to, to consent to, to adhere to (a treaty, a convention); (der Meinung eines Anderen) to assent to.

Beitritt m accession, consent, adhesion; ~surkunde f act of accession.

Beischlitten m supplementary sledge.

Beiwagen m supplementary vehicle, by-coach, extra-coach, by-chaise; (der Bahnpost) supplementary mail van; (Eisenb.) by-waggon.

Bekanntmachung f (Aushänge im Schalterflur) notice to the public; (in den Zeitungen) advertisement.

bekiesen v to gravel, to cover with gravel; eine Straße ~, beschottern to ballast a road.

bekleben v to paste, to paste on; (Packete mit Zetteln) to label (parcels).

Beklebezettel *m* (auf Packeten) label, facing slip; f. auch Aufgabezettel u. Leitzettel; ~ für das Gepäck (Eisenb.) way bill.

Beklebung *f* labelling (of parcels).

bekleiden *v* (Techn.) to cover, to serve (a wire with guttapercha).

Bekleidung *f* (Techn.) covering, serving (of a wire); (Belegung *f*) coating (of a Leyden jar, of a condenser).

beladen *v*, to load, to lade, to freight, to charge; ~e Hin-, ~e Rückfahrt f. Hin... u. Rück...

Beladung *f* load, charge.

Belag, Beleg *m* (Techn.) covering, lining; (Elektr.) coating; (Beweisstück) voucher; als Ausgabe-~ dienen to serve as voucher for expenses incurred.

belasten *v* (Mech.) to weight, to load; (ein Gestänge) to load a t. structure; (mit einer Schuld) to debit, to place to one's debit, to charge to one's account.

Belastung *f* (Mech.) load, weight; ~ des T.-Gestänges load on the t. structure; (Beanspruchung) strain; f. auch a. T. unter load; (eines Grundstücks) encumbrance of a real estate; (mit einer Schuld) debit.

belaufen *v*, sich ~ auf... to amount to, to come to; die zu erstattende Summe beläuft sich auf... the sum to be refunded amounts to...

belegen *v* (Techn.) to line, to cover, to serve; (darthun) to support by documents & vouchers; (einen Platz) to secure (a seat); (mit Arrest) to seize; (mit Strafe) to inflict a punishment on *smb*; to set a fine upon *smb*.

Belegung *f* (des Kondensators, der Leydener Flasche) inner & outer coating of the condenser, of the Leyden bottle (jar); layer (of mica ob. tinfoil).

beleuchten *v* to illuminate; (fig.) to throw light upon, to view closely.

Beleuchtung *f* illumination, lighting; (elektrische) electric lighting.

bemerken *v* to mark, to note; ich möchte in dieser Sache noch ~ I would remark (observe) in this matter.

bemerkenswerth *a* worth remarking; deserving notice.

Bemerkung *f* remark, observation, notice; die Spalte „~en" the column for observations.

bemessen *v* to determine.

Bemessung *f* der Zahl der Elemente determining the number of elements.

benachbart *a* neighboring, adjacent (countries).

benachrichtigen *v* to inform, to advertize *smb* of *smt*.

Benachrichtigung *f* information; (schriftliche) notice in writing; um ~ bitten to ask for information; ohne vorgängige ~ without previous information; ~szettel (*T P*) notice; letter of advice.

benachtheiligen *v* to prejudice, to injure.

Benachtheiligung *f* prejudice, harm, detriment; ohne ~ der Dienstzucht without prejudice to the discipline.

berauben *v* (die Post) to rob; (Briefe) to steal from a letter, to rob letters of their contents.

Beraubung *f* (eines Briefes, Packets) spoliation (of a letter, package); (der Post) depredation (of mails); Post Office robbery.

berechnen *v* to calculate, to estimate; die Taxe wird berechnet nach.. the rate of tax is calculated according to...

Berechnung *f* calculation, estimation.

berechtigen *v* to entitle, to give a right (power), to authorize; der berechtigte Empfänger the authorized recipient.

Berechtigung *f* authorization, right, privilege; ~sgrund *m* title; ~skarte *f* pass card.

Bereich *m* district, province; das liegt nicht in meinem ~ that comes not within my province.

bereit *a* ready for, prepared for, resolved on; sich ~ finden lassen to be ready.

Bereitschaft *f* readiness.
Berg *m* mountain; ~ab *adv* down hill; ~auf *adv* up hill.
Bericht *m* report, relation; einen ~ abstatten, vorlegen to make, to submit a report; statistischer ~ statistical return; tabellarischer ~ tabular statement; unverzüglich ~ erstatten to report at once (without delay); laut ~ (kaufm.) as per advice; ~sjahr *n* year under report.
berichten *v* J—m über etw. to report to *smb* on *smt*.
berichtigen *v* to correct, to rectify.
Berichtigung *f* correction, rectification; ~sbogen *m* correction sheet; ~stelegramm telegram of correction (rectification), correcting telegram; ~ der angekündigten Wortzahl correction of the number of words signalled.
Bernstein *m* amber, yellow (mineral) amber, electron, electrum; schwarzer ~ jet.
berücksichtigen *v* to take notice of, to mind, to take into consideration.
Berücksichtigung *f* consideration; in ~ der Verhältnisse considering the circumstances.
Berufung *f* appeal; unter Vorbehalt der ~ reserving an appeal.
Beruhigungs= (Richt=) Magnet *m* directing magnet.
berühren *v* to touch, to come into contact; die Drähte ~ sich the wires are making contact.
Berührung *f* (der Drähte) contact; (in Folge schlechten Wetters) weather contact; die Drähte stehen mit der Erde in ~ the wires make contact with the earth.
Berührungs ... ~elektricität *f* galvanism, voltaism, galvanic electricity; ~fläche *f* surface of contact.
beschädigen *v* to damage, to spoil; ein beschädigter Apparat an instrument out of order; vom Seewasser beschädigt sea-damaged.
Beschädigung *f* injury, damage.
Beschaffenheit *f* state, quality, condition, temper; je nach ~ der Umstände according to the nature of circumstances.

Beschaffung *f* purveyance, providing for, furnishing with; ~s=kosten *pl* first cost, prime cost.
beschäftigen *v* sich mit etwas to occupy one's self with *smt*, to engage one's self; J—n ~ to employ *smb*.
Beschäftigung *f* business, occupation, employment; ~sort *m* location; ~szeugniß *n* certificate of employment.
Bescheid *m* answer, order; bis auf weiteren ~ until further orders; (mit rechtlicher Wirkung) award.
bescheiden *v* to inform *smb* about (of) *smt*; sich ~ to acquiesce in, to be contented, to concede.
bescheinigen *v* to attest, to certify, to approve; (den Empfang von Geld) to give a receipt; (den Empfang eines Telegramms) to acknowledge the receipt.
Bescheinigung *f* certificate, acquittance, receipt.
beschicken *v* (eine Batterie [*T*]) to fill, to charge (a battery).
Beschlag, in ~ nehmen to arrest, to attach, to seize.
Beschlagnahme *f* arrest, seizure (of mail matter); ~ von Postsendungen in Porto-Uebertretungsfällen seizure of correspondence on account of embezzlement of postage.
beschleunigen *v* to accelerate, to quicken; beschleunigte Bewegung *f* increasing motion.
Beschleunigung *f* acceleration, rate of variation of the velocity (speed).
beschließen *v* (sich vornehmen) to determine, to resolve upon; etwas mit einander ~ to agree upon.
Beschluß *m* resolution, conclusion, determination.
beschneiden *v* to dress, to pare; (die Zweige) to clip, to lop (the branches).
beschottern *v* to ballast, to gravel.
Beschotterung *f* ballast, gravelling, ballasting.
beschränken *v* to confine, to limit, to bound, to restrain; (sich auf etwas) to restrict one's self to

smt; der Spielraum des Hebels wird durch Kontraktschrauben beschränkt the motion of the lever is limited by screw-stops; beschränkte Wortzahl *f* limited number of words.

Beschränkung *f* limitation, restriction.

Beschwerde *f* complaint; ~ erheben to complain of; eine ~ zurückweisen to reject a complaint; ~buch *n* complaint book.

beschweren *v* (mit etw.) to load, to charge; beschwerter Brief *m* insured letter; sich bei J— m über etw. ~ to make complaints to *smb* about *smt*.

beseitigen *v* eine Leitungsstörung (*T*) to remove a fault.

besetzen *v* einen Apparat to staff an instrument.

besetzte Linie *f* (*T*) occupied line; stark ~ ~ line crowded with business, busy circuit.

besichtigen *v* to inspect, to examine.

Besichtigung *f* inspection, examination; ~sreise *f* tour (journey) of inspection.

besolden *v* to pay, to have (to keep) in pay, to give wages (salary).

Besoldung *f* pay, salary, wages.

besonder *a* particular, peculiar; ~e Einnahmen *f/pl* (im Etat) extra receipts; ~ Leitung special wire; ~es Telegramm special telegram.

besonders *adv* separately.

Bespannung *f* number of horses put to a carriage; (Gespann) team; die ~ der Post ist einem Unternehmer übertragen the horsing of the mail is done by contract.

bespinnen *v* to spin, to cover; mit Seide besponnener Draht silk-covered wire.

bestallen *v* to appoint to a place, to invest with an employment.

Bestallung *f* appointment, commission; ~surkunde *f* commission, letters patent.

Bestand *m* (Kassen=~) balance, remainder; (Güter ꝛc.) stock; Baar= ~ *m* balance in cash, clear amount; ~ an Freimarken value of postage stamps in store; (~ der sich aus dem Kassenabschluß ergiebt) due balance, balance in hand; (~ an Packeten in der Packkammer) amount of parcels in hand (in store); ~ aufnehmen (einer Kasse) to prepare the balance, (von Waaren) to take stock of; eiserner ~ s. eisern; ~theil *m* (Chem., Phys.) element, constituent, component, ingredient.

beständig *a* constant, uniform; ~e Größen *f/pl* (Math.) constant quantities; ~er Preis *m* fixed, (standard) price; ~es Wetter *n* settled weather.

bestätigen *v* (probeweis beschäftigte Beamte) to confirm the appointment of probationers; (einen Vertrag) to ratify a treaty (convention).

Bestätigung *f* confirmation, ratification; legislative ~ eines Gesetzentwurfs consolidation.

bestehen *v* to exist; (eine Prüfung) to pass an examination; (auf etw.) to insist upon, to persist in; (aus etw.) to consist of; ~de Linien *f/pl* (*TP*) lines in existence.

Bestell ... ~amt *n* delivering office (s. auch Bestellung); ~bericht *m* (des Briefträgers) remark of the letter carrier showing to whom any article has been delivered or stating cause of non-delivery of each article returned; ~bezirk *m* (der Postanstalt) delivery district; (des Briefträgers) beat, round, walk; innerhalb des ~bezirks within the delivery; ~buch *n* (des Briefträgers) postman's (letter carrier's, rural carrier's) register, delivery book; ~dienst *m* delivery service; ~einrichtung *f* delivery system; ~fahrt *f* delivery trip; ~gang *m* delivery, walk of the (rural) post messenger; ~gebühr *f*, ~geld *n* porterage, carrier's (messenger's) fee; ~schein *m* (~zettel *m*) form of requisition; ~schreiben *n* requisition; ~zeit *f* time (term) of delivery; ~zettel *m* requisition.

bestellen *v* (Briefe) to deliver; (Auftrag geben) to order; (einen Auftrag ausrichten) to acquit one's self of a commission (an order); (Plätze) to secure, to order (a seat in the mail coach); (kommen lassen) to send for; (einen wohin) to make an appointment with *smb* to meet at a certain place.

Bestellung *f* (s. auch Bestell...), (Aushändigung) delivery; (Auftrag) order, commission; (Besorgung) execution; (Verabredung) appointment; ~ von Zeitungen s. Zeitungs... falsche ~ mis-delivery; ~sbezirk *m* district of delivery, delivery; (des Briefträgers) beat, round, walk; ~sbogen *m* form of requisition; ~sbuch *n* (des Briefträgers) register; ~sdienst, ~sgeschäft delivery service (business); ~skarte *f* delivery docket; ~sstempel *m* delivery stamp; ~svermerk letter carrier's remark (note).

bestimmen *v* (anordnen) to order, to dispose; (festsetzen) to determine, to fix; (ausersehen) to designate; quantitativ ~ (Chem.) to estimate; zu bestimmten Stunden at fixed hours.

Bestimmung *f* (dienstliche) regulation (of the service); (Ort) destination; ~samt *n* office of destination, terminal office; ~sland *n* country of destination; ~sort *m* place of destination; ~postanstalt *f* arrival office.

bestrafen *v* to punish; (mit Geldstrafe belegen) to fine, to inflict a fine.

Bestrafung *f* punishment; (Geldstrafe) fine.

betheiligt *a*, die ~en Parteien the parties interested.

Betheiligung *f* participation (in).

Beton, Grundmörtel *m* concrete, beton; ~unterlage *f* concrete foundation.

Betracht *m* consideration, regard, respect; in ~ considering; in ~ ziehen to take into consideration; es kommt nicht in ~ it is out of the question.

Betrag *m* amount, sum, sum total; (eines Antheils) quota; netto ~ net, clear amount; roher, (brutto) ~ gross amount.

betragen *v* to amount to.

Betreff *m* regard, respect, consideration; in ~ (eines Dinges) with respect to, in consideration of.

betreffen *v* to concern; was Ihr Verlangen betrifft concerning your request; die ~den Personen the respective persons.

Betrieb *m* working, management; Postwege *m/pl*, Telegr. Linien *f/pl* im ~e mail routes, t. lines in operation; in ~ setzen to put in operation; ~ der Leitungen working of the lines; ~sanlage *f* plant; ~sbureau *n* Office for the traffic, manager's office; ~sdienst *m* technical service; ~sdirektor *m* managing officer, manager; ~sfähig *a* in working condition; ~sjahr *n* fiscal (financial) year; ~skapital *n* stock, funds, capital employed; ~skosten *pl* working expenses, working cost; ~sleitung *f* management; ~smaterial (*T P*) stock in trade; ~smittel *n/pl* working stock; ~spersonal *n* personnel, staff for the traffic.

beurkunden *v* to attest by documents, to verify, to certify, to authenticate.

Beurkundung *f* verification, authentication.

beurlauben *v* to grant leave of absence; beurlaubt *a* absent on leave.

Beurlaubung *f* leave of absence; (der Soldaten) furlough; (s. Urlaub).

Beutel *m* bag, (*Am*) sack; Brief-~ *m* mail bag; Geldbrief-~ *m* bag containing insured letters; Versteck-~ *m* transit bag, (*Am*) inner sack; ~messer knife for opening mail bags; ~schloß *n* mail lock; ~stück *n* parcel enclosed in the mail bag; den ~ wenden to turn a bag inside out; fehlgesandter ~ a bag out of course; ein schadhafter ~ a bag out of repair, a damaged bag.

bevollmächtigen *v* to authorize, to empower.

Bevollmächtigte *m* attorney, authorized agent; (Inhaber einer Prokura) proxy, procurist.

Bevollmächtigung *f* authorization.

bewachen *v* to watch, to guard.

Bewachung *f* supervision (of t. lines).

bewältigen *v* (eine Arbeit) to finish; geeignete Maßregeln treffen, daß die vorliegende Korrespondenz noch vor Dienstschluß aufgearbeitet wird (*T*) to take such measures (to dispose so) that the whole business is finished before the official closing hour.

bewandt *a*, bei so ~en Umständen under the existing (given) circumstances.

Bewandtniß *f* condition, state, case; es hat damit folgende ~ the case is this; es hat damit eine andere ~ the case is quite different.

bewegen *v* to move; sich ~ to move; sich auf und nieder ~ to work up and down; ~de Kraft *f* motive power.

beweglich *a* mobile, moveable; ~e Rolle *f* moving pulley.

Bewegung *f* motion, movement; in ~ setzen to actuate; (Masch.) to throw into gear; auf- und niedergehende ~ up and down motion; beschleunigte ~ increasing motion; rotirende ~ rotary motion, rotation.

Beweis *m* (im gesetzlichen Sinne) evidence, proof, instance; ~stück evidence, article of proof; zum ~e dessen in support (proof) of *smt*; in witness whereof.

beweisen *v* to show, to prove, to demonstrate.

bewenden *v*, es dabei ~ lassen to acquiesce in.

Bewenden *n* rest, end; dabei hat es sein ~ there the matter may rest.

bewerben *v* sich um etw. to sue for, to apply for, to solicit.

Bewerber *m* candidate, applicant, suitor.

Bewerbung *f* um ein Amt application (for a place).

bewickeln *v* to wrap in, to inwrap (a parcel with paper); to wind (the coils of an electromagnet).

bewirken *v* (Mech.) to effect, (Chem. Techn.) to exert.

bezahlen *v* to pay; (abschlägig) to pay in part (on account); (baar) to pay down (in cash); (die Kosten) to defray the expenses; (eine Rechnung) to settle an account; (terminweise) to pay by instalments; (im Voraus) to pay in advance.

bezeichnen *v* to mark, to mark out, to distinguish by a mark, to denote, to indicate, to designate; (benennen) to denominate.

Bezeichnung *f* marking, denotation, designation, denomination.

bezetteln *v* to label.

Bezettelung *f* labelling.

beziehen *v* (Zeitungen) to subscribe to; (Waaren) to be provided (supplied) with, to get, to have; (Gehalt) to receive salary; (einen Wechsel auf J—d) to draw a bill upon *smb*; sich ~ auf to refer to, to relate to.

Beziehung *f* mit J—m unterhalten to sustain a relation with *smb*; in dieser ~ in that respect; die telegraphischen ~en *f/pl* the telegraphic relations.

Bezirk *m* division, district; ~s-Aufsichts-Beamter *m* officer (official agent) of the district; ~s-Leitungs-Revisor *m* (*T*) supervising officer of the district; ~s-Materialien-Magazin *n* store; ~s-Materialien-Verwalter *m* storekeeper; ~sverfügung *f* post office circular.

Bezug *m* (von Zeitungen) subscription to; ~sdauer *f* term (period) of subscription; ~squelle *f* place (or office) through which one is provided (supplied) with *smt*; ~ nehmen auf *v* to refer to.

Biege . . . ~maschine *f* (Draht) bending machine; ~zange, Drahtzange *f* wire plyers.

biegen *v* to bend, to curve.

biegsam *a* pliant, flexible.

Biegsamkeit *f* flexibility.

Biegung *f* bent, curve, inflection; (einer Straße, eines Flusses) bend, turn of a street, of a river.

Biegungs . . . ~elasticität und Festigkeit, relative Elasticität und Festigkeit *f* elasticity and strength of flexure; ~festigkeit strength of flexure, transverse strength; ~moment *n* moment of flexure (flexion).

Bifilar-Magnetometer *n* bifilar magnetometer.

bildlich *a* figurative, pictorial; ~e Darstellung *f* figurative (pictorial) representation.

Bildung *f* formation, shape; (Erziehung) education, learning; ~stufe *f* degree of learning.

Billet s. Fahrschein.

billig *a* equitable, just, fair; (wohlfeil) cheap, low; zu einem ~en Preise at a moderate price; auf dem ~sten Wege *m* on the cheapest (least expensive) way.

billigen *v* to approve of, to allow, to grant.

Billigung *f* approbation, approval.

Binde . . . ~draht *m* binding wire; ~kraft, Atombindekraft *f* atomicity, atomic combining-capacity; ~kraft *f* (z. B. des Mörtels ꝛc.) binding power; ~mittel *n* medium; ~strich *m* hyphen; durch ~strich verbundene Worte words coupled by hyphens.

binden *v* (einbinden) to bind a book; (absorbiren [Chem., Phys.]) to absorb, to take up; (mit einander ~ [Chem.]) to combine, to unite, to bind chemically; (festbinden) to bind (the wire to the insulator).

Binder *m* (*T*) binder.

Bindfaden *m* pack thread, string, twine; ~abfälle *m/pl* waste twine; ~kapsel *f* twine box; ~rolle *f* pack thread reel.

Bindung *f* (der Drähte) binding, binder; (chemische Verbindung, Vereinigung) combination, combining.

binnen *adv* within; ~ 8 Tagen within a sennight; ~ jetzt und 6 Monaten within a six-month from now (from this).

Binnen . . . ~gewässer *n* inland water; ~hafen *m* inner harbour, inner port, wet dock; ~handel *m* inland (domestic, home) trade; ~transport *m* inland transportation; ~verkehr *m* inland traffic.

bis *adv* till, until; bis zu to, up to; bis zu 5 kg einschließlich up to 5 kg inclusively.

Bitte *f* entreaty, request, solicitation; ~ zu wiederholen (*T*) please (to) repeat.

Bittersalz *n* bitter salt, Epsom salt, sulphate of magnesia (magnesium).

blank *a* blank, white, bright, clean; ~er Draht *m* uncovered wire; den Draht ~ machen to scour the wire.

Blank *n* (im Typenrad des *HA*) blank portion, blank space; ~taste *f* blank key; Buchstaben-~taste letter blank key; Zahlen-~taste figure (cipher) blank key.

Blase *f* (v. Flüssigkeiten) bubble; ~n entwickeln to bubble; Luft-~ air bubble.

Blatt *n* leaf; ein ~ Papier a sheet of paper; (einer Säge) blade of a saw; (einer Schaufel) blade (pan) of a shovel; (eines Tisches) table leaf; ~feder *f* plate spring.

Blätter . . . ~Kondensator *m* condenser made of alternate sheets of thin (silver) paper saturated with paraffin and tin foil (*Varley*), condenser insulated with sheets of specially prepared gutta percha (*Smith*); ~-Magnet *m* laminated magnet, lamellar magnet.

Blau . . . ~schreiber *m* s. Apparat; ~stift *m* blue chalk.

Blech *n* sheet metal, sheets, sheet, plate; gewalztes ~, Walzblech rolled plate; (gewelltes) corrugated plate; schwarzes ~, Schwarzblech black iron-plate; weißes ~, Weißblech white iron-plate, tin plate; ~büchse *f* tin can; ~magnetkerne *m/pl* sheet-iron cores; ~scheere *f* plate shears, cutters.

Blei *n* lead; ~kabel, ~rohrkabel *n* lead covered cable; ~loth *n* lead plummet; ~loth, (Löthung) soldering with lead, lead-solder; ~platte *f* lead plate; ~rohr *n* lead pipe, leaden tube; ~rohrkabel *n* s. Bleikabel; ~stift *m* pencil, black lead-pencil; ~verschluß *m* (Plombe) lead; schwefelsaures ~oxyd *n* sulphate of lead.

bleibende Zeichen gebender Apparat (*T*) recording instrument.

Blitz *m* lightning; ~schlag *m* lightning (electric) discharge; ~schutzvorrichtung *f* arrangement to protect the instruments etc. against lightning; ~strahl *m* flash of lightning.

Blitzableiter *m* lightning conductor, lightning rod, lightning discharger, lightning arrester, lightning guard, paratonnere; 𝔓latten~ plate lightning protector, grooved plate discharger; Schneiden~ lightning protector with serrated plates (with an arrangement of opposite edges cut out in the form of sharp teeth); Spindel~ lightning protector with fusible wire; Spitzen~ lightning protector with opposing points, point discharger, comb protector; Stangen~ lightning protector to poles; Vacuum~ rarefied air lightning discharger, ſ. auch Vacuum a. T.; ~iſolator *m* lightning rod insulator; ~pult *n* stand for the lightning protectors; ~ſeil *n* wire strand serving as lightning conductor.

Block *m* block, log (of wood, of stone etc.); (Mech.) block, pulley, system of pulleys; ~ſignal *n* block signal; ~ſignalſyſtem *n* block system.

bloß *a* naked, bare, uncovered; ~legen *v* to bare.

bloßgehende Packete *n/pl* separate parcels.

Blutlaugenſalz *n* (rothes) ferricyanide of potassium; (gelbes) ferrocyanide of potassium.

Bock *m* (*T*) ſ. Doppelſtänder; (einer Kutſche) coach box, driver's seat.

Boden *m* (Erde) ground, earth, soil; (eines Gefäßes) bottom.

Boden ... ~beſchaffenheit *f* nature of the ground; ~bildung *f* geological formation; ~einſchnitt *m* cutting in the ground; ~erhebung *f* rising ground; ~platte, Grundplatte *f* einer Maſchine, eines Apparats bottom plate, bed plate, floor plate; ~ſatz *m* sediment, deposit, residuum; einen ~ſatz ablagern to deposit.

Bogen *m*, der Volta'ſche ~ voltaic arc; ~licht *n* arc light; ~lichtlampe *f* arc lamp.

Bohle *f* plank, thick board.

Bohr ... ~knarre, Bohrratſche *f*, Ratſchbohrer *m* ratchet drill, rock drill; ~kurbel *f* brace, crank brace, brace drill; ~zeug *n* boring implements, boring tools.

bohren *v* to bore, to drill, to perforate; (in Stein) to jump; (einen Tunnel) to cut, to drive a tunnel.

Bohrer *m* borer, bore, drill, auger, tap; (Bergbohrer) terrier, jumper; (großer ~ zur Stangen-Unterſuchung [*T*]) auger; (Schraubenbohrer) twisted auger, screw auger, screw drill.

Boie, Boje *f* buoy, beacon; wachende ~ floating buoy.

Bolzen *m* bolt, pin; flachköpfiger ~ flat-headed bolt; ~ mit flachrundem Kopf round-headed bolt; ~ mit Splint ob. Vorſtecker, Splintbolzen *m* eye bolt; ~ mit verſenktem Kopf countersunk-headed bolt; ~ mit viereckigem Kopf square-headed bolt; ~ mit Widerhaken barb bolt, rag bolt, jag bolt.

Bolzen ... ~blech *n*, Unterlagſcheibe *f* collar, washer; ~kopf *m* bolt head; ~ring *m* shackle.

Börſe *f* exchange, change; ~nzelle *f* (*F*) call box on change.

Borſtenpinſel *m* painting brush (made of bristles).

Böſchung *f* slope, sloping talus, acclivity; ~smauer *f* scarp, steep.

Boſſekel *m* ſ. Poſſekel.

Bote *m* messenger, carrier, postman; ~namt *n* messenger's office, office whence messengers use to be despatched; ~ndienſt *m* service of a messenger; ~gang *m* a messenger's walk (errand); ~lohn *m* messenger's fee (pay); ~meiſter *m* inspector of messengers; ~poſt *f* a mail conveyed by a messenger (runner); Brief~ *m* postal (post office) messenger; eigener (beſonderer) ~ special messenger, express; Eil~ *m* special messenger; Fuß~ *m* foot messenger, runner; Hülfs~ *m* auxiliary (temporary) messenger (carrier); Telegraphen~ *m* telegraph messenger.

Bottich *m* trough, tub, tun.

Bouſſole *f*, **Kompaß** *m* compass.

Brack *m* refuse.

Braunkohle *f* brown coal; ~ntheer *m* brown-coal tar.

Braunſtein *m* peroxide of man-

ganese, pyrolusite; ~element *n* Leclanché element.
Brecheisen *n*, **Brechstange** *f* iron bar, crow, crow bar.
brechen *v* to break; Papier ~ to fold a sheet of paper; sich ~ to be refracted (said of rays of light).
Breite *f* breadth; (Geogr.) latitude.
Breithacke *f* mattock, broad axe.
Brems... ~buchse *f* (*HA*) brake box; ~feder *f* brake spring; ~hebel *m* lever of the braking mechanism; ~kolben *m* friction pad; ~kugel *f* pendulum (ball) of the braking mechanism; ~ring *m* brake ring; ~stange *f* vibrating tongue, rod of the braking mechanism; ~trommel *f* brake box; ~vorrichtung *f* braking mechanism, (beim Abrollen des Kabels) friction brake; ~wagen *m* (Eisenb.) brake van.
Bremse *f* brake, stopper, carriage stopper.
bremsen *v* to stop a movement, to brake.
Bremser *m* braker, brake man.
brennbar *a* combustible; ~, entzündlich *a* inflammable.
brennen *v* (Thon) to bake (clay); (Ziegel) to burn (bricks, tiles); to bake.
Brennmaterial *n* fuel.
Brennstempel *m* brand iron; marking iron.
Brett *n* deal, board, shelf, thin plank; mit ~ern belegen to board, to plank.
Brief *m* letter; (gewöhnlicher) ordinary letter; (eingeschriebener) registered letter; (einfacher) single letter; (frankirter) prepaid letter; (unfrankirter) unpaid letter; (unzureichend frankirter) insufficiently prepaid letter; (portofreier) letter exempt from postage; (unbestellbarer) dead letter.
Brief... ~abfertigung *f* despatch of mails, (als Dienststelle) despatch department; ~abgangszettel *m* bag list; ~abholung *f* calling (asking) for letters; ~aufgabe *f* posting of a letter; ~aufgabestempel *m* date stamp; ~ausgabe *f* window delivery, delivery at the counter; ~bestellung *f* delivery of letters; ~beutel *m* mail bag, letter bag; ~beutelmesser *n* knife for opening mail bags; ~bogen *m* sheet of letter paper; ~bund *n* letter packet, bundle of letters; ~einwurf *m* (am Briefkasten) letter drop, slip of the letter box; ~entkartung *f* checking of the letter bill; ~fach *n* (eines abholenden Korrespondenten) private box, (des Sortirspinds) sorting case, pigeon hole; ~geheimniß *n* secrecy & inviolability of letters; ~karte *f* letter bill; ~kartenschluß *m* closed mail, letter mail; ~kasten *m* letter box, (*Am*) mail box; Haus=~kasten *m* door letter box; Pfeiler=(Säulen-)~kasten pillar (letter) box; (in der Wand angebrachter, Mauerbriefkasten) wall (letter) box; beweglicher ~kasten portable letter box; fester ~kasten fixed letter box; aus dem ~kasten (amtl. Vermerk) „posted in box", „found in the letter box"; ~kasteneinwurf = Briefeinwurf; ~kastenleerer *m* letter box clearer; ~kastenleerung *f* clearance; ~kasten=Sammelsack *m* collecting bag; ~kasten-Stundenplatte *f* notice plate; ~korb *m* letter basket; ~marke u. Zusammensetzungen s. Postwerthzeichen; ~packet *n* closed bundle of letters, letter packet, letter package, (direktes) direct mail, (geschlossenes) closed mail; ~porto *n* postage; ~post *f* letter post, letter mail; ~postsendung *f* article of the letter post; ~postsendungen *pl* correspondence; ~posttarif *m* table of (domestic & foreign) postage; ~posttitelschild *n* label affixed to a mail bag; ~taube *f* carrier pigeon; ~taxe *f* rate of postage for letters; ~träger *m* letter carrier, postman; ~trägertasche *f* pouch, letter carrier's pouch; Land=~träger s. Land...; Ober=~träger head letter carrier; ~umschlag *m* envelope, cover; gestempelter ~umschlag stamped envelope; ~waage *f* letter balance; ~wechsel *m* correspondence.
brieflich *adv* by letter, in writing.
bringen *v* (Aufforderung zum

Bronze — Bund

(Geben von Telegrammen) „G" (d. h. „go on").

Bronze *f* bronze, brass, metal (alloy of copper and tin); (gefirnißte, unechte) varnished bronze; Phosphor-~ phosphor bronze.

bronziren *v* to bronze, to bronze over.

Bruch *m* (eines Drahtes) break, breakage, fracture, rupture; (Arithm.) fraction; (Waare, Materialien) broken ware, rubble, fragments; echter, unechter, gemischter ~ (Arithm.) proper, improper fraction, mixed number; faseriger ~ (Eisen) fibrous fracture; körniger ~ granular fracture; krystallinischer ~ crystalline fracture; schiefriger, blätteriger ~ foliated fracture; splitteriger ~ splintry fracture.

Bruch ... ~belastung *f* breaking load; ~belastungsgewicht *n* breaking weight; ~festigkeit, relative Festigkeit, Biegungs- oder relative Elastizität und Festigkeit *f* strength of flexure, elasticity and strength of flexure, transverse strength, resistance to breaking strain; ~fläche *f* fracture; ~kraft, ~spannung *f* breaking strain; Probe *f* auf ~kraft breaking test; ~stelle *f* (Fehler) break, position of the break; ~strich *m* bar; ~stück *n* fragment, shred; ~theil *m* fraction, ~theil eines Pfennigs fractional pfennig, fraction of a pfennig.

Bruch ..., Moor ... ~boden *m* swamp earth, moor earth; ~land *n* marsh.

bruchfrei *a* free from fracture.

brüchig *a* brittle (said of metals); fragile; short (said of wood).

Brücke *f* bridge; Wheatstone'sche ~ Wheatstone's bridge (balance); (bewegliche) moveable bridge; (feste, stehende) fixed (permanent) bridge; (fliegende) flying (ferry) bridge; (hängende, Hängebrücke *f*) suspension bridge; (mit Hängewerk) truss bridge.

Brücken ... ~aufseher *m* overseer of the bridge; ~bahn, ~fahrbahn *f* breadth, surface, bridge road; ~balken, ~träger *m* beam, balk, baulk, girder; ~belag *m* flooring, planking of a bridge; ~durchlaß *m* cut, opening for the passage of ships; ~geländer *n* hand rails, railing, parapet, balustrade; ~= (Meß-) Methode *f* bridge method; ~öffnung *f* aperture; ~pfeiler *m*, ~joch *n* (von Holz) pile, pier; ~spannung *f* im Lichten width of the bay; ~waage *f* portable weighing machine; ~weite *f* span of a bridge.

brutto *a* brutto, brute, gross.

Brutto ... ~betrag *m* gross amount; ~einnahme *f* gross amount of receipts; ~ertrag *m* gross revenue, brutto proceeds, brutto profit; ~fracht *f* gross freight; ~gewicht *n* gross weight.

Buch *n* book, register; (ein ~ Papier) quire of paper; (gebundenes ~) volume; ~halter *m* book keeper, accountant; ~halterei *f* book keeper's office; ~halterei-Rechnung *f* special annual account of the Cash office; ~haltung *f* bookkeeping, (einfache, doppelte) by single, by double entry; ~ führen, ~ halten to keep the books; Bücher anlegen to arrange a set of books.

buchen *v* to book, to enter (into the books), to carry into the books.

Bücher ... ~post *f* book post; ~postgebühren *f/pl* book post rate of postage; ~zettel *m* open printed order for books and other articles of the book trade.

Buchhaltereirechnung s. Buch ..

Büchse, Buchse *f* box, shell; (Masch.) socket, bush, bushing; pipe (HA); ~, Achsbüchse, Lagerbüchse *f* (Masch.) journal box.

Buchung *f* booking, entering (carrying) into the books.

Buchstabe *m* letter, type, character; ~n-Blank *n* (HA) letter blank; ~n-Blanktaste *f* letter-blank key; ~n-Wechsel change from letters to figures and signs.

buchstäblich *a* literal; *adv* literally, literatim.

Budget *n* estimate (of receipts and expenditures).

Bügel *m* a curved small piece of metal, bridle, shackle, strap.

Bund *n* (Briefe ꝛc.) bundle,

packet; (Bündniß n) union, confederation; der deutsche ~ the German confederation; ~esrath m federal council.

Bündel n (Draht-~) bundle of wire; ~magnet, Magnetstab~ m compound magnet.

Bureau n office; ~assistent m assistant clerk; ~ausgaben f/pl costs of stationery and necessary incidentals; ~beamter m clerk; ~materialien n/pl supplies, equipment of an office; s. auch Amts...

Bürgersteig m, Trottoir n foot path, side walk.

Bürgschaft f security, bail; ~ leisten to give security; ~sschein m bond, bail bond; ~sleistung, Kaution f bond.

Bürste m brush.

Büschel m, elektrischer Strahlen-~ electric brush, electric aigret (aigrette).

Butte, Bütte f coop, trough.

C.
Siehe auch K und Z.

Carcel-Lampe f Carcel lamp, clock-work lamp, mechanic lamp.

Cement m cement.

Cementation f, **Cementirungsprozeß** m des Eisens cementation, conversion of iron into steel by cementation.

Cementir... ~ofen m cementing furnace.

Cent... s. Z..

Chaussee f high road, main road, turnpike road, causeway; ~aufseher m overseer; ~graben m ditch along both sides of a high road; ~stein m number stone.

Chemie f chemistry.

chemisch a chemical; ~er Telegraph chemical telegraph.

Chiffer f cipher; in ~n schreiben to cipher; ~nschrift f cipher system; ~nschlüssel m cipher key; ~telegramm n cipher telegram.

Chiffreur m cipherer, cipher operator.

Chlor n chlorine; ~gas n chlorine gas; ~natrium n chloride of sodium; ~wasser chlorine water.

Chlorid n, **Chlorverbindung** f,

Chlormetall n chloride, perchloride.

Chlorsäure f chloric acid.

Chrom n chrome; ~alaun m chrome alum; ~eisenstein m chromate of iron; ~oxyd n chromic oxide; ~sauer a chromic; ~saures Salz n chromate; ~saures Kali n chromate of potassa, potassium chromate; ~säure f chromic acid.

Chronograph m chronograph.

Circular s. Zirkular.

Cisterne s. Zisterne.

Coak, Coaks s. Koke.

Coëfficient s. Koeffizient.

Coupé n compartment, coupé; (des Postwagens) front compartment.

coupiren v (ein Billet) to punch (a ticket).

Coupirzange f hand punch.

Couvert n, **Briefumschlag** m envelope.

Croquis n eye sketch, rough sketch.

Cylinder m cylinder, barrel, tube; (Dampfmasch.) steam cylinder.

cylinderförmig a cylindric, cylindrical.

D.

Dach n roof; über die Dächer der Häuser geführte Linie (T) over house line.

dachförmig abgeschrägte Stange (T) pole scarfed (cut) roof-like (wedge-like) at top.

Dachs... ~beil n, Dächsel m addice, adze.

Damm m dam, dike; (Eisenb., Straßenb.) bank, embankment; (Aufschüttung f von Erde) filling.

Dampf m steam; ~boot, ~schiff n

steam boat, steam vessel, steamer; ~heizung *f* steam heating; ~kessel *m* steam boiler, boiler; ~ventil *n* steam valve; ~wagen *m* steam carriage, steam waggon, locomotive engine.

Dämpfapparat *m* steamer; (für Holz) steam tank.

dämpfen *v* (abschwächen) to damp, to check (the swinging of a needle); (Holz) to steam (timber), to treat with steam.

Dämpfer *m* (magnetischer) damper; (Tondämpfer *m*) damper, sordine.

Darstellung *f* (Chem.) preparation, (Math.) construction; (Darlegung) statement; (gedrängte) succinct (concise) statement.

datiren *v* to date; falsch ~ to misdate; später ~ to postdate; vor~ to antedate.

Datum *n* date; ein Schreiben vom 10. a letter d. d. (de dato) 10th; ohne ~ undated; ~stempel date stamp.

Dauer *f* duration, length of time, term; (Dauerhaftigkeit, Reaktion gegen Formveränderung) durability, permanence, rigidity, stiffness.

dauerhaft *a* durable, lasting.

dauern *v* (dauerhaft sein) to endure, to last.

Daumen *m* (Wellenzahn, Masch.) cam; ~ der Auslösung des Einstellhebels (*HA*) detent cam; ~welle *f* cam shaft; Druck~ (*HA*) printing cam; Korrektions-~ (*HA*) correcting cam; Papierführungs-~ (*HA*) paper-moving cam.

Debet, Debit *n* debit; im ~ stehen to be on the debtor side.

debitiren *v* to charge to one's account, to place to the debit.

Decharge *f* s. Entlastung.

Dechiffreur *m* decipherer.

dechiffrirbar *a* decipherable.

dechiffriren *v* to decipher.

Deck ... ~laberaum *m* (*P*) receptacle for parcels on top of the mail coach; ~leiste *f* rod, tringle covering a joint; ~platte *f* top plate, covering plate; ~platte des Stiftgehäuses (*HA*) contact plate; ~schicht *f* covering, layer (of gutta-percha etc).

Decke *f* cover, covering; (des Zimmers) ceiling; (Wagendecke) tilt; (des Kutscherbocks) hammer cloth; (eines Pferdes) horse cloth, rug.

Deckel *m* cover, lid.

decken *v* (einen Defekt) to reimburse, to cover (a deficiency); (einen Wechsel) to answer a bill.

Deckung *f* reimbursement, remittance.

Defekt *m* deficiency; ~beschluß *m* s. Beitreibungsbeschluß.

Defizit *n* deficience, deficit.

Deflagration *f* (Chem.) deflagration.

Deflagrator *m* calorimotor.

dehnbar *a* (Phys.) extensible; ~ durch Druck ob. Stoß, hämmerbar (Metall) malleable; ~ durch Ziehen, streckbar (Metall) ductile.

Dehnbarkeit *f* extensibility, malleability, ductility (vergl. d. Vorst.).

dekartiren s. entkarten.

deklariren *v* (Waaren beim Zollamt) to enter, to report goods at the custom house; zu wenig ~ to enter short.

Deklination *f* der Magnetnadel variation, declination of the magnetic needle.

Deklinationsbussole *f*, **Deklinatorium** *n* declination compass.

Depesche *f*, telegraphische ~ despatch, dispatch; s. Telegramm.

depeschiren *v* to send a (telegraphic) despatch, to telegraph, to wire.

Depolarisation *f* depolarization.

depolarisiren *v* to depolarize; ~der Körper *m* depolarizer.

Deponent, Hinterleger *m* depositary, deponent, bailor.

deponiren, hinterlegen *v* to depose, to deposit, to lay down.

Deponirung *f* consignation.

Depositar, Aufbewahrer *m* consignatary, trustee, bailee.

Depositen ... ~bank *f* circulating bank, bank of deposits, transfer bank; ~kasse *f* trust funds; ~schein *m* receipt for a deposit.

Depositorium *n*, **Niederlegung** *f* depository.

Depositum *n* deposit, trust, bailment.

Depot *n* depot.
Desinfektion, Desinfizirung *f* disinfection; ~apparat *m* disinfector, disinfecting apparatus; ~smittel *n* disinfectant, disinfecting agent.
desinfiziren *v* to disinfect.
deutlich *a* clear; ~e Handschrift *f* legible hand-writing; ~ schreiben to write distinctly.
deutsch *a* German; ~er Aufsatz (als Prüfungsaufgabe) German composition; das ~e Reich *n* the German Empire; ~-französisch *a* franco-german; deutsch-österreichischer Postvertrag German-Austrian postal convention.
Dezimal ... ~bruch *m* decimal fraction; ~maaß *n* decimal measure; ~system *n* decimal system; ~waage *f* decimal balance.
Dextrin *n*, **Stärkegummi** *m* dextrine, British gum, starch gum; ~lösung *f* mucilaginous dextrine.
Diagonalstrebe *f* diagonal brace.
Diagramm *n*, **Figur** *f* diagram, figure.
diamagnetisch *a* diamagnetic.
Diamagnetismus *m* diamagnetism.
Diaphragma *n* (poröse Scheidewand *f*) diaphragm, diaphragma.
Diamantfarbe *f* grey minium (varnish protecting iron against rust).
Diäten *pl* daily allowance.
dicht *a* compact, solid, substantial.
dichten, dicht machen *v* to make close, to make tight; Röhren ~ (in denen Kabel verlegt sind) to caulk tubes.
Dichtigkeit *f* compactness, solidness; ~, spezifisches Gewicht specific gravity (weight); Wasser im Zustande seiner größten ~ water at its maximum density; ~smesser *m* instrument for measuring the density of bodies; dasymeter, hydrometer, areometer.
dick *a* thick; ~flüssig thickly liquid; ~flüssig, zähe viscid.
Dicke *f* thickness, dimension.
Dienst *m* duty, service; außer ~ (dienstfrei) off duty; (verabschiedet) retired from service; im ~ on duty; im Interesse des ~es in the interest of the service; J—n des ~es entlassen to dismiss *smb* from the service; seinen ~ thun to do duty; den ~ versagen to refuse working (said of materials); ununterbrochener ~ (*T P*) the Office is open day & night; ~ablösung *f* relief, change, substitution; ~ablösungszeit *f* relieving time; ~alter *n* seniority; ~angelegenheit, ~sache *f* matter relating to official business (to the public service); in ~angelegenheiten (letter) on official business; ~anweisung *f* instructions, regulations; ~bedürfniß *n* the requirements (exigencies) of the service; dem ~bedürfniß genügen to meet the requirements of the service; ~betrieb *m* transaction of business; ~buch *n* der Postillone (+) book in which the testimonials of service etc. are entered; ~eid *m* oath of office; ~enthebung *f* suspension; ~entlassung *f* (freiwillige) discharge, (unfreiwillige) dismissal from service; ~eröffnung *f* official opening hour; ~fähig *a* fit for service; ~fähigkeit *f* fitness for service; ~frei *a* off duty, (vom Militär) exempt from military service; ~führung *f* service, official conduct; ~habend *a* on (upon) duty; ~jahr *n* a year spent in the service; anrechnungsfähiges ~jahr a year to be counted as spent in the service; ~kleidung *f* uniform clothing, uniform; ~korrespondenz *f* correspondence on official business; ~leistung *f* service; zur militärischen ~leistung eingezogener Beamter officer called out for military service (for training); ~lich *a* official; ~liche Angaben *f/pl* (*T*) service instructions; ~mütze *f* uniform cap; ~notiz *f* service notice; ~papiere *n/pl* official papers; unbrauchbare ~papiere waste paper; ~pflicht *f* obligation to perform certain services; der militärischen ~pflicht genügen to serve his time in the army; aus ~pflicht (etwas thun, v. Beamten) by virtue of office; ~pflichtig *a* liable to (perform) certain services; ~plan *m* s. Geschäfts-

ordnung; ~räume m/pl, miethsweise Beschaffung der ~räume renting of a building (parts of a building) to establish an office therein; ~reise f travelling in the service, official trip; Vergütung für eine ~reise allowance; ~rock m uniform coat; ~sache f s. ~angelegenheit; ~schluß m (T P) official closing hour; ~schuld f amount which an officer is held to refund to the postal Treasury; ~siegel n office seal, official seal; ~stelle f place; ~stunden f/pl hours of attendance; während der gewöhnlichen ~stunden during the regular hours of business; ~stundenplan m order of working; ~tauglich a able to perform (military) service, fit for service; ~telegramm n service message; ~thätig a effective (im Gegensatz zu den ausgeschiedenen Beamten); ~thuend a performing service (duty); ~übergabe f transfer of duty; ~übernahme f taking over the service; ~unfähig, ~untauglich a not fit (unfit) for service, incapacitated, (Militär) disabled; ~unfähigkeit f incapacity, disability; ~verbrechen n crime committed in office; sich eines ~verbrechens schuldig machen to prevaricate; ~vergehen n breach of duty, misdemeanor in office; ~vermerk m service notice; ~vernachlässigung f delinquency; ~verrichtung f duty; I—m die ~en untersagen to suspend smb from duty; ~weg m, auf dem ~wege etwas beantragen to apply for smt through the proper channels; ~wohnung f official residence; ~wohnungsinhaber m officer whom an official residence is assigned to; ~zeit f time spent in the service; anrechnungsfähige ~zeit vgl. ~jahr; ~zeugniß n testimonial (of service); ~zucht f discipline; ~zweig m branch of service; Nacht-~ night service; Tages-~ day service; beschränkter Tages-~ limited day service; verlängerter Tages-~ service prolonged till midnight; voller Tages-~ day service; Tag- und Nacht-~ permanent service, the office is open day & night; s. auch Amt...

Differenzial .. differential; ~galvanometer n differential galvanometer; ~-Methode (= Schaltung) f differential method (principle).

direkt a direct; ~er Verkehr m, ~stellung f (T) direct communication, through traffic (the stations are in direct communication, the way stations cut out).

Direktion f direction, management; a body (board) of directors.

Direktor m director, manager; technischer ~ chief engineer.

dirigiren v to direct, to manage.

Disciplin u. s. w. s. Disziplin.

discontiren v to discount.

Disconto n discount.

Dispensation f dispensation from..., dispense, special licence.

dispensiren v to dispense from..., to grant a dispensation from...

Disponibilität f disponibility, disposal.

Distanz, Entfernung f, Abstand m distance; ~messer m ambulator, range finder.

Distrikt m district.

Disziplin f discipline.

Disziplinar... ~hof m (+) Supreme Court (Court of appeal) for the trial of civil service officers; ~kammer f (+) court for the trial of civil service officers; ~maßregel f disciplinary measure; ~strafe f disciplinary punishment (penalty); ~verfahren n (+) disciplinary proceeding.

Divergenz f divergence.

divergiren v to diverge.

Dokument n document, deed.

Doppel, Duplikat n double, duplicate, duplicate copy; als ~ befördern to transmit a duplicate copy.

Doppel... doppel... ~gängig a, ~gewinde (einer Schraube) double thread (of a screw); ~geleis (Eisenb.) double way, double track; ~gestänge n (T) double set of poles; poles coupled by struts; ~glocke f double bell, double bell insulator; ~klemme f double binding screw; ~konsole f double bracket; ~krone f piece of 20 marks; ~mutter f thimble; ~nadeltelegraph m double needle telegraph; ~punkt m colon; ~schreiber m s. ~stiftapparat; ~sprechen *n* simultaneous transmission of two communications in the same direction; ~sprecher, ~sprechapparat *m* diplex telegraph, apparatus for simultaneous transmission in the same direction; ~ständer, Bock *m* A pole (so called from its form); coupled poles forming a triangular frame; ~stange *f* double pole; ~stiftapparat *m* double style apparatus; ~ T Eisen *n* double T iron, H iron; ~taste *f* double key; ~währung *f* double standard-currency; ~winkeleisen *n* double angle-iron.

doppelt *a* double, duplex, two-fold, *adv* double; ~e Eintragung *f* double entry; ~ wirkend double-acting; ~ chromsaures Kali *n* bichromate of potash; ~ kohlensaures Natron *n* bicarbonate of soda; ~ schwefelsaures Natron *n* bisulphate of sodium.

Dorn *m* (Stachel) spur; (eiserner Stift) iron pin, spike; (Metallbohrer) punch, prick punch.

Dosenrelais *n* box relay.

Draht *m* wire; (ausgeglühter) annealed wire; (besponnener) covered wire; (gewalzter) rolled wire; (gezogener) drawn wire; (primärer, sekundärer) primary, secondary wire; ~ader *f*, ~bund *n* bundle of wire; ~aufleger *m* shear legs (f. a. T.) ~berührung *f* contact; ~bruch *m* break, breakage, rupture; ~bündel *n* fagot of wires (for the cores of an electro magnet); ~bürste *f* wire brush; ~eisen, Zieheisen *n* draw plate, drawing plate; ~fabrik *f* wire works; ~gabel *f* (f. ~aufleger); ~gitter *n* wire trellis, wire grating; ~haspel *m* u. *f* reel; ~hülle *f* wire covering; ~klemme *f* clamp; ~klinke *f* f. ~leere; ~lager *n* (auf der Strecke) depot; oberes, seitliches ~lager des Isolators top groove, side groove (neck) of the insulator; ~leere *f* instrument for measuring the diameter of wires; wire guage vernier; die Birmingham ~leere Birmingham wire gauge (B. W. G.), calliper; ~leier *f* (Scheibe zum ~ziehen) wire drum, winding drum, coiling drum; ~leitung *f* wire;

~litze *f* strand wire, stranded wire; ~mühle, ~zieherei *f* wire-drawing mill; ~öse *f*, ~reiter *m* eye, eye splice; ~ring *m* coil of wire; ~rolle *f* = ~ring *m*; ~seele *f* core (of a cable); ~seil *n* wire rope, guy wire; ~spirale *f* spiral wire; ~stift *m* wire tack; ~strang *m* strand of wire; ~träger *m* (des Baum-Isolators) wire carrier, wire hook; ~umspinnungsmaschine *f* armoring machine; ~verschlingung *f* cross; ~wagen (*MT*) wire waggon; ~winde *f* wire straightening tool, vice with drum & ratchet; ~windung *f* coil, convolution; ~zieher *m* wire drawer; ~zieherei *f* wire-drawing mill, wire mill; ~zug *m* strain, tensile strain; ~ ziehen *v* to draw wire, to wiredraw; Binde-~ tie wire, binding wire; Wickel-~ thin (fine) wire; den ~ auf die Stangen aufbringen to erect, to run the wire, to wire the poles.

Draisine *f* trolly.

Dreh ... u. **dreh** ... ~achse, Umdrehungsachse *f* axis of rotation, axis of revolution; ~bank *f* lathe, turning engine; ~bar *a* turning; ~bar um eine Achse turning on an axis; ~brücke *f* swivel bridge, turning bridge; ~punkt, Stützpunkt *m* centre of motion, (eines Hebels, einer Waage) fulcrum of a lever, of a balance; ~waage *f*, einfaches Torsionspendel *n* torsion balance, torsion rod.

drehen *v* to turn, to revolve; sich ~ to turn, to rotate.

Drehung *f* turn, turning, revolution, rotation; ~ der Kurbel stroke of the crank.

Drehungs ... ~achse *f* axis of revolution; ~elastizität u. Festigkeit *f* elasticity and strength of torsion; ~festigkeit *f* torsional strength, resistance to torsional strain; ~geschwindigkeit *f* speed of rotation; ~moment, Trägheitsmoment *n* momentum of inertia; ~punkt *m* centre of motion, centre of rotation, fulcrum; ~vermögen *n* rotary power.

Dreieck *n* triangle; gleichschenkliges isosceles, gleichseitiges equi-

lateral, rechtwinkliges right angled, spitzwinkliges oxygon, stumpfwinkliges amblygon, ungleichseitiges scalene triangle; ~sverbindung *f* lateral fastening (in the shape of a triangle).
dreieckig, dreiseitig *a* triangular, three angled.
Dreifuß *m* (gußeis. Fußpunkt der eis. Stangen) triangular iron base plate.
dreikantige Feile *f* three square file, triangular file.
Drillbohrer *m* drill, drill borer, wimble.
drillen, bohren *v* to drill.
dringend *a* urgent; ~ bitten *v* to urge.
dringlich *a* urgent.
Dringlichkeit *f* urgency; ~serklärung *f* declaration of urgency.
Druck *m* pressure, pression, compression; (auf die Taste T) depression of the key; (Buchdruck) printing, impression; (Luftdruck) atmospheric pressure, pneumatic pressure; voller ~, Volldruck *m* full pressure.
Druck ... ~achse *f* (HA) printing axis (axle, shaft); ~apparat *m* printing telegraph; ~daumen *m* printing cam; ~festigkeit *f* compressive strength; ~formular *n* printed form; ~hebel *m* (HA) printing lever (forked in two prongs); ~kolben, ~stempel *m* piston, forcer; ~kraft *f* force of pressure; ~pumpe *f* forcing pump; ~rohr, Steigrohr *n* einer ~pumpe column lift, rising main; ~rolle, ~walze *f* (M & HA) impression roller, printing roller, platen; ~sache *f* printed paper, printed matter, (Formular) blank form; ~schraube *f* binding screw, pressing screw, press screw, attachment screw; ~schrift *f* printed work, type, types; ~telegraph *m* printing telegraph, type printing telegraph; ~vorrichtung *f*, ~werk *n* (HA) printing mechanism; ~walze *f* s. ~rolle; ~welle (HA) printing axis.
drücken, niederdrücken *v* (die Taste T) to press, to depress, to close (the key).
Dübel, Diebel *m* peg, dowel,

tree nail, wooden pin; (Keil *m*) key.
dübeln, verdübeln *v* to peg, to dowel.
dünn *a* thin; (of fluids) dilute, weak; (of the air) rare.
Duplex-Apparat *m* duplex apparatus.
Duplikat *n* s. Doppel.
durchbiegen, sich ~ *v* to sag.
durchbohren *v* to pierce, to pierce through.
durchdringen *v* to permeate, to penetrate.
Durchdruckpapier *n* carbonic paper.
durchfließen *v* (vom elektr. Strom) to traverse, to flow through.
Durchgang *m* passage, transit; ~aufnahme-Formular *n* transmitted telegram form; ~stelegramm *n* transmitted telegram, transit message; ~sverkehr *m* transit.
durchgleiten *v* (vom Draht) to run, to run back.
Durchhang *m* dip (s. a. T. chain curve).
Durchkreuzung *f* (Eisenb.) crossing, crossing point.
Durchlaß *m* passage, opening; ~brücke *f* culvert bridge, wicket; ~öffnung *f* bei Brücken opening of a bridge.
durchlaufen *v* s. durchfließen.
durchlochen *v* s. lochen.
Durchlocher *m* s. Locher.
Durchmesser *m* diameter; äußerer ~ *m* diameter outside; innerer ~, ~ im Lichten diameter inside.
durchscheuern *v* (Kabel auf felsigem Untergrund) to scour (said of a cable on rocky ground).
durchschneiden *v* to cut through, to cut across, to intersect.
Durchschnitt *m* (Durchschneidung *f*) cutting through, cut, intersection; (Zeichnung) section, profile, diagram; (mittleres Ergebnis) average, mean; ~sgehalt *n* mean salary; ~sgewicht *n* average weight; ~szahl *f* mean number.
durchschnittlich *a* on an average.
durchsehen, prüfen *v* to revise, to read over, to examine.

Durchsicht f revision, examination.

durchsprechen v (T) to communicate through.

Durchsprechstellung f (T) the terminal stations are in direct (through) communication.

durchstechen v to cut, to make a cutting (into a dike etc.).

Durchstich m (Eisenb., Straßen-, Wasserbau) cutting, excavation, thorough-cut.

durchstreichen v to blot out, to cross, to cancel.

durchwachsen v to penetrate through the porous partition (said of the decomposed copper in a Dan. cell).

Dynamik f dynamics s. pl.

dynamisch a dynamical.

dynamo-elektrische Maschine f dynamo-machine, dynamo-electric machine.

Dynamometer, Kraftmesser m dynamometer.

E.

eben a plain, flat level, smooth; ~ machen to make plain, to level.

Ebene f level country, flat country; (Geom.) plane; schiefe, vertikale, wagerechte ~ inclined, vertical, horizontal plane.

ebenen v to level, to make even, to smooth.

Ebenholz n ebony, ebony wood; aus ~ ebony.

Ebonit n, **Hartgummi** m ebonite, hard rubber, indurated caoutchouc, vulcanite; ~glocke f (T) ebonite insulator; ~hülse f ebonite cylinder; ~rohr n, ~röhre f ebonite tube; ~unterlage f ebonite base, ebonite washer.

Echtheit f der Unterschrift authenticity of the signature.

Eck ... ~säule, ~stange f, ~ständer m corner post; ~schaltung f (T) way station with arrangements of translation between a closed & an open circuit, arrangement for cross-connecting two through lines at a junction; ~station f (T) way station with more than one wire; ~stein, Prellstein m stone stud.

Ecke f corner, edge, angle, nook; (abgestumpfte) blunt angle, obtuse corner; (aus-, herausspringende) salient angle, jet, cant; (räumliche ~, Kante f) solid angle; (scharfe) shot; (spitze) sharp edge; (stumpfe) obtuse corner, blunt edge.

eckig a cornered, angular, angulous.

Effektivbestand m effective balance of cash.

ehemalig a & adv formerly, before now.

Ehren ... ~peitsche f (P) whip of honor; ~posthorn n (silver) post-horn of honor.

eichen v s. aichen.

Eiche f oak, oak tree.

eichen a (von Eichenholz) oaken.

Eid m oath; einen ~ leisten to take an oath; J—m einen ~ abnehmen to administer an oath to smb; durch einen ~ erhärten to affirm by an oath (upon oath); ~esleistung f taking an oath; ~esmündig a capable of having an oath administered.

eidlich a by oath, upon (with an) oath; ~es Zeugniß n affidavit.

eigen a, auf ~e Kosten at one's own cost; für ~e Rechnung for (on) one's own account; der ~e Apparat m (T) the home apparatus; ~händig a with (in) one's own hand; ~händig zu bestellen to be delivered into the addressee's own hands; ~händiges Schreiben n autograph letter; ~thümlich a (Phys.) specific.

Eigenname m proper name.

Eigenschaft f quality, capacity, property, attribute.

Eil ... ~bestelldienst m special delivery service; ~bestellgeld n (fee) porterage for special delivery; ~bote m special messenger, courier, express; ~brief m letter to be delivered by special messenger; ~

briefzettel *m* (+) slip denoting (showing) a letter to be delivered by special messenger; ~fracht *f*, ~gut *n* conveyance of despatch; ~sendung *f* mail matter to be delivered by special messenger; ~zug, Kurierzug *m* express train; ~zugswagen *m* express carriage, mail carriage.

Eile *f* haste, speed, despatch.

eilen *v* to hasten, to make haste; die Arbeit eilt the work is pressing.

eilig *a* hasty, speedy; *adv* hastily, speedily; ein ~er Brief an urgent letter.

einadriges Kabel *n* cable with one (a single) conductor.

einbaggern *v* to imbed by means of a dredging machine.

einbegreifen *v* to include; die Bestellgebühr einbegriffen the cost of delivery included.

einberufen *v* (zur Probedienstleistung) to call in a candidate for a probationary appointment; (zu einer militärischen Dienstleistung) to call out officials for military training.

einbetten *v* to imbed (a cable into the ground).

einbiegen, sich *v* (v. Draht, v. Balken u. dergl.) to sag, to bend inwards.

einbleien f. verbleien.

einbrennen *v*, den Stempel in Stangen (*T*) to mark poles with the branding iron.

Eindruck *m* (Techn.) impression, mark, stamp.

Einfahrts-Signal *n* (Eisenb.) distance signal, station signal, up-signal.

Einfall ... ~haken *m* (*HA*) detent; ~loth *n* normal line, normal, perpendicular; ~rad *n* escape wheel; ~winkel *m* (Opt.) angle of incidence.

einfordern *v* (Geld) to call in money, to demand payment; (Kosten für Bestellung eines Telegramms) to collect charges for porterage.

einführen *v* (Waaren) to import (goods); (einen Gebrauch) to introduce, to establish (a practice); (in ein Amt) to install, to inaugurate (*smb* in an office); (Drähte ins Amt [*T*]) to lead wires into the office.

Einführung *f* importation, introduction, installation, leading-in; (vgl. das Vorstehende).

Einführungs... ~brett *n* (*T*) board for the leading-in wires; ~drähte *m/pl* (*T*) leading-in wires; ~glocke *f* (*T*) insulator outside the office to which the line wire is fastened; ~kasten *m* (*T*) box; ~rohr *n* tube of india rubber with the outer end inclined downwards; ~stütze *f*, ~träger *m* (*T*) bracket for the leading-in wires.

Eingang *m* (Kopf des Telegramms) top, preamble; (der Post) arrival of the mail; ~svermerk *m* (auf Schriftstücken) notice of receipt; ~zettel *m* (*P*) bag list; ~zoll *m* customs inwards *pl*; ~zollamt *n* custom house.

eingängige Schraube *f* single-thread screw.

eingehen *v* (von Schriftstücken) to come in, to come to hand, to receive; (ankommen) to arrive; eine Leitung, (ein Amt) ~ lassen (*T*) to discontinue a line, (an office).

eingelassen *a* (Techn.) sunken, trimmed, mortised.

eing(e)leisige Bahn *f* single line, railway line with a single set of tracks.

eingeschlossen, einliegend *a* subjoined, enclosed (in a letter).

eingeschriebener Brief *m* registered letter.

eingezahnt *a* indented.

eingießen *v* to cast in; (zuschmelzen) to seal.

eingipsen *v* to plaster, to fasten with plaster.

eingraben *v* to cut in, to carve; Pfähle ~ to drive in poles.

eingreifen *v* in die Zähne eines Rades to catch, to gear into the teeth of a wheel.

Eingreifen *n* catch, toothing, bite.

eingreifendes Rad *n* (Masch.) match wheel.

eingrenzen *v* (einen Fehler [*T*]) to locate, to localise (a fault).

einhaken v to hook in, to fasten with a hook.

einhalten v (in der Bewegung) to check, to stop; (mit der Arbeit) to discontinue, to pause, to leave off; (das Gehalt) to stop, to arrest, to withhold the salary.

Einhalten n einer Gehaltszulage stoppage (arrest) of increment.

Einheit f unity; (Gleichförmigkeit, Uebereinstimmung f) uniformity, conformity; ~ der elektr. Kapazität unit of capacity, farad; ~ des elektr. Widerstandes unit of resistance, ohm; ~ der elektromotorischen Kraft unit of electromotive force, volt; ~ der mechanischen Arbeit od. Leistung dynamical unit, unit of work.

einheitlich a (s. auch Einheits..); ~e Taxe f uniform tax; ~e Verwaltung f centralized administration.

Einheits... ~porto n uniform rate of postage; ~taxe f uniform tax.

einigen, sich v to agree, to come to terms.

einkassiren v to get in, to cash, to collect.

Einkassirung f encashment, collection.

Einkaufspreis m first cost, prime cost, purchase money; zum ~ verkaufen to sell at first cost.

einkehlen v (Techn.) to channel, to hollow.

einkeilen v to wedge, to wedge in, to fasten with wedges.

einkerben v to indent, to notch.

Einkerbung f notch.

einkitten v s. kitten.

Einlageschein m des Post-, Spar- und Vorschußvereins m depositor's book for the Postal Saving & Loan Association.

einlassen v (Techn.) to sink, to trim; (in eine Mauer) to trim in.

Einlauge-Bottich m cask (vessel, tub) for impregnating poles with bichloride of mercury.

einlaugen v Stangen to immerse poles in the vessel.

Einlaugezeit f duration of the immersion of poles to be kyanized.

einlegen, einhüllen v to imbed; (einschließen) to enclose (a letter); Geld ~ to [deposit money (in the Saving Fund).

einleiten v to bring about, to manage; to take the preliminary steps (measures), to take the initiative.

einliefern v (zur Post) to hand in, to hand over the counter, to deliver; (Briefe in den Briefkasten) to mail letters, to drop letters at the post office.

Einlieferung f (zur Post) handing in, delivery; ~samt n (P) office of origin; ~sgebühr für Spätlingsbriefe (P) late letter fee; ~schein m receipt, acknowledgment of deposit; (für Packete) certificate of posting.

einliegend a enclosed (inclosed).

einlothen v to plumb, to set vertical.

einlöthen v to solder in.

einmauern v to immure, to wall in, to imbed in a wall.

Einmauerung f eines Eisenstabes in ein Mauerwerk chain bond.

einmeißeln v to work in with a chisel.

Einnahme f revenue (of the post office); (Ertrag) receipt, receipts; etatsmäßige ~ s. Etat; ~ u. Ausgabe receipts & expenditures pl; in ~ nachweisen to account for as received; in ~ stellen to book as received; ~journal n book of receipts.

einnehmen v (Geld u. dgl.) to receive; (von Schiffen, die die Post ~) to pick up the mail.

Einnehmer f receiver, collector (of money).

einölen v to oil, to rub with oil.

einpacken v in Holzasche to imbed in wood ashes; Waaren ~ to pack, to embale goods.

einpassen v to fit in, to trim in.

Einpassung f adjustment.

einpfählen v to pale in, to fence with pales.

einrammen v to ram in, to drive in (a pile).

Einrammen n pile driving.

einräumen, zugestehen v to concede, to grant, to allow, to permit.

einrechnen v to take into (to include in) an account.

einreichen v, ein Gesuch to hand in an application, to apply, to present a petition.

Einreichung f, eines Gesuchs presentation of a petition.

einrichten v (Techn.) to adjust; (ein Amt *TP*) to fit up an office; (elektr. Beleuchtung) to install electric lighting.

Einrichtung f fitting up, installation.

einsammeln v to collect (letters).

Einsammlung f collection (from the letter boxes).

Einsatzkasten m, des Briefkastens (removable) inside case of a letter box.

einsaugen v to suck in, to suck up; (absorbiren [Chem.]) to absorb.

einschalten v (Apparat, Widerstand) to place in circuit, to join up in circuit, to insert in circuit, to cut in; (mehrere Aemter in eine Leitung) to place several offices upon one wire, to fix several stations on one (Morse) circuit; sich wieder ~ to enter again in circuit; ein Wort ~ to intercalate, to interpolate, to insert (a word).

Einschalter m s. Umschalter.

Einschaltung f (*T*) joining up in circuit, insertion (of an apparatus, a shunt etc.); interpolation (of a word).

einscheeren v to reeve (the end of a rope).

einschlagen v (Techn.) to drive in, to beat in, to knock in; (vom Blitze) to strike.

einschlägige Bestimmungen f/pl instructions relating to ...

einschleifen v (ein Amt in eine Leitung [*T*]) to loop in (a station).

einschließen v (zur Aufbewahrung) to lock in, to lock up (for safe keeping); (enthalten) to include.

einschmieren v to grease; (mit Oel) to oil.

einschneiden v to cut into, to incise; (einkerben) to notch, to score.

Einschnitt m cutting, incision, notch, slot; (im Terrain) cut, cutting, excavation; (zackiger) toothed (indented) incision; mit ~en versehen v to notch.

einschränken v to restrain, to restrict.

Einschränkung f restriction.

Einschreib(e)... ~brief m registered letter; ~gebühr f registration fee; ~sendung f registered mail matter, registered article.

einschreiben v to enter, to book; einen Brief ~, rekommandiren to register a letter; eingeschriebener Brief registered letter.

Einschreibung f entering, booking; (eines Briefes) registration; (der Postreisenden) booking.

einschreiten v, gegen J—n to proceed against smb.

einsenden v to send, to send in.

Einsenkung f eines Kabels in die See submersion of a cable.

Einsicht f von etw. nehmen to look closely into smt, to take insight.

Einsprache f, **Einspruch** m objection, protest; ~ thun, erheben to raise an objection, to protest against; ohne ~ nem. con., nem. dis. (nemine contradicente, nemine dissentiente).

Einspruchsfrist f time allowed for protesting (against any order).

einspurig s. Eisenbahn.

einstampfen v (Erde) to ram down, to beat down the earth; (Papier) to put into the paper stamp, to reduce paper.

einstehen, Bürgschaft leisten v für J—d to answer for smb, to be security for smb.

Einstell... ~hebel m (HA) zero-adjusting lever; ~schraube f (HA) adjusting screw.

einstellen v (die Arbeit, den Betrieb) to stop the work; (streifen) to strike; (den Apparat) to adjust the instrument; sich ~, sich richten (vom Magneten) to set.

Einstellung f (der Arbeit) leaving off, stopping; (der Stangen [*T*]) erection; (des Apparats) adjustment; ~sjahr n year of erection; ~stiefe f depth to which poles are inserted in the ground (in Germany generally one fifth of their length).

einstemmen *v* to make holes with the chisel, to mortise, to fix in a mortise.

Einstreichfeile, Schraubenkopf=feile *f* slitting file, screw-head file.

einstreichen *v* to slit (to make a deep and narrow stroke with a file).

Einströmung *f* (Dampf) admission, introduction of steam; (der Elektrizität in das Kabel) first rush of current into the cable; ~ u. Ausströmung (im Kabel) flux & reflux.

eintauchen *v* to dip, to steep, to immerge, to plunge in.

Eintauchung *f* dipping, immersion.

eintheilen *v* to divide, to distribute; (in Grade) to graduate.

Eintheilung *f* division, distribution; (in Grade) graduation.

eintragen *v* (einschreiben) to enter, to register, to make an entry; (einbringen) to yield, to bring in.

einträglich *a* profitable, lucrative.

Einträglichkeit *f* profitableness.

Eintragung *f* entry, entering, registration.

eintreiben *v* (Techn.) to drive in; (die Gebühr vom Empfänger) to collect the tax from the adressee.

Eintreibung *f* (Techn.) driving in; (Einziehung) collection, exaction.

eintreten *v* to enter upon (an office); (sich ereignen) to happen; ein Zwischenamt ~ lassen (*T*) to call in a way station.

Eintritt *m* entering upon (an office); ~skarte *f* card of admission, pass card.

eintrocknen *v* to dry up.

Einvernehmen *n* understanding; sich mit J—m in ~ setzen to communicate with *smb*.

einverstanden sein mit J—m to agree with *smb*, to be agreed in (on), to concur with *smb* in *smt*.

Einverständniß *n* agreement, understanding; zu einem ~ kommen to come to an agreement (to an understanding, to terms).

Einwand *m* einen ~ erheben to raise an objection to.

einweichen *v* to steep, to soak.

einwenden *v* to object to, to demur to, to oppose; es ist nichts dagegen einzuwenden there is nothing for it.

Einwendung *f* objection; ~ machen gegen etw. to demur to, to except to (against) *smt*; ~en unterworfen exceptionable.

einwilligen *v* to consent to, to assent to, to permit.

Einwilligung *f* consent, assent, permission.

Einwirkung *f* (des elektr. Stromes) action.

Einwohner=Meldeamt *n* intelligence office (where the residences of a town's inhabitants may be enquired for).

Einwurf *m* (des Briefkastens) letter drop, slip (of the letter box), aperture.

einzahlen *v* to pay in.

Einzahlung *f* payment, the sum paid in; ~sschein *m*, Quittung *f* receipt, acknowledgment.

einzahnen, einzacken *v* to indent, to notch, to jog.

Einzahnung *f* indentation.

einzeichnen *v* (die Maaße) to write (to draw) the dimensions into a design.

einzeln *a* single, separate, isolated (abbreviations must be counted as *single* words ob. combinations of two words are to be charged for as *separate* words).

Einzeltransit *m* (*P*) open mail transit.

einziehen *v* (Techn.) to haul in, to draw in (a cable into a pipe [tube]); (Geld) to collect, to gather money; (ein Amt) to discontinue an office; (Erkundigungen) to enquire, to make enquiries about *smt*.

einziehbar *a* which can be collected (said of fees, charges etc.).

Einziehung *f* (vgl. einziehen) hauling in, drawing in; collection, gathering; discontinuance; ~s=kosten *pl* (Geld) collecting charges.

Eisen *n* iron; (zu Bauzwecken) structural iron; (brüchiges, faulbrüchiges) short iron; Entkohlung des ~s decarburization of iron; (Flacheisen) flat iron; (feinkörniges) fine grained (cast, wrought) iron; (galvanisirtes, verzinktes) galvanized iron; Frischeisen fined (refined) iron; (gegossenes) cast iron; (gehämmertes ~, Hammereisen) hammered (beaten, tilted) iron; (geschmiedetes ~, Schmiedeeisen) forged (wrought) iron; (geschweißtes) welded iron; (gezogenes, gestrecktes, gewalztes) drawn (out, rolled, bar) iron; (grobkörniges) coarse grained iron; (hämmerbares, schmiedbares, geschmeidiges) malleable (soft, ductile) iron; (heißbrüchiges) hot short iron; (kaltbrüchiges) cold short iron; (körniges) grained (granular) iron; (reines, chemisch reines) pure iron; (rohbrüchiges) dead short iron; (rohes ~, Roheisen, Gußeisen) pig iron, cast iron; (rothbrüchiges) red short iron, red sear iron; (Rundeisen, Handelseisen mit kreisförmigem Querschnitte) round iron, round bar iron, rod iron; (schieferiges, blätteriges) scaly iron; (übergares) kishy pig iron; (unganzes) weak (flawy) iron; (verzinktes, s. galvanisirtes); (viereckiges, vierkantiges Quadrateisen) square iron; (weiches) soft (malleable) iron; (weißbrüchiges) lamellar (laminated) iron; (schlackiges) slaggy (drossy) iron; (schmiedbares s. hämmerbares); (schwarzbrüchiges) black short iron; (sehniges, zähes, faseriges) stringy, fibrous iron; (Spiegeleisen) specular cast iron, speculum iron, crystalline pig, spiegel iron; (sprödes) brittle iron, white short iron; (Winkeleisen) angle iron.

Eisenbahn *f* railway, railroad; (atmosphärische) atmospheric (pneumatic) railway; (im Bau befindliche) railway in progress; (im Betriebe befindliche) railway in operation; (breitspurige) broad gauge railway; (durchgehende) through line, (in Amerika) trunk road, trunk; (einspurige) railroad with a single set of tracks, single way; (schmalspurige) narrow gauge railway; (unterirdische) underground railway; (zweispurige, doppelspurige) railroad with double way, with two sets of tracks, double way; s. auch Bahn.

Eisenbahn... ~bau *m* construction of railways; ~beamte *m* officer (employee) of a railway; ~betrieb *m* working of a railway; ~coupé *n* railway compartment; ~diensttelegramm *n* message sent on the business of a railway Company; ~fahrkarte *f*, Billet *n* railway ticket; ~fracht *f* carriage by railway; ~gesellschaft *f* railway company; ~gesetz *n*, ~verordnungen *f/pl* railway laws & regulations; ~läutewerk *n* bell, alarum; ~linie *f* railway line, set of tracks; ~polizei-Reglement (+) railway laws & regulations; ~post s. Bahnpost; ~reglement *n* regulations on the railway service; ~schiene *f* rail; ~schwelle *f* railway sleeper; ~signal *n* railway signal; ~station *f* railway station; ~strecke *f* section of a railway; ~tarif *m* railway tariff; ~-Telegr.-Leitung *f* line wire; ~-Telegraphenstation *f* railway telegraph station; eine dem öffentl. Verkehr dienende ~-Telegr. Station railway station open for the collection and delivery of telegrams, delivering railway station; ~transport *m* railway conveyance, transportation by railway; ~unterbau *m* earthworks & viaducts; ~uhr *f* railway clock; ~verkehr *m* railway traffic; ~waggon *m* (in England) railway carriage, railway waggon, (in Amerika) railway car; ~wärter *m* railway watchman; ~zug *m* railway train; ~zug mit Post mail train; s. auch Bahn.

Eisenblech *n* iron plate, sheet iron; (gehämmertes) hammered sheet iron; (gewalztes) rolled sheet iron; (gewelltes) corrugated sheet iron.

eisenhaltig *a* ferriferous, ferruginous.

eisern *a* iron, made of iron; ~er Bestand *m* (an Postwerthzeichen) credit stock (of stamps), full stock, standing stock; ~er Vorschuß *m* standing stock on credit.

4

elastisch *a* elastic, elastical; ~es Harz, Federharz, Gummi elasticum, Kautschuk *n* elastic gum, India rubber, caoutchouc.

Elastizität, Federkraft, Schnellkraft, Spannkraft *f* elasticity; ~ u. Festigkeit *f* gegen Abscheeren elasticity and strength of shearing; absolute ~, (Zug-~) u. Festigkeit elasticity and strength of extension; relative ~, (Biegungs-~) u. Festigkeit elasticity and strength of flexure; rückwirkende ~, (Druck-~) u. Festigkeit elasticity and strength of compression.

Elastizitäts ... ~grenze *f* elastic limit, limit of elasticity; ~modul *m* modulus of elasticity.

Elektriker, Elektrotechniker *m* electrician, electro-technician.

elektrisch *a* electric, electrical; ~, mit Elektrizität geladen electrified; ~ abgeschieden, galvanisch niedergeschlagen electro-deposited; ~e Beleuchtung *f* electric lighting; ~e Entladung *f* electric discharge; ~er Lichtbogen *m* electric arc; ~es Geläute, ~es Läutewerk *n* electro-magnetic ringing apparatus; ~er Haustelegraph *m* electric annunciator; ~e Induktion *f* electrification by induction; ~es Kohlenlicht *n* carbon light; ~e Ladung *f* electric charge; ~er Strom *m* electric current, current of electricity; ~er Telegraph *m* electric telegraph.

elektrisiren *v* to electrify.

Elektrisirmaschine *f* electrical machine.

Elektrisirung *f* electrization, electrification.

Elektrizität *f* electricity; (dynamische, in Bewegung befindl., strömende) dynamical electricity, electricity in motion, Voltaic electricity, galvanism; (galvanische ~, Galvanismus *m*) galvanism; mit ~ geladen, elektrisirt *a* electrified, charged with electricity; die ~ leiten to conduct the electricity; Leiter *m* der ~ conductor; (negative, Harz-~) negative (resinous) electricity; (positive, Glas-~) positive (vitreous) electricity; (statische) statical electricity, electricity of tension.

Elektrizitäts ... ~ableiter *m* arrester; ~entwicklung *f* electric excitement; ~erregung *f* electromotion; ~leiter *m* conductor of electricity; ~messer *m* electrometer; ~sammler, (Kondensator) collector, condenser; ~verlust *m*, Entweichen *n* von ~ leakage, escape; ~vertheilung *f* distribution of electricity; ~waage *f*, ~zeiger *m* electrometer, electroscope.

Elektro ... ~chemie *f* electro-chemistry; ~chemisch *a* electro-chemical; ~dynamik *f* electro-dynamics; ~dynamometer *m* electro-dynamometer; ~lyse *f* electrolysis; ~lyt *m* electrolyte; ~lytisch *a* electrolytical; ~magnet *m* electro-magnet; ~magnetismus *m* electro-magnetism; ~magnetisch *a* electro-magnetic; ~meter *m* electrometer; ~motor *m* electro-motor; ~motorische Kraft electro-motive force (EMF); ~negativ *a* electro-negative; ~positiv *a* electro-positive; ~phor *m* electrophorus; ~skop *n* electroscope, rheoscope; ~statisch *a* electrostatic; ~technik *f* electrotechnics; ~techniker *m* electrician, electro-technician.

Elektrode *f*, **elektrischer Pol** *m* electrode.

Element *n* element; (galvanisches) pair of plates, cell, galvanic cell, cell of a galvanic battery (pile); (Volta'sches) Voltaic cell; ein Zink-Kohlen-~ a zinc-carbon element (s. auch Batterie).

Elfenbein *n* ivory; ~platte *f* ivory plate.

Elimination, Ausscheidung *f* elimination.

eliminiren *v* to eliminate.

Elle *f* ell, yard.

Emballage *f* packing, package, casing, charges for packing.

emballiren *v* to bale, to embale, to pack up.

Emblem *n* emblem, badge.

Empfang *m* receipt; ~s-Anerkenntniß *n* der Auswechslungs-Postanstalt verification of the exchange office; ~s-Anzeige (über ein aufgegebenes Telegramm) certificate of receipt; (über ein befördertes Telegramm) acknowledgment of receipt; ~sapparat m (T) receiver, receiving instrument; ~sberechtigt a (TP) authorized to receive (to receipt for) mail matter, telegrams; ~bescheinigung f bezahlt (T) notice of delivery paid; ~bestätigung f acknowledgment of receipt; ~zettel m (+) label.

empfangen v to receive.

Empfänger m (Adressat) addressee, receiver; (eines Betrages) recipient; (einer Postanweisung) payee; (T Apparat) receiving instrument.

empfindlich a (Techn.) sensitive, delicate (said of instruments, galvanometers etc.).

Empfindlichkeit f (Techn.) sensitiveness.

End... ~amt n, ~station f (T) terminal office (station); ~geschwindigkeit f final (terminal) velocity; ~klemme f terminal screw; ~los a endless; ~punkt m terminal point; ~stück n end piece, butt end.

Ende n end; (eines Zeitraums) expiration; ein kurzes ~ Kabel a short length of cable.

enden v to end, to finish, to conclude.

endgiltig a definitive; adv definitively.

Endosmose f endosmosis.

Energie, lebendige Kraft, Kraft f energy (the capacity of doing work); aktuelle ~ actual (kinetic) energy; potentielle, mögliche ~, mögliche Arbeit f potential energy; verwendbare, verwandelbare ~ available energy.

entbinden v (vom Dienste) to excuse from duty; (disziplinarisch) to suspend from duty; (Gase) to evolve (to disengage) gases.

enteignen v to expropriate.

Enteignung f expropriation.

Entfernung f distance; ~ aus dem Amte removal, dismissal from office; ~starif m tariff of distances.

entgegengesetzt a contrary, inverse, reverse; Ströme ~er Richtung inverse currents, currents of opposite directions.

entgleisen v to slip off, to get off the rails, to run off the line.

Entgleisung f derailment, leaving the rails.

enthalten v to hold; (in sich begreifen) to include.

enthärten v (Stahl) to anneal steel.

entheben v von Dienstgeschäften to suspend from office.

Enthebung f von Dienstgeschäften suspension from office.

entkarten v (P) to deal with the received mail.

Entkartungsstelle f (P) place where the received mail is dealt with.

entkohlen v (Roheisen) to decarburize, to decarburate, to decarbonize pig (cast) iron.

Entkohlung f decarburization, decarbonation.

entkuppeln v to disconnect, to throw out of gear.

entladen v to discharge.

Entlader m discharger, discharging rod.

Entladung f discharge; (allmähliche) gradual (successive) discharge; (augenblickliche) instantaneous discharge; (fortlaufende, kontinuirliche) continued discharge; (intermittirende) intermittent discharge; (langsame) slow discharge; (Neben-~, Seiten-~) lateral discharge; (oszillirende) oscillatory discharge.

Entladungs... ~batterie f discharging battery; ~strom m discharging current.

entlassen v to dismiss from..., to discharge from...

Entlassung f dismission, dismissal, discharge; ~sattest n certificate of discharge.

entlasten v (das Telegr.-Gestänge [T]) to relieve the telegraph structure, to diminish the load bearing on the poles; (Decharge ertheilen) to discharge; (von Porto [P]) to discharge, to put on the monthly statement of uncollectible postage.

Entlastung f (T) des Telegr.-Gestänges relieving the telegraph structure; (Decharge) discharge; (von

Porto [*P*]) discharge, vgl. entlaften;
~**§farte** *f* (+) monthly statement of uncollectible postage on re-directed or returned correspondence.

entmagnetifiren *v* to demagnetize.

Entreprife, *f*, in ~ in contract; in ~ vergeben to let out in contract.

entrichten *v* to pay, to discharge.

entrinden *v* to decorticate, to bark (a tree).

entschädigen *v* to indemnify, to re-imburse, to compensate.

Entschädigung *f* indemnification, indemnity, re-imbursement, compensation; ~ gewähren bis zu ... to grant (to give) compensation to the amount of ...

entscheiden *v* to decide; sich ~ to resolve upon.

Entscheidung *f* decision, determination; eine Frage zur ~ bringen to bring a question to the issue.

Entschließung *f* resolution, determination.

entsetzen *v* des Amtes to dismiss (to remove) *smb* from office.

Entsetzung *f* dismissal, removal.

entstellen *v* to alter, to mutilate (a telegram); ein entstelltes Telegramm (*Am*) a bulled message.

Entstellung *f* alteration, mutilation (of a telegram).

Entwässerungsgraben *m* drain, draining ditch.

entweichen *v* to escape (said of gases); nicht ~ lassen to retain.

entweichender Dampf *m* waste steam.

Entweichung *f* escape (of gases); ~röhre, Abzugsröhre *f* eduction pipe; ~ventil *n*, ~sklappe *f* escape valve, delivery valve.

entwerfen *v* (einen Plan u. dergl.) to project, to plot, to trace, to design; (ein Konzept) to draft (a letter).

entwerthen *v* to cancel, to obliterate, to deface (stamps).

Entwerthung *f* (v. Waaren) depreciation; (v. Marken) cancellation, obliteration of stamps; ~stempel *m* defacing (obliterating) stamp.

entwickeln *v* to develop; (von Gasen) to evolve, to develop, to issue; sich ~ to develop.

Entwicklung *f* development (of electricity, of the postal business etc.).

Entwurf *m* projected plan, delineation, sketch, conception; (flüchtiger) rough sketch; (Konzept) draft.

entzündbar, entzündlich *a* inflammable, combustible.

Entzündbarkeit *f* inflammability.

Erd ... ~arbeit *f* earth work; ~arbeiter *m* digger; ~bohrer *m* terrier, earth-boring auger; ~damm *m* earth bank, embankment; ~draht *m* (*T*) ground wire; ~fehler *m* (*T*) earth, earth fault; ~hacke *f* hoe, pickaxe, bill; ~kabel f. Kabel; ~harz, ~pech *n*, Asphalt *m* bitumen, mineral pitch, asphalt; ~klemme *f* (*T*) earth terminal; ~leitung *f* (*T*) ground wire, ~leitungsstange *f* (*T*) iron rod; ~loch *n* hole, excavation, (gebohrtes) bored hole, (gegrabenes) dug hole, (gesprengtes) blasted (sprung) hole; ~magnetismus *m* earth (terrestrial) magnetism; ~platte *f* (*T*) earth plate, ground plate; ~plattenstrom *m* earthplate current; ~rohr *n* tube, pipe; ~schleife *f* (*T*) two underground wires looped, loop; ~schließung *f*, ~schluß *m* earth, ground connection; die Leitung hatte ~schluß the wire was found to be to earth; ~schraube *f* (bei eis. Stangen) earth screw; ~schraube mit Muffe versehen screw earth tube; ~stampfe *f* earth ram, rammer; ~strom *m* earth (terrestrial) current; ~verbindung *f* ground connection; ~winde *f*, Göpel *m* crab, field capstan.

Erde *f* (*T*) earth, ground; die Linie an ~ legen to ground the line, to put the line on a ground; an ~ liegen to be to earth; mit ~ verbinden to connect to earth, to make (to find) earth; mit der ~ in Verbindung stehen to make earth.

erforderlich *a* necessary; ~en Falls in case of exigency.

Erfordernisse *n/pl* des Dienstes requirements of the service.

ergänzen *v* to complete.

Ergänzungs ... ~porto supplementary postage; ~zettel (zur Geld-karte [+]) supplementary money-letter bill.

ergeben *v*, es ergiebt sich it results from ...

Ergebniß *n* result.

erhalten *v* (im Stande) to maintain; die guten Beziehungen ~ to preserve the good relations; Erlaubniß ~ to obtain permission; Löhnung ~ to receive wages.

Erhaltung, Unterhaltung *f* maintenance (of t. lines); preservation, conservation (of energy).

erheben *v* (Steuern) to levy taxes; (Gehalt) to touch (to receive) salary; (im Voraus) to receive in advance; (die Gebühren bei der Bestellung) to collect the charges on delivery.

erhebliche Verzögerung *f* (in der Beförderung) considerable retardation.

Erhebung *f* von Gebühren collection of charges.

erhöhen *v* die Gebühr to raise the tax.

Erholungsurlaub *m* leave of absence (granted in regular turns to all officials).

Erhöhung *f* elevation (of temperature); raising (of taxes).

erinnern *v* an die Erledigung eines Schreibens to remind *smb* (to bring to the remembrance of *smb*) that a letter has not been answered yet.

Erinnerungs ... **~schreiben** *n*, **~verfügung** *f* monitory letter.

erkrankter Beamter *m*, der ein ärztliches Zeugniß eingereicht hat officer absent under medical certificate (*M. C*).

Erlaß *m* (des General-Postmeisters) ordinance of the Postmaster General; (einer Strafe) remission of a fine.

erlassen *v* to enact (an ordinance); to remit (a penalty).

erledigen *v* (Geschäfte) to despatch business.

erleuchten *v* to light, to illuminate (with gas, with electric light).

Erleuchtung *f* lighting, illumination.

ermangeln *v* to fail, to want of *smt*; ich werde nicht ~ I shall not fail to ...

Ermangelung *f* want; in ~ dessen for want of which, in default whereof.

ermächtigen *v* to authorize.

Ermächtigung *f* authorization.

ermäßigen *v* to moderate, to lessen, to reduce (the price).

Ermäßigung *f* moderation, reduction.

ermitteln *v* to ascertain, to discover, to find out; Adressat nicht zu ~ addressee not to be found.

Ermittelung *f* ascertaining, enquiry; nach den angestellten ~en as far as has been ascertained.

erneuern *v* to renew.

Erneuerung *f* renewal.

eröffnen *v* (eine Telegr.-Linie) to open a t. line to traffic; der Telegr.-Verkehr wird eröffnet mit ... telegraph business will be transacted with ...; J—m etw. ~ to let *smb* know, to communicate *smt* to *smb*.

Eröffnung *f* opening; (Mittheilung) communication.

erregen *v* to excite; **~de Flüssigkeit** *f* exciting liquid (solution); excitant.

erreichbar *a*, die nächste ~e Station (*T*) the next available station.

erreichen *v* to reach, to attain, to get.

errufen *v*, ein Amt (*T*) to gain the attention of an office.

Ersatz *m* compensation (for loss & damage), indemnification; replacing (of materials); **~betrag** *m*, **~leistung** *f* compensation, indemnification, indemnity; **~verbindlichkeit** *f* responsibility; **~verfahren** *n* proceeding with the view to grant a compensation.

Erscheinung *f* (Phys.) phenomenon.

erschöpfen *v* (einen Gegenstand) to sift a matter thoroughly.

erschweren *v* to aggravate; ~de

Umstände *m/pl* aggravating circumstances.

ersehen *v* to see, to perceive; es ist hieraus zu ~ it is to be seen from this.

ersetzen *v* to repair, to replace, to remove (bad pieces of wire etc.); to re-imburse (the cost); to compensate (for loss & damage); Jn ~ to replace *smb*.

erstatten *v* (Kosten) to refund, to compensate, to repay; Bericht ~ to report.

Erstattung *f* restitution, reimbursement; ~sanweisung *f* order for re-imbursement (refunding).

ersteigen *v* to climb up, to ascend, to scale.

ersuchen *v* to entreat for..., to request for...

Ersuchen *n* entreaty, request.

Ertrag *m* proceeds, profit, revenue.

Ertrags... ~ergebniß *n* financial result; ~fähigkeit *f* produce, revenue.

erwähnen *v* to mention; der oben erwähnte the above mentioned; ~swerth *a* worth being mentioned.

Erwähnung *f* mention; ~ thun to mention, to make mention.

erweichen *v* to soften.

erweisen, sich ~ als... *v* to prove.

erweitern *v* to widen, to enlarge, to dilate, to expand; sich ~, ausladen to gain.

Erweiterung *f* widening, dilation; ~, Ausladung *f* gain; ~sbau *m* enlargement of a building.

Erz *n* ore; (Bronze *f*) hard brass, bronze.

erzeugen *v* to produce, to generate.

Erzeugniß *n* production, produce, products.

Erziehungsgeld *n* sum paid for the education of children.

erzielen *v* (einen Gewinn) to realize a gain.

Estafette *f* horse express; ~npost *f* special courier line.

Etapen = Telegraphen = Abtheilung *f* intermediate detachment of military telegraphists.

Etat *m* der Post- u. Telegr.-Verwaltung estimate of the amount required in the (fiscal) year 18.. to defray the charges of the Post Office & Telegraph Department; den ~ vorbereiten to prepare the estimate etc.; ~sersparnisse *f/pl* savings in the expenditures; ~sjahr *n* fiscal (financial) year; ~smäßig *a* consistent with the estimate of charges fixed by the legislature; ~smäßige Einnahmen u. Ausgaben *f/pl* receipts & expenditures according to the estimate of charges; ~smittel *pl* funds appropriated by the legislature; ~süberschreitung *f* expenses in excess of the estimated charges; ~sverhältnisse *n/pl* financial condition; die ~sverhältnisse gestatten es it is consistent with the financial condition.

etwa *adv* (ungefähr) about, circa (ca.).

etwaig *a* eventual; *adv* eventually.

Examen *n* examination; ~ machen to pass an examination; s. auch Prüfung.

Examinations = Kommission *f* board of examiners.

Exemplar *n* specimen (of postage stamps); copy (of a book, print etc.).

Exhaustor *m* exhaustor, blast governor.

Expansion, Ausdehnung *f* expansion, dilation; ~smaschine *f* engine with expansion, expansion engine; ~shebel *m* detent lever; ~skraft *f* expansive (elastic) force; ~sstange *f* expansion rod, detent rod; ~sventil *n* expansion valve, cut-off valve; ~svorrichtung *f* expansion gear; ~swelle *f* cut-off shaft.

expediren *v* to despatch.

Expedition *f* Office; ~svorsteher *m* directing clerk, head clerk.

Experiment *n*, **Versuch** *m* experiment; ~e anstellen to conduct experiments.

explodiren *v* to explode; ~de Stoffe *m/pl* explosives.

expreß *a* express.

Expreß... ~bote *m* special mess-

Extra — fallen

enger; ~kosten *pl* fee for special messenger; s. auch Eil...

Extra... ~post *f* extra mail coach; ~postpferde *n/pl* horses of the extra mail coach; ~postreisende *m/pl* travellers by extra mail coach; ~strom *m* (*T*) extra current.

Exzentrik *n* eccentric, (*HA*) curved end of the pusher; ~rad *n* eccentric wheel; ~stange *f* eccentric rod; ~welle *f* eccentric shaft.

exzentrisch *a* eccentric, eccentrical.

Exzentrizität *f* eccentricity.

F.

Fabrik *f* factory, manufactory, works, establishment.

Fach *n* (Abtheilung) compartment, case, partition; (im Schreibtisch) pigeon hole; Abholungs=~ auf der Post box; (verschließbares Abholungs=~) lock box, (wenn die Thür von Glas ist) glass box; ~gebühr *f* box rent, box rental.

Fachwerk *n* frame work, bay work, panel work.

Fackel=Telegraphie *f* telegraphy by means of torches.

Facsimile *n* facsimile, autotype.

Factage=Gebühr *f* delivery fee.

Faden *m* thread; (Maaß zur See) fathom; ~=Telephon *n* toy telephone (the means of communication being a tightly stretched string).

Fahlerz, Kupferfahlerz *n* grey copper, fahl ore, fahlerz.

Fahne *f* (am Packet od. Gepäckstück) label.

Fahr... ~bahn *f* (Brücke) bridge road, carriage way; (Eisenb.) railway line, road way; (Schiffe) channel, track for vessels; (Straße) cart way; ~bar *a* practicable, navigable, sailable; ~damm *m* high road, causeway; (Deich) dike practicable for carriages; ~gelegenheit *f* conveyance; ~geleise *n* track; ~karte *f*, ~billet *n* ticket; ~plan *m* time table, itinerary; ~planmäßiger Eisenbahnzug *m* train starting at fixed hours previously notified, ordinary railway train; ~post *f* stage coach, (Packetpost) parcels post; ~postbeutel *m* s. Geldfahrpostbeutel; ~poststücke *n/pl* articles of the parcels post; ~posttarif f. Packetposttarif; ~posttaxe *f* s. Packetposttaxe; ~schein *m* ticket; ~weg *m* carriage road, carriage way.

Fähr... ~boot *n* ferry boat; ~brücke *f* rising scaffold bridge; ~geld *n* ferriage, compensation established (paid) for conveyance over a river (lake) in a boat; ~haus *n* ferry house; ~mann *m* ferry man; ~schiff *n* wherry.

Fähre *f* traject, ferry; (fliegende) flying bridge, swing bridge.

fahren *v* to ride in a carriage; (mit der Post) to go by post (by the stage coach).

Fahrlässigkeit *f* des Absenders (*P*) carelessness (negligence) of the sender.

Fahrt *f* ride, drive; (auf der Eisenb.) journey, passage; (zu Schiff) voyage; (des fahrenden *P* Beamten) trip, journey, (*Am*) run; (Bahnhofsfahrt *P*) trip to the railway station; ~ und Ueberlagergebühren (*P*) allowance for travelling & stopping; ~bericht *m* (*P*) report of trips; ~geld *n* (*P*) trip money, travelling allowance.

Faktur *f* invoice; ~ geben to invoice; laut ~ as per invoice; ~enbuch *n* invoice book.

Fall *m* (Phys. Mech.) fall, descent; (Fallen *n*, Neigung *f*) descending, gradient, fall, declivity, incline; (Begebenheiten) case, accident, event; in diesem ~ in this case; in keinem ~ at no events, by no means; erforderlichen ~s in case of exigency; im ~ der Noth in case of necessity (emergency).

Fall... ~geschwindigkeit *f* velocity of fall; ~rohr, Abfallrohr *n* gutter pipe, rain pipe, down pipe; ~scheibe Signalscheibe *f* (Eisenb.) colored glass.

fallen *v* to fall, to descend; (vom Preise) to go down; (nieder=

fallen [Chem.]) to settle down, to separate, to precipitate.
fällen v to fell, to cut, to hew down (trees); (Chem.) to precipitate.
fällig a due, payable; ~ werden to become due; (von Wechseln) to expire.
Fälligkeitstermin m term of expiration.
Fällzeit f der Bäume proper season for felling trees.
falsch, unecht, nachgemacht a false, counterfeit, adulterated; ~es Geld n base coin; ~er Wechsel m forged bill of exchange; ~ schreiben to misspell, to write incorrectly.
fälschen v to forge, to adulterate, to falsify.
Fälscher m defaulter, falsificator.
Falschmünzer m forger of base coin, counterfeiter of coin.
Fälschung f forging, adulteration, falsification.
Falz m fold, lap; ~bein n folder, folding bone.
Fang ... ~-Apparat m (Bahnpost) mail bag apparatus, catching apparatus, mail catcher; ~leine f (Techn.) painter; ~rad n striking wheel; ~spitze f (am Blitzabl.) point; ~winkel m hook.
Farbe f color, dye; (M & HA) ink, blue liquid, colored liquid; ~gefäß n ink reservoir, ink well; ~rädchen n printing wheel, inking disc; ~radträger m arbor of the printing wheel; ~rolle, ~walze f impression roller; ~scheibchen n printing roller, inking disc; Buchdrucker-~ ink.
Farbschreiber m ink writer, inker, printer (f. auch Apparat); (ohne Relais arbeitender) direct ink writer.
Faser f fibre, filament; (Pflanzenfaser, Holzfaser f) lignine, woody fibre; hänfene ~ hempen fibre.
faserig a fibrous, filamentous (f. auch Eisen).
Faß n cask, barrel.
fassen v (ergreifen) to seize, to lay hold on; (enthalten) to hold, to contain; (von Nägeln 2c.) to put on, to take.

faul a rotten, putrid, decaying.
faulbrüchig f. Eisen.
Fäule f rot; (trockene) dry rot; (nasse) wet rot; f. auch Fäulniß.
faulen v to rot.
Fäulniß f putrescence, putrefaction, rottenness; ~ befördernd septic, septical; ~widrig antiseptic; ~ widrige Mittel n/pl antiseptics.
Faustpacket n (P) small parcel.
Feder f (Mech.) spring; ~aufzug f. Apparat mit ~aufzug; ~bolzen m spring bolt; ~gehäuse n, ~trommel f spring box, spring barrel; ~harz, Kautschuk n elastic gum, India rubber, caoutchouc; ~hebel m spring lever; ~unterbrechung f (T) spring break (of the automatic circuit breaker); ~waage f spring balance, spring yard.
federhart a hard and elastic like a steel spring, springy; ~ machen to hammer-harden.
federn v to spring, to be elastic; ~der Kontakt m spring contact.
Fehlbetrag m deficiency; f. auch Forderung.
fehlen v to fail, to miss; es fehlt ein Packet one parcel is missing, to be short of one parcel.
Fehler m fault, defect; (Irrthum m) mistake, error, blunder; einen ~ begehen to commit a fault, to make a mistake; ~quelle f source of error; ~stelle im Draht fault, location of a fault, a faulty wire, a leak; ~ beim Ablesen der Winkelgrade error of parallax in reading the angular measures.
fehler ... ~frei a faultless; ~haft a faulty.
fehlleiten v (T P) to missend, to misdirect, to put out of course; fehlgeleitete Post mail arriving out of course.
Fehlleitung f (T P) missending, misdirection.
Feiertag m holiday, blank day (day on which the ordinary mail service is wholly or in part suspended, in England: Sunday, Christmas day & Good Friday).
Feil ... ~kloben, Handkloben, Handschraubstock m vice, hand vice,

pin vice, filing vice; ~kluppe sloping clamp of a vice; ~späne pl, ~staub m file dust, filings, limature.

Feile f file; (doppelrunde) cross file; (dreikantige) three square (triangular) file; (dünnflache) pillar file; (flache, Ansatz~, Hand~) flat file; (halbrunde) half-round file; (runde) round file; (große runde, Stroh~) rough file; (kleine runde, Rattenschwanz m) rat tail file; (scheibenförmige) turning file; (spitzflache) taper-flat file; (vierkantige) square file.

feilen v to file.

feinkörnig s. Eisen.

Feld n field, ground, land; (magnetisches) magnetic field; (des Zeiger=Apparats) division; ~batterie f (T) field battery; ~isolator m field insulator; ~messer m surveyor; ~meßkunst f art of surveying, land survey; ~post f war postal service; ~postbeamter m officer of the war postal service; ~postdienstordnung f regulations (instructions) for the war postal service; ~seite f (der Stange [T]) back of the pole; ~telegraph m military telegraph; ~telegraphen=Abtheilung f field telegraph train, military telegraph detachment; ~telegraphenlinie f field telegraph line; ~telegraphenstange f field telegraph pole; ~telegraphenstation f telegraph station in the field; ~weg m field way, field path; ~zulage f extra allowance for field service.

Felge f felloe, felly, jaunt.

Felleisen n bag, knapsack, valise; Post=~ s. Post.

Fels m rock; ~boden m rocky ground; ~geröll n pebble, rubble, boulder; (Geol.) detritus.

ferne Station f, **fernes Amt** n (T) distant (remote) station.

Fernglas n perspective glass, telescope, spy glass, spying glass.

fernhalten, sich v vom Amte (unter Vorlegung eines ärztlichen Zeugnisses) to be absent under medical certificate; (unerlaubt) to be absent without leave; (unter Vorschützung von Krankheit) to malinger.

Fernhaltung f ~ vom Amte (unter Vorlegung eines ärztlichen Zeugnisses) absence under medical certificate; (unerlaubte) absence without leave; (unter Vorschützung von Krankheit) malingering.

Fernrohr n telescope.

Fernsprech ... ~=Anlage f telephone plant; ~anschluß m connection; ~betrieb m telephone service, telephony; ~leitung f telephone wire; ~linie f telephone line; ~stelle f call room, call station, (öffentliche) public call room, public telephone; ~verbindungslinie f telephone trunk line; Stadt=~vermittlungsamt n local telephone exchange, telephone office; ~zelle f call box.

Fernsprechen n **auf weite Entfernungen** long distance telephony.

Fernsprecher m, **Telephon** n telephone.

fertig a (zum Geben [T]) ready.

fest a solid, strong, fast, firm, fixed, rigid; ein Eisenstab ist ~ mit der Schiene verbunden an iron rod is rigidly attached to the metallic bar; ~ angestellter Beamter m established servant of the department; ~e Gebühr f fixed tax (rate); ~er Körper m solid body.

fest ... ~binden v to bind (the insulator); ~halten, unterhalten to support; die Nägel halten ~ the nails stick fast; ~kleben (an den Polen des Elektromagneten) to adhere (to stick) to the poles; ~klemmen to cramp; den Draht mit einer Klemmschraube ~klemmen to attach the wire by a binding screw; ~machen to fasten; ~punkt m fixed point; ~nageln to nail, to spike; ~rammen, ~stampfen to ram, to beat down the earth around the pole; ~schrauben to screw; ~setzen to plant (a pole), to fix (the tax); (Bestimmungen) to establish (to lay down) regulations; ~setzen auf ... to put it at ...; ~setzung f regulation, stipulation, law; gegenseitige ~stellung f der Rechnungen mutual settling of accounts.

Festigkeit f tenacity, strength,

rigidity, stiffness; absolute ~, Zug-~ tensile strength, strength of extension; relative ~, Bruch-, Biegungs-~ strength of flexure, transverse strength; rückwirkende ~, Druck-~ strength of compression, compressive strength; Scheer-~ shearing strength; Torsions-~, Drehungs-~ torsional strength; ~smodul m breaking load.

Festtag s. Feiertag.

feucht a humid, moist, damp.

Feuchtigkeit f moisture, moistness, dampness, humidity; ~ einsaugen to imbibe moisture; ~sgehalt m moisture; ~smesser m hygrometer.

Feuer n fire; ~ fangen to take fire; ~alarmapparat m fire alarm; ~fest a fire proof, refractory; ~fester Schrank fire-proof safe; ~gefährlich a liable to catch fire; ~melder m fire alarm; ~wehrtelegraph m fire telegraph, fire-alarm telegraph.

Feuerung f, **Brennmaterial** n fuel, combustibles.

feurig a (Phys.) igneous; ~er Schwaden m, schlagendes Wetter, Grubengas n fire damp.

Fiber, Faser f fibre, filament; (des Holzes) grain of wood.

Fichte f pine, pine tree, fir pitch pine (Pinus abies); Sprossenfichte spruce (Nord-Amerika).

fichten a pine, of pine wood of fir wood.

Figur f figure, diagram.

Figurenwechsel m (HA) change from letters to figures & signs (& vice versâ).

Filialstation f sub office, branch office.

Filz m felt; ~band n ribbon of felt; ~bekleidung f felt covering; ~rolle, ~walze f felt roller.

Finanzjahr s. Etatsjahr.

Firmenregister n commercial register.

Firniß m varnish; ~überzug m coat of varnish.

firnissen v to varnish.

First, Dachfirst m ridge, top of a house; (einer Mauer) coping of a wall.

fiskalisch a fiscal; ~es Eigenthum n Government property.

Fixpunkt m fixed point, station.

flach a flat, plain, level, even; ~es Dach n flat roof.

Flach... ~bohrer m flat auger; ~eisen s. Eisen; ~feder f flat spring; ~feile f flat file; ~land n, Ebene f flat plain; ~pinsel m flat brush; ~zange f flat plyer, flat-nosed plyers.

Fläche f level, flat, plane, smooth surface, area; (ebene, horizontale) plane surface, dead level; (geneigte, schiefe) inclined plane.

Flächen... ~inhalt, ~raum m superficial content, area; ~maaß n square measure, superficial measure; ~messung f planimetry.

Flagge f flag, standard.

flaggen v to hoist the flag, to display the colors.

Flaggen... ~signal n flag signal; ~signalsystem n code of flag signals; ~stange f, ~stock m flag staff.

Flammbogen m (galvanischer, elektrischer) Voltaïc arc, electric light-arc.

Flamme f flame.

Flansch m, **Flansche, Flantsche** f flange, collar; ~, Latsche f eines Eisenträgers flange of an iron girder; mit ~n versehen a flanged.

Flantschen... ~röhre f flange pipe; ~verbindung f in Röhren flange joint.

Flasche f bottle, flask; ~, Kloben m (Gehäuse mit Rollen) block, tackle, system of pulleys; Leydener ~ Leyden jar, electric jar.

Flaschenzug, Rollenzug, Klobenzug m set of pulleys, pulley block, tackle, block and tackle, fall and tackle; (doppelter) double purchase; (mit nur einer Rolle) simple purchase; (an einem Hebebock ob. Krahn angebrachter) gin block.

fliegende Brücke, Fähre, fliegende Fähre f flying bridge, swing bridge.

Fließpapier n blotting paper, sinking paper.

Floß n raft, float.

flößen v to float (wood).

Flucht *f* (Techn.) straight line, line of direction, alignment.
Fluchtlinie *f* line of direction.
flüchtig *a* (Chem.) volatile; ~es Oel *n* volatile oil, ethereal oil; ~ arbeiten to work carelessly (hastily).
Flug ... ~schrift *f* pamphlet, print; ~sand *m* quicksand, shifting sand, moving sand.
Flügel *m* (Techn.) wing, branch, part; (Zeichengeber des opt. Telegr.) wing, indicator, receiver; (d. Schiffsankers) fluke of the anchor; (einer Flügelthür) leaf of a folding door.
Flügel ... ~bremse (*HA*) wing regulator; ~klemme *f* winged clamp; ~mutter *f* thumb nut, finger nut, fly nut, nut of a winged screw; ~rad *n* (Masch.) flying pinion; ~regulator *m* wing regulator; ~schraube *f* winged screw, thumb screw; ~telegraph *m* semaphore.
Fluidum *n* fluid; (elektrisches, magnetisches) electric, magnetic fluid.
Fluorescenz *f* fluorescence.
Fluß *m* river; (das Fließen) flow, flowing; (Flußmittel *n*) flux (to melt metals).
Fluß ... ~bett *n* river bed; ~damm *m* river dike, quay; ~eisen *n* ingot iron; ~kabel *n* river cable; ~krümmung *f* bend, sinuosity of a river; ~mittel *n* flux.
flüssig *a* (Phys.) fluid; (Chem.) liquid; elastisch-~ gaseous, aëriform; tropfbar-~ liquid; ~er Zustand *m* fluidity.
Flüssigkeit *f* (Phys.) fluidity, fluid; (Chem.) liquid, liquor; ~smaaß *n* fluid-measure; ~smesser *m* fluid-meter.
Flut *f* flood, tide; Anfang *m* der ~ setting tide; (Ebbe *f* u. ~ ebb & flood, flux & reflux; niedrigste ~ neap tide.
Folge *f* consequence; (Reihe) series, order; (Fortsetzung) continuation; (Aufeinanderfolge) succession, sequence; (Wirkung) result, effect, consequence; in ~ according to, in consequence of; dem zu ~ in pursuance of which, consequent on (upon); in der ~ hereafter; ~ leisten to comply, to obey; ~punkte *m/pl* (Phys.) intermediate poles, consequent poles, consequent points (of a magnet).
folgen *v* to follow, to succeed; (aus etw.) to follow, to be the consequence of ...; und lautet wie folgt and runs as follows; ~d, auf einander folgend successive; ~de Woche next week.
foliiren *v* (ein Buch) to page a book, to mark the pages.
Folio *n* folio; ein Buch in ~ a folio volume; ~papier *n* foolscap paper.
Fond *m* fund, funds, capital, stock; ~sbörse *f*, ~smarkt *m* stock exchange; ~smakler *m* stock broker.
fordern *v* to demand, to ask, to call for.
Forderung *f* demand, claim; (Guthaben) credit; in ~ stellen to enter (an item) on the payment side of an account, to recover; eine Postanstalt stellt sich einen Fehlbetrag in der Frachtkarte in ~ a post office *recovers* the difference by raising the amount to its credit in the way bill; ~skonto *n* credit account; ~snachweis *m* liquidation, amount to be paid.
Form *f* form, figure, shape; (Gießform) mould, cast; in gehöriger ~ in due form; der ~ wegen for form's sake.
Formalien *f/pl* formalities.
Format *n* size.
Formel *f* formula (*pl* formulae).
formell, förmlich *a* formal, *adv* formally.
Formular *n* form, (*Am*) blank; Telegramm-~ message (telegram) form, (für Ursprungs-Telegramme) forwarded telegram form, (für Durchgangs-Telegramme) transmitted telegram form, (für Orts-Telegramme) received telegram form, (für bezahltes Antworts-Telegramm) form of authority; ~heft *n* pad.
fort ... ~bewegen *v* to feed (the paper slip); ~laufend *a* continued, continuous; ~pflanzen *v* to propagate (the electricity); ~schaffen *v* to convey, to transport; ~schaffungsmittel *n/pl* means of conveyance; ~satz, Ansatz *m* (Techn.) continuation.

Fracht ... ~brief *m* bill of carriage, letter of conveyance, way bill; ~buch *n* (+) book containing the receipts of the mail guard, driver etc. for the mail; ~frei *a* freight-free, carriage-free; ~fuhre *f* land carriage; ~gut *n* load, freight, goods; ~gut=Expedition *f* (Eisenb.) goods office; ~karte (*P*) parcel bill; ~kartenschluß *m* (*P*) parcel mail; ~lohn *m* freightage, cartage; ~stück *n* parcel, article of the parcels post.

Frage *f* question, interrogation; ~bogen *m* schedule of interrogatories; ~zeichen *n* sign (note) of interrogation.

frankiren *v* to prepay, to send postpaid; baar ~ to pay the postage in money; mit Freimarken ~ to prepay by means of postage stamps.

frankirt *a* prepaid.

Frankirung *f* prepayment; ~s=zwang *m* compulsory prepayment.

Franko *n* prepaid postage; (baar erhobenes) postage paid in ready money; f. auch frei; ~couvert *n* stamped envelope; ~=Einnahmenachweisung *f* (+) statement of the amount of prepaid postage not accounted for in postage stamps; ~=Gegennachweisung *f* (+) duplicate of the statement of the amount of prepaid postage not accounted for in postage stamps (for controling purposes); ~stempel *m* date stamp denoting prepayment; ~vermerk *m* notice of prepayment; ~zettel *m* (+) advice of postage due.

frei, portofrei *a* post-paid, post-free; ~ versenden (mittels eines zur Portofreiheit berechtigenden Vermerks z. B. Postdienstsache, von Parlamentsmitgliedern u. dgl.) to frank.

Frei ... ~billet *n* free pass; ~exemplar *n* (einer Zeitung) copy of a newspaper free of cost; ~fahrt *f* (auf Eisenb.) free ride, (*Am*) dead head; ~fahrtschein *m* (*P*, Eisenb.) free pass; ~händige Vergebung, (Verdingung) *f* giving out work, striking a bargain (for any work to be done) without any preliminary formalities; ~gepäck *n* luggage free of charge; ~gewicht *n* weight of luggage allowed; ~marke *f* postage stamp, f. auch Postwerthzeichen; ~=paß *m* (Eisenb.) free pass.

Friktions ... ~rad *n* friction wheel, fine toothed wheel; ~sperrklinke *f* ratchet of the friction wheel; ~walze *f* friction roller.

Frischeisen f. Eisen.

Frist *n* space of time, set term, delay; ~ bewilligen to grant delay; ~ erbitten to ask for time; ~sache *f* (*P*) report to be handed in at a fixed time; ~verlängerung *f* prolongation of a term.

Froschklemme *f* dutch tongs, draw tongs, devil's claw.

Fuchsschwanz *m*, **Blattsäge** *f* pad saw, whip saw, saw without a frame.

Fuhr ... ~lohn *m*, ~kosten *pl* freight, cartage, price of conveyance; ~mann *m* carrier, carter; ~werk *n* vehicle, carriage, cart, conveyance; ~wesen *n* the business of transporting goods by carriages; f. auch Postfuhrwesen.

Fuhre *f* cart load, waggon load.

Führung *f* (Techn.) management, direction, carrying on; (der Bücher) keeping of the books; (des Papiers) f. Papier ...; ~attest *n* conduct certificate; ~rolle *f* (*MA*) guide pulley, roller; ~sstift *m* (*MA*) guide pin.

Füll ... ~apparat *m* (Dampfmasch.) feeding apparatus, (*T*) replenisher; ~material *n* (um den Fußpunkt der Stange) filling (packing, stuffing) material.

Fundament *n* foundation, basis, base; (gemauertes) foundation walling.

fungiren *v* to perform the duties of one's office, to officiate.

Funken *m* spark; ~sammler *m* spark condenser.

Funktionszulage *f* supplementary allowance, duty pay, pay for the performance of special duties.

Fuß *m* (Techn.) foot, foundation, base.

Fuß ... ~bote *m* foot messenger; ~boden *m* floor, flooring; ~brett,

Trittbrett *n* pedal; ~gestell *n* pedestal, socle; ~platte *f* foot plate; ~punkt *m* basis, base; ~steig *m* foot path, path.

Futter *n* (Techn.) lining, case, inner covering, coating; (Vieh=~) food, feed, forage; ~kosten *pl* expenses for food; ~kostenzuschuß s. Zuschuß; ~mauer *f* revetment wall.

G.

Gabel *f* fork, prong.
gabelförmig *a* forked, furcate, bifurcate.
gabeln, sich gabeln *v* to bifurcate (said of the electric current, of t. lines, railroads etc.).
Gabelung *f* bifurcation.
galvanisch *a* galvanic; ~e Batterie, Säule *f* galvanic battery, galvanic (voltaic) battery (pile); ~es Element *n* galvanic element (cell).
galvanisiren *v* to galvanize.
Galvanisirung *f* galvanizing, galvanization.
Galvanismus *m* galvanism, voltaism, voltaic electricity.
Galvanomagnetismus *m* galvano-magnetism, electro-magnetism.
Galvanometer *n* galvanometer; (astatisches) astatic g.; (mit Nebenschließung) shunted g.; (Differential=~) differential g.; (Marine=~) marine g.; (Spiegel=~) mirror g., reflecting g.; (Universal=~) universal galvanometer.
Galvanoskop *n* galvanoscope, detector.
Gang *m* way, passage; (Arbeit) work, working, motion, action (of an apparatus, of a machine); thread (of a screw); (des Briefträgers) walk, s. auch Bestellgang; (der Posten) post routes, mail routes; den ~ der Posten regeln to regulate the post (mail) routes; außer ~ (in Unordnung) sein (Masch.) to be out of gear; außer ~ setzen (in Unordnung bringen) to throw out of gear; im ~ (Masch.) at work, in gear; im ~ sein, arbeiten to be at work; in ~ setzen to start.
Gang . . . ~höhe *f* eines Schraubengewindes pitch; ~werk *n* (einer Uhr 2c.) work.

gangbar *a* practicable (road); current (money); marketable, saleable (article).
Gangbarkeit *f* good working (of an apparatus).
Gänsefüßchen *n/pl* signs of quotation, inverted commas, carets („ ").
ganz *a* entire, whole; *adv* entirely, wholly.
gar *a* good, refined; ~es (gargefrischtes) Eisen *n* fined iron, iron which has come to nature; ~es Kupfer *n* refined copper.
Gas *n* (Phys., Chem.) gas; (Leuchtgas) gas. lighting gas; ~beleuchtung *f* gas lighting; ~brenner *m* gas burner; ~flamme *f* gas light; ~förmig *a* gaseous, gasiform, aëriform; ~förmig, elastisch flüssig aëriform; ~förmige Körper *m/pl* elastic fluids; ~kohle *f* gas carbon, coke; ~leitung *f* conduit of gas, gas conduit, gas supply; ~licht *n* gas light; ~messer, Gasometer *m* gas meter; ~rohr *n*, ~röhre *f* gas pipe.
Gasse *f* lane.
Gatter, Gitter *n* grate, lattice, railing; ~thor *n*, ~thür *f* grated door, barrier gate, barrier; ~werk *n* lattice work, grated work, trellis work.
gattiren *v* to mix (the ores); ~ und möllern to mix the ores and fluxes.
Gattirung *f* mixing of ores, mixture of ores, mixed ores.
Gattung *f* class.
geben *v* (abgeben) to forward, to transmit (telegrams).
Geber *m* s. Apparat.
Gebiet *n* district, territory; (Gerichts=~) jurisdiction; (Herrschaft) command, government.
Gebirge *n* chain of mountains, ridge of hills, mountainous district.

Gebirgs ... ~gegend *f* mountainous region; ~paß *m* mountain pass, gorge; ~schlucht *f* canyon, canon.

Gebläse *n* bellows; ~luft *f* blast, blast air; ~ofen *m* blast furnace; ~vorrichtung *f* blast, blast apparatus, blower.

geböschte Mauer s. Böschungsmauer.

Gebot *n* (bei Auktionen) bid, bidding, offer.

gebrannt *a* burnt (gypsum, plaster of Paris), baked (earthenware).

Gebrauch *m* use, employment, custom, practice; ~ von etw. machen to make use of *smt*; außer ~ kommen to fall into desuetude.

gebrauchen *v* to use, to make use of, to employ; (nöthig haben) to need, to want; wie lange gebraucht der Zug bis ...? how long does it take the train (to go) to ...?

gebräuchlich *a* usual, customary, current; ~ sein to be in use; nicht mehr ~ sein to be out of use.

gebrochener Hebel *m* lever consisting of 2 parts, compound lever, recording lever for open and closed circuits.

Gebühr *f* charge, rate, fee; (Porto) rate upon letters (parcels etc); (für Postanweisungen) commission on money orders.

Gebühren ... ~einnahme *f* receipts; außergewöhnliche ~einnahme extraordinary receipts; ~frei *a* (*PT*) free of charge; ~freiheit *f* exemption from postage; ~pflichtig *a* subject to charge (postage); ~quittung *f* (*T*) receipt for the charges.

gediegen *a* native (metals).

Geding *n*, **Gedingarbeit** *f* work agreed upon, job; im ~ arbeiten to work by the job; Arbeit in kleinen ~en piece work.

geeignet *a* appropriate, proper (to take such measures which may be deemed proper).

Gefahr *f* danger, risk; auf ~ des Absenders at the risk of the sender; Gegenstände, deren Beförderung mit ~ verbunden ist articles the conveyance of which is likely to entail risk or injury to the ordinary contents of the mail.

Gefälle *n* slope, inclination, incline; head, hight (of water); fall (of a river); (Einnahmen *f/pl*) revenues; steiles ~ steep (heavy) gradient.

gefälliges Schreiben *n* (Your) honored letter, favor.

gefrischt *a*, ~s Eisen *n* fined iron; mit Holzkohlen ~es Eisen *n* charcoal iron; ~er Stahl *m* fined steel, charcoal steel, furnace steel.

Gefüge *n*, **Struktur** *f* construction, structure, (the wire, when broken must disclose a fibrous structure).

gegen *prp*, ~ Abzug der Kosten upon deduction of the costs; ~ eine Gebühr at a rate; ~ das Licht against the light.

Gegen ... ~batterie *f*. Batterie; ~beweis *m* counter proof, counterevidence; ~buch *n* (Porto-Konto) book of control, control; ein ~buch unterhalten to keep a book of control; ~forderung *f* set-off, offset; ~gewicht *n* counter-poise, counterweight, balance weight; ~kraft *f* counter-force, opposed force; ~mutter, Stellmutter *f* jam nut, check nut, lock nut, pinching nut; ~rechnung saldirt balanced in account; ~schaltung *f* (*T*) opposition method; ~schraube *f* counter screw; ~schuld *f* reciprocal debt; Schuld u. ~schuld active and passive debts; ~seitig *a* reciprocal, mutual; ~seitigkeit *f* reciprocity, reciprocality; ~sprechen *n* (*T*) simultaneous transmission of two communications in opposite directions, duplex telegraphy; ~sprechen und Doppelsprechen quadruplex system; ~sprecher, ~sprech-Apparat *m* apparatus arranged for the simultaneous transmission of two communications in opposite directions, duplex apparatus; ~strom *m* counter current, reverse current, extra current; ~theil *n* contrary; im ~theil on the contrary; ~über *prp* over-against, opposite to; ~über stehend (liegend) *a* opposite;

~vorstellung *f* remonstrance; ~werth *m* equivalent, counter value; ~wirken f. entgegen.

Gehalt *n* (für Dienstleistungen) salary, pay, wages, allowance; (Inhalt) contents, capacity; ~abzug *m* deduction from the salary, stoppage of pay (for a day, for two days); ~abzugsverfahren *n* proceeding with the view to cause regular reductions made from an official's salary; ~aufbesserung *f* increase of salary (wages); ~sliste *f* pay roll; ~stufe *f* scale of pay; ~szulage *f* increment, augmentation (increase) of salary (wages); ~zuschuß *m* supplement of salary.

gehämmert ... es Stabeisen *n* forged bar-iron; ~er Stahl, Gerbstahl *m* tilted steel.

gehärtet *a* hardened, tempered (steel); nicht ~ unhardened, soft.

Gehäuse *n* case, casing, box, shell; (eines Flaschenzugs) shell, cheek of a pulley, pulley case, pulley frame; (des *MA*) frame, brass case.

geheim *a* secret, private; ~er Rath *m* private (privy) counsellor, f. auch Rath; ~e Sprache *f* (*T*) secret language (words not to be found in a standard dictionary of the language), conventional language, code language.

Geheim ... ~schrift *f* cipher, cryptography; ~secretair *m* privy secretary; ~sprache f. geheime Sprache.

gehen *v* (Mech.) to go, to work; vor Anker ~ to cast anchor; an's Land ~ to go ashore; leer ~ (said of a screw) to have end-play; in See ~ to go to sea; (vom Strome) to flow (to earth).

Gehilfe, Gehülfe *m* helper, assistant.

Gehör *n* hearing; ~ geben to give (*smb*) a hearing; nach dem ~ aufnehmen (*T*) to read (a despatch) by the sound; Aufnahme nach dem ~ sound reading, acoustic reading.

Gehwerk *n* movements, wheel work.

gekröpft *a* bent at right angles; ~es Schneidwerkzeug *n* edged tool with gullet-end.

gekuppelte Stangen *f/pl* coupled poles.

geladen *a* (mit Stückgütern) laden in parcels; (mit Stürzgütern) laden in bulk: tief ~ deep loaden.

Gelände *n* tract of land.

Geländer *n* rail, railing, side rail, hand railing; mit einem ~ versehen *v* to rail; ~pfosten *m* rail post.

Geläute *n* (elektrisches Läutewerk) electro magnetic ringing apparatus, electric bell.

Geld *n* money, coin; ~ anlegen to invest money (capital); (baares) ready money, cash; (falsches) coined money; (gangbares) current money; (schlechtes, geringwerthiges) base money (coin); (umlaufendes, zirkulirendes) running cash.

Geld ... ~abgangszettel, ~eingangs- und ~übergangszettel *m* (+) list of mail bags containing articles with value declared; ~ablieferungsschein *m* (+) receipt for articles with value declared; ~anlage *f* investment of funds; ~anweisung *f* money order; ~ausgabekonto *n* account book of expenditures for office supplies; ~bedarf *m* requisite cash, money required; ~beutel *m* money bag; ~brief *m* (+) letter containing money, letter with value declared, money letter; ~briefbeutel *m* (+) bag containing letters with value declared (money letters); ~bund *n* money packet; ~eingangsbuch *n* (+) book of cash-receipts; ~eingangszettel *m* f. ~abgangszettel; ~einlieferungsschein *m* receipt; ~fahrpostbeutel *m* (+) mail bag containing articles with value declared, (*Am*) coin mail bag (for sending money order funds in coin); ~fahrpostpacket *n* (+) money-letter mail; ~faß *n* cask containing money, cask (barrel) of money; ~karte *f* (+) money-letter bill; ~kartenschluß *m* (+) money-letter mail; ~kasse *f* money box; ~kiste *f* money chest, strong chest, strong box; ~kurs *m* rate of exchange for money; ~packet *n* money packet; ~rolle *f* roll of money; ~sack *m* money bag; ~schrank *m* money chest; feuerfester

~schrank safe, banker's safe, fireproof safe; ~sendung *f* (+) letter (packet) containing money; (t. Handel) remittance (of money), ~sorte *f* sort (denomination) of coin; ~strafe *f* fine; J—n in eine ~strafe nehmen to inflict a fine on *smb;* ~stück *n* piece of money, coin; ~übergangszettel *m* f. ~abgangszettel; ~umlauf, ~verkehr *m* circulation of money, ~verkehr durch die Post use made of the post office for the transmission of money; ~vorschuß *m* cash advance; ~waage *f* money balance, money scales; coin balance; ~wechsel *m* exchange of money, banking business; ~wechsler *m* money changer, banker; ~werth *m* value of the money, value in money; ~werthe Papiere *n/pl* stocks; ~zählapparat *m* coin counter.

Geleis, Geleise *n* track, track way, line, railway line; (Fuhrw.) track, rut; durchgehendes ~ (Eisenb.) thorough line, through line, main line; einfaches (doppeltes) ~ single (double) line, single (double) set of tracks.

Gelenk, Glied *n* (Techn.) link, joint; (Scharnier) hinge, hinge joint, turning joint; ~band, Scharnierband *n* joint hinge; ~hebel *m* joint lever.

gemäß *a* u. *adv* conformable (to, with), conformably, according to . . .

Gemäßheit *f* conformity; in ~ in conformity with . . (to . .).

Gemeinde *f* commonalty, community, parish; ~behörde *f* municipality, common council; ~bezirk *m* commune, rural district; ~weg, Bizinalweg *m* parish road.

gemeinsame, gemeinschaftliche Batterie f. **Batterie**.

gemeinschaftlich *a*, auf ~e Rechnung on joint account; ein Unternehmen auf ~e Rechnung joint operation, mutual operation.

Gemenge, Gemisch *n* mixture; (Beschickung) mixture of ores & fluxes (f. Gattirung).

genau *a* fine, true, accurate; ~ um 4 Uhr at four o'clock precisely; (passend) fitting close, fitting exactly; ~e Abschrift *f* true copy; ~este Preis *m* the lowest price.

genehmigen *v* to approve, to sanction, to ratify, to allow, to accept.

Genehmigung *f* granting, permission, approbation, sanction.

geneigt *a* inclined; ~e Ebene *f* (Bahn *f*) inclined plane; ~e Fläche *f* gradient, slope.

General . . . ~-Abrechnung *f* general settlement of accounts; ~bevollmächtigter *m* attorney-in-fact, one who holds a full & absolute power of attorney, (als Minister, Gesandter u. dgl.) plenipotentiary; ~direktion *f* general direction, besser department (of Posts, of Telegraphs) ob. in Adressen „an die ~direktion" to the Director General etc.; ~erlaß *m*, ~ordre *f* general order; ~postdirektor f. Post; ~postkasse f. Post; ~postmeister f. Post; ~Quittung *f* receipt in full, acquittance in full for all demands; ~Versammlung *f* general meeting, general assembly; ~vollmacht *f* general power of attorney.

Gepäck *n* baggage, luggage; das ~ aufgeben to book the baggage; ~annahme *f* booking of luggage; ~ausgabe, ~-Expedition *f* luggage office; ~schein *m* luggage bill; ~wagen *m* baggage van, luggagewaggon, box waggon, (*Am*) baggage car.

gerade *a* straight, right; ~ Linie (Math.) straight line; ~ Isolatorstütze *f* bolt; ~ machen (strecken) to straighten (the wire).

geradlinig *a* rectilinear, right lined.

Gerinne *n* channel, trough, water course; (der Straße) gutter, side gutter, kennel.

gerinnen *v* (Chem.) to coagulate, to congeal, to curdle; ~ lassen to coagulate, to curdle.

Geröll e, Geschiebe *n* rubble, boulder, pebble.

Gerüst *n* frame, framing, truss, trestle, rack, scaffold; ein ~ aufschlagen to erect a scaffold; fliegendes ~ flying scaffold; hängendes ~ hanging scaffold.

Gesammt ... ~anlage *f* (die fertige Anlage) the whole plant, plant; ~bestand *m* (der Kasse) balance of cash; ~ergebniß *n*, ~ertrag *m* total return; ~gebühr *f* the entire rate; ~gewicht *n* net weight in bulk; ~inhalt *m* (einer Schrift ꝛc.) total contents, summary, (von Fässern ꝛc.) contents, capacity, (Geom.) area; ~zahl *f* total number.

Geschäft *n* business, employement, function, charge, occupation; (Handel) business, transaction.

Geschäfts ... ~bereich *m* jurisdiction, sphere of business, department (of business); das liegt außer meinem ~bereich that is out of my line (sphere), that is not (in) my province; ~bericht *m* der Post-Verwaltung Report of the management of the P. O.; ~betrieb *m* management of affairs, managing of business; ~eintheilung *f* arrangement of business, distribution of the several duties; ~führung *f*, vorschriftsmäßige proper management; ~gang *m* course of business, in den ~gang geben to hand in (a letter, an application) to be dealt with in the regular way of business; ~kreis *m* dasf. wie ~bereich; ~ordnung *f*, ~plan *m* einer Verkehrsanstalt order of working; ~papiere *n/pl* (P) commercial papers.

Geschirr *n*, (Pferde~) harness.

geschmeidig *a* (Techn.) pliant, soft, flexible; (v. Metallen) ductile, malleable, soft.

Geschmeidigkeit *f* (v. Metallen) malleability, ductibility; (Biegsamkeit) flexibility.

Geschoß, Stockwerk *n* story, floor.

Geschwindigkeit *f* speed, velocity; größte (volle) ~ full speed; mit einer ~ von at a rate of ...; mittlere (Normal-)~ mean speed, proper speed, mean velocity; die ~ vermindern to slacken the movement; ~smesser *m* speed indicator, speed recorder, velocimeter.

Gesichts ... ~feld *n* (Opt.) field of view; ~linie *f* visual line; ~punkt *m* point of view; ~winkel *m* visual (optic) angle.

Gesperre *n* ratchet with catch; (Sperr-, Sperrradvorrichtung) click and ratchet wheel.

Gestänge *n* (T) telegraph structure, set of poles.

Gestell *n* support, stand; (des HA) frame; ~platte *f* (vordere, hintere) front plate, hind plate.

Gesuch *n* application, petition; ein ~ stellen to apply, to make an application, to hand in a petition.

getrennt *a* separate; ~er Strich *m* (bei der Magnetisirung) separate touch.

Getriebe *n* (Mech.) motion, gear, gearing, driving gear, pinion; (Gesammtheit des Räderwerks) machinery, machine work; (Uhrm.) work, wheel work; aus dem ~ bringen to uncouple, to ungear, to throw out of gear; konisches ~ bevelled gear, bevel gear; laufendes ~, Triebwerk *n* in Bewegung running gear.

gewachsener Boden *m* grown earth.

Gewalt *f*, im Fall höherer ~ in case of „force majeure", in a case beyond control.

Gewähr leisten *v* to warrant, to guarantee, to give bond, to become surety.

Gewährleistung *f* warranty, security, bail.

Gewahrsam *m* custody, keeping, safe keeping.

Gewährsmann *m* warranter, guarantee.

Gewicht *n* (Phys. Mech. Techn.) weight, gravity; (leichtes) troy weight; (schweres) avoir dupois weight; (spezifisches) specific gravity (weight); (todtes) dead weight.

Gewichts ... ~abgang *m* deficiency in weight; ~abweichung *f* difference in the weight; ~angabe *f* declaration of the weight; ~aufzug *m* f. Apparat mit ~aufzug; ~berechnung *f* calculation of the weight; ~bestimmung *f* (Chem.) determination of the weight, analysis of the weight; ~ermittlung *f* ascertaining the weight; ~grenze *f* limit of weight; ~satz *m* set of weights: ~stück, Gewicht *n* weight; ~tarif *m* tariff of weight; ~verlust *m* loss in weight (in the weight, of weight).

Gewinde *n* (Schrauben-~) thread, fillet of a screw; doppeltes, dreifaches, einfaches, mehrfaches, scharfes ~ double, triple, single, multiplex, angular thread.

Gewinn *m* gain, profit, proceeds; ~ abwerfen (bringen) to turn to account (advantage), to leave a profit; ~ aus dem Kurs gain of exchange, profit by exchange; reiner ~ clear gain, net proceeds; ~antheil *m* dividend, contingent, premium.

Gewinnung *f* winning, gaining; ~ der Metalle (aus den Erzen) extraction of the metals.

Gewitter *n* thunderstorm; ~luft *f* electricity of the atmosphere; ~stellung (des Telegr. Apparats) cutting the apparatus out of the circuit, putting the lightning protector in.

gewöhnlich *a* ordinary (telegram); (gebräuchlich) usual, customary.

gezahnt *a* toothed, jagged, notched, denteled, dentated.

Gicht *f* (obere Oeffnung eines Hochofens) top, mouth, throat of a furnace; (Beschickungsmaterial *n*) charge, smelting charge.

Giebel *m* gable; ~balken *m* top beam; ~dach *n* gable roof, gable end; ~fassade, ~front *f* frontispiece, face; ~mauer *f* gable wall; ~seite *f* gable side, frontispiece of a house.

gießen *v* (v. Metallen) to cast, to found (by means of a mould).

Gießen *n* casting, cast.

Gießerei *f* foundry, casting house.

Gießform *f* (der Zinkringe) mould, casting mould.

Giltigkeitsdauer *f* eines Fahrscheins (Eisenb.) the time which a railway ticket has to run.

Gips, Gyps *m* gypsum, hydrated sulphate of lime; gebrannter ~ burnt gypsum, plaster; ~bewurf *m* plaster coat; ~stuck *m* stucco.

gipsen *v* to plaster, to cover (to fasten) with plaster.

giriren *v* to put in circulation, to circulate; (einen Wechsel indossiren) to endorse a bill of exchange.

Giro *n* endorsement; ~ in Blanko blank endorsement; sein ~ geben to endorse; ~bank *f* (Depositenbank) bank of deposits, bank of circulation, transfer bank.

Gitter *n* grate, grating, trellis, lattice, railing, rails; ~brücke *f* lattice bridge, (*Am*) truss bridge; ~werk *n* lattice work, trellis work, grating.

Glanzleinwand *f* glazed linen, buckram.

Glas *n* glass; gehärtetes ~, Hartglas hardened glass; ~ballon *m* (im Meidinger Element) glass balloon; ~bohrer *m* glass drill; ~Elektrizität *f* vitreous (positive) electricity; ~glocke *f* glass bell, glass insulator; ~röhre glass tube.

gläsern *a* of glass, glassy, vitreous.

glasiren *v* to glaze.

Glasur *f* glaze, glazing, gloss.

glatt *a* (Techn.) smooth, even, sleek.

Glätteisen *n* smoothing iron, tooling iron.

Glättfeile *f* smoothing file.

glaubwürdig *a* authentic (report), worthy of belief (witness).

gleich *a* equal, even, level, flush; ~ machen to equalize, to even, to make flush, to level; ~ weit entfernt equidistant.

Gleich ... u. **gleich** ... ~artig *a* (Phys.) homogeneous (structure), similar (poles); ~artigkeit *f* homogeneousness, similarity; ~gerichtete Ströme *m/pl* currents of the same direction; ~gewicht *n* equilibrium, balance; im ~gewicht sein to be in equilibrium; stabiles, labiles ~gewicht stable, unstable equilibrium; für ~lautende Abschrift for the conformity of the copy with the original; ~mäßige Bewegung uniform motion; ~mäßiges Brennen *n* steady burning (of the electric light); ~schenkliges Dreieck isosceles triangle; ~seitig *a* equilateral; ~winklig *a* equiangular; ~werthig *a* equivalent; ~zeitig *a* (Mech.) synchronous, isochronous; ~zeitigkeit *f*, Synchronismus *m* synchronism, isochronism.

Gleit ... ~bahn f slide, slide way; ~block, ~klotz m slide block, guide block; ~kontakt m sliding (glide) contact; ~schiene f. ~bahn.

gleiten v to slide; (v. Drahte, durchgleiten) to run.

Glied n member; (Math.) term; link (of a chain).

Gliederkette f link chain, ring chain, chain with flat links; Gelenk einer ~ flat link of a chain.

Gliederung f organization.

Glimmer m mica, glimmer, glist; ~blättchen n/pl scales of mica; ~kondensator m condenser made with tin foil & sheets of thin mica coated with paraffin or shell lac; ~schiefer m mica slate.

Glocke f bell; Doppel~ double bell; recipient (of an air pump).

Glocken ... ~apparat m electric bell; ~förmig a bell shaped; ~isolator m bell insulator.

Glüh ... ~hitze f glowing heat; rothe ~hitze red heat; weiße ~hitze white heat; ~lampe glow lamp, incandescent lamp; ~licht n glow light, incandescent light; ~ofen m heating furnace, annealing furnace; ~span m iron scale, scale, hammer slag, hammer scale.

glühen v to anneal (the wire), to glow, to heat, to calcine.

Glühen n des Roheisens unter Luftzutritt glowing, heating of pig iron with access of atmospheric air, annealing (of wire).

glühend a glowing; roth ~ red hot; weiß ~ white hot, incandescent.

Gnaden ... ~gesuch n application for pardon (for crimes committed); ~monat m, Bewilligung des Gehalts eines verstorbenen Beamten für den ~monat allowance granted to the widow (orphans etc.) of a defunct officer (for the period of one month etc); the monthly (quarterly) salary is granted to the widow etc.

Gold n gold; ~blatt n leaf of gold, gold leaf; ~blatt-Elektroskop n gold-leaf electroscope; ~münze f gold coin; ~papier n gold paper, gilt paper.

Göpel m whim, whim gin, capstan; (Dampfgöpel m, Fördermaschine f) whim gin, winding engine, drawing engine; ~kreuz n, ~steg m wooden cross of a whim; ~welle f axle tree of a whim.

Goudron, Mineraltheer, Braunkohlentheer m mineral tar, brown-coal tar, mineral goudron.

graben v to dig, to ditch, to trench, to excavate.

Graben m (Erdarb.) ditch, trench; (Wasserb.) drain, gullet; (Eisenb.) trench; einen ~ ziehen to cut (to sink) a ditch; ~böschung f, äußere, innere counterscarp, escarp; ~kante f edge, crest of a ditch; ~sohle f bottom, sole of a ditch.

graduiren v to graduate.

Graduirung f graduation.

Gramm n gram, gramme (the weight of a cubic centimeter of water at a temperature of 4^0C.) = 15,43 grains.

Gran n grain (0,0648 grammes).

Granit m granite.

granulirtes Metall n metal in grains.

graphisch a graphic; ~e Darstellung f mapping-out.

Graphit graphite, plumbago, black lead; ~stift, Bleistift m black-lead pencil; ~widerstand m graphite resistance.

Gravitation, Schwerkraft f gravitation.

Gravitations ... ~-Batterie f gravitation battery; ~gesetz n law of gravitation.

Grad m degree; (Rang) rank; in ~e theilen, graduiren to graduate; ~ der Beschleunigung rate of acceleration; (Maaßeinheit) measure; ~abtheilung, ~eintheilung f division into degrees, graduation, scale; ~bogen m limb, graduated arc; ~kreis m circle divided into degrees.

greifen v to touch (said of a file), to catch; in einander ~ (Mech.) to catch, to gear together (said of wheelwork).

Grenz ... ~ausgangspostanstalt f exchange post office; ~graben m ditch marking the boundary, boundary ditch; ~kontakt m limiting stop;

~land *n* frontier land, (angrenzendes Land) adjacent country; ~pfahl *m* post marking a boundary; ~punkt *m* point in a boundary line, point on the frontier; ~scheide *f* boundary; ~stein *m* land mark, boundary stone; ~verkehr *m* (*P*) exchange of mail between the frontier districts of neighboring countries; ~zoll *m* transit duty; ~zollamt *n* custom office on the frontier.

Grenze *f* border, boundary, frontier; (Spielraum) margin.

grenzen *v* to border, to be contiguous (adjacent) on (upon).

Griff *m* (der Taste) handle; (einer Feile) tang of a file; (eines Messers) haft of a knife; s. auch Handgriff.

Gros *n* gross (twelve dozen = 144); ~gewicht *n* brutto weight.

Groß ... u. groß ... ~plattig = nebeneinander geschaltet; ~quartformat *n* large quarto.

Größe *f* largeness, magnitude, size; ~ der Abweichung amplitude of aberration; bekannte ~ (Math.) known quantity; halbe ~ half size; natürliche ~ full (real, natural) size.

Grund *m* (Techn.) ground, bottom; (Baugrund *m*) ground, soil; ~ und Boden premises; (Beweisgrund) argument, reason, cause; (Beweggrund) motive; auf ~ in (by) virtue of; ~brett *n* (*MA*) wooden base, base board; ~buch *n* register of landed property; ~einheit *f* fundamental unity; ~fläche *f* base, basis, bottom, sole; ~lage *f* foundation; ~legung *f* foundation, laying of a foundation; ~linie *f* (Techn.) basis, base line, (Standlinie) base line, datum line; ~riß *m* sketch, outline, (Plan) ground plan, horizontal section; ~stück *n* estate; ~taxe *f* (*T*) fixed tax, fundamental tax; ~wasser *n* underground water; ~wasserspiegel, ~wasserstand *m* underground water-level.

Gruppe (Techn.) set, suit, group; (*P*) section of the map showing railway & postal connections.

gruppenweise *adv* in groups.

gruppiren *v* to group.

Gruppirung *f* grouping.

gültig *a* valid, lawful binding (contract), current (postage stamps); ~e Münze *f* current and passable coin; ~ sein to be valid, to be in force; ~ machen to make valid.

Gültigkeit *f* validity, legality.

Gummi *n* u. *m* gum; (Pflanzenschleim *m*) mucilage; ~ arabicum gum-arabic; Lösung von ~ arabicum mucilage; ~ elasticum, Kautschuk *m* elastic gum, gum-elastic, caoutchouc, India rubber, (in Tafeln) sheet-India rubber, sheet-rubber, sheet-caoutchouc; vulkanisirtes, geschwefeltes ~ elasticum vulcanized India rubber; hornisirtes ~ elasticum, Hartgummi, Horngummi, Ebonit *m* hardened caoutchouc, ebonite, vulcanite.

Gummi ... ~band *n* caoutchouc band, India-rubber band; ~harz *n* gum resin; ~lack, Schellack *m* shellack; ~schlauch *m* elastic tube, India rubber hose.

gummirte Freimarke *f* (*P*) adhesive stamp.

Gürtelbahn *f* encircling railway.

Guß *m* casting, founding; ~eisen *n* pig iron, crude iron; (gegossenes Eisen) cast iron; ~form *f* mould, casting mould; ~stahl *m* cast steel; ~stahldraht cast steel wire.

Gut ... ~achten *n* report, statement; ~haben *n* credit (balance) in favor; ~ sagen *v* to answer for, to be security for.

Güter *n/pl* goods, merchandise, wares; bewegliche ~ movable goods, movables; sperrige ~ bulky goods.

Güter ... ~abfertigung *f* freight service; ~abgangsbureau *n* outwards office; ~anmeldezettel *m* notice of goods to be conveyed by railway; ~annahmebureau *n* inwards office; ~bahnhof *m* goods station, freight station; ~expedition *f* office where goods are forwarded, goods office, depot; ~post *f* goods line; ~postfahrt *f* trip of the baggage waggon; ~postwagen *m* baggage waggon, van; ~schuppen *m* goods depot, freight depot; ~transport, ~verkehr *m* transport, conveyance of goods, merchandise traffic, carrying traffic; ~versicherung f insurance of goods; ~wagen m baggage waggon, van; ~zug m goods train, baggage train, freight train.

Guttapercha f gutta percha (the inspissated juice of Isonandra gutta); ~presse f covering machine; ~überzug m gutta percha covering, gutta percha sheath (envelope).

Gyrotrop m gyrotrope.

H.

Haar ... ~besen m hair broom; ~förmig a capillary; ~nadel-Galvanometer n galvanometer with a magnet in the shape of a hair pin; ~röhrchen n capillary tube; ~röhrchen-Anziehungskraft, Kapillarität f capillary attraction, capillarity.

Hacke f (Techn.) pick, pick-axe; (Ackb.) hoe, mattock.

hacken v (Techn.) to chop, to hack; (Ackb.) to hoe; (Holz) to cleave wood.

Hafen m harbor, port; in den ~ einlaufen to put into port; ~damm m mole, pier, quay; ~dock n basin, wet dock; ~gebühren f/pl harbor dues, harborage, port dues, (Ankergeld) anchorage; ~lootse m harbor pilot; ~meister m harbor master, port warden; ~zoll m port dues, port duties.

Haft f prison, arrest, durance, imprisonment; ~befehl m warrant of arrest; ~pflicht f liability; ~pflichtgesetz n liability law.

haften v für etw. to answer, to be answerable for smt, to be held responsible.

Hahn m (Techn.) cock, stop cock.

Haken m hook, crook, clasp, catch, hasp, claw; ~bolzen m hook bolt, rag bolt; ~eisen n iron rod with a hook, winder; ~förmig a aduncous, hooked; ~förmige Schraubenstütze f curved iron bracket, swanneck bracket; ~nagel m clasp nail, hook nail, spike; ~öse f der Isolatorstütze eye, ear, hook, eyelet hole.

hakiger Bruch m (des Eisens) hackly fracture.

halb a half; eine ~e Stunde half an hour; ~ ein Uhr half past twelve, 30 minutes to one (o'clock); um ~ 5 (Uhr) at half past four (o'clock); ~durchmesser, ~messer m semi-diameter, radius; ~feile f square file; ~flach a half-flat; ~insel f peninsula; ~jährig a of six months, lasting six months; ~jährlich a half yearly, semi-annual, adv every half year, semi-annually; ~kreis m semicircle; ~kugel f hemisphere; ~kugelförmig a hemispheroidal; ~mondförmig a semilunar, crescent shaped; ~part m halves, ~part machen to go halves; ~rund a half-round, semicircular; ~runde Feile f half-round file; ~scheid f one half; ~scheidliche Theilung f des Portos equal division of postage; ~wegs adv half-way; ~zirkel m half-circle, hemicycle.

halbirbar a which can be halved.

halbiren v to halve, to divide into two equal parts; (Geom.) to bisect.

Halbirung f halving; (Geom.) bisection; ~spunkt m point of bisection.

Hälfte f the (one) half; über die ~ more than half; um die ~ größer (kleiner) more (less) by half; bis auf die ~ to the middle.

Hals m (Techn.) neck (of an insulator); collar (of a bolt); ~band, Reifen m collar, hoop; ~eisen n iron collar, hoop.

Halt m, **Anhalten** n stop, stopping, halt.

Haltbarkeit f (Mech.) strength.

Halte ... ~platz m (Eisenb.) station, station of secondary order; ~scheibe f (Eisenb.) stopping signal disk; ~stelle f (P) way station; ~zeichen n (Eisenb.) block signal.

halten v to stop; (enthalten) to hold, to contain; Buch ~ to keep accounts, to keep the books; doppelt (einfach) Buch ~ to keep books by double (single) entry; (festhalten) to hold, to stick fast (said of nails), to keep (to continue) in a good condition.

Halter *m* (am Fernsprecher) holder.

Hammer *m* hammer; (großer schwerer) sledge hammer; (hölzerner) mallet; (elektrischer, Wagner'scher) make and break; ~schlag *m* iron scale, hammer slag.

hammerbar *a* malleable; nicht ~ immalleable.

hämmern *v* to hammer; (schmieden) to hammer, to forge.

Hand ... ~arbeit *f* manual labor, handwork, handiwork; ~arbeiter *m* laborer; ~bagger *m* hand drag; ~beil *n* hatchet, hacket; ~betrieb *m* hand working; ~bohrer *m* small borer, gimlet; ~geld *n* advance money, hand money; ~gepäck *n* small luggage; ~griff *m* handle, (beim Einnadel-Telegraph) drop handle, crutch handle; ~habe *f* einer Kurbel crank handle; ~karren *m* hand cart, hand barrow, wheel barrow; ~langer *m* helper; ~säge *f* hand saw, arm saw; ~schlag *m*, I—n mittels ~schlags verpflichten to engage *smb* (to do his duty faithfully) by the joining of hands; ~wagen (*P*) hand cart; ~zeichen *n* (unter Ablieferungsscheinen [*P*]) mark; ~zeichnung *f* drawing, sketch, design; freie ~zeichnung free-hand drawing.

Handel *m* commerce, trade, traffic, business; ~ mit dem Auslande foreign trade; inländischer ~ inland (home) trade; einen ~ schließen to strike (to make) a bargain; ~ treiben to trade.

Handels ... ~amt, ~departement *n* board of trade; ~firma *f* firm; ~genossenschaft, ~gesellschaft *f* company, joint-stock company, partnership; ~gericht *n* court of commerce (trade); ~gewicht *n* avoirdupois weight; ~kammer *f* commercial board, chamber of commerce; ~minister *m* minister of commerce; ~ministerium *n* ministry of commerce; ~register *n* commercial register.

Hanf *m* hemp; ~strang *m*, ~trense *f* hempen strand; ~umwickelung *f* packing (serving) of hemp; getheerte ~fäden *m/pl* tarred hemp threads.

Hänge ... ~brücke *f* suspension bridge; ~gerüst *n* hanging scaffold; ~glocke *f* (T) suspended bell; ~lampe *f* suspended lamp; ~leiste *f* (T) bracket.

hängen *v* to hang, to suspend; (hangen) to be suspended; an einander ~ to stick together.

hantiren *v* to handle, to work.

Hantirung *f* working.

hart *a* hard.

Hart ... ~gummi *m*, Ebonit *n*, Vulkanit *m* hardened caoutchouc, ebonite, hard rubber, vulcanite; ~guß *m*, eigentl. Bronze *f* hard brass; ~loth, Schlagloth *n* hard solder; ~löthen *v* to braze, to hard-solder; ~löthung *f* hard soldering; ~werden *n* induration, hardening.

Härte ... ~grad *m* (v. Eisen 2c.) temper; ~mittel *n* temper.

härten *v* to harden, to temper; (durch Einsetzen in Schalen) to case-harden; (durch Zuführung eines kalten Luftzugs) to chill-harden.

Härtung *f* hardening, tempering.

Harz *n* resin; ~elektrizität *f* resinous (negative) electricity; ~kitt *m* resinous cement; ~öl *n* oil of rosin, rosin oil; ~tanne *f* Norway fir, common pitch-fir; ~theer *m* resinous tar.

harzig *a* resinous, resiniferous.

Haspe *f* cramp, cramp iron, clink, clamp.

Haspel *m* whim, windlass, winch, windle, crab; Kabel~ (Winde) reel; ~baum *m* beam of a windlass, cheek of a crane.

Haufen *m* heap, pile; in ~ setzen *v* to pile up.

Haupt ... ~abrechnung *f* general account; ~achse *f* principal axis; ~ansicht *f* face plan; ~ausstattungsgegenstände *m/pl* s. Ausstattungs...; ~bahnlinie *f* (Eisenb.) main line, main road, main track; ~buch über gestundetes Porto ꝛc. principal account of postage credited; ~buchhaltereirechnung *f* special annual account of the General Postal Treasury; ~geleise *n* (Eisenb.) main line; ~hahn *m* (Masch.) main cock; ~kasse *f* s. Kasse; ~ladezettel *m* bei Posten auf Landwegen (+) statement form showing the whole load (bags & loose articles) of the mail coach; ~linie *f* (Eisenb. u. *T*) main line, trunk line; ~niederlage *f* principal magazine, staple place; ~postamt *n* head post office; ~quittung *f* annual receipt; ~rad, Triebrad *n* main wheel; ~rechnung *f* annual account; ~summe *f* sum total; ~wagen *m* (*P*) principal mail coach; ~zollamt *n* principal toll office.

Haus . . . ~briefkasten *m* door letter box, private letter box; ~diener *m* domestic servant; ~suchung *f* searching of a house by the magistrates, domiciliary visit; ~suchung halten to search the house; ~suchungsbefehl *m* search warrant.

Hebe . . . ~arm *m* lever; ~baum *m*, Brechstange *f* hand spike, heaver, lever, crow bar; ~bock *m* winding engine, gin; ~kraft *f* lifting power; ~vorrichtung *f* lever, apparatus for lifting weights, purchase.

Hebel *m* lever, jack; Doppel-, zweiarmiger ~ lever of the first kind; einarmiger ~ lever of the second kind; gebrochener ~, Winkel~ bent lever, angle lever; gerader ~ straight lever; rechtwinkliger ~ rectangular lever.

Hebel . . . ~achse *f* axis of the lever; ~arm *m* lever arm; ~bewegung *f* motion of the lever; ~kraft *f* leverage, leverage power; ~stütze, ~unterlage *f* fulcrum, prop of a lever, hypomochlium; ~winde *f* lever jack.

heben *v* to lift, to raise; mit dem Hebebaum ~ to take a purchase with the lever; (beheben) to remove (a fault).

Heber *m*, **Heberzeug** *n* (Mech.) lever, raiser; (Saugheber) siphon, syphon; ~barometer *m* siphon barometer; ~röhre *f* siphon tube; ~schenkel *m* leg of a siphon; ~Schreibapparat *m* (*T*) siphon recorder.

Heft *n* writing book, copy book; (Griff *m*) hilt, handle; (Lieferung *f*) number.

Heft . . . ~ahle *f* stitching awl; ~faden *m* basting thread; ~nadel *f* stitching needle; ~stift *m* pointel; ~weise erscheinende Schriften *f*/*pl* writings published by numbers; ~zwirn *m* basting thread.

heften *v* (ein Aktenstück) to stitch, to put papers on file by stitching.

Heimfall *m* von Ruhegehalt discontinuance of the payment of pensions.

Heiß . . . ~brüchiges Eisen s. Eisen; ~wasserheizung *f* hot-water heating, hot-water stove.

Heiz . . . ~apparat *m* heating apparatus, heater; ~kraft *f* calorific power, heating power; ~material *n* fuel, combustibles; ~rohr *n* fire tube, flue, tube flue.

heizbares Zimmer *n* room containing a fire place (stove).

heizen *v* to heat, to fire; mit Holz, Kohlen ~ to burn wood, coal.

Heizer *m* fireman, furnace man.

Heizung *f* heating, firing; ~smaterial *n* fuel; ~vorrichtung *f* heating apparatus, (in einem Eisenb.-Wagen) car heater, (unter dem Fußboden) hypocaust.

Hemm . . . ~feder *f* stopper; ~rad *n* escape wheel, scapement wheel; ~schraube *f* stop screw; ~schuh *m* skid, wheel drag; ~vorrichtung *f* stop.

hemmen *v* to stop; (Fuhrw.) to skid, to lock.

Hemmung *f* stopping, (im Uhrwerk) scapement, escapement; freie ~ free escapement; ruhende ~ repose escapement; schleifende ~ dead-beat escapement; selbstthätige ~ an Aufzügen ꝛc. (Mech.) safety stop.

Hemmungs . . . ~lappen *m* pallet of the escapement; ~rad, Steigrad *n* scapement wheel, escape wheel, swing wheel, balance wheel.

Henkel, Griff *m*, **Oehr** *n* handle, ring, ear, lug; ~ einer Glocke cannon of a bell.

herab ... ~gehen *v* (Preise) to go down; ~setzen to reduce, to lower the tariff, the price etc.; ~setzung *f* der Gebühren reduction of the rates.

Hergang *m* event, circumstance; ~ einer Sache way in which a thing has come to pass.

herrenlos *a* unclaimed (article); ~es Gut *n* waif.

herrühren *v* von etw. to come, to flow, to proceed from.

herstellbar *a* reparable, that may be repaired.

herstellen *v* to produce, to raise, to establish; (eine T. Linie) to construct, to build; wieder ~ to repair, to restore, to re-establish.

Herstellung *f* restoration, reparation, re-establishment; ~skosten *pl* cost of reparation.

herumdrehen *v* to turn round (about).

hervorragen *v* to project, to jut (out), to stick out, to stand out.

hervorspringen *v* to spring forth, to leap forth (out).

hervorstehen *v* to stick out, to stand out.

hervortreten *v* to come out.

hier *adv*, hierorts (auf Briefen) „city"; ~bei herewith, with this.

Hilfs ... f. Hülfs ...

Hin ... ~ u. her *adv* to & fro; Fahrkarte *f* für ~- u. zurück (Retourbillet *n*) ticket to go & return, return ticket; ~fahrt *f*, beladene (unbeladene) the trip out with (without) passengers or mails; ~kunft *f* arrival at the place; ~leitung *f* (T) the line going; ~ u. Rückfahrt *f* going & returning.

Hinterbliebene *pl* family (left behind).

hinter einander schalten *v* to connect (cells) one after the other (in single series).

hinterlassen *v* (TP) to leave (a notice at the dwelling place of the addressee).

hinterlegen *v* to deposit (a sufficient amount), to make a deposit (for porterage).

hinterziehen *v* to embezzle (postage).

Hinweis *m* pointing at, hint, direction, reference to.

hinweisen *v* auf to point towards (to), to point at, to refer to, to allude to.

Hinweisungsvermerk *m* (in Kassenbüchern ꝛc.) reference to ..

Hirn ... ~holz *n* wood cut across the grain, cross grain, end grain; ~fläche *f* cross way, endway of the grain.

Hitze ... ~messer *m* pyrometer; ~zeiger *m* pyroscope.

Hobel *m* plane; (mit doppeltem Eisen, Doppel~) double plane, double-ironed plane; (mit einfachem Eisen) single-ironed plane.

hobeln *v* to plane.

Hoch ... ~bahn *f* elevated railway; ~bau *m* building above ground; ~druck *m* (Dampfmasch.) high pressure; ~ofen *m* high furnace, blast furnace.

Hochachtung *f* esteem, respect; Genehmigen Sie u. f. w. (am Schluß von Briefen) I am, Sir, Very (Most) Respectfully, Your Obedient Servant.

Höhe *f* hight, elevation; lichte ~ headway; ~ eines Ortes latitude.

Hoheitsrecht *n* government monopoly, royalty.

hohl *a* hollow, concave; (ausgehöhlt) cored; ~feile *f* round hollow file; ~kehle *f* (Techn.) hollow, furrow, groove, channel; ~leiste f. Hohlkehle; ~trieb *m* hollow pinion.

höhlen, aushöhlen *v* to hollow, to excavate.

Hollundermarkkügelchen *n* ball made from the pith of elder.

Holz *n* wood, timber (f. auch a. T. timber); lufttrockenes ~ seasoned timber; ~ fällen to cut wood, to fell timber (trees); frisches ~ green wood; geradstämmiges ~ straight timber; hochstämmiges ~ tall (lofty) timber; knorriges ~ knotty (knaggy) wood; ~ auf dem Stamme standing tree; unbehauenes ~, Stammholz uncleft timber, round (unhewn) timber; verfaultes ~ decayed (rotten) wood; vertrocknetes ~ dead wood; windbrüchiges ~ rolled timber, wind fall;

Holz — Hypothek

windschiefes ~ back-sided timber; wurmstichiges ~ worm-eaten wood; wurzelfaules ~ root-rot wood.

Holz ... ~anstrich *m* paint for wood; ~bekleidung *f* timber lining; ~bestand *m* amount (stock) of wood; ~block *m* log of wood; ~bohrer *m* (der in die Brustleier gesteckt wird) duck's-bill bit; ~bohrer (Zimmermannsbohrer) auger, gimlet, wimble; ~brücke *f* timber bridge, wooden bridge; ~dübel *m* wooden pin, treenail; ~faser, ~fiber *f* fibre of wood, woody (ligneous) fibre, (Chem.) cellulose; ~fäule *f* dry rot, druxey; ~floß *n* raft, float of wood; ~flöße *f* wood-floating place; ~gerüst, ~gestell *n* wooden frame; ~hacke *f* wood axe; ~handel *m* trade in wood, timber trade; ~händler *m* dealer in wood; ~hauer *m* wood cutter; ~hof, Zimmerplatz *m* timber yard, wood yard; ~imprägnir=Vorrichtung *f* wood-preserving apparatus; ~kern *m* (der Morserolle) roller; ~kohle *f* vegetable coal, wood charcoal, charcoal; ~leiste *f* batten; ~nagel *m* wooden peg, treenail; ~pflock *m* wooden pin, peg; ~ring, Jahresring *m* annual ring; ~scheibe *f* wooden disc; ~schraube *f* (Eisenschraube zum Schrauben in Holz) wood screw, screw nail; ~schraubengewinde *f* thread of a wood screw; ~schwamm, Hausschwamm *m* dry rot, wood fungus; ~schwelle *f* (Eisenb.) timber sleeper; ~theer *m* wood tar, vegetable tar; ~theeröl *n* oil of wood tar; ~werk *n* timbering, timber work, framework; ~zapfen *m* wooden plug, peg, treenail.

hölzern *a* wooden, wood, of wood; ~er Hammer *m* mallet; ~er Nagel *m* treenail; ~e Schraube *f* wooden screw.

homogen, gleichartig *a* homogeneous.

Hör=Instrument *n* (*F*) ear tube.

Horizont, Gesichtskreis *m* horizon; wahrer ~ (Feldm.) true level; scheinbarer ~ visual (visible, apparent) horizon.

horizontal, wagerecht *a* horizontal, level; ~e Fläche *f* dead level; ~e Linie *f* horizontal line; ~e Strecke *f* (Eisenb.) level length.

Horizontal ... ~ebene *f* level; ~waage, Wasserwaage *f* water level.

Horngummi s. Gummi.

Hub *m* lift, lifting, raising; ~höhe *f* length of the stroke, stroke.

Hufeisen *n* horse shoe; ~magnet *m* horse-shoe magnet.

Hülfs ... ~anspänner *m* hand (helper) of the mail contractor; ~arbeiter *m* helper, assistant; ~batterie s. Batterie; ~bote *m* auxiliary messenger; ~elektromagnet *m* relay magnet; ~schreiber *m* aid (for copying), temporary writer.

Hülle *f* cover, envelope; Schutz~ protecting cover (of the cable).

Hülse *f* shell; (Buchse) socket, shell, collar, box.

Humus *m* vegetable soil, humus.

hundert ... ~gradig *a* centigrade (thermometer); ~theilig *a* centesimal (scale).

Hydrant *m* hydrant, stand pipe; street washer.

Hydraulik, Mechanik der flüssigen Körper *f* mechanics of fluids; (Lehre von der Bewegung der flüssigen Körper) hydraulics.

Hydrodynamik *f* hydrodynamics.

hydroelektrisch *a* hydro-electric.

Hydrogen *n* hydrogen.

Hydromechanik s. Hydraulik.

Hydrometer *n* hydrometer; (Senkwaage) areometer, gravimeter (for ascertaining the specific gravity of liquids & solids).

Hydrostatik *f* hydrostatics.

hydrostatisch *a* hydrostatic; ~e Waage *f* hydrostatic balance.

Hydroxygengas, Knallgas *n* oxyhydrogen gas.

Hydroxygen=Licht *n* oxyhydrogen light, calcium light.

Hypothek *f* mortgage, security; Geld auf ~ nehmen to raise money on mortgage.

J.

identisch *a* identic, identical, the same; *adv* identically.

Identität *f* identity; die ~ erweisen to identify.

immerwährend *a* permanent (magnet).

Immobilien *s. pl* immovables; (Immobiliar-Vermögen) immovable (real) estate, dead stock.

imprägniren s. zubereiten.

Imprägnirung s. Zubereitung.

Inangriffnahme *f* beginning of the work.

inbegriffen *a* included, inclusive; *adv* inclusively.

Inbetriebnahme *f* putting in operation.

Index, Zeiger *m* index, hand.

indiziren *v* to indicate; indizirte Leistung *f* (Masch.) indicated effect (power); indizirte Pferdekraft *f* indicated (nominal) horse power.

Indifferenz ... ~linie, neutrale Mittellinie *f* eines Magneten neutral line; ~punkt, magnetischer *m* point of indifference.

Indossament, Indossement *n* endorsement, indorsement.

Indossant *m* endorser.

Indossat *m* endorsee.

indossiren, giriren *v* to endorse; wieder ~ to reindorse.

Induktion *f* induction; elektrische ~ electrical induction; elektromagnetische ~, Strominduktion electro-magnetic, electro-dynamic induction, induction of currents; elektrostatische ~, Influenz *f* electro-statical induction, influence.

Induktions ... ~apparat s. ~maschine; ~elektrizität *f* electricity by induction; ~kapazität *f* inductive capacity; ~rolle *f* induction coil, bobbin; ~maschine *f* inductive machine; ~spule *f* directing coil; ~störung *f* inductive disturbance; ~strom *m* inducing current, induction current, inductive circuit; ~taster *m* induction key.

induziren *v* to induce.

Ineinander ... greifen *v* (von Rädern) to gear; ~greifen *n* gear, gearing.

Influenz, elektrostatische Induktion *f* influence, electrostatical induction.

Ingangsetzung *f* starting (of an apparatus, of an engine, of the wheelwork [*HA*]), release.

Ingenieur *m* engineer; ~wesen *n* engineering.

Inhaber *m* holder (of an office, of a bill); auf den ~ lautende Werthpapiere bills payable to bearer.

Inhalt *m* (Techn. Geom.) contents; (einer Fläche) area of a surface; (körperlicher) solid content, cubical content, volume; (einer Schrift) contents, tenor, purport, substance.

Inhaltsangabe *f*, **Inhaltsverzeichniß** *n* index, table of contents; (einer Postsendung) declaration of contents.

inhibiren *v* to stop (the sending of a telegram).

Inkandeszenz s. Glühlicht.

Inkasso *n* encashment.

Inklination *f* inclination; ~s-bussole *f* inclination compass; ~s-nadel *f*, **Inklinatorium** *n* inclinatory needle, dipping needle; ~swinkel *m* angle of inclination.

inklusive *adv* inclusively.

Inkraftsetzung *f* putting in vigor.

inkrustiren *v* to incrust, to incrustate.

innere *a* interior; ~ Arbeit, (Wirkung) *f* internal work (action); ~r Dienst (*PT*) indoor duty; ~r Tarif *m* inland rate; ~ Korrespondenz inland (domestic) correspondence; ~r Verkehr *m* domestic service; ~r Widerstand *m* (einer Batterie) internal resistance; Minister des ~n minister (secretary) of the interior.

innerhalb *adv* (2er Monate von heut) within (two months from this date).

inquiriren *v* to examine, to question, to interrogate.

Inserat *n* insertion, advertisement.

inseriren *v* to insert, to advertize (in a paper).
Inserirung *f* insertion.
Insertionskosten *pl* costs of insertion.
insgemein *adv* (beim Kostenanschlag) miscellaneous.
Inspektion *f* survey.
Inspektor *m* inspector, superintendent, surveyor; Post-~, Telegraphen-~ s. Post u. Telegraphie.
instandhalten *v* to maintain.
Instandhaltung *f*, **Instandhaltungsarbeiten** *f/pl* maintenance; ~skosten *pl* costs of maintenance.
instandhalten *v* to repair.
Instandsetzung *f* repair; ~sarbeiten *f/pl* repairs; ~skosten *pl* costs of repairs.
Instanz *f* resort, instance; letzte ~ last resort; an die letzte ~ gehen to join issue.
Instanzenweg *m*, den ~ einhalten to go through the proper instances; ein Gesuch auf dem ~e einreichen to make an application through the immediate superior officer.
Instradeur *m* clerk charged with noting down the viâ (route) on telegrams to be forwarded.
instradiren *v* (*TP*) to direct, to despatch telegrams (mail articles) on their proper routes.
instruiren *v* to instruct.
Instruktion s. Dienst-Instruktion.
Intensität *f* intensity.
intensiv *a* intense, intensive.
interimistisch *a* ad interim, provisional.
intermittirend *a* intermittent.
Inkurssetzung *f* von Werthpapieren putting stocks in circulation.
intern *a* interior; ~es Telegramm *n* inland telegram; ~er Verkehr *m* inland traffic.

international *a* international; ~es Telegramm *n* foreign telegram; ~er Telegraphen-Vertrag *m* International Telegraph Convention.
Interpunktion *f* punctuation; ~szeichen *n* mark of punctuation.
interpunktiren *v* to interpoint.
Intervall *n*, zwischen zwei Stangen (*T*) distance between two poles.
Inundationsgebiet *n* land liable to inundation (to be flooded).
irrthümlich *a* erroneous; *adv* erroneously.
Irrungszeichen *n* (*T*) „erase" signal.
isochron, isochronisch *a* isochronous, isochronal (uniform in time); ~e Linie *f* isochronal line, isochronous curve.
isoklinische Linien *f/pl* isoclinical lines.
Isolation *f* insulation; ~s-Vermögen *n* insulating capacity; ~s-Vorrichtung *f* insulator; ~s-Widerstand *m* insulating resistance.
Isolator *m* insulator; ~kopf *m* top; ~mantel *m* cap, cup; ~stütze *f* bracket, bolt, arm; ~ von Porzellan porcelain insulator; ~ Abspann-~ shackle; gußeiserner Spann-~ *m* mit Porzellantülle tightening apparatus of cast iron with porcelain socket; Doppelglocken-~ *m* double-bell insulator; Pendel-~, Baum-~ swinging insulator suspended by an iron rod from the eye of an iron carrier; Untersuchungs-~ insulator for testing purposes; ~ kleiner Form small insulator.
Isolir... ~anschlag, ~kontakt *m* insulating stud; ~feder *f* insulating spring; ~glocke *f* insulating bell; ~schicht *f* insulating layer.
isoliren *v* to insulate.
Isolirung *f* insulation; ~swiderstand *m* insulating resistance.

J.

Jahres... ~bericht *m* annual report; ~rechnung, Hauptrechnung *f* annual account; ~quittung, Haupt- quittung *f* annual receipt; ~schlußbilanz *f* annual balance; ~ring *m* (im Holze) annual ring.

jedesmalig a then being, for a certain case, at the time; die ~en Umstände the circumstances in such a case.

Joch n (Brückenjoch) bay, arch of a bridge; ~brücke f pile bridge.

Journal, Tagebuch n day book, journal, waste book; in das ~ eintragen, journalisiren v to journalize.

Judenpech n, **Asphalt** m jew's pitch, mineral pitch, asphaltum.

Jute, Jutehanf m jute, yute, hemp, chinese reed; Seil aus ~ jute rope; getheerte ~umhüllung f tarred jute covering.

K.

Kabel n cable; Erd-~ n (unterirdisches) underground cable; Fluß-~ river cable; See-~ (unterseeisches) submarine (ocean) cable; ~ader f the insulated conductor; ~gesellschaft f cable company; ~graben m trench; ~halter m cramp, clincher (contrivance to hold the shore end of a cable), (beim Löthen) wooden compressor; ~haspel m reel; ~kasten m box; ~lageplan m plan of the cable (containing the route of the wires carefully surveyed & drawn, the positions of the joints being marked); ~lager n path of the cable; ~legung f laying of a cable; ~linie f underground line, cable line; ~litze f wire in a strand; ~löthstelle (spleißung) f splice, joint; ~muffe f pipe, covering tube; ~rinne f (im Tunnel) wooden trough; ~schiff n cable ship; ~schrank m cable box; ~schutzdrähte m/pl guards, sheath, sheathing armor, protecting wires; ~schutzkonferenz f cable protection conference; ~seele f core of a cable; ~sonde f grapnel; ~untersuchungsbrunnen m test box, testing box; ein ~ abrollen to pay out a cable; ein Land-~ legen to lay an underground cable; ein See-~ legen to submerge a cable.

Kaiser-Wilhelm-Stiftung f für Angehörige der Reichs-P. u. T. Verwaltung Emperor William fund for the improvement of the position of Postal & Telegraph Officers.

Kalender m calendar, almanac; ~jahr n calendar year.

kalfatern v to calk, to caulk.

Kali n protoxide of potassium; chlorsaures ~ chlorate of potassium; doppelt chromsaures ~ red chromate (bichromate) of potash; einfach chromsaures ~ yellow chromate of potassium; kohlensaures ~ carbonate of potassium; doppelt kohlensaures ~ bicarbonate of potash; salpetersaures ~ nitrate of potassium, saltpetre; schwefelsaures ~ (bi)sulphate of potash.

Kaliber n caliber; ~maaßstab m (sliding) calliper scale, vernier calliper.

kalibriren v to size; eine Walze ~ to cut the grooves of a roller, to groove a roller.

Kalibrirung f taking the size; ~ der Walzen grooving.

Kalium n potassium, s. auch unter Kali.

Kalium ... ~eisencyanid, rothes Blutlaugensalz n ferricyanide of potassium; ~eisencyanür, gelbes Blutlaugensalz n ferrocyanide of potassium.

Kalk m (Chem. Miner.) lime; ~, Kalziumoxyd n, Kalkerde f lime, oxide of calcium; ungelöschter ~ quick lime; abgelöschter ~ wet lime; ~licht n (Drummond'sches) calcium light, lime light, Drummond's light; ~wasser n lime water.

Kalkulator m verifier of accounts, examiner.

kaltbrüchig s. Eisen.

Kamm ... ~masse f ebonite; ~rad, Zahnrad n (von Holz od. Eisen) cog wheel, toothed wheel, cogged wheel; ~rad, Kronenrad n crown wheel, face wheel; eisernes ~rad iron jack.

kanneliren, auskehlen v to channel, to chamfer, to flute, to groove.

Kannelirung f flute, channeling.
Kant ... ~**feile** f three square file; ~**haken** m cant hook; ~**holz** n squared (square) timber.
Kante f edge, cant; hohe ~ edge, narrow side (of a board); auf der hohen ~ edgeways, on edge; auf die hohe ~ verlegt laid edgeways.
kanten v to cant; (viereckig behauen) to square (timber).
kantig a angular, edged, canted.
Kanzlei f copying office.
Kanzlei ... ~**arbeit** f copying work; ~**papier** n short demy paper, medium paper; ~**tinte** f record ink.
Kanzlist m copying clerk, clerk of the ruler.
Kaolin m, **Porzellanerde** f kaolin, porcelain earth, porcelain clay, china clay.
Kapazität f (Techn.) capacity; Messung der ~ capacity test.
kapillarförmig, haarförmig a capillary.
Kapillarität, Haarröhrchenkraft f capillarity, capillary attraction.
Kapital n (einer dekorativ behandelten Stange) capital, chapiter.
Kappe f (Techn.) cap, hood; (des Isolators) rain cap; gußeiserne ~ hood of cast iron (to protect the insulator).
kappen v to head, to lop off (trees).
Kapsel f (HA) case, box.
Kariol n, **Kariole** f gig, a light two-wheeled carriage, cariole, mail cart; ~**fahrt** f cariole trip; ~**post** f cariole line.
Karre f, **Karren** m cart, barrow, wheel barrow.
karren v to cart, to transport by means of a cart or barrow.
Karren m cart, dray cart; ~**transport** m carting, cartage.
Karte f (Landkarte) map; (Seekarte) chart, sea chart.
Karten ... ~**brief** m card letter; ~**postanweisung** f. Postanweisung; ~**schluß** m mail; ~**schluß-Postanstalt** f post office of exchange; ~**wechsel** m exchange of mails; ~**zeichner** m draftsman: Brief-~schluß f. Brief;

Fracht-~schluß f. Fracht; Geld-~schluß f. Geld.
kartiren v (P) to make up the mail, to enter the mail articles in the bill.
Karton m cartoon, fine pasteboard, pasteboard box.
Kasse f (PT) cash office, treasury, fund; Haupt-~ eines P.- od. T.-Amts (+) cash office; Post-~ f. Post; General-Post-~ f. Post; Ober-Post-~ f. Post.
Kassen ... ~**abschluß** m, täglicher ~abschluß eines P.- od. T.-Amts daily balance of a post (telegraph) office; ~**anweisung** f (Schein) treasury bill, (Zahlungsanweisung) order of payment; ~**ausfall** m deficiency, deficit; einen ~ausfall haben to be short of cash; ~**ausweis** m balance, cash account; ~**bestand** m balance of cash (in cash); ~**bilanz** f balance; die ~bilanz ziehen to strike the balance; ~**buch** n cash account; ~**einnehmer** m receiver, collector; ~**führer** m cash-keeper, cashier; ~: u. Rechnungswesen n bookkeeping; ~**revision** f inspection of the cash; ~**revisionsverhandlung** f protocol of an inspection of the cash; ~**revisor** m examining clerk (officer); ~**tagebuch** n daily cash account; ~**vorschuß** m advance of cash.
kassiren v to get in (to collect) money; (für ungiltig erklären) to annul (a telegram), to cancel; einen Beamten ~ to dismiss an officer.
Kassirer m cashier, cash-keeper, treasurer; Post-~ f. Post ...
Kastellan m (eines Posthauses) housekeeper.
Kathode f, **negativer Pol** m cathode, kathode, negative pole.
Kaufwerth m, **Waarenproben** 2c. mit ~ samples (patterns) containing merchandise of intrinsic value.
Kaution f (P) bond, caution; eine ~ bestellen to give bond, (seitens der Briefabholer) to deposit an amount; ~**sbestellung** f giving bond; ~**sempfangsschein** m receipt for bond given; ~**snachschuß** m supplementary bond; ~**ssumme, ~shöhe** f amount of bond; ~**sverschreibung** f

declaration of a third party to give bond for an officer.

Kautschuk m, **Federharz, Gummi elasticum** n caoutchouc, gum elastic, India rubber; (vulkanisirter, geschwefelter) vulcanized caoutchouc, vulcanized rubber; ~band n caoutchouc band; ~packung, ~umhüllung f India rubber packing.

Kegel m (Techn.) cone; (Sprungkegel) detent; abgestumpfter ~ truncated cone; gerader, senkrechter ~, Umdrehungskegel right (upright) cone; schiefer, ungleichseitiger ~ oblique (scalene) cone.

Kegel ... ~achse f axis of a cone; ~förmig a conical, coniform, tapering; ~rad n bevelled wheel, bevel wheel, conical wheel; ~radgetriebe n bevel pinion.

kehlen v to chamfer, to channel, to flute, to groove.

Kehlung f channel, chamfer, chamfering.

Keil m wedge; (Splint) peg, splint; (Dübel) key; (Vorstecker eines Bolzens) forelock, forelock key; ~bolzen m (Bolzen mit Vorstecker) eye bolt with key, forelock bolt; ~förmig a cuneate, cuneiform, wedge shaped; ~hacke f pick, pick axe; ~haue f mattock; ~verbindung f keying.

Kenntniß f information, knowledge, cognizance, notice; ~ haben to be informed; ~nahme (the act of) taking cognizance of smt; zur ~nahme u. Beachtung for guidance in similar cases, to be guided by the foregoing instructions; ~ nehmen to take cognizance of smt; J–n in ~ setzen to inform smb of smt.

Kerbe f indent, notch, scarf, nick.

kerben v to notch, to indent.

Kern m core, heart; (Elektromagnet) ~e cores of an electro magnet; ~ von weichem Eisen core of soft iron; ~ des Holzes pith of wood; ~holz n heart wood; ~rissiges Holz shaken wood.

Kerze f taper, candle; ~nstärke f candle power.

Kessel m kettle, caldron; (Dampfkessel) boiler; ~haus n boiler house; ~zubereitung f impregnation of timber effected in closed boilers (from which the air is exhausted, after which creosote oil is forced into the timber by pressure).

Kette f chain; (Meßkette) measuring chain; ~ ohne Ende, geschlossene ~ endless chain, chain without an end; galvanische ~ galvanic circuit.

Ketten ... u. **ketten** ... ~förmig a catenarian; ~gelenk, ~glied n link of a chain, chain link; ~getriebe n chain gear; ~linie f chain curve, catenary (f. a. T.); ~rad n chain wheel; ~schriftgeber m (T) automatic chain-transmitter.

Kiefer f fir tree, pine; (rothe oder schottische) red pine, Scotch fir.

kieferne Stange f pine pole.

Kilogramm n kilogram, kilogramme (a thousand grams = 15 432 grains).

Kilogrammmeter, Kilogrammometer, Meterkilogramm n kilogrammeter (the power necessary for lifting in one second a kilogram to a height of one meter).

Kilometer n kilometer (a thousand meters, about $5/8$ of an English statute mile); ~stein m kilometer stone.

Kink m (Schlinge in einem Tau) kink, nip; ~e bekommen to kink.

Kissen n (Stempelkissen) pad; (der Elektrisirmaschine) rubber; (elastisches ~, Luftkissen) elastic (air) cushion.

Kiste f chest, trunk, box, case.

Kitt m cement, mastic.

kitten v to cement, to glue, to join together by means of putty.

Klammer f cramp, cramp iron, clamp, clinching iron; (Einschlußzeichen) parenthesis; eckige ~n pl (Buchdr.) brackets; ~ des Löthkolbens cramp handle for soldering; ~band n tie piece, brace, strut.

Klang m sound, ring; ~farbe f stamp; ~lehre f acoustics; ~welle f wave of sound.

Klappe f clack, flap; (eines Briefes) cover; (eines Fensterladens) back flap of a window shutter;

Klappen — Kohlen 79

(am Klappenschrank [*F*]) annunciator, indicator.
Klappen ... ~schrank *m* (*F*) switch board, indicator board; ~system *n* switch system.
Klappleiter *f* folding ladder.
Klasse, Wagenklasse *f* (Eisenb.) class; Wagen erster ~ first-class carriage.
Klassifikation *f* classification.
klassifizieren *v* to classify, to rate.
Klauenfett *n* neat's foot-oil, grease boiled out of the feet of oxen and sheep.
Klaviatur, Tastatur *f* (*HA*) key board.
kleben *v* to glue (a telegram to the form); to stick, to cling to (the armature sticks to the cores of the electro magnet).
Klebpfosten *m* brace (to prop a pole).
Kleiderkasse *f* clothing fund.
Klemm ... ~backen *f/pl* holdfast; (der Froschklemme) cheeks, jaws, claws, s. Froschklemme; ~leiste *f* clamp board; ~schraube *f* binding screw, attachment screw, clamp, connector, terminal; ~vorrichtung *f* zum Festhalten des Drahtes draw tongs.
Klemme *f* holdfast jam; Batterie-~ *f* clamp, binding screw, terminal; Draht-~ *f* clamp; Pol-~ *f* s. Batterie ...; Tisch-~ *f* clamp, terminal; Verbindungs-~ *f* connector.
Kletter ... ~eisen *n/pl* climbing irons; ~stange *f* climbing pole; ~vorrichtung *f* climbing contrivance, climbing appliance.
Klingelwecker *m* bell, call bell, electric bell.
klingende Münze *f* ready money, cash, metallic currency.
Klink ... ~bolzen *m* (Techn.) clinch-bolt, clinched bolt, rivetted bolt; ~haken *m* (Techn.) hook (staple) which receives the latch.
Klinke *f* (der Thür) latch, clink.
Kloben *m* a log of wood; (Feilkloben *m*) hand vice; (Rolle *f*) pulley, block; (Rollkloben, einfacher Rollkloben) single block; (Rollengehäuse, Flasche eines Flaschenzugs)

block, block pulley, shell, tackle; ~ des Hebebocks gin block; ~ einer Waage cheeks of a balance; ~gehäuse *n* pulley frame, case, shell; ~schraube *f* shackle jack; ~seil *n* pulley rope; ~zug, Flaschenzug *m* tackle, set of pulleys, pulley block, block and tackle, tackle of pulleys.
Klopf ... ~apparat *m* (*T*) sounder; ~signalapparat von Morse Morse sounder; Magnet *m* eines ~signalapparats sounder magnet.
Klöppel *f* einer Glocke clapper of a bell, bell clapper, hammer.
Kluppe *f* (Zange) pincers, clamp; (Schneidkluppe, Schraubenkluppe) screw stock, die stock, stocks.
Knagge *f* bracket, peg.
Knallgas *n* oxyhydrogen gas, inflammable gas.
Knebel *m* stick, iron stick, picket (to strain the wire stays).
Kneifzange s. Beißzange.
Knick *m* break, brisure; (im Draht) nip; ~hebel s. Kniehebel.
Knie ... ~hebel, Winkelhebel *m* knee lever, joint lever, bent lever; ~rohr *n*, ~röhre *f* knee pipe, elbow pipe, bent pipe; ~verbindung *f* elbow joint.
knistern *v* (Chem.) to crepitate, to crack.
Knopf *m* (an der Taste [*MA*]) button, knob; (*F*) electric bell push; ~leiste *f* board with terminals.
Knorren *m* (im Holz) knot, knag, snag in wood.
knorriges Holz *n* knotty (knobby, snaggy) wood.
Knoten *m* knot; (im Drahte) hitch, bend; (Seemeile) knot (the sixtieth part of a degree of longitude); ~punkt *m* (Eisenb.) railway center.
Kohle *f* coal; (Braunkohle) brown coal; (Holzkohle) charcoal; (Steinkohle) mineral coal.
Kohlen ... ~asche *f* ashes of coal; ~element *n* carbon element; ~oxyd *n* carbon monoxide; ~säure *f* carbon dioxide; ~stab *m* rod of carbon; ~stift *m* (am Mikrophon) pencil of carbon; ~stoff *m* carbon; ~zylinder *m* carbon cylinder.

Kokon m, **Gallette** f (Seidenzeug) cocoon; ~faden m ohne Torsion cocoon fibre devoid of torsion.
Kolben m (Mechanik) piston, sucker; (Löthkolben) soldering iron.
kollationiren f. vergleichen.
Kollo n (pl Kolli) package, bale of goods.
Kolophonium n colophony, rosin, common rosin, colophonium.
kommissarisch a temporary, for the time being; adv temporarily; einem Beamten wird die kommissarische Verwaltung einer höheren Stelle übertragen an officer, before being definitely promoted to a higher rank, has to serve a probational term.
Kommissions-Sitzung f meeting of the Committee.
Kommissorium n temporary (preliminary) employment.
Kommunalstation f (T) municipal telegraph office.
Kommutator m commutator, switch.
Kompaß m, **Bussole** f compass, ~nadel f compass needle, magnetic needle.
Kompensation f compensation; ~smethode compensation method; ~sstrom compensation current.
Kompetenz f competence, competency; (Einkünfte) allowance.
Komponente f. Seitenkraft.
komprimirte Luft f compressed air; Maschine durch ~ ~ getrieben compressed air engine.
Kondensator m condenser.
kondensiren v to condense.
Konduktéur m (P, Eisenb.) guard; Eisenbahn-~, Schaffner m railway guard; Post-~ m mail guard; f. auch Post ... u. Bahnpost ...
Konduktor m conductor.
konisch, **kegelförmig** a conical, coniform, tapering; ~es Getriebe n (Masch.) bevelled gear; ~es Rad n bevelled (conical) wheel; ~e Verzahnung, Winkelverzahnung f cone gear, conical gearing.
Konkurs m bankruptcy; der an eine in ~ befindliche Person gerichtete Brief (P) bankrupt's letter.
Konsole f console, bracket; gußeiserne ~ bracket of cast iron (let into the masonry to form a point of attachment).
konstant a constant (battery, current); ~e Zahl f constant.
konstruiren v to construct.
Konstruktion f construction, structure.
Konsumverein m supply association.
Kontakt m contact; ~feder f contact spring; ~feile f contact file; ~hebel m contact lever; ~schraube f contact screw, set screw; ~stift m stud, platinum stud, steel stud, contact pin; Arbeits-~ front (working, closing) contact; fester ~ fixed contact; gegenüberstehende ~e contacts facing each other; oberer, unterer ~ upper, lower stop; Schrauben-~ screw stop; Ruhe-~ back (rear, resting) contact.
kontinuirlich a continuous (line, current).
Konto n account; ~ über gestundetes Porto ꝛc. (+) daily account (day book, journal) of postage credited; ein ~ eröffnen to open an account; ~ geben to give credit; ein ~ haben to keep a deposit account; ~ nehmen to take credit; ~buch n account book.
Kontor, **Comptoir** n counting house, office.
Kontrahent m contracting party.
kontrahiren v to contract, to stipulate; Schulden ~ to incur debts.
Kontrakt m contract, agreement; ~mäßig a according to contract; ~widrig a contrary to contract.
kontraktlich a und adv as contracted, per contract; ~ verpflichtet bound by contract.
Kontroll ... ~apparat m controlling apparatus, signal repeater; ~beamter m controlling officer; ~büreau n für Postanweisungen money order department; ~rad, Stellrad n, Aufzugkontrolle f spring stop, stop wheel; ~zahn m cam of the spring stop, stop finger.
kontroliren v to control, to tally.
Kontrolle f control, check.

Konventionalstrafe f. s. Verzugs=
strafe.
konventionell a conventional; ~e
Zeichen n/pl (T) conventional signs.
Konvolut n bundle (of papers).
Konzept n minute, sketch, draft
(draught); ~papier n ordinary paper,
common foolscap.
Konzession f patent, license,
grant, concession; Ertheiler einer
~ granter; ~sbedingungen f/pl conditions on which a concession is
granted.
Konzessionär m concessionary,
grantee.
Kopf m head, top; (eines Bolzens) head of a bolt; (des Isolators)
top; (eines Nagels) nail head; (des
Telegramms) preamble; ~bahnsta-
tion f terminal station, reversing
station, cusp station; ~bohrer m
trepan; ~bolzen, Schraubenbolzen m
set bolt; ~schraube f round-headed
screw.
Kopialien pl copy money.
Kopie f copy, duplicate.
Kopir... ~maschine f copying
machine, copying press; ~papier n
copying paper; ~telegraph m copying
telegraph, autographic telegraph;
~tinte f copying ink.
kopiren v to copy, to transcribe,
to take a copy of.
Kordenschraube f screw with a
milled head.
körnig a granular; ~, gekörnt
(v. Metallen) granulated, shotted;
~es Eisen n crystalline iron.
Körper m (Phys., Chem., Mech.,
Techn.) body, substance; (Geom.)
solid, body; (des Schlittens [HA])
contact maker; ~schiene f (der Taste
[MA]) axis of the key.
Korrektion f correction; ~s-
daumen m (HA) correcting cam;
~splatte f (HA) correcting plate;
~srad n (HA) correcting wheel;
~sstift m (HA) correcting pin.
Korrektur f correction, revision,
reading; ~bogen m proof sheet,
proof; die ~ besorgen, lesen to correct the errors of the press, to
look over.
Korrespondenz f correspondence; ~austausch m (P) postal exchange; ~karte f postal card.
korrespondiren v to correspond
(by telegraph, by letter).
Kosten pl charges, cost, expenses; außerordentliche, unvorhergesehene ~ additional, extraordinary
charges; die ~ bestreiten, tragen to
bear the charges.
kosten v to cost, to bear a price.
Kosten... ~anschlag m estimate
(of the costs of the work), statement,
bill of cost; ~ansatz m im Einzel-
nen, Rechnungsposten m item; ~-
frei a u. adv cost-free, free of
charge; ~rechnung f note, account
of charges; ~verzeichniß n list of
expenses.
Kraft f force, power, strain;
~ der Bestimmung by virtue of the
rule; in ~ sein to be in force; in
~ setzen to execute; ~einheit f dynamical unit, unit of force (power);
~messer m dynamometer; ~moment n
momentum of a force; ~übertra-
gung f transmission of force.
Krahn m crane, sheer legs; ~-
balken m horizontal beam, neck,
gibbet of a crane; ~gestell n crane
frame; ~pfosten m post, upright
post of a crane.
Kramme, Krampe f cramp,
cramp iron, clincher, staple; mit
~n befestigen to cramp.
Kreis m circle; (Geschäftskreis)
sphere of business; (Bezirk) county,
district; sich im ~e bewegen to turn
round; sich im ~e drehend rotatory.
Kreis... ~bewegung f rotary
(rotatory) motion, rotation; ~bo-
gen m arc of a circle; eingetheilter
~bogen (Gradbogen) m graduated
sector; ~bohrer m borer with a circular bit; ~förmig a circular; ~-
lauf m circulation, circular motion,
circuit; ~linie f circle, circular line;
~säge f circular saw; ~scheibe f circular disc.
Kreosot n creosote, creasote.
Kreosotiren n, **Imprägniren mit
Kreosot** n creosoting.
Kreosotöl n creosote oil.
Kreuz... ~band n (Techn.) cross
piece, cross beam, diagonal brace,

diagonal stay; ~band n, Umschlag m (P) wrapper; ~bandsendung f (P) article sent under wrapper; ~holz n (Befestigung von Stangen im Erdboden) cross feet, cross bar; ~strebe f cross stay.

kreuzen, sich v to cross.

Kreuzung f (Eisenb.) crossing, railway crossing; (von Telegraphendrähten) cross; ~spunkt m crossing point, (Niveauübergang) level crossing.

Kriegs ... ~jahr n, Anrechnung der ~jahre bei Festsetzung des Ruhegehalts counting the years of actual warfare in calculating the pension due; ~telegraphie f military telegraphy, field telegraphy; ~zulage f extra allowance for field service.

Kronrad n (Techn.) crown wheel.

kröpfen v to bend at right angles, to bend, to curve.

krumm a crooked, bent, curved; ~ werden, sich werfen v (vom Holz) to warp.

krümmen v to curve, to crook, to bend.

Krümmung f curve; (eines Flusses) bend, sinuosity, winding of a river; (eines Weges, einer Eisenb.) curvature.

Kubik ... ~fuß m cubic foot; ~inhalt m cubature, cubic contents; ~maaß n cubical measure; ~meter n cubic metre.

Kugel f (Techn.) ball; (Erd-, Himmelskugel) globe; (Math.) sphere; (eines Baro- od. Thermometers) bulb; ~förmig a spherical, globular.

Kuhfuß m, **Brechstange** f crooked crow-bar.

Kummet n collar, horse collar; ~geschirr n collar harness.

kund a, J—m etwas ~ thun to make known, to notify smt to smb; ~ werden to get abroad.

kündigen v (einen Dienst) to give warning, to give smb notice to leave; (ein Kapital) to give notice that a capital is to be returned; (einen Vertrag) to give notice that a contract shall not be renewed.

künftig a future, next, to come; ~e Woche next week; ~ adv in future, henceforth, hereafter.

Kunststraße f high road, causeway.

künstlich a artificial (resistance).

Kupfer n copper; Gar~ refined copper; ~draht m copper wire; ~niederschlag, ~schlamm m deposit, precipitate of copper, (in der Batterie) black mud; ~oxyd n black oxide of copper; ~oxydul n red oxide of copper; ~platte f copper plate; ~pol m copper pole; ~stahldraht m compound wire; ~streifen m strap of copper; ~strom m copper current; ~vitriol m blue vitriol, sulphate of copper.

kuppeln, koppeln v to couple, to connect; (HA) to look together by a detent; gekuppelte Stangen f/pl coupled poles.

Kuppelung, Verkuppelung f coupling.

Kurbel f crank, handle, winch handle; (des Umschalters) switch bar, lever; ~achse f cranked axle, crank axle, crank shaft; ~handgriff m winch handle; ~umschalter m (T) lever switch, switch with levers (switch arms).

Kurierzug m express train.

Kurs m (Postkurs) mail route, mail line; (Geldkurs) exchange rate, course of exchange; in ~ setzen to put in circulation; außer ~ setzen to put out of circulation, to call in.

Kurs ... ~änderung f alteration of mail routes; ~ausstattungsgegenstände m/pl mail equipment; ~berechnung f (des Geldkurses) calculation of exchanges; ~bericht m (Geld) exchange advice; ~buch n Official Railway Guide, time tables, itinerary; ~büreau n Office of mail routes, (Am) Office of the Topographer; ~einrichtungen f/pl mail routes, arrangement of mail routes; ~habende Papiere n/pl current stocks; ~karte f map of mail routes; ~materialien n/pl supplies of mail routes; ~notirung f (Geld) quotation of exchange; ~postanstalt f. Postanstalt; ~sack m mail sack; ~stempel m date stamp of the travelling post office; ~uhr *f* watch (in keeping of mail guards & drivers); ~verluſt *m* loss on the exchange, loss by exchange; ~wagen *m* any postal vehicle for use on high roads; ~werth *m* value in exchange; ~weſen *n* (*P*) mail routes, post routes; ~zettel *m* (des Geldkurſes) list, statement of exchanges, exchange list, stocks list, printed exchange.

Kurſus *m* course; (auf Univerſitäten) term.

Kurve *f* curve, curveline; Stromankunfts-~ (*T*) curve of the in-coming current.

kurz ſchließen *v* (*T*) to short-circuit; die Apparate ſind kurz geſchloſſen the instruments are joined up on short circuits ob. are short-circuited.

Küſte *f* coast, shore; ~n-Kabel *n* (Ende des Kabels) shore end; ~n-Telegraph *m* coast telegraph, semaphore.

Kyaniſirung *f* kyanization (preservation of timber by steeping it in a solution of corrosive sublimate).

kyaniſiren *v* to kyanize timber.

L.

Lack *m* lac, gum lac; (Lackfirniß) varnish; (Siegellack) sealing wax.

lackiren *v* to lacker; (mit Firniß überziehen) to varnish.

Lade . . . ~brief, Verladungsſchein *m* bill of freight; ~platz *m* (Eiſenb.) platform; ~rampe *f* ascent, carriage landing; ~ſchein, ~zettel *m* (*P*) list of bags & parcels, (Eiſenb.) bill of lading.

laden *v* (Techn.) to load, to charge; (Elektr.) to charge (a battery).

Ladung *f* charge; ~sbatterie *f* secondary battery; ~s-Koeffizient *m* coëfficient of charge; ~splatz *m* place of lading; ~sſtrom *m* current of charge, charging current.

Lage *f* situation, site; (Techn.) layer, coat, coating; (Schicht) layer, stratum; (Farbeſchicht) couch; ~plan *m* plane of site.

lagernde Vorräthe *m/pl* stores in depot.

Lager *n* (Techn.) couch, layer, bed; (Handl.) stock, stock on hand, store house, warehouse, magazine; (Zapfen-, Wellen-, Achslager) bearing, journal bearing; Poſtſtück *n* auf ~, Lagerſendung *f* (*P*) mail article not yet delivered.

Lager . . . ~buch *n* (*P*) statement of mail articles arrived; ~friſt *f* (*P*) period of retention; ~gebühr *f* (*P*) demurrage, demurrage charge;

~platz *m* place for storing goods, depot; ~ſendung ſ. Lager.

lagern *v* (liegen) to lie, to be stored; es ~ 100 Packete in der Packkammer there are 100 parcels in the parcels room; ~de Sendung ſ. poſtlagernd.

Lamelle *f* (*HA*) vibrating spring.

Lampe *f* lamp.

Land *n* land, country; (Topogr.) land, ground, soil; auf dem ~e wohnen to live in the country; ~beſtellbezirk *m* (*P*) rural delivery, district, walk; ~beſtelldienſt (*P*) *m* duty of the rural messenger; ~beſtellfahrt *f* (*P*) rural delivery drive; ~beſtellgang (*P*) *m* rural delivery trip, walk, route; ~beſtellungseinrichtung *f* (*P*) rural post arrangements, driving letter carrier service; ~briefbeſtellung *f* rural postal service; ~briefträger *m* rural post messenger, rural postman, rural letter carrier; fahrender ~briefträger *m* mounted rural messenger, driving rural letter carrier, mail driver; ~esgeſetze *n/pl*, Landrecht *n* laws of the land, common law; ~eswährung *f* the standard currency of a country; ~karte *f* map, land chart; ~linie *f* (*T*) land line, air line, aërial line, pole supported line, overhead line, (*P*) high road, ſ. auch ~poſtkurs; ~poſt *f* rural post; ~poſtkurs *m* mail-coach line, postal line in the country, postal line

6*

on high roads; ~straße *f* high road, public road, main road; ~transit *m* (*PT*) territorial transit; ~transport *m* land carriage, land conveyance; ~üblich *a* usual in a country, customary; ~weg *m* high road, road; Postbeförderung *f* auf~wegen mail conveyance by road.

landen *v* to land, to disembark; in einem Hafen ~, um die Post einzunehmen to touch at a port to pick up the mails.

Landung *f* landing, coming on shore; ~spunkt *m* landing place, landing point (of a cable).

Länge *f* length; (Astron.) longitude; der ~ nach at length, lengthwise.

Längen... ~ausdehnung *f* linear expansion, linear dimension; ~durchschnitt *m*, ~profil *n* longitudinal section; ~faser *f* grain, streak, fibre of wood; ~maaß *n* long measure, linear measure, instrument for measuring lengths; ~nachweisung *f* (*T*) list showing the length of the lines.

länglich *a* oblong; ~rund oval.

Längsfäden *m/pl* (der Kabelschutzhülle) longitudinal threads.

Lärche *f*, **Lärchenbaum** *m* larch, larch tree, larch fir.

Lasche *f* (Techn.) flap, lappet; (Masch.) fish, fish plate.

Laschung, Blattung, Holzverbindung *f* scarf.

Last *f* load, burden, weight; (Gewicht) last (etwa 4000 Pfund); todte ~ (Mech.) dead weight; zur ~ legen to impute to; J—m zur ~ schreiben to put down to a person's account, to charge against *smb*.

Lauf *m* run, course; (Schifff.) way, course, rate; (Geschwindigkeit) speed, velocity; (eines Flusses) course, bed of a river; ~achse *f* trailing axle; ~bahn *f* (Techn.) run; (Lebens~bahn) career, life; ~brett, Trittbrett *n* foot plank; ~gewicht *n* sliding weight, running weight; ~werk *n* wheel work; ~zettel *m* (*P*) note (form, letter) of inquiry.

laufen *v* (Mech.) to turn (said of a wheel); to run (said of a ship); to work, to run (said of an apparatus); (umlaufen) to circulate, to be in circulation (said of a bill, a rumor, a letter etc.).

laufend..., ~ numeriren *v* to number consecutively; ~e Geschäfte *n/pl* the business of the day, the course of affairs; ~es Jahr *n* the current (present) year; ~er Kurs *m* course of exchange; ~es Meter *n*, ~er Fuß *m* meter run, foot run; ~e Nummer *f* number, consecutive number; ~er Preis *m* current price, market price; sich auf dem ~en halten to keep one's self well posted.

Läufer (*HA*) s. Schlitten.

laugen s. auslaugen.

Läute... ~apparat *m* bell telegraph, alarum telegraph, alarm; zweigloctiger ~apparat, Doppelläuter *m* Bright's bells; ~induktor *m* ringing inductor; ~stellung, ~schaltung *f* des Apparats alarm bell in circuit; ~taste *f* bell key; ~werk, electrisches Geläute *n* electro magnetical ringing apparatus; ~werklinie *f* bell line.

Lebensversicherung *f* life insurance, insurance of the lives (of officers); ~s-Gesellschaft *f* Life Insurance Company.

Leder... ~tasche *f* (der Briefträger) pouch; (der Briefkastenleerer) bag; ~überzug *m* (auf dem Verdeck der Packetpostwagen) cover, leather cover.

leer *a* empty; (Phys.) vacuous; ~er Bogen Papier clear sheet of paper; ~es Papier blank (paper); ~er Raum *m* (Phys.) vacuum; ~ lassen to leave in blank.

Leer... ~karte *f* (*P*) bill left blank; ~nachweisung *f* vacat return, blank return (in England mit dem Worte „nil" [nihil] beschrieben).

Leere *f* (Phys.) vacuum, s. auch Drahtleere.

leeren *v* to empty, to evacuate; (den Briefkasten) to clear (the letter boxes), to collect (the letters from the boxes).

Leerung *f* (der Briefkasten) clearance, collection; (the hours of collection are printed on the letter boxes).

legen v (einen Brief in den Briefkasten) to post a letter, to drop a letter in the letter box; (eine Leitg. an Erde [T]) to put a wire to earth, to ground a wire; (ein Kabel) to lay a cable.

legiren v to alloy.

Legirung f alloy, alloying, alloyage, mixture (combination) of metals.

Legitimation f legitimation, evidence, proof; ~spapiere n/pl papers to prove the identity.

Lehm m loam, common clay; ~boden m clayey soil, clayey ground.

Lehre s. Drahtleere.

leicht a light, not heavy; ~e Leitung f (T) light wire, thin wire.

Leier, Drahtleier s. Draht.

Leim m glue; flüssiger ~ clear cole, liquid glue; ~farbe f size color, glue-water color, distemper color.

leimen v to glue.

Leinöl n linseed oil; in ~ getauchter Hanf hemp steeped in linseed oil; ~firniß m boiled linseed oil, oil varnish, linseed oil varnish.

Leinpfad m tow (towing) path, track road.

Leiste f ledge, border, strip.

leisten v to do, to make; (Mech.) to perform.

Leistung f (einer Kraft, Maschine eines Apparats) effect of a power, work done by a machine, laboring force; (mechanische) performance, work performed, work done, mechanical effect, services to be rendered; ~en des Postfuhrunternehmers (P) services to be rendered (work to be done) by the mail contractor; ~en der Postverwaltung operations of the postal administration; ~ im Nebenverdienst service rendered in addition to . . .

Leistungs . . . ~fähigkeit f (höchstmögliche einer Maschine) load; (Befähigung) capacity; ~nachweisung f specification of the services (to be) rendered, report (return) of work (to be) done, (des Postfuhrunternehmers) statement of the regular trips to be performed by the mail contractor.

Leit . . . ~ort m, ~postanstalt f post office upon which the mail is to be directed; ~rolle, feste Rolle f (an Flaschenzügen 2c.) guide pulley, (zur Führung des Papierstreifens) guide roller; ~stange f conducting rod, guide rod; ~übersicht f (P) circulation list, sorting list, table of despatch; ~vermerk m (P) notice indicating the route by which mail articles are to be forwarded; ~zettel m (P) label indicating the route by which parcels are to be forwarded.

leiten v (führen) to guide, to lead, to conduct; (beaufsichtigen) to direct, to conduct; (Telegramme, Briefe u. s. w.) to send, to direct (telegrams, letters etc); (anderswohin) to turn off.

leitende Verbindung f, **leitender Körper** m conductor.

Leiter m (Phys.) conductor; guter, schlechter ~ good, bad conductor; (einer Fabrik 2c.) director, manager.

Leiter f ladder; (mit Gelenken versehene, zusammenlegbare) folding ladder, jointed ladder; (mit Verlängerungsstück) extension ladder; ~baum m, ~stange f ladder beam, cheek, side piece; ~sprosse f ladder step.

Leitung f (T) wire, s. auch Telegraphen . . . ; (P) despatch, conveyance, forwarding, route; (Masch.) direction, guide; (eines Unternehmens) management, direction; (Chem.) passage, ohne Rücksicht auf die ~ der Sendung (P) without destination of route; (eine besondere ~ ist für die Staats-Korrespondenz bestimmt (T) a special wire is assigned for government telegrams; zwei Orte durch eine direkte ~ verbinden (T) to establish direct telegraphic communication between two places; die ~ wiederherstellen (T) to repair the wire; unrichtige ~ eines Briefes, Telegramms mis-sending, mis-directing.

Leitungs . . . ~-Aufseher m (T) inspector of linemen, foreman; ~draht m (T) conductor, conducting wire, main wire; ~fähigkeit f (T) conductivity, (Phys.) conductibility; ~führung f (T) lead; ~geräthe pl (T) tools for telegraph construction; ~klemme f (T) clamp to which the line wire is attached; ~kraft f (Phys.) conducting power; ~materialien n/pl (T) materials for telegraph construction; ~litze f (T) strand, eine siebenadrige Kupferdrahtlitze a strand of seven copper wires; ~netz n (T) network of lines; ~platte f plate to which the line wire is attached; ~revisionsbezirk m (T) division (district) of the supervising officer; ~revisor m (T) supervising officer; ~rohr n (für unterirdische Führung [T]) conduit, conduit pipe; ~schnur f (F) telephone cord; ~störung f (T) disturbance; ~unterbrechung f (T) disconnection, interruption; ~vermögen n conducting power, spezifisches ~vermögen specific conductivity; ~widerstand m resistance (to the passage of the electric current).

Leucht ... ~gas n coal gas, illuminating gas, lighting gas; ~schiff n floating light, light boat, light ship; ~thurm m light house, beacon.

Leydener Flasche f Leyden jar, Leyden phial, electric jar; Batterie von ~ ~n Leyden battery.

Libelle, Wasserwaage f level, water level.

Licht n (Phys.) light; (farbige Signalscheibe [Eisenb.]) colored glass; ~bogen m luminous arc; elektrischer ~bogen m voltaic arc; ~bündel, ~büschel n electric brush; ~stärke f candle power.

Lichte n clear; im ~n in the clear, clear.

lichte Weite f diameter inside; s. auch Lichten ...

Lichten ... ~breite f einer Oeffnung breadth, width of the day; ~höhe f height of the day, day height, head way; ~weite f intermediate space, open space, (einer Oeffnung) width, (eines Raumes) width in the clear, inside width.

lidern, ledern, beledern v (Masch.) to pack, to leather (a piston etc.).

Liderung f (Masch.) packing, leathering; mit ~ versehen a packed; ~ mit Hanf, Hanfliderung hemppacking.

Lieferant m purveyor, contractor, furnisher.

liefern v to furnish, to purvey, to supply; die Batterie liefert schwachen Strom the battery gives a weak current.

Liefer ... ~schein m bill of delivery; ~zeit f term for delivery.

Lieferung f supplying, purveying, supply, purveyance; (Ablieferung f) delivery; (Heft) number; ~sangebot n tender; ~sbedingungen f/pl specification (z. B. of the wire); ~spreis m contracting price; ~svertrag m contract for supplying, contract of delivery; ~szeit, Lieferzeit f term for delivery.

liegen v to lie; (von Orten) to be situated; der Zinkpol liegt an Leitung (T) the zinc pole is to line.

liegende Güter n/pl immovable goods, immovables, landed estate, landed property.

Linie f line; gerade ~ straight line; krumme ~ curve line, curve; senkrechte, lothrechte, vertikale ~ vertical line, plumb line.

Linien ... ~ausstattungsgegenstände m/pl (T) appliances for telegraph construction; ~batterie s. Batterie; ~karte f des Bezirks (T) map of lines of the district; ~materialien n/pl (T) construction stores, line equipment; ~schreiber, Monostychograph m (T) apparatus printing in one line; ~umschalter s. Umschalter; ~wechsel m switch.

Lippe f (Techn.) lip; ~ des Schlittens (HA) steel plate of the chariot.

Liste f list.

Litze f (verseilte Drähte) strand; siebenadrige ~ strand of seven wires; ~ndraht m wire in a strand.

Loch n (für die Stangen [T]) s. Stange; (des Zieheisens) die.

lochen v (T) to perforate, to punch.

Locher, Schriftlocher m, **Locheisen** n (T) perforator; (Handlocher) hand perforator; (Tastenschriftlocher) keyboard perforator.

Löffelbohrer *m* (Bergb.) spoon bit, wimble; ~, **Hohlbohrer** (Zimm.) shell auger, shell bit.

Lohn *m* (Techn.) hire, wages, pay; ~arbeiter *m* hired workman; ~auszahlung *f* payment of wages; ~berechnung *f* amount to be paid (as wages); ~liste *f* register of wages; ~satz *m* rate of wages; ~tag, Zahltag *m* pay day.

Lokal ... ~batterie *f*. Batterie; ~behörde *f* local authority; ~korrespondenz *f* (TP) local traffic; ~schaltung (= stellung) *f* (T) a way station has its connections so arranged that it works as a terminal station; ~strom *m* (T) local current, local circuit; ~verhältnisse *n/pl* local concerns, local circumstances; ~zug *m* (Eisenb.) stopping train; ~zulage *f* increase of salary on account of being stationed in a certain place.

Lori, Lowry *f* (Eisenb.) lowry, truck, truck waggon.

lose *a* loose, movable; ~ Rolle *f* movable pulley; ~ Waaren *f/pl* loose articles, articles in bulk.

lösen *v* (Chem.) to dissolve, to solve; (lose machen) to loosen, to untie; (eine Aufgabe) to solve (a problem).

Lösung *f* (Chem.) solution, dissolution; (einer Aufgabe) solution (of a problem).

Loth *n* (als Gewicht) half an ounce (the 32th part of an old German pound); (senkrechte Linie *f*) perpendicular; (Senkblei, Bleiloth *n*) plumb, plumb line, plummet, lead; (Löthmittel *n*) solder.

loth ... ~recht, senkrecht *a* perpendicular, vertical; ~rechte Ebene *f* perpendicular plane; ~rechte Linie *f* perpendicular line, vertical line, plummet line.

Löth ... ~bahn *f* bevel of the soldering iron; ~bock *m* soldering support; ~büchse *f* soldering box; ~eisen *n*, ~kolben *m* soldering iron, (um kleine Gegenstände zu löthen) copper bit, (um große Gegenstände zu löthen) iron bolt; ~haken *m* joint hook; ~kluppe *f* clamp; ~korn, ~kupfer *n* link for soldering; ~lampe *f* soldering lamp; ~löffel *m* soldering groove (spoon); ~mischung *f*, ~mittel *n* solder, link for soldering; ~muffe *f* tube (the wire being held by hooks or wedges); ~ofen *m* soldering furnace (stove); ~pfanne *f* soldering pan; ~rohr *n* blow pipe; ~stelle *f* joint, Britannia ~stelle, Wickel-~stelle Britannia joint, Würge-~stelle twisted joint, bell hanger's joint; ~stellentrog *m* trough for testing the joints; ~wasser *n* flux; ~zange *f* pincers *pl*, hawk bill, hawkbill plyer; ~zinn *n* tin solder.

löthen *v* to solder.

Löther *m* jointer, solderer; ~zelt *n* tent for (the protection of) the jointers.

Löthung *f* soldering, s. auch Löthstelle.

Lowry s. Lori.

Luft *f* (atmosphärische) air, atmospheric air; verdickte, verdünnte ~ compressed (rarefied) air; ~abzug *m* air exhauster; ~blase *f* air bubble; ~bremse *f* atmospheric brake; ~dicht *a* air proof, air tight, hermetical; ~dichtigkeitsmesser *m* manometer; ~druck *m* atmospheric (pneumatic) pressure; ~druckeisenbahn *f* atmospheric railway; ~druckmaschine *f* air-pressure engine, atmospheric (pneumatic) engine; ~druckpumpe *f* atmospheric pump; ~elektrizität *f* atmospherical electricity; ~förmig *a* aëriform, gaseous; ~hahn *m* (Absperrhahn einer Röhrenleitung) stop cock of a wind (air) pipe; ~heizung *f* hot air heating, heating buildings by transmission of hot air; ~heizungsofen *m* air stove, air heater; ~kabel *n* (T) air cable, aërial cable, overhead cable; ~leer *a* exhausted, void of air, ~leer pumpen to exhaust of air; ~leere *f*, ~leerer Raum *m* vacuum; ~leitung (T) *f* air span, aërial wire, overland wire; ~linie (T) *f* aërial line, overland line, overhead line; ~pumpe *f* air pump, pneumatic pump; ~trocken, an der Luft getrocknet *a* air dry, air dried; ~trocknes Holz *n* seasoned timber;

~verdichtungspumpe *f* condensing air-pump; ~verdünnung *f* rarefaction of air.

lüften, der Luft aussetzen *v* to air, to weather, to expose to the air. Lüften, Ventiliren *n* airing.

M.

Maaß *n* measure; eingeschriebenes ~ in Zeichnungen figured dimension; ~band, Meßband *n* tape measure, measuring tape; ~einheit *f* unit of measure, unit of measurement; nach ~gabe in proportion to, according to; ~latte *f* gage, gauge, measuring rod; ~regeln (danach) treffen to take measures (accordingly); ~stab *m* scale, nach dem ~stabe zeichnen to draw to a scale, natürlicher ~stab plain scale, full size, vergrößerter ~stab enlarged scale, verjüngter, verkleinerter ~stab reduced scale; ~stock *m* measure rule, scale rule.

Magazin *n* magazine; magnetisches ~ magnetic magazine, compound magnet.

Magnesia *f* magnesia, oxide of magnesium; schwefelsaure ~, Bittersalz *n* sulphate of magnesia (magnesium), bitter salt, Epsom salt.

Magnet *m* magnet; einen ~ armiren, bewaffnen to arm (to cap) a magnet; bleibender, immerwährender ~ permanent magnet; künstlicher ~ artificial magnet; ~-Armatur *f* furniture of a magnet; ~bündel *n* compound magnet; ~eisen, ~eisenerz *n*, ~eisenstein *m* magnetic iron, magnetic iron-ore, magnetic iron-stone, load stone; ~elektrisch *a* magneto electric; ~-Induktion *f* magnetic induction; ~-Induktionsstrom *m* magneto-electric current; ~nadel *f* magnetic needle, compass needle, needle; ~stab *m*, ~stange *f* bar magnet; ~system *n* magnetic system.

magnetisch *a* magnetic; ~ machen to magnetize; ~e Abstoßung *f* magnetic repulsion; ~e Anziehung *f* magnetic attraction; ~er Aequator *m* magnetic equator, aclinic line, line of no dip; ~e Batterie *f*, ~es Magazin *n* magnetic battery, magnetic magazine; ~e Induktion s. Magnet-Induktion; ~es Residuum *n*, remanenter Magnetismus *m* residual magnetism.

magnetisiren *v*, durch den einfachen, doppelten Strich, durch Vertheilung to magnetize by the method of the single touch, double touch, by influence.

Magnetisirung *f* magnetizing; galvanische ~ magnetizing by the galvanic current; ~ durch Induktion magnetic induction, magnetic influence; ~s-Konstanten *f/pl* magnetic elements; ~s-Spirale *f* magnetizing coil.

Magnetismus *m* magnetism; Erd-~ terrestrial magnetism; freier ~ free magnetism; gebundener ~ bound magnetism; Nord-~ north magnetism; remanenter ~ residual magnetism; Süd-~ south magnetism.

magneto-elektrisch *a* magneto electric.

Magneto-Elektrizität *f* magneto-electricity, electro-magnetism.

Magnetometer *m* magnetometer (an instrument for measuring the intensity of a magnetic current).

Makulatur *f* waste paper.

Mal *n* mark; (Grenzmal) boundary.

Mangan *n* manganese; ~eisen *n* manganesian iron, ferro-manganese; ~oxyd *n* manganic oxide; ~superoxyd *n* (Braunstein *m*) pyrolusite, manganese-ore.

mangelhaft *a* defective, incorrect, faulty.

Manko *n* deficiency (of measure, of weight).

Manometer *n*, **Dampfdruckmesser** *m* steam gage, steam gauge, steam pressure gauge, manometer; geschlossenes ~ closed manometer, compressed steam gauge; offenes ~ open manometer, steam gauge with open leg.

Manöver-Postordnung s. Post-ordnung.

Manschette f (Masch.) leather collar.

Mantel m (Techn.) (äußere Bekleidung f) mantle, case, casing; (Hülle [Masch.]) case; (des Isolators) cap, cup, outer cup; (eines Kegels) surface of a cone; (einer Pyramide) convex surface of a cone.

Marine . . . ~angelegenheit f marine business, official business of the imperial navy; ~brief m marine letter, letter addressed to a private seaman (mariner) of the imperial navy; ~ministerium n ministry of naval affairs, (in England) Admiralty; ~postanweisung f marine money order; ~postbüreau n marine post office.

Mark f mark (of silver or gold, eight ounces); mark (a German coin); die ~ Banko banco mark.

Markir . . . ~pfahl m, ~stange f mark pile; ~pfählchen n (T Bau) peg, stave.

markiren v to mark.

Material n material; liegendes ~ (Eisenb.) railway plant; rollendes ~ (Eisenb.) rolling stock.

Materialien n/pl materials, stores; ~bankett n banquette for piling up the materials; ~bedarfsnachweisung, ~berechnung f specification of the materials or stores to be used; ~lager, ~depot, ~magazin n depot, store; ~transporttabelle f specification of the mode of conveyance of the materials or stores; ~verwalter m controler of stores, storekeeper; ~verwendungsnachweisung f specification of the materials used.

Matrize f (Schraubenmutter f) female screw, nut screw.

Mauer f wall; ~bock, ~bügel m bracket (of cast iron) let into the masonry to form a point of attachment, wall attachment; ~briefkasten s. Briefkasten; ~werk n masonry; ~ziegel m brick.

Maximaltarif m maximum tariff (the highest rate which must not be overstepped).

Mechanik f mechanics.

Mechaniker m mechanician, mechanist.

mechanisch a mechanic, mechanical; ~e Arbeit, Leistung f work done, work performed; Einheit der ~en Arbeit unit of work, dynamical unit.

Mechanismus m mechanism, contrivance.

Megafarad n megafarad (a million of farads).

Megaveber m megaveber (a million of vebers).

Megavolt n megavolt (a million of volts).

Megohm n megohm (a million of ohms).

mehr a more; jedes Wort ~ 6 Pfg. every *additional* word 6 pfennig.

Mehr n (bei Kassenabschlüssen) cash over; ~arbeit f additional labor, extra work; ~ausgabe f excess of expenditures; ~betrag m an Baar in der Kasse (P) excess in the official cash; ~betrag beim Kassenabschluß (P) cash over; ~- od. Minderbeträge aus den Kassenabschlüssen cash over or cash short; ~bespannung f (P) supplementary teams of horses; ~einnahme f surplus of receipts; ~gewicht n excess of weight.

Meile f mile, league; deutsche geographische ~ German geographical mile (7420,428 metres = the fifteenth part of a degree of longitude of the equator = 4 English geographical miles); englische ~ English mile, statute mile (1609,315 metres = 1760 yards); ~, Seemeile sea mile (1855 metres = the sixtieth part of a degree of longitude); metrische ~, deutsche Reichsmeile German metrical mile (7500 metres = 4,66 English miles).

Meilen . . . ~geld n allowance; ~stein, ~zeiger m mile mark, mile stone, mile post.

Meißel m chisel.

Meist . . . ~betrag m maximum amount; ~bietend verkaufen to sell at public auction, to sell to the highest bidder; ~gebot *n* the highest bidding; ~gewicht *n* maximum weight.

Melde ... ~amt *n* intelligence office f. auch Einwohner=~amt; ~zettel *m* der Bahnpoſtwagen (P +) notice (of parcels loaded into, & unloaded from, a supplementary luggage waggon).

Membrane *f* membrane, diaphragm.

Menge *f* quantity, a great number; ~einheit *f* unit of quantity.

Mennige *f*, **rothes Bleioxyd** *n* red lead, red oxide of lead, minium.

Meridian, Mittagskreis *m* meridian; magnetiſcher ~ magnetic meridian.

Merk ... ~buch *n* der Regiſtratur minute book of the Registrar's office; ~zeichen *n* mark pile, mark.

Meß ... ~band *n* measuring tape, tape measure; ~brücke *f* Wheatstone's balance (bridge); ~inſtrument *n* measuring instrument, (für Drahtſtärke) ſ. Drahtleere, (elektriſches) testing instrument; ~kette *f* measuring chain; ~kunde *f* (als Titel) instructions for testing, electrical testing; ~latte *f* surveying staff, measuring staff; ~leine *f* measuring line; ~ordnung *f* instructions for electrical testing; ~ſtab *m*, ~ſtange *f* surveyor's pole, surveying rod, perch; ~tiſch *m* surveyor's table, plane table; ~werkzeug *n* measuring (surveying) instrument; ~zelt *n* tent for testing purposes.

meſſen *v* (Techn.) to measure, to survey; (mit den Augen) to sight; (mit der Kette) to chain, to survey by the chain; (Elektr.) to measure (the insulation etc.), to test (a cable).

Meſſing *n* brass, latten (alloy of copper and zinc); ~blech *n* sheet brass, brass plate; dünnſtes ~blech *n* shaven latten; ~draht *m* brass wire, latten wire; ~hebel *m* brass lever, (um das Laufwerk des MA auszulöſen u. anzuhalten) brake, friction brake; ~klemme *f* brass clamp; ~platte *f* brass plate; ~rahmen *m* brass frame; ~ſchiene *f* brass bar; ~ſchlagloth *n* brass solder, zinkreiches

~ſchlagloth *n* spelter solder; ~ſtäbchen *n* brass strip; ~ſtänder *m* brass pillar; ~ſtift *m* brass pin; ~ſtange *f* brass rod; ~wange *f* (des Apparats) brass side.

Meſſung *f* test, testing, measuring, vgl. meſſen.

Metall *n* metal; (Bronze) bronce, brass; (Kompoſition) yellow metal; (edles) noble metal; (unedles) base (ignoble) metal; ~gelb *n* specie; ~gewinnung *f* aus Erzen metallurgy, extraction of metals from ores; ~hebel *m* metallic lever; ~ſtift *m* metallic pin; ~ſtreifen *m* strip of metal; ~überzug *m* plating, metallic covering; ~verbindung *f* (Chem.) metallic compound.

metalliſch *a* metallic, metalline, metalliferous; ~blank bright & clean; ~e Berührung *f* metallic contact.

Meter *n* metre; ~maaß *n* metric measure.

Meterkilogramm ſ. **Kilogrammmeter.**

Meteor ... ~eiſen *n* meteoric iron; ~ſtahl *m* meteor steel; ~ſtein, **Meteorolith** *m* meteoric stone, meteorite, aerolite.

Meteorologie *f* meteorology.

meteorologiſch *a* meteorologic.

Miethe *f* hire, rent; in die ~ geben to let (out); in die ~ nehmen to hire, to rent; in der ~ haben to have on hire, to hold on lease.

miethen *v* to hire, to rent; (ein Schiff) to charter a ship.

Miether *m* hirer, renter; (eines Hauſes) tenant, lodger.

Mieths ... ~entſchädigung *f* indemnification on (reimbursement of, sum of indemnity for) rent paid; ~geld *n* (Handgeld *n*) earnest money; ~poſtgebäude *n* rented post office building; ~preis *m* rent, amount of rent, rental; ~ſteuer *f* tax on the rent; ~vertrag *m* agreement (contract) between hirer and letter (out), lease; ~weiſe *a* u. *adv* by lease, on hire, ~weiſe Uebertragung *f* einer T. Leitung letting out a wire (to a newspaper); ~zeit *f* hiring time.

Mikrofarad *m* microfarad (the millionth part of a farad).

Mikrohm *m* microhm (the millionth part of an ohm).

Mikrometer *n* micrometer (an instrument for measuring very small distances); ~schraube *f* micrometrical screw, micrometer screw, regulating screw.

Mikrophon *n* microphone (an instrument for making audible very slight sounds); ~geber *m* microphone transmitter.

Mikrovolt *n* microvolt (the millionth part of a volt).

Mikroweber *m* microveber (the millionth part of a veber).

Militär . . . ~angelegenheiten, Portofreiheit in ~angelegenheiten exemption from postage in matters relating to the army (on military business); ~anwärter *m* (*TP*) military (naval) pensioner (one who is entitled to an appointment in the civil service in consideration of his military (naval) services; ~dienstzeit *f*, Anrechnung der ~dienstzeit counting the time passed in active military service (in the army); ~ehrenzeichen *n* military medal (given as a reward for meritorious conduct in the field); ~invaliden-Pension *f* military pension (granted to disabled soldiers & officers of the army & navy); ~telegraphie *f* military telegraph (telegraphy), field telegraph; ~verhältnisse *n/pl* der Beamten military obligations of civil service officers.

Millimeter *n* millimeter; ~maaßstab *m* scale graduated in millimeters.

Minder . . . ~bespannung *f* (*P*) putting less horses to the mail coach than stipulated by contract; ~betrag *m* deficiency, den ~betrag decken to cover (to make good, to refund, to reimburse) the deficiency; ~betrag beim Kassenabschluß cash short; ~beträge s. Mehr- u. ~beträge; ~einnahme *f* deficiency in the receipts; ~ertrag *m* deficiency in the proceeds; ~gewicht *n* deficiency in weight.

Mindest . . . ~betrag *m* minimum amount, (Minimalsatz) minimum rate; ~fordernder *m* he that asks the least (lowest) price, lowest contractor; dem ~fordernden den Zuschlag ertheilen to adjudge to him that asks the lowest price; ~gehalt *n* minimum pay; ~gewicht *n* minimum weight.

Minimal . . . s. Mindest . . .

mischen *v* to mix.

Mischung *f*, **Gemisch, Gemenge** *n* mixture, compound; (Chem.) composition, combination; ~sgewicht *n* (Chem.) equivalent, atomic weight; ~strog *m* (beim Kyanisiren) reservoir; ~sverhältniß *n* combining proportion.

mißbräuchliche Benutzung *f* misuse (of the mails, of the telegraph).

Mit . . . ~arbeiter *m* fellow worker, assistant; ~hören *n* (*F*) in einer Leitung, was in einer andern gesprochen wird overhearing from wire to wire; ~lesen *v* (*T*) to read through messages (in order to exercise a control); ~nahme *f*, heimliche Aufnahme von Personen bei den Posten (*P*) taking up stray passengers on the road clandestinely (under hand); ~nehmer, Daumen *m*, Knagge, Nase *f* (Techn., Masch.) cam, catch, cog, tappet; ~sprechen *n* (*T*) contact, leakage of one wire into another; ~sprechen *v*(*T*) to be in contact (the current from another wire leaks into the home wire & confuses the signals passing.

Mittel *n* middle, midst; means, expedient; (Math.) mean, medium; (Durchschnitt *m*) average, ratio.

mittel *a* middle, mean.

Mittel . . . ~band *n* (*T* Bau) middle brace; ~größe *f* middling size; ~kraft *f* resultant, resulting force; ~linie *f* (Geom.) axis, centre line, (neutrale, indifferente [Magn.]) neutral line, medial line; ~punkt *m* centre, central point, einen gemeinschaftlichen ~punkt habend concentric, verschiedene ~punkte habend eccentric, ~punkt der Masse, Schwerpunkt centre of gravity; ~riegel *m* middle cross bar; ~schiene, Körperschiene *f* (der *MA* Taste) axis of the key; ~stück *n* middle piece, middle part.

mittels *adv* by means of, by the aid of.
mittheilen *v* to communicate, to impart (*smt* to *smb*).
Mittheilung *f* communication.
Mobilmachung *f* mobilization, putting the army (& navy) on the war establishment.
Modul, Modulus module, modulus; **Festigkeits-~** modulus of strength; **Sicherheits-~** modulus (factor) of safety.
Molekül *n* molecule, atom.
Molekular ... ~anziehung *f* molecular attraction; ~gewicht *n* molecular weight.
Moment *n* momentum; statisches ~, Kraft~ momentum of a force.
momentan *a* (Mech.) instantaneous.
Monat *m* month; der laufende ~ instant; ~e lang for months; ~s-abschluß *m* monthly settlement; ~s-bericht *m* monthly report; ~srech-nung *f* monthly account; in ~sfrist *f* within a month; ~sweise *adv* by the month.
monatlich *a* monthly.
Moor moor, turf moor, bog, swamp; ~b den *m* swampy ground, moory soil.
moorig *a* moory, marshy.
Morse ... ~-Apparat *m* Morse writer (telegraph); ~klopfer *m* Morse sounder; ~streifen *m* Morse slip; ~-Taster *m* Morse key.
Muff *m*, **Muffe** *f* (Masch. Techn.) socket joint, socket; (übergeschobene Hülse zur Verbindung zweier Röhren) muff; Röhre *f* mit ~, **Muffenrohr** *n* socket pipe, faucet pipe; (*T*) tubular connector (for making wire joints or splices).
Muffenverbindung *f* (des T.-Drahtes) covering tube; (der Röhren) joint with nozzle and socket, socket joint of pipes.
Multiplex-Apparat *m* multiplex.
Multiplikator *m* multiplier; ~-rolle *f* bobbin of a multiplier.
mundiren *v* (in's Reine schreiben) to copy fair.
Mundstück *n* (*F*) mouth piece.
Münz ... ~einheit *f* unity of coinage; ~fuß *m* standard of coinage.
Münze *f* coin, money, small money, change; (Münzstätte *f*) mint; (falsche) false coin, counterfeit coin; (klingende) ready money, cash.
muschelig *a* conchoidal (said of the fracture of iron wire).
Muster *n* (Techn.) pattern, design; (Probe) sample, specimen; (Modell) model; (Norm, Typus) standard, type.
Mutter, Matrize *f* (Techn.) matrix, matrice; (Schraubenmutter) nut, female screw, inside screw, screw nut; vierkantige ~ square nut; ~bohrer, Gewindebohrer *m* tap, screw tap; ~bolzen *m* screw bolt; ~gewinde *n* female thread of a screw; ~schraube *f* female screw, screw nut.
Myriagramm *n* myriagram (10 000 grams).
Myriameter *n* myriameter (10 000 meters).

N.

nachachten *v* to observe.
Nachachtung *f*, zur ~ which is to observe.
Nachanschlag *m* supplementary estimate.
Nachbar ... ~land *n* neighboring land (country); ~staat *m* neighboring state; ~volk *n* neighboring people.
Nachbezahlung *f* subsequent payment.

nachbohren *v* (Techn.) to bore up, to rebore, to ream, to rime.
Nachdruck *m* (betrügerischer) fraudulent impression, counterfeiting (of postage stamps).
nachdrucken *v* to imprint fraudulently, to counterfeit.
Nachforderung *f* after-claim, after-account; (v. Porto [*P*]) collection of supplementary postage.

nachforschen v to search into, to enquire after.

Nachforschung f searching, enquiry; ~en einleiten to make enquiries, to institute an enquiry (with the view to trace a letter).

Nachfrage f enquiry; ~ halten, in ~ ziehen to enquire, to make an enquiry after.

Nachfrist f f. Fristverlängerung.

nachfüllen v to fill up (earth around the poles).

nachgeben v (Techn.) to yield, to give way (said of the soil).

nachgefragt a in demand, enquired for.

nachgemacht a (Techn.) counterfeit, artificial, factitious, imitated.

nachglühen v to soften (the steel), to temper, to anneal.

Nachkosten pl after-cost.

Nachlaß m (einer Gebühr, einer Strafe u. f. w.) remission; (Abzug) deduction, abatement.

nachlassen v (vom Preise ꝛc.) to remit; (Techn.) to slacken (a spring).

Nachlieferung f von Zeitungen supplementary order of newspapers.

nachliegende Zeichen n/pl kommen an (HA) the letters (signs) on the receiving instrument go backwards in the alphabet.

Nachnahme f (P [+]) reimbursement (re-imbursement), collection; Postpacket mit ~ (+) parcel on the delivery of which a certain amount is to be collected; ~betrag m amount to be collected; ~gebühr f (P [+]) collection fee; ~postanweisung (+) money order by which the amount of re-imbursement is transmitted; ~sendung f (P [+]) value-payable article; ~verfahren n (P [+]) value-payable system.

nachnehmen v (P [+]) to reimburse one's self.

Nachprüfung f (subsequent) control, double checking, re-examination.

nachrechnen v to reckon after (again), to reckon up, to examine, to control (an account).

Nachrutsch m slip of the soil.

Nachschuß m after-payment; ~=bestellgeld n (TP) supplementary porterage (delivery fee); ~porto n (P) surcharge, supplementary postage.

nachsenden v to send after; (Briefe, Telegramme) to re-direct, to re-transmit; nachzusendender Brief m letter to be re-directed; nachzusendendes Telegramm n telegram to be re-transmitted, telegram to follow.

Nachsendung f (PT) (v. Briefen, Telegrammen ꝛc.) re-direction, retransmission.

nächst a, mit ~er Post with next post.

nachtaxiren v (PT) to charge additional postage (upon a letter).

Nachtdienst m (PT. Eisenb.) night duty, night service; Tag= u. ~ f. Dienst; ~=Entschädigung f special payment for night duty.

nachtelegraphiren v to re-transmit, to re-direct (from one town to another town).

Nachtelegraphirung f re-transmission, re-direction; ~sgebühr f re-direction fee.

Nachtrag m supplement, addition; ~=Artikel m additional article; ~starif m supplementary tariff; ~=Verzeichniß n additional (supplementary) list, supplement.

nachtragen v (ergänzen) to supply, to add.

nachträglich a u. adv supplementary, by way of addition.

Nachtransport m (P) supplementary mail.

Nachurlaub m prolongation of leave of absence.

Nachweis m, **Nachweisung** f (Beweis) record, proof; (nähere Auskunft) information, reference; (Liste) list, return, statement; ~ der Werth- u. Einschreibesendungen (P) record of insured & registered letters.

nachweisen v (beweisen) to prove, to establish, to make good; (einen Brief [P]) to record a letter; (einen Brief durch Laufzettel [P]) to trace a letter; nachzuweisende Postsendungen f/pl correspondence to be recorded.

Nachweisung *f* (Liste) list, return; f. auch Nachweis; ~sbüreau *n* office of address.

nachwiegen *v* to weigh again, to verify the weight.

nachzahlen *v* to pay afterwards; (eine Gebühr [*PT*]) to pay a supplementary charge.

Nachzahlung *f* (*P*) supplementary payment.

Nadel *f* needle; ~apparat, Doppel-~apparat, ~telegraph u. s. w. s. Apparat.

Nagel *m* nail; (Bolzen) spike, spike nail; (hölzerner) treenail, peg, wooden pin; versenkter ~ driven-in nail; ~ mit versenktem Kopf countersunk nail; ~auszieher, ~zieher *m* nail claw, nail drawer, claw, claw wrench; ~bohrer *m* gimlet, piercer, nail passer.

Näherungswerth *m* approximative value.

Naht *f* suture, seam; Briefsack ohne ~ (*P*) seamless mail bag.

Namens ... ~stempel *m* stamped signature; ~unterschrift *f* signature; ~verzeichniß *n* list of names, nomenclature.

Napf *m* bowl, cup.

Naphtha *f*, Erdöl, Steinöl, rohes Petroleum *n* naphtha, mineral oil, rock oil.

Nase *f* (Masch. *HA*) cam, catch, tappet, lifter; (einer Röhre) spout of a pipe; (Ansatz) projection.

naß *a* wet; ~e Fäulniß *f* wet rot.

Nationalflagge *f* national flag, flag of one's country; (auf See) ancient.

Natrium *n* sodium, (the metallic base of soda); ~bikarbonat, doppeltkohlensaures Natron *n* sodium bicarbonate, bicarbonate of soda; ~karbonat, neutrales kohlensaures Natron *n* carbonate of sodium, soda; ~sulphat, schwefelsaures Natron, Glaubersalz *n* sodium sulphate, sulphate of soda, Glauber's salt.

Natron, Natriumoxyd *n* protoxide of sodium, sodium oxide.

Naturereigniß *n*, Verlust von Postsendungen in Folge eines ~es loss of mail articles in consequence of a natural event.

natürliche Größe *f* real size, full size.

Neben ... ~achse *f* secondary axis; ~amt *n* (*P*) branch office, (Nebenbeschäftigung) other than postal office); ~arbeit work to be done by postal officials in addition to their proper duties, incidental work; ~ = Ausstattungsgegenstände s. Ausstattungsgegenstände; ~bahn *f* (Eisenb.) junction railway, secondary line; ~bezüge *m/pl* perquisites, casual (accidental) emoluments; ~draht, induzirter Draht *m* secondary wire; feine Batterie, Elektromagnete ~einander schalten *v* (*T*) to join up a battery in quantity, to make up a quantity battery, to connect elements side by side, to join up electromagnets in quantity; ~einanderschaltung *f* (*T*) connection (of elements) side by side, joining up of (electromagnets) in quantity; ~einkünfte *pl* fees, occasional perquisites; ~fuhrdienst *m* (*P* [+]) supplementary vehicles; ~fuhrkosten *pl* (*P* [+]) expenses for supplementary vehicles; ~gebühr *f* für die vom Landbriefträger eingesammelten Postsendungen collection fee; ~geleise *n* (Eisenb.) side track, shunt, siding; ~geräusch *n* (*T*) interrupting noise; ~kariol *n* supplementary cariole; ~kosten *pl* bei Beförderung der Posten od. für Beiboten (*P* [+]) expenses for supplementary vehicles or auxiliary messengers; ~leitung, ~linie *f* (*T*) secondary line; ~linie *f* (Eisenb.) branch line, siding line, secondary line; ~materialien *n/pl* (*T* Bau) small stores; ~schließung *f*, ~schluß *m* (*T*) leakage, eine Leitung mit ~schluß a leaky line, a line subject to (heavy) escapes, ~schließung in Folge von Drahtverschlingung (*T*) cross leakage, künstliche ~schließung, Shunt *m* shunt, eine künstliche ~schließung anlegen to shunt; ~schlußlampe *f* shunt lamp; ~spirale, Spirale mit induzirtem Strom *f* secondary coil; ~straße *f* by-street, by-road, branch road; ~verbindungen *f/pl* (*PT*) cross country connections; ~vergütung *f* für Telegraphen-Geschäfte extra payment (pecuniary allowance) for telegraph business; ~zollamt n branch office of a custom house.

negativ a negative; ~e Elektrizität f negative (resinous) electricity; ~e Elektrode f negative electrode, copper electrode; ~er Pol m negative pole, terminal zinc plate (of a galvanic battery).

nehmen v (ein Billet) to take (a ticket); (ein Telegramm) to receive; (eine Ltg. auf Morse) to join up a Morse printer in circuit.

neigen, sich ~ v (Techn.) to incline, to stoop.

Neigung f (Topogr.) inclination, incline, slope, declivity, descent; (Gefälle bei Eisenb. 2c.) ascent, gradient, fall; (der Magnetnadel) dip of the needle, inclination; ~sanzeiger m (Eisenb.) gradient post, indicator of gradient; ~sfläche f gradient; ~swinkel m angle of inclination, (Böschungswinkel [Eisenb.]) angle of slope, gradient of a sloped wall.

Nenner m eines Bruches denominator of a fraction.

netto a net, netto, neat, clear.

Netto ... ~betrag m net amount, clear amount; ~bilanz f, reiner Saldo m net balance; ~ertrag, Reinertrag m net proceeds, net produce; ~gewicht n neat weight, net weight, suttle weight; ~preis m real price, fixed price, short price.

Netz n net, network (of telegraph lines); ~punkt, Richtpunkt m station; ~riegel, Schußriegel m pullock, putlog.

Neu ... ~anlage f (T) new line, establishment of a new line; ~baufonds m (T) funds for the erection of a new line; ~bauten m/pl new buildings, construction (erection) of new buildings; ~silber n German silver, nickel silver, white brass, white metal.

neutral a neutral; ~e Salze n/pl neutral salts, middle salts.

neutralisiren v to neutralize, to render neutral.

nicht ... ~amtlich a non-official; ~ angenommen (von Briefen 2c.) refused; ~ankunft f non-arrival; ~annahme f non-acceptance, refusal; ~bestellung f non-delivery; ~leiter m (der Elektriz.) non-conductor, dielectric body, insulator; ~übereinstimmung f (der HA) want of synchronism, (the instruments are) not working synchronously, not keeping time; ~vorhandensein n nonexistence; ~zulassung f non-admission.

nichtig a null, void; null und ~ s. null.

Richtigkeit f nullity.

Nickel n u. m nickel; ~überzug m, Plattirung mit ~, Vernickelung f nickel plating.

Nieder ... ~druck m (Dampfm.) low pressure; ~drücken v (T) to depress (the key); ~druckcylinder m low-pressure cylinder; ~legen v (vermahrlich) to deposit, (ein Amt) to resign; ~schlag, Absatz, Bodensatz m (Chem.) deposit, sediment, (niedergeschlagenes Präzipitat n) precipitation, precipitate, (elektrogalvanischer Niederschlag [Phys.]) voltatype, electrotype, electro-galvanic deposit; ~schlagen, sich ~schlagen v (Chem.) to precipitate; eine Gebühr ~schlagen (PT) to remit an overcharge; eine Strafe ~schlagen to remit a penalty.

Niederung f low country, low ground.

niedrig a low; ~ im Preise stehen to bear a low price.

Niet n, **Niete** f, **Nietnagel** m rivet; ~ mit halbrundem Kopf cup rivet; versenktes ~ countersunk rivet, flush rivet; ~blech, Schraubenmutterblech n rivet plate, bur; ~bolzen m riveted bolt, clinched bolt; ~eisen n rivet iron, riveting stock; ~hammer m rivet hammer; ~keil, ~stift m rivet pin; ~kluppe f rivet tongs; ~nagel m riveting nail; ~ u. nagelfest a clinched and riveted; ~ und nagelfeste Gegenstände m/pl fixtures.

nieten, vernieten v to rivet, to clinch.

Niveau n, horizontale Ebene f level; auf gleichem ~ mit etw. sein to be on the level with smt; ~fläche, ~höhe f level; ~übergang m

(Eisenb.) level crossing, (rechtwinkeliger) level crossing in right angle, (schiefwinkeliger, schräger) level crossing on the skew.

niveauständige Stellung *f* der Isolatoren the insulators are fixed (fitted) in the same horizontal plane on opposite sides of the pole.

Nivellement *n*, **Nivellirung** *f* levelling; auf's ~ nehmen to take the level.

Nivellir ... ~instrument *n* levelling instrument; ~kreuz *n* boning rod, boning stick; ~maaßstab *m* levelling rule; ~waage *f* air level, spirit level.

nivelliren *v* to level; (abnivelliren, abwägen) to take the level; (einnivelliren) to bring to the level.

Nivellirung *f* levelling.

Nocken *f/pl* (der Wickellöthstelle) extremities of the two wires joined by a Britannia joint.

Nonius, Vernier, Zehntelzeiger *m* nonius, vernier (f. Kaliber).

Nord ... ~licht *n* north light, polar light, aurora borealis; ~lichtstrom *m* polar light current; ~magnetismus *m* north magnetism; ~pol *m* north pole, arctic pole.

Norm *f* norm, rule, model, pattern.

normal *a* normal, standard; die Leitungen wieder ~ schalten (*T*) to restore the ordinary arrangement of the wires.

Normal ... ~bespannung *f* (*P*) normal team; ~draht *m* standard wire; ~gewicht *n* standard weight; ~kerze *f* (französ.) carcel lamp, (engl.) London spermaceti candle, (deutsche Vereins-~kerze) paraffin candle; ~profil f. Profil; ~uhr *f* standard clock.

Nota, Note, Rechnung *f* note, account, memorandum.

Notat *n* des Rechnungshofes note, memorandum.

Noth ... ~adresse f. Adresse; ~brücke *f* temporary bridge; ~signal *n* (Eisenb.) danger signal, danger whistle, (Schifff.) signal of distress; ~stange *f* (*T*) temporary pole.

Notizbuch *n* memorandum book, diary; (*T Bau*) field book.

Null *f* nought, zero; ~ und nichtig *a* null and void; ~punkt, Gefrierpunkt *m* zero mark, freezing point; ~strich *m* zero mark.

numeriren *v* to number, to mark with a number.

numerisch *a* numeral, numerical.

Numerirung *f* numbering; ~smaschine *f* numbering machine.

Nummer *f* number; ~folge *f* numerical order; ~pfahl *m* (Eisenb.) number peg; ~schild *n* der Postwagen plate (indicating the number of the mail coach); ~stein *m* number stone, kilometer stone; ~stempel *m* numbering stamp, (der Briefträger [+]) stamp of the rural postman showing the number of his walk; ~zettel *m* (auf eingeschriebenen Briefen) gummed label (containing the letter „R").

Nuß *f*, **Nußgewinde** *n* (Techn., Mech.) nut.

Nuth *f* groove, furrow, channel; mit einer ~ versehen *v* to groove.

Nutz ... ~barmachung *f* utilization; ~effekt *m*, ~arbeit, ~leistung *f* einer Maschine useful effect, effective power; ~holz, Bauholz *n* timber; ~kraft *f* effective power; ~loser Effekt *m* lost effect.

Nutzung *f* using, usufruct, produce, revenue; ~swerth *m* efficiency.

O.

Oberansicht *f* plan view.

Oberaufsicht *f* (*TP*) superintendence; ~sbeamter *m* superintendent on duty.

Oberbau *m* building above ground, superstructure; (Eisenb.) permanent way.

Oberbriefträger *m* chief letter carrier.

Oberfläche *f* (Geom., Techn.) surface, superficies; x Quadratfuß ~ x superficial feet; ~nleitung *f* (Elektriz.) surface conduction.

oberflächlich *a* superficial.

Obergeschoß *n* upper story, over-story.

Obergestell *n* (eines Wagens) part of a carriage above the fore- & hind-carriage.

Ober=Ingenieur *m* head engineer, chief (first) engineer.

Ober=Inspektor *m* (in England) Surveyor, (*Am*) Chief (Post Office) Inspector, f. auch Post=Inspektor.

oberirdisch *a* (*T*) aerial; ~e Leitung, Luftleitung *f* aerial line, overland line, overhead line, pole supported wire, terrestrial line, air line.

Oberkante *f* (der Eisenb.=Schiene) edge.

Oberlicht *n* sky light, full sky light; ~fenster *n* sky light window; ~saal *m* hall with sky light.

Ober=Post ... f. Post.

Ober = Rechnungshof f. Rechnungshof.

Ober=Regierungsrath f. Regierungsrath.

Oberriegel *m* (Techn.) upper brace.

Oberschaffner *m* (Eisenb.) upper guard.

Oblate *f* wafer.

Obligation *f*, **Obligationsschein** *m* bond, debenture.

Observatorium *n*, **Sternwarte** *f* observatory.

Ofen *m* oven, stove; (Brennofen, Darrofen) kiln; (Chem.) furnace; (Hüttenofen) furnace; (eines Dampfkessels) furnace of a steam boiler; ~herd *m* hearth of a stove (furnace).

offen *a* open (the office is open at night), every post office must be open for general business from 7 A. M. to 10 A. M.; ~es Geschäft *n* retail business; ~e Rechnung *f* open account, running account; ~er Wechsel *m* letter of credit.

öffentlich *a* public, open; ~es Amt *n* public office; ~ bekannt machen *v* to post, to advertize in a public place.

Öffentlichkeit *f* publicity, openness.

öffnen *v* to open.

Öffnung *f* (Techn.) opening, aperture, mouth: (am Briefkasten) aperture, letter slit; ~s=Induktionsstrom *m* induced current on breaking contact (flowing in the same direction as the battery current).

Ohm *n*, **Widerstandseinheit** *f* Ohm, unit of electrical resistance; das ~sche Gesetz *n* Ohm's law; (Maaß) hogshead.

Ohr *n* (Techn.) ear, handle; (eines Beils) eye of a hatchet; (eines Schlüssels) handle of a key; ~bolzen *m* bolt with a ring & a hook.

Öhr *n* ear, handle, shackle, eye, catch (for a hook); (Henkel) handle; (eines Bolzens) eye of a bolt; (einer Nadel) eye of a needle.

Oktav, Oktavformat *n* octavo; breites ~ crown-octavo; ~band *n* octavo-volume, book in octavo.

Oktroi *n*, **Verzehrsteuer** *f* excise; ~erheber excise-man.

Öl *n* oil; mit ~ schmieren to oil, to lubricate, to grease; ~anstrich *m* oil paint; ~farbe *f* oil color, oil paint; ~kännchen *n* oil cup, oil feeder; ~lampe *f* oil lamp.

Omnibusleitung, Omnibuslinie *f* (*T* [+]) omnibus line, telegraph line of minor importance.

Optik, Lichtlehre *f* (Phys.) optics.

Optiker *m* optician.

optisch *a* optic, optical; ~e Achse *f* optic axis; ~e Instrumente *n/pl* optic instruments; ~er Telegraph, Flügeltelegraph, Semaphor *m* optic telegraph, semaphore.

Ordinate *f* (Geom.) ordinate, offset.

Ordnung *f* order; betriebsfähige ~ (Techn.) working order; ~, Grad *m* einer Kurve (Geom.) order (degree) of a curve; Alles in ~! (*T*) cleared out & all right.

Ordnungsstrafe *f* penalty, fine; ~ für Zuspätkommen in den Dienst (*TP*) fine for late attendance; einen Beamten in eine ~ nehmen to subject the officer to a fine, to inflict a penalty upon an official.

organisch *a* organic; ~e Chemie *f* organic chemistry; ~e Säuren *f/pl* organic acids.

Orts ... ~batterie = Lokalbatterie

f. Batterie; ~behörde *f* local authority; ~beschreibung *f* topography; ~bestellbezirk *m* town postal delivery; ~briefträger *m* town letter-carrier; ~polizei *f* local police authorities; ~postanstalt *f* town post office; ~sendung *f* local letter, (*Am*) drop letter; ~üblich *a* usual in a place, customary; ~veränderung *f* locomotion; ~verhältniß *n*, je nach den besonderen ~en according to the special local circumstances.

Ortschaftsverzeichniß *n* list of places.

Öse *f* (Techn.) ear, hook, lug, eye; (eines Bolzens) eye of a bolt; (Schleife eines Taues) loop, bend, eye of a rope; Haken u. ~n *pl* hooks and eyes; ~, Ring *m* von Metall (Techn.) link, loop, hoop.

Oszillation, Schwingung *f* (Phys.) oscillation, vibration.

oszilliren *v* to oscillate, to vibrate; ~der Zylinder *m* oscillating cylinder; ~e Welle *f* (Masch.) rock shaft, rocker shaft, rocking shaft.

Oxyd *n* oxide; ~haltig *a* containing oxides, oxidiferous, oxidic; ~hydrat *n* hydrated oxide.

Oxydation, Oxydirung *f* oxidation (the act of combining with oxygen).

oxydationsfähig *a* oxidable.

Oxydationsstufen *f/pl* degrees of oxidation.

Oxydirbarkeit *f* oxidability.

oxydiren *v* to oxidize, to oxygenate, to oxygenize.

Oxydul, Protoxyd *n* protoxide (lowest degree of oxidation, oxide with one atom of oxygen).

Oxygen *n*, **Sauerstoff** *m* oxygen.

Ozon *n*, **aktiver Sauerstoff** *m* ozone, active oxygen; ~messer *m* ozonometer.

P.

paarweise *adv* in couples.

Pack *m* (Techn.) pack, bag, bale; ~ Schriften bundle (file) of papers.

Pack ... ~haus *n*, ~hof *m* warehouse, bonding warehouse, custom house; ~kammer *f* (P) parcel room; ~kammerbuch *n* (+) daily balance of parcels (containing a specified statement of all parcels received, delivered & forwarded during the day); ~kiste *f* packing case; ~korb *m* (P) basket for parcels; ~lack *m* pack wax, packing wax, bag sealing wax, sealing wax used in packing; ~leinwand *f* pack cloth, packing cloth; ~materialien *n/pl* packing materials; ~meister *m* surveyor of packers; Post~meister *m* mail guard (having received that title after long faithful service); ~papier *n* wrapping paper; ~raum *m* packing room; ~tisch *m* packing table; ~wagen *m* (Güterwagen) baggage waggon, van, (Passagiergepäckwagen) luggage van, passenger's luggage van.

Packer *m* packer; ~lohn *m* packing, package, casing.

Päckerei ... ~beiwagen *m* (P) supplementary parcel vehicle; ~dienst *m*, ~geschäft *n* (P) transmission of parcels; ~verkehr *m* (P) exchange of parcels.

Packet *n* (Postpacket) parcel; (Briefpacket) packet, bundle; ~abgangsverzeichniß *n* (P) list of parcels to be forwarded (transmitted); ~abgangszettel *m* list of parcels; ~adresse *f* (+) despatch note; ~annahmebuch *n* (P) counter list of parcels; ~annahmeschalter *m* counter for the transaction of parcels post business; ~aufgabenummer *f*, ~aufgabenummerzettel *m*, ~aufgabezettel *m* label (indicating the registered number & the office of origin); ~beiwagen *m* (Güterwagen) supplementary luggage van; ~beförderung *f* transmission of parcels; ~besteller *m* (Postschaffner im Packetbestelldienst) mail guard employed on parcel delivery duty; ~bestellfahrt *f* parcel delivery trip; ~bestellkarte *f* list of parcels turned over to the mail guard for delivery; ~bestellung *f* delivery of parcels; ~bestellwagen *m* van; ~boot *n* packet boat, mail boat; ~eingangsverzeichniß *n* (+) list of parcels arrived (to be prepared for each mail); ~eingangszettel *m* (+) list of parcels; ~handkarren, ~karren *m* hand cart; ~karte *f* parcels bill; ~korb *m* basket for parcels; ~lagerbuch *n* (+) list of parcels received at a P. O. for delivery; ~leitzettel f. Leitzettel; ~post *f* parcels post, parcel post; ~posttarif *m* parcel post tariff, parcels post regulations; ~posttaxe *f* rate of postage for parcels; ~sendung *f* parcel; ~träger *m* porter; ~trägerstelle *f* situation of porter; ~trägertasche *f* pouch, porter's pouch; ~verkehr *m* exchange of parcels; ~waage *f* scales.

paginiren *v* to mark the pages with numbers, to page.

Papier ... ~bogen *m* sheet of paper; ~brücke *f* (*HA*) paper guide; ~führung *f* (*HA*) guide pulley; ~führungshebel *m* (*HA*) paper moving lever; ~führungswalze *f* (*H & MA*) paper roller; ~geld *n* paper money, paper currency; ~rolle *f* (*H & MA*) paper roll, roller; ~streifen *m* (für die Telegr.-Apparate) slip, paper tape, paper ribbon; ~träger *m* (*H & MA*) paper carrier, reel.

Pappdeckel *m* pasteboard.

Pappelement *n* Siemens Halske's element (the porous partition is substituted by specially prepared paper pulp).

Parallaxe *f* parallax f. auch Fehler.

parallel schalten, zwei Leitungen ~ ~ *v* (*T*) to loop two wires.

Parallelschaltung *f* der Elemente einer Batterie (*T*) joining up a battery in series; ~ zweier theilweise gestörter Leitungen zu einer Leitung (*T*) looping two wires.

Passagier *m* passenger; blinder ~ stray passenger taken up on the road in violation of the mail driver's instructions; ~büreau *n* (*P*) mail coach office, (Eisenb.) booking office; ~gut *n* passengers' luggage, (herrenlos gebliebenes) unclaimed luggage; ~stube *f* passengers' room, waiting room.

Patent, Erfindungspatent *n* patent, letters patent; ein ~ anmelden to apply for a patent; ein ~ ertheilen to grant a patent; ein ~ nehmen to take out a patent; ~anwalt *m* patent agent; ~inhaber *m* patentee; ~register *n* record of patents.

Paus ... **Pause** ... ~papier *n* calking paper, tracing paper; ~zeichnung *f* calking, counterdrawing.

Pause *f* (Zeichnung) tracing, calking, pouncing; ~, Karton *m* pricked drawing.

pausen *v* to trace, to calk, to pounce, to counterdraw.

Pedal *n* (*HA*) pedal.

Peil ... ~loth *n* lead, plummet; ~stange *f* gauge rod, sounding rod.

peilen *v* to sound.

Peilung *f* sounding, bearing.

Pendel *n* pendulum; ~isolator *m* (*T*) insulator suspended by an iron rod from the eye of an iron carrier screwed into a tree; ~kugel *f* (*HA*) pendulum ball; ~schwingung *f* vibration, oscillation of the pendulum; ~stange *f* (*HA*) vibrating spring, pendulum rod.

Pensionirung f. Versetzung in den Ruhestand.

pensionsfähig *a*, ~es Diensteinkommen *n* salary according to which the pension is calculated; ~e Zulage *f* allowance included in the calculation of the pension.

Peripherie *f* circumference.

Perron *m* (Eisenb.) platform of a station, passenger's platform, landing; ~karren, ~wagen *m* (*P*) hand cart (for the transportation of parcels).

Personal *n* staff, ~akten *pl* personal papers; ~büreau *n* Office for the staff, (in England) appointment branch.

Personen ... ~beförderung *f* (*P*) conveyance of passengers; ~beiwagen *m* (*P*) supplementary passenger vehicle; ~fuhrwerk *n* (*P*) passenger

7*

vehicle, vehicle for the conveyance of passengers; ~geld n fare charged to passengers; ~post f mail coach; ~postkurs m line for the conveyance of passengers by mail coach; ~tarif m tariff for the conveyance of passengers; ~verkehr m passenger traffic; ~wagen m (P) passenger vehicle, (Eisenb.) passenger carriage, (Am) passenger car; ~zettel m (+) bill showing the number of passengers by mail coach & the fare charged.

Petition f petition.

Pfahl m pale, post; ~ramme f pile rammer, pile driver.

Pfanne f pan, kettle, boiler; (Lagerpfanne, Lagerfutter) pillow, bush, brass; (Halslager) collar; (Zapfen-, Wellen-, Achslager) bearing, journal bearing, carriage; (einer horizontalen Welle) plumber block, pillow block, carriage.

Pfeil m arrow; ~höhe f (Durchhang des Drahtes [T]) dip.

Pfeiler m pillar, post; (Brückenpfeiler) pier, buttress.

Pfennig m pfennig.

Pferde ... ~bahn f railway worked by horses, horse tramway, tramway, (Am) cars, horse cars; ~gelder n/pl bei Extraposten (+) amount charged to passengers by extra mail coach for the horses; ~kraft, ~stärke f horse power ([HP] = 75 Kilogrammeter = 550 Engl. foot pounds).

Pflaster n pavement; das ~ aufreißen to unpave a street; ~stein m paving stone.

pflastern v to pave, to make a pavement; mit Fliesen ~ to flag.

Pflock m wooden nail, tree nail, pin, peg; (Absteckpfahl) picket.

Pfosten m post; (Ständer) wooden pillar, standard, upright; (Thür-, Fenster-~) jamb, side post; (aus Gitterwerk) trellis post.

Phänomen n, **Naturerscheinung** f phenomenon.

phonisches Rad n (T) phonic wheel.

Phonograph m phonograph, phonautograph.

Phosphor m phosphorus; ~bronze f phosphor bronze; ~bronzedraht m phosphor bronze wire; ~eisen n phosphide of iron, phosphorized iron; ~entfernung f aus dem Eisen dephosphorizing of pig iron; ~haltig a phosphoric, phosphatic.

Photometer, Lichtmesser m photometer.

Physik f physics, natural philosophy.

physikalisch a physical.

Physiker m physicist.

Picke f, **Pickel** m pick, pick axe.

Pinne f the feather-shaped edge of several objects; (Stift) pin, peg, tack.

Pinsel m pencil, brush, paint brush.

Plakat n placard.

Plan m plan, ground plan; (farbiger) colored plan; ~karte f plane chart; ~spiegel m plane mirror; ~zeichnen n plan drawing.

plan, eben a plain.

Plane f (Leinwanddecke für Wagen) tilt.

planiren, einebenen v to level, to even, to lay flat.

Planke f (starkes Brett) plank, thick board.

Planum n (Eisenb.) surface of the formation, formation level; (einer Straße) form, bed, soil of a pavement.

Platin n platinum; ~schwamm m spongy platinum, platinum sponge; ~spitze f platinum point; ~-Zink-Element n Grove's element.

platiniren v to cover with platinum, to platinize, to platinate.

platt a flat, plain, even; (abgeglichen mit der Oberfläche) flush; ein ~es Dach a flat roof.

Platte f plate; ~ am Briefkasten f. Stundenplatte; gereifelte ~ grooved plate; Guttapercha-, Kautschuk-~ sheet of guttapercha (caoutchouc); Metall~ metal plate; Kupfer~ copper plate; Tisch~ slab (leaf) of a table, table board.

Plattenblitzableiter f. Blitzableiter.

Plenarsitzung *f* plenary sitting.
Plinthe f. Sockel.
Plombe *f* lead.
Plusbetrag f. Mehrbetrag.
Pneumatik *f* pneumatics.
pneumatisch *a* pneumatic; ~er Aufzug *m* pneumatic lift; ~e Klingel *f* pneumatic bell; ~e Post *f* pneumatic post.

Pol *m* pole; gleiche ~e *pl* like poles; gleichnamige (feindliche) ~e poles of the same name, similar poles; magnetischer, negativer, positiver ~ magnetic, negative, positive pole; ungleiche ~e unlike poles; ungleichnamige (freundschaftliche) ~e poles of contrary names; Einheits-~ unit pole; Nord-~ marked pole, north seeking pole, north pole; Streich-~ touch pole; Süd-~ unmarked pole, south seeking pole, south pole; der Zink-~ liegt an Leitung the battery is connected up with its zinc pole directed to line.

Pol ... ~fläche *f* polar surface; ~klemme *f* connection binder; ~platte *f* polar plate; ~schuhe *m/pl* pole pieces; ~spannung *f* polar tension; ~wechsel *m* change of poles.

polar *a* polar.
Polar ... ~licht, Nordlicht *n* polar light; nördlicher, südlicher ~kreis *m* arctic, (antarctic) circle.
Polarisation *f* polarization.
polarisiren *v* to polarize.
Polarität *f* polarity.
Police *f* policy of insurance; Inhaber einer ~ policy holder.
Pore *f* pore.
porös *a* porous; ~es Batteriegefäß *n* porous cell.
Porosität *f* porosity.
Porto *n* postage; Hauptbuch über gestundetes ~ f. Hauptbuch; Konto über gestundetes ~ f. Konto; ~abgangsbuch *n* (+) statement recording unpaid postage entered in the despatched letter bills; ~ankunftsbuch *n* (+) account of unpaid postage entered in the received letter bills; ~ansatz *m* charging of postage; ~antheil *m* share of postage; ~einnahme *f* revenue from postage; ~ermäßigung *f* reduction of postage; ~frei *a* exempt from postage, free of charge, prepaid, postpaid; ~freiheit *f* exemption from postage, franking privilege; ~freiheitsbezeichnung *f* certificate that correspondence is exempt from postage; ~gefälle *pl* revenues from postage; ~hinterziehung *f* embezzlement of postage; ~kosten *pl* amount of postage; ~marke *f* (Taxmarke) postage due stamp; ~nachschuß *m* supplementary postage; ~pflichtig *a* liable to postage, ~pflichtige Dienstsache *f* official correspondence liable to postage; ~satz *m* rate of postage; ~stundung *f* crediting the postage due; ~stundungsbuch *n* register of the postage due; ~stundungsgebühr *f* rate to be charged for crediting the postage due; ~taxe *f* table of (foreign & colonial) postage; ~theilung *f* division of postage; ~übertretung *f* defraudation (embezzlement) of postage; ~vergünstigung *f* (für Soldaten) privilege granted to private soldiers (seamen) & non-commissioned officers to receive their correspondence free of postage or at a reduced rate, (für Behörden) privilege granted (by law) to government authorities to designate their correspondence as official but liable to postage, whereupon the addressees pay only a single rate of postage; ~zuschlag *m* additional charge; ~zuschreibung *n* (+) hand-to-hand receipt for the amount of postage due on unpaid correspondence.

Porzellan *n* porcelain, china, china ware; ~-Doppelglocke *f* double bell of porcelain; ~-Isolator *m* porcelain insulator.

positiv *a* positive; ~e Elektricität *f* positive (vitreous) electricity.

Possekel, Poseckel, Poßhammer *m* large hammer, sledge hammer.

Post *f* post; (Postamt) post office (P. O.); fahrendes ~amt travelling post office (T. P. O.); railway post office (R. P. O.); ~agent *m* postal agent; ~agentur *f* postal agency; ~annahmestempel *m* date stamp, dated stamp; ~anstalt *f* post office (P. O.), (nur zeitweilig geöffnete) summer post & telegraph office (on mountain peaks & other remote & solitary spots visited by tourists); Abrechnungs-~anstalt head post office; Absatz-~anstalt (+) P. O. of sale; Absendungs-, Aufgabe-~anstalt office of origin; Auswechselungs-~anstalt exchange office (of two neighboring countries); Bestimmungs-~anstalt P. O. of destination; Grenz-~anstalt exchange office, office of exchange; Kartenschluß-~anstalt exchange office; Kurs-~anstalt P. O. situated on the mail route; Leit-~anstalt P. O. upon which the mail is to be directed; Orts-~anstalt local post office; Ueberweisungs-~anstalt head post office; Umleitungs-~anstalt re-transmitting office, forward office, s. auch a. T. forward; Verlags-~anstalt (+) office of publication; eine ~anstalt aufheben to discontinue a P. O.; eine ~anstalt einrichten to establish a P. O.; eine ~anstalt schließen to close a P. O.; eine ~anstalt umwandeln to change a P. O.; ~anwärter m candidate; ~anweisung f money order, (in England für Beträge von 1—20 sh) postal order; Karten-~anweisung f card money order; verfallene ~anweisung void (lapsed) money order; ~anweisungsamt n money order office; gestempeltes ~anweisungsformular n stamped money order form; ~anweisungsgebühr f money order rate, commission; ~anweisungsschalter m (+) counter for the transaction of money order business; ~anweisungsverfahren n money order business; ~anweisungsverkehr m exchange of money orders; ~armenkasse f postal relief fund; ~assistent m (+) postal assistant, festangestellter ~assistent m (+) definitively appointed postal assistant; ~auftrag m zu Bücherpostsendungen (+) order for the collection of bills accompanying articles of the book post; ~auftrag zur Geldeinziehung (+) postal collection order, postal order for collecting money on bills through the agency of the P. O.; ~auftrag zur Einholung von Wechselakzepten (+) postal order for the collection of bills of exchange; ~auftragsbrief m (+) letter containing an order for collecting money; ~auftragsformular n collection order form; ~auftrags-Postanweisung f money order by which the amount collected is transmitted; ~ausgabebuch n für Soldatensendungen (+) delivery book of mail articles addressed to private soldiers (seamen) & non-commissioned officers; ~ausgabestempel m stamp of distribution; ~baurath m special architect of the P. O. department; ~beamter m postal officer, postal clerk, employee of the P. O., (Beamter im Fahrdienst) railway mail clerk; ~beförderung f auf Landwegen conveyance of mails by road; ~beförderungsdienst m mail service, conveyance of the mails; ~begleitdienst m travelling mail service; ~begleiter m mail guard; ~bericht m (Aushang) regulation notice; ~betrieb m auf Eisenbahnen railway mail service; ~betriebsdienst m technical postal service; ~bezirk m postal district; ~bon m postal order; ~bote m postman, messenger, letter carrier; ~botentasche f letter carrier's pouch; ~briefkasten m letter box, s. auch Briefkasten; ~buch n official postal guide, P. O. Handbook; ~dampfschiff n mail packet; ~debit n sale by the post; ~diebstahl m mail depredation; ~dienst m postal service, mail service, post office duty; ~diensträume m/pl rooms for the transaction of P. O. business; ~dienststunden f/pl hours of business (of attendance); ~direktor m (+) postal director; Ober-~direktion f (+) chief (principal) postal direction, provincial direction of posts; Ober-~direktionssekretär m (+) clerk of the chief postal direction; Ober-~direktor m (+) chief (principal) director of posts; ~druckformular n printed form; ~einlieferungsbuch n (+) book containing the receipts for mail articles handed in at the office window; ~eleve *m* (+) apprentice; ~felleisen *n* pouch; ~flagge *f* postal flag; ~freiheit *f* franking privilege; ~freimarke f. ~werthzeichen; ~fuhrbetrieb *m* horse post service; ~fuhrordnung *f* horse post regulations; ~fuhrstation *f* horse post station; ~fuhrunternehmer *m* mail coach service contractor, horse post contractor, contractor for the carriage of mails by passenger conveyance; ~fuhrvergütung *f* compensation (price of conveyance) allowed to the mail contractor for services rendered; ~fuhrvertrag *m* contract for the conveyance of mails by road; ~fuhrwesen *n* horse post service; ~fußbote *m* foot messenger, ständiger (nicht ständiger) ~fußbote *m* foot messenger, whose whole time is (not) given to the service; ~gebäude *n* post office building, reichseigenes ~gebäude Government P. O. building; Mieths-~gebäude rented P. O. building; ~gebühr *f* postal charge; ~gebühren, ~gefälle *pl* revenues of the P. O.; ~gehülfe *m* (+) auxiliary, applicant for the situation of postal assistant; ~gesetz *n* P. O. (protection) bill, postal law; ~gewicht *n* weight found by the receiving P. O.; ~halter *m* (+) mail contractor, station master; ~halterei *f* (+) horse post station; ~halterei Ausstattungsgegenstände *m/pl* (+) equipment of the horse post station; ~haltereigebäude *n* (+) station house; ~halterei-Revision *f* (+) (ordinary, extraordinary) revision of the horse post service; ~haltereiwagen u. Schlitten *m/pl* (+) coaches & sledges to be kept & maintained by the mail contractor; ~haus *n* postal building, f. auch ~gebäude; ~hausdiener *m* porter; ~hausneubau *m* erection of a new postal building; ~hausschild *n* P. O. sign; ~hauswächter *m* watchman; ~hof *m* yard (court yard) of the P. O. building; ~horn *n* posthorn, postal bugle; Ehren-~horn honorary bugle; ~hülfsbote *m* auxiliary subaltern postal official; ~hülfsstelle *f* auxiliary P. O.; ~inspektor *m* (in Engld.) surveyor, (*Am*) P. O. inspector, (in Deutschland) Government P. O. inspector; ~karte *f* postcard, (Landkarte) post route map, map of the mail routes; ~kasse *f* postal cashier's office; Ober-~kasse *f* (+) chief postal cash office; General-~kasse *f* General postal treasury; ~kassirer *m* cashier, (als Aufsichtsbeamter) (*P* [+]) controlling officer, representative of the postmaster; ~-Kleiderkasse *f* postal clothing fund; ~kongreß *m* postal congress; ~kontravention *f* infringement of the postal laws; ~koupé *n* (im Eisenb.- u. Postwagen) compartment; ~kurs f. Kurs; ~lagernd *a* to be called for, poste restante; ~lagernde Sendung *f* mail matter addressed to a P. O. to be kept till called for, poste-restante article; ~leitheft *n* schedule of arrivals & departures of mails; ~leitvermerk f. Leitvermerk; ~mandat f. ~auftrag; ~marke *f* postage stamp, f. auch ~werthzeichen; ~meister *m* postmaster; General-~meister *m* Postmaster General; ~monopol *n* monopoly, government monopoly; ~nachnahme *f* postal reimbursement; ~nachnahmeverkehr *m* exchange of articles with reimbursement; ~ordnung *f* postal regulations; Manöver-~ordnung postal regulations for the field manoeuvres; ~packet *n* parcel, f. auch Packet; ~packetboot *n* mail packet; ~packetdienst *m* parcel post service; ~packmeister *m* mail guard, f. Pack...; ~papier *n* letter paper, post paper; ~praktikant *m* (+) clerk without definitive appointment; ~rath *m* (+) councilor (counselor*) of posts; Ober-~rath *m* (+) superior councilor of posts; Geheimer ~rath *m* (+) privy councilor of posts; Geheimer Ober-~rath *m* (+) privy superior councilor of posts; Wirklicher Geheimer Ober-~rath *m* actual privy superior councilor of posts; ~regal *n* postal privilege; ~schaffner *m* mail guard, (im Bahnpostdienst) railway

---

*) councilor: a member of a council; counselor: one who gives counsel.

mail guard, (im Begleitungsdienst) guard in charge of the mail, (im Bestellungsdienst) mail guard employed on outdoor duties; (im innern Dienst) mail guard employed on indoor duties; ~schalter *m* office window, counter; ~schalterdienst *m* duty at the counter; ~schein *m* receipt, acknowledgment of receipt, (für Packete) certificate of posting; ~schiff *n* packet, mail packet; ~sekretär *m* (+) postal clerk; Ober-~sekretär *m* (+) chief clerk; ~sendung *f* mail matter, postal article; ~sparkasse *f* P. O. savings bank, s. auch Spar...; ~sparkasseneinlage *f* deposit; ~Spar- u. Vorschuß-Verein *m* Postal Saving & Loan Association; ~statistik *f* postal statistics; ~stempel *m* office stamp; ~sterbekasse *f* fund for the assistance of the families of deceased officers; ~straße *f* post road, post route, high road; ~stunden *f/pl* hours of business; ~tag *m* post day; ~tasche *f* private bag; ~taxgesetz *n* law regulating the rates of postage; ~transport *m* conveyance of mails; ~transport-Ordnung *f* regulations concerning the conveyance of mails; ~übertretung *f* infringement of the postal laws; ~uhr *f* (Kursuhr) watch, (am Posthause) P. O. clock; ~unterbeamter *m* subaltern postal official, (in England) official of the minor establishment, s. a. T. establishment; ~unterstützungskasse *f* postal relief fund; ~verbindungen *f/pl* einrichten to establish connecting services (f. i. between railway stations & post offices); ~verein *m* Postal Union; allgemeiner ~verein, Welt-~verein Universal Postal Union; ~vereinsvertrag *m* Postal Union Convention; ~verkehr *m* postal traffic, (mit dem Auslande) postal relations (with foreign countries); ~vermerk *m* (auf Postanweisungen) official particulars concerning the booking etc. of money orders, space for the official particulars; ~vertrag *m* postal convention; ~verwalter *m* (+) postal manager; ~vorschuß s. ~nachnahme; ~wagen *m* (allgemein) any vehicle used in the P. O. service, (im besonderen) mail coach, stage coach, (Eisenb.) P. O. carriage, mail car, (*Am*) postal car; ~wagenwascher *m* porter employed in cleaning postal vehicles; ~wartezimmer *n* passengers' room, waiting room; ~werthzeichen *n* postage stamp; amtliche ~werthzeichen-Verkaufsstelle *f* official place for the sale of postage stamps, (Inhaber einer solchen) official stamp seller; ~-Zeitungsamt *n* (+) P. O. mediating the subscription to newspapers, Postal Newspaper Office; ~zeugamt *n* office of equipments; ~zug *m* (Eisenb.-Zug, der die Post mitführt) mail train; ~zwang *m* postal privilege.

Posten *m* (Rechnungsposten) item, entry in an account; (Partie) parcel, lot of goods.

Postillon *m* postilion, postillion, mail driver; ~trinkgeld *n* drink-money, gratuity, remembrance (to the driver).

Potential *n*, **Spannung** *f* potential (property of electricity which determines its motion from one point to another).

Potenz *f* (Masch.) power; mechanische ~, einfache Maschine mechanical power, simple machine; zweite, dritte ~ einer Zahl second, third power.

Pottasche *f* potash.

Prahm *m* pram, praam (a flat bottomed boat, a sort of lighter).

pränumerando *adv* to be paid beforehand.

Pränumeration *f* paying beforehand, subscription.

Präparatur s. Zubereitung.

Präzedenzfall *m* precedent.

Prell ... ~pfahl *m* fender; ~stein *m* guard stone, stone stud.

Preßschraube *f* screw, nut of a press, clamp screw, screw clamp, press screw.

Primawechsel *m* first of exchange, prima bill.

primär *a* primary; ~er Draht *m* primary wire, primary inducing wire (in which the inducing current circulates); ~e Spule *f* primary coil; ~er Strom *m* primary current.

Priorität f. Vorrang.

Privat ... ~gesellschaft *f* private company; ~eigenthum *n* private property; ~gehülfe *m* (P) auxiliary clerk engaged & paid by the postmaster; ~personenfuhrwerk *n* vehicle belonging to private persons, private vehicle, (P) stage coach kept by a carrier & used for the conveyance of the mails; ~telegramm *n* private telegram; ~unterbeamte *m* porter engaged & paid by the postmaster.

Probe *f* (Techn.) trial, proof, experiment (the act of testing); sample, specimen, pattern (the substance to be tested); (Waarenprobe) sample, pattern; der auf ~ angenommene Beamte probationary appointee; ~ machen (Arithm.) to prove; nach Probe (Handl.) according to sample.

Probe ... ~dienstzeit, ~zeit *f* probationary term, the period during which a candidate is under trial with a view to test his qualifications for the appointment to which he has been nominated; einem Beamten wird die ~weise Verwaltung einer Dienststelle übertragen an official has to serve a probationary term before he is definitely appointed to a place; ~weise Verwendung *f* probationary employment.

probiren *v* (Techn.) to prove, to test, to try.

Profil *n*, **Durchschnittsansicht** *f* profile, section, side projection; Normal-~ des lichten Raumes (Eisenb.) standard width in the clear.

Prokura *f* procuration.

Prokurist *m* authorized agent, managing clerk.

Proportion *f* proportion.

proportional *a* proportional.

Proportionale (mittlere) *f*, **geometrisches Mittel** *n* mean proportional.

Protest *m* protest; (wegen Nichtannahme) protest for non-acceptance; (wegen Mangels an Zahlung) protest for non-payment; einen ~ notiren to make a note of protest; mit ~ zurückkommen to return dishonored; sofort zum ~ to be protested at once; ~instrument *n* instrument of protest; ~kosten *f/pl* protest charges.

protestiren *v* to protest (a bill).

protestirter Wechsel *m* dishonored bill.

Protokoll *n* protocol, verbal process; ein ~ aufnehmen to protocol, to draw up a protocol, to draw up the minutes of a transaction.

protokolliren *v* f. ein Protokoll aufnehmen.

provisorisch *a* temporary.

Prozent *n* per cent; ~e *pl* percentage.

prüfen *v* (Techn.) to prove, to examine, to test; Draht in der Fabrik ~ und abnehmen to test, examine & receive wire; (eine Beschwerde) to investigate a complaint; (eine Leitung) to test a wire.

Prüfer, Batterieprüfer *m* (T) battery detector.

Prüfung *f* (Techn.) trial, examination; ~ für Anstellung im Staatsdienst civil service examination; eine ~ bestehen to pass an examination; ein Kandidat, der die ~ bestanden hat a successful candidate; sich einer ~ unterziehen to undergo an examination; nach ~ der Akten after examination of the papers on file; mündliche ~ oral examination; schriftliche ~ written examination; ~apparat *m* (T) apparatus for testing purposes, testing apparatus; ~rath *m* board of examiners, examining committee; ~zeugniß *n* certificate of examination.

Puddel ... ~eisen *n* puddled iron; ~ofen *m* puddling furnace; ~prozeß *m* puddling process; ~stahl *m* puddle steel, semi-steel; ~werk *n* puddle iron works, forge.

Pult f. Schreibpult.

Pumpe *f* pump.

Punkt *m* (Geom.) point; (Buchdr.) period, full stop; fester ~ (Mech.) fixed point; todter ~ dead centre;

(des Morse-Alphabets) dot; ~e und Striche (des Morse-Alphabets) m/pl dots & dashes.
Punktationsverhandlung f engagement between two parties (preliminary to the conclusion of a contract but with legally binding power).
Putzwolle f engineer's waste, woolen refuse.
Pyrolusit m (Braunstein) pyrolusite, peroxide of manganese.

Q.

Quadrant, Viertelkreis m quadrant.
Quadranten-Elektrometer n quadrant electrometer.
Quadrat n (Geom., Techn.) square; (zweite Potenz f) second power, square; ~centimeter m square centimeter; ~eisen, vierkantiges Eisen n square bar iron; ~fuß m square foot; ~kilometer m square kilometer, myriare; ~meile f square mile; ~meter n square meter, centiare; ~wurzel f square root; ~zahl f square, square number; ~zoll m square inch.
quadratisch a square, in the square; ~e Gleichungen f/pl quadratic equations, quadratics.
quadriren v to square; (eine Zahl) to square a number, to raise a number to the second power.
Quadruplex-System n (Gegensprechen und Doppelsprechen [T]) quadruplex system.
Qualität f quality; (die beste) best kind.
qualitative Analyse f qualitative analysis.
Quantität f quantity; (der Elektrizität) amount (magnitude) of electricity present.
quantitativ a quantitative.
Quantum n quota, share, portion.
Quarantäne f quarantine; ~ halten to make (to perform) quarantine; die ~ aufheben to release (to discharge) from quarantine; ~hafen m place (port) of quarantine.
Quart n quart (a liquid measure); ~, ~format n quarto; Buch n in ~, ~band m book (volume) in quarto, quarto; ~bogen m a sheet divided into four parts.

Quartal s. Vierteljahr.
Quecksilber n mercury, quicksilver; ~chlorid, ~sublimat n perchloride of mercury; ~-Element n (T) Marié-Davy element, quicksilver element; ~faden m mercury thread'; ~legirung f, Amalgam n alloy of mercury; ~säule f mercurial column (of a barometer tube); ~verbindung f compound of mercury; Knall-~ n fulminating mercury.
quer a diagonal, traverse, transversal; adv across, diagonally.
Quer... ~achse f transverse axis; ~balken m cross beam; ~gasse f cross lane; ~geschnittenes Holz n cross grained wood, wood cut across the grain; ~holz, Hirnholz n s. ~geschnittenes Holz; ~latte f cross lath, cross batten; ~profil, Breitenprofil n cross section, transverse section, traverse section; ~riegel m cross rail; ~schiene f transverse bar, stretcher; ~schnitt, Durchschnitt m cross section, lateral section, sectional elevation, diagram; ~schnitt, Flächeninhalt m sectional area; ~straße f cross road; ~stück n cross piece; ~träger m (an Telegr.-Stangen v. Holz) cross arm, arm, (v. Eisen) bracket; ~verbindung f lateral fastening.
quittiren v (eine Rechnung) to acquit, to receipt an account, to give a receipt for an account; (ein Telegramm) to acknowledge the receipt; (einen Wechsel) to discharge a bill.
Quittung f receipt, acknowledgment of receipt, quittance, acquittance; eine ~ ausstellen to give a receipt; ~sbogen m receipt-sheet; ~sbuch n receipt-book; ~sstempel m, ~smarke f receipt-stamp; ~szeichen n

(T) the „cleared out all right" signal.

Quote, Quota *f* quota, proportional share.

quotiren *v* to quote, to divide proportionally.

R.

Rabatt *m* rebate, abatement, deduction, discount.

Rad *n* wheel; zylindrisches ~, Stirnrad cylindrical wheel, spur wheel, s. auch Stirnrad; gekuppeltes ~ coupled wheel; gezahntes ~, Zahnrad toothed wheel; konisches ~ conical wheel; Kron~ crown wheel, face wheel; phonisches ~ phonic wheel; ~ mit innerer Verzahnung annular wheel.

Rad... ~achse *f* shaft of a wheel, wheel shaft; ~bremse *f* wheel lock; ~felge *f* felloe, felly; ~speiche *f* spoke, arm; ~spur *f* wheel track; ~welle *f* wheel & axle, arbor wheel; ~zahn *m* tooth of a wheel, cog; ~zapfen *m* trunnion, pivot, nut of a wheel.

Radehacke *f* mattock.

Räderwerk *n* (Masch.) wheel work, gearing; (konisches) wheel work with conical gearing, bevel (bevelled) gear.

radial *a* radial.

radiren *v* to rase, to erase.

Radirung *f* (v. Worten, Zahlen u. s. w.) erasure.

Rahmen *m* (Gestell) frame; (eines Fensters) casement, chase; (einer Thür) frame, framing; (Einfassung) frame.

Ramme *f* ram, pile driver, pile driving engine.

Rampe *f* (Eisenb.) ascent, sloping terrace.

Ramsch *m*, im ~ in the lump, by the bulk.

Rand *m* (Techn.) border, edge, rim; (eines Buchs) margin; ~bemerkung *f* (~verfügung) marginal note; ~leiste *f* border.

Rändeleisen *n* milling iron.

rändeln *v* to mill; gerändelter Schraubenkopf milled head of a screw.

Rändelung *f* milling.

Rang, Stand *m* rank, degree, grade; order, class; degree of dignity; ein Beamter im ~e eines Postmeisters a government official holding the rank of postmaster; ~ordnung *f* regulation concerning rank or dignity; ~stufe *f* order, rank.

rangiren *v* to rank; der Postrath rangirt mit den Majors der Armee the postal councilor ranks with the majors of the army.

Rath *m* councilor (a member of a council); counselor (one who gives counsel); Geheimer ~ privy councilor; wirklicher Geheimer ~ (+) actual privy councilor; Regierungs-~ councilor of the government.

Rapport *m* report.

Rasen *m* turf, sward, grass plot; ~eisen *s.* Sumpfeisen; ~narbe *f* sod.

Raspe, Raspel *f* rasp, grater; ~feile *f* rasp file; ~meißel *m* rasp chisel; ~späne *m/pl* rasping chips, rasp chips.

Rate *f* rate, instalment; in ~n bezahlen to pay by instalments; ~n-zahlung *f* payment by instalments.

ratifiziren *v* to ratify.

Ratifizirungs-Urkunde *f* document of ratification.

Ratschbohrer *m* ratchet drill, rock drill.

Raum *m* (Phys.) space, room; (Buchdr.) space, distance between the lines; (Rauminhalt) volume, content; (im Lichten) area, clearance; (luftleerer) vacuum, vacuous space; (zwischen den Zeichen des Morse-Alphabets) space.

Räumlichkeiten *f/pl* (eines Hauses) parts.

Raupe *f* (des Drahts am Abspann-Isolator) twist.

Raute *f* (Rhombus *m*) rhomb, lozenge.

rautenförmig *a* rhomboidal, lozenge shaped.

Rechenschaftsbericht *m* der Postverwaltung Report on the operations of the Post Office Department.

rechnen *v* (Math.) to calculate; (berechnen) to reckon; als ein Wort ~ (*T*) to count as one word.

rechnerisch prüfen *v* eine Rechnung to examine an account as to the correctness of the figures, to check an account, to audit accounts, (= bescheinigen) to verify an account.

Rechnung *f* calculation, reckoning, account; eine ~ aufstellen to cast an account; seine ~ einreichen to bring in one's account; eine ~ schließen to close (to settle) an account; in ~ stehen to have an account with *smb*; in ~ stellen to carry to account; gemeinschaftliche ~ joint account; offene, laufende ~ account current; Monats-, Vierteljahres-, Jahres-~ monthly (quarterly, yearly) account.

Rechnungs... ~ablegung *f* rendering of accounts; ~abschluß *m* close of accounts, balance of accounts, account agreed upon; ~arbeiten *f/pl* accounts; ~auszug *m* abstract, extract, statement of accounts; ~beleg *m* voucher; ~büreau *n* office of accounts, accountant's office; ~Fehl- und Vergütungs-Beträge *m/pl* (*P*) errors discovered in the primary examination of the monthly accounts, Nachweisung der ~-Fehl- und Vergütungs-Beträge (*P*) balance of errors; ~führer *m* accountant, keeper of accounts; ~führung *f*, ~wesen *n* keeping of accounts, bookkeeping, accounts; ~gebühr *f* bei Extraposten (+) extraordinary fee (connected with the extra mail coach service); ~geschäfte *n/pl* accounts, bookkeeping; ~hof *m* audit-office, chamber of accounts, (*Am*) office of the Auditor of the Treasury, Ober-~hof des deutschen Reiches audit-office of the German empire; ~jahr *n* financial year; ~kammer *f* dass. wie ~hof; ~legung *f* rendering of accounts; ~rath *m* (+) titular councilor (counselor); ~revi- for *m* controller (comptroller) of accounts; ~stellung *f* dass. wie ~legung; ~vergütungsbeträge s. ~-Fehl- und Vergütungs-Beträge; ~wesen *n* accounts.

recht *a* right; (richtig) correct; ~er Winkel right angle.

Recht... u. recht ... ~eck *n* rectangle; ~eckig *a* right-angled; ~fertigen *v* to justify; ~mäßig *a* lawful, legal; ~mäßigkeit *f* legality, validity; ~winkelig *a* rectangular; ~zeitig *a* in due time.

rechts *adv*, nach ~ gewickelter Elektromagnet electromagnet wound with the right-handed helix.

Rechts ... ~anspruch *m* title, legitimate claim; ~beistand *m* solicitor, legal assistance; ~grund *m* legal argument, plea, claim; ~kräftig *a* legal, valid; ~mittel *n* legal means; ~streit *m* law-suit, einen ~streit mit J—b anfangen to bring a law-suit (to enter an action) against *smb*; ~verhältnisse *n/pl* der Reichsbeamten rights & duties of the civil service officers of the German Empire; ~widrig *a* contrary to law; ~wohlthat *f* benefit of the law.

recken, anspannen *v* to stretch (the wire).

Reduktion *f* reduction; ~stabelle *f* table of reduction.

reduziren *v* to reduce.

reflektiren *v* (zurückwerfen) to reflect, to throw back.

Reflektor *m* reflector.

Reflexions ... ~ebene *f* plane of reflection; ~galvanometer *n* reflecting galvanometer; ~winkel *m* angle of reflection.

Regal *n* (Hoheitsrecht *n*) royalty (which pertains to a king as his right); (Monopol *n*) government monopoly; (Gestell *n*) stand, shelves *pl*; ~brett *n* shelf; ~papier *n* royal paper.

Regel *f* (Math.) rule; ~bespannung *f* (*P*) normal team; ~mäßige Leistungen des Posthalters s. Leistungen.

Regierung *f* government, administration, executive power; ~srath *m* councilor of the government, f. auch Rath.

Register *n* register, index, table of contents; (eines Ofens) damper, register.

Registrator *m* keeper & registrar of official papers.

Registratur *f* registry.

Registrir...~-Apparat *m* counter, indicator, recorder, registering apparatus; (Thomson's) syphon recorder.

registriren, aufzeichnen *v* to record.

Reglement *n* Regulations.

reglementarische Bestimmung *f* regulation.

Regreß *m* the act of having loss or injury repaired; seinen ~ an J–d nehmen to cause *smb* to repair loss or injury done, to recover (an amount) against *smb*; ~klage *f* law-suit with the view to obtain an indemnification for loss or injury sustained; ~nahme *f* obtaining an indemnification for loss or injury; ~pflicht *f* obligation (enforced by law or administrative measures [proceedings]) to repair loss or injury caused; ~pflichtig *a* bound to repair loss or injury.

Regulator *m* (Masch.) regulator, (HA) governor (for maintaining the synchronism); (Schwungrad *n*) regulator; ~kugel *f* governor ball; ~stange *f* regulator rod.

Regulir...~apparat *m* regulating apparatus, regulator; ~bar *a* adjustable; ~vorrichtung *f* adjusting device.

reguliren *v* (den Durchhang) to regulate (the sag or dip of a wire), to adjust (the strain of the wire); (einen Apparat) to adjust; (den Gang der Posten) s. Gang; sich selbst ~d *a* (Techn.) self-regulating, self-acting.

Reib...~kissen *n* (der Elektrisir-Masch.) rubber, cushion; ~klotz *m* (HA) fly wheel break, friction pad.

reiben *v* (Techn.) to rub, to clean by friction; (abnutzen) to grind.

Reibung *f* friction, rubbing; ~selektrizität *f* frictional electricity; ~srad, Friktionsrad *n* friction wheel; ~swiderstand *m* frictional resistance.

Reich *n*, das deutsche ~ the German Empire; ~sangehörig *a* belonging to the empire, citizen of the empire; ~sbeamte *m* person holding an office of the empire, government agent; ~sbehörde *f* government authority, department of the empire; ~sdienstangelegenheit *f* official business (of the empire); ~sdruckerei *f* imperial printing office; ~sgesetzblatt *n* official publication of the laws; ~skanzler *m* chancellor of the empire; ~spost *f*, ~spostamt *n* Post Office Department of the German Empire; ~sregierung *f* imperial government; ~stag *m* diet of the empire, Reichstag; ~sverfassung *f* constitution of the empire; ~swährung *f* standard of the empire.

reifeln *v* (auskehlen) to channel, to groove.

Reifelung *f* channeling, grooving.

Reihe *f* range; (Aufeinanderfolge) series; (Arithm., Math.) progression; in ~n arbeiten (T) to work in batches, abwechselndes Arbeiten in ~n (T) up & down working.

Reihenfolge *f* der Beförderung (T) (each telegram is to be sent in its) proper turn.

Rein...~einnahme *f* net revenue; ~ertrag *m* clear (net) profit, net proceeds; ~gewicht *n* net weight; ~gewinn *m* net profit; ~schrift *f* fair copy.

reinigen *v* to clean, to purify, to scour.

Reinigung *f* cleaning.

Reise...~bedürfnisse, kleine ~bedürfnisse *n/pl*, Handgepäck *n* (P) small luggage; ~gepäck *n* passengers' luggage, travelling baggage; ~kosten u. Tagegelder *pl* allowance for removal & sojourn; ~kostenvergütung *f* allowance for travelling expenses; ~stipendium *n* allowance for official trips; ~tag *m* day passed in travelling.

Reitpost *f* mounted messenger line.

Reklamation f complaint; eine ~ erheben to make a complaint.

Rekognoszirung f. Auskundung.

Relais n (T) relay; ankerloses ~ relay without an armature; Induktions- (polarisirtes, Siemens'sches) ~ polarized relay; liegendes, stehendes ~ relay with upright (with horizontally disposed) electromagnet.

rekommandiren s. einschreiben.

Rekurs m appeal (removal of a cause from an inferior to a superior judge (authority) for examination or review); ~gesuch n appeal; ~instanz f superior (judge, authority); ~verfahren n proceeding by which the removal from an inferior to a superior judge (court, authority) is effected.

relative Elastizität f elasticity of flexure; ~ Festigkeit, Biegungsfestigkeit, Bruchfestigkeit f strength of flexure, transverse strength, resistance to breaking strain.

Reliefschreiber m embosser, embossing register.

remanenter Magnetismus m residual magnetism.

Rendant m der Ober-Postkasse (+) cashier of the chief postal cash office.

rentabel a profitable.

Rentabilität f profit; ~sberechnung f calculation of the profits.

Rente f rent, annuity.

Rentenanstalt, Rentenbank f annuity office, insurance company for granting annuities.

rentiren v to be profitable, to yield, to pay.

Reparatur f reparation, repair; ~kosten f/pl cost for repair, expenses of maintenance; ~werkstätte f repairing shop.

repariren v to repair.

Repartition f repartition.

Requisiten-Wagen m (MT) Telegraph wagon.

Reserve ... ~fonds m reserve fund; ~leitung f (T) spare wire; ~stück n reserve piece.

Reservoir n reservoir, cistern, tank.

Residuum n residuum, residue; elektrisches ~ residual electricity; magnetisches ~ remnant of magnetism, residual magnetism.

Resonanz f resonance; ~boden m resounding board; ~kasten m resounding box.

Ressort n jurisdiction, province, department; das gehört nicht zu meinem ~ this comes not within my jurisdiction.

ressortiren v to fall within (to belong to) the jurisdiction, to be within the competence.

Rest ... ~materialien n/pl remaining materials; ~nachweisung f list of the official correspondence (reports etc.) not yet acted upon; ~zahlung f final payment.

restiren, im Rückstand sein v to be in arrear.

Resultante, resultirende Kraft, Mittelkraft f resultant, resulting force.

Retention f retention; ~skraft f coercive force; ~srecht n lien.

Retour ... ~billet n ticket to go and return, return ticket; ~brief f. Rück...; ~fracht f return freight; ~waaren f/pl returns.

Rettungs ... ~boot n safety boat, life boat; ~station f life saving station.

revidiren v to revise, to look over; (eine Rechnung) to examine (to verify) an account.

Revier n quarter, district; ~briefträger m postman of the walk.

Revision f (PT) eines Amtes durch den Inspektor visit; (einer Kasse) examination, revisal; (eines Prozesses) new trial; (einer Rechnung) audit (of an account); (zollamtliche) custom's examination.

Revisions ... ~bericht m (PT) report, examiner's report; ~kostenberechnung f (T) report after completion of the work (stating totals & results; the actual & estimated costs of the work are compared, explanation given of excess or defect of the estimated costs, departures from the original design are described & explained); ~verhandlung f protocol.

Revisor *m* controller, comptroller, supervisor, f. auch Leitungs-~; (v. Rechnungen) verifier of accounts, controller, comptroller.

Rheochord *n* rheochord (for measuring the electro-magnetic resistance).

Rheomotor *m* rheomotor (apparatus for producing an electric current).

Rheostat *m* rheostat (apparatus for regulating an electric current).

Rheotom *m* rheotome, circuit breaker.

Richt . . . ~blei, Bleiloth *n* plummet and line, plummet; ~kraft *f* der Erde directing force of the earth; ~linie *f* guide line; ~loth f. ~blei; ~magnet *m* governor (directive) magnet; ~punkt *m* station; ~schnur *f* plumb line, chalk line.

richten *v* (die Stangen [*T*]) to dress, to adjust; sich ~, sich einstellen to set (said of a magnet).

richtig *a* true, correct.

Richtung *f* (Techn., Mech.) bearing, direction; ~slinie *f* line of direction; ~swechsel *m* (*T*) up & down working (each station sends alternately one [up & down working proper] or several messages).

Riegel *m* bolt, bar, (beim Doppelgestänge) cross bar, oberer ~ upper bar, mittlerer ~ middle bar, unterer ~ lower bar; ~ mit Vorstecker fox bolt.

Riemenscheibe *f* pulley, band pulley, strap wheel.

Ries Papier *n* a ream of paper (20 quires, 480 to 500 sheets).

Rille *f* (Techn.) chamfer, flute, groove.

rillen *v* to chamfer.

Rimesse *f* remittance; ~n machen to make remittances.

Rinde, Borke *f* bark, cortex; die ~ abschälen to bark, to decorticate.

Ring *m* ring, (Oese) ring, loop, collar, (Kettenglied) link; ~bahn *f* circular railway, encircling line; ~förmig *a* annular, annulate, annulated.

Rinne *f* (Gerinne *f* Wasserlauf *m*) channel, gutter, water course; (Furche, Kerbe, Nuthe in einer Walze) groove, channel; (Kabelrinne) trough, wooden trough.

Riß *m* (Techn.) rent, cleft, crack; chink (in wood), crack (in the wire); (Grundriß *m*) tracing, plan, drawing.

rissig *a* cracked, sprung, chinked; ~ werden, Risse bekommen *v* to crack, to spring.

roh *a* raw, crude, unwrought; ~er Betrag *m* gross amount; ~es Eisen f. Roheisen.

Roh . . . u. **roh** . . . ~brüchig f. Eisen, ~eisen, Gußeisen *n* pig iron, cast iron, unwrought iron; ~material *n* raw material.

Rohr *n* (Techn.) tube, pipe, (eines Thermometers) stem, (einer Rohrpost) pneumatic tube, (spanisches) rattan, Bengal cane; ~blech, Röhrenblech *n* sheet iron for making tubes, flue iron; ~flantsche *f* flange of a pipe; ~post *f* tubular post, pneumatic post, pneumatic despatch; ~postbrief *m* tubular letter; ~posteinrichtung *f* pneumatic despatch; ~postkarte *f* tubular post card; ~ständer *m* von Schelleisen getragen gas pipe with collar bands.

Röhre *f* pipe, tube, conduit, conduit pipe; ~ mit Muffe faucet pipe, socket pipe; ~nstrang *m* range of pipes; ~nzweig *m* branch pipe.

Roll . . . ~fuhrunternehmer *m* contractor for carting goods; ~geld *n* cartage; ~kloben *m* pulley block, pulley sheave, pulley; ~kutscher *m* drayman; ~material *n* (Eisenb.) rolling stock; ~scheibe *f* sheave of a pulley; ~wagen *m* cart, dray, go-cart, (Eisenb.) truck, truck carriage, (zur Abfuhr von Gütern von der Bahn) lowry, lory.

Röllchen *n* roller, rowel.

Rolle *f* (Techn., Mech.) roll, roller; (Walze) cylinder, roll, drum; (Papierrolle [*M & HA*]) paper roll; (feste Leitrolle an Flaschenzügen u. s. w.) guide pulley, fixed pulley; (einfache) single-purchase pulley; (lose, bewegliche) moveable pulley, loose sheave; (zum Aufwickeln v. Draht ꝛc.) reel.

rollen v to roll.
Rosettenkupfer n rose copper, rosette copper, copper in disks.
Rost m rust (of iron).
rosten v to rust, to gather rust, to get rusty.
rostig a rusty; aeruginous.
Rotation f rotation, revolution.
Roth ... u. **roth** ... ~brüchig f. Eisen; ~fichte, ~tanne f red fir, spruce fir (pinus abies); ~stift m red pencil, red chalk pencil.
rotiren v to rotate, to turn; (um seine Achse) to turn about its axis.
rotirend a rotatory, rotary, revolving.
Rubrik f head, heading, column.
rubriziren v to head, to put into a column.
Ruck m fit, jerk, start; ~weise adv by jerks, by starts.
Rück ... ~antwort f reply; ~bewegung f retrograde motion; ~brief m returned letter; ~datiren v to post-date; ~fahrt f the trip home, beladene (unbeladene) ~fahrt the trip home with (without) passengers & mails; ~forderung f reclamation, claim made, ~forderung von Postsendungen demand that a letter may be returned to the writer (sender); ~fracht f return freight, home freight, return cargo; ~frage f enquiry; ~fuhre f f. ~fahrt; ~gabe f return, unter Beding der Rückgabe with the request for return, gegen gefällige ~gabe with the (respectful) request for the ultimate return; ~gängig machen v to annul (a bargain); ~karte f (P [+]) bill on which a P. O. certifies the amount of unpaid postage due on re-directed or returned mail matter; ~kauf m repurchase, redemption; ~leiter m, ~leitung f (T) return wire, the line returning; ~melden v (P) to notify by means of a verification certificate; ~meldung f (P) verification certificate; ~porto n postage charged on returned correspondence; ~schein m (P) advice of receipt, return receipt for registered articles; ~scheingebühr f (P) fee on advice of receipt; ~schlag m (Techn.) back stroke; ~seite f back side, reverse; ~sendungen f/pl (P) returns; ~signal n back signal; ~sprache f nehmen to confer with smb; ~stand m residuum, residue, remains, (Techn.) refuse; im ~stande sein to be in arrears, to be behind; ~stoß m (Phys.) repulsion; ~strahlung f radiation; ~strom m (T) return current; ~vergütung f re-imbursement; ~wand f back wall; ~wärtsbewegung f retrograde motion; ~wirken v to react; ~wirkende Elastizität, Druckelastizität f elasticity of compression; ~wirkende Festigkeit, Druckfestigkeit f strength of compression; ~wirkung f reaction; ~zahlbar a returnable; ~zahlung f return of payment, paying back; ~zoll m drawback of duty.
Ruf m (T) call; ~taste f (F) bell push, bell key; ~zeichen n (T) call signal, (in England) code, code letters, f. a. T. code.
rufen v to call (an office three times in succession).
Ruhe f (Phys., Mech.) rest, repose; ~gehalt n pension; ~gehaltsanspruch m title to pension; ~kontakt m (MA) back stop, rest stop, receiving anvil, resting contact; ~kontaktschiene f (MA) back stop piece; ~lage f (MA) position of rest; ~punkt, Stützpunkt, Drehpunkt m fulcrum, fulcre; ~schaltung f (T) position of rest; ~schiene f (MA) back stop piece; ~stand m (Versetzung in den ~stand) granting a pension, Versetzung in den einstweiligen ~stand to dispense with an official's services for the time being; zwangsweise Versetzung in den ~stand f. Pensionirung; ~stellung f. ~schaltung; ~strom m constant current, closed circuit; Apparat (Linie) für ~strom apparatus (line) worked on the closed circuit plan; ~stromtaste f key of the instrument worked on the closed circuit plan; ~tag m day of rest; ~zustand m der Telegr.-Linie is the t. line in the state of repose, the t. line being idle.
rühren, sich rühren v to stir.
rund a round, spherical, glo-

bular; eine ~e Zahl f a round number.
Rund ... ~eisen s. Eisen; ~feile f round file, rat's tail file; ~holz n round timber, round wood, in Stücke gesägtes ~holz dimension lumber; ~reise f (Eisenb.) circular trip; ~reisebillet n circular tourist-ticket; ~schreiben n circular, circular letter; ~zange s. Zange.
Rußschreiber m carbon recorder.

S.

Saal m hall, large room; Apparat·~ m (T) instrument room.
Sachverständiger m competent party, competent judge, expert.
Sack m bag, sack, pouch; ~leinwand f sack cloth, sacking, bagging; in geschlossenen Säcken (P) in closed bags.
Saft m juice; (Pflanzensaft) sap; die Bäume stehen im ~ the trees are in sap; ~holz n sap wood.
saftig a succulent.
Säge f saw; ~bank f saw bench; ~blatt n saw blade; ~block, ~klotz m saw block; ~spänbatterie f (Elektriz.) sawdust battery (a Daniell battery with sawdust supplanting the porous cylinders); ~späne m/pl sawdust.
sägen v to saw; der Länge nach ~, trennen to cut.
säkulare Variationen f/pl der Magnetnadel secular variations of the needle.
saldiren v eine Rechnung to balance, to settle, to strike (an account).
saldirt a durch Gegenrechnung balanced in account, counter-balanced by ...
Saldirung f balancing, settling, liquidation.
Saldo m balance of an account current; ~ zu unseren Gunsten balance in our favor; den ~ ziehen to strike the balance; ~betrag m amount of balance; ~zahlung f payment per appoint.
Salmiak m, **Chlorammonium**, **salzsaures Ammoniak** n sal ammoniac, ammonium chloride, hydrochlorate (muriate) of ammonia.
Salon m (große Kajüte f) state room, great cabin, saloon; ~wagen m (Eisenb.) saloon carriage, (Am) palace car.

Salpeter, **Kalisalpeter** m saltpetre, nitre, nitrate of potassium; ~sauer a nitric; ~säure f nitric acid; rauchende ~säure red fuming nitric acid, nicht rauchende, gewöhnliche ~säure ordinary nitric acid.
salpetrige Säure f nitrous acid.
salpetrigsaures Salz n (Nitrit) nitrite.
Salz n salt; ~haltig a saliferous, saliniferous; ~lösung f solution of salt; ~sauer a muriatic, hydrochloric; ~säure f muriatic acid, wässerige ~säure f aqueous (commercial) muriatic acid.
salzig a saline, saltish.
Sammel ... ~amt s. ~stelle; ~apparat m (Phys., Techn.) collector; ~behälter m reservoir, receiver; ~beutel m (P) collecting (collection) bag; ~briefe m/pl a bundle of letters sent under cover to a post office to be delivered by the same; ~linse f collective lens, condensing lens; ~punkt m, ~spitze f (Elektriz.) collecting point; ~sack s. ~beutel; ~stelle f (P) office where transit correspondence (mostly parcels) is dealt with; office for intermediary service.
Sand m sand; ~batterie f (T) sand battery; ~boden m sandy ground; ~stein m sandstone, grit stone.
sättigen v (Phys.) to saturate.
Sättigung f saturation; eine Lösung in den Zustand der ~ versetzen to bring a solution to the point of saturation; ~sfähigkeit f, ~svermögen n capacity of saturation; ~sgrad m degree of saturation; ~spunkt m point of saturation.
Satz m (Techn.) set; (Lehrsatz) principle, theorem; (Bodensatz) deposit, sediment, dregs; (Grammatik) sentence, period; (Handel) price,

rate; ein ~ Gewichte a set of weights; ein ~ Werkzeuge a set of tools & implements.

sauer *a* sour, acid, (Chem.) acid.

säuerlich *a* sourish, (Chem.) acidulous, acidulated.

säuern *v* to sour, (Chem.) to acidify, to acidulate.

Sauerstoff *m* oxygen; ~ entziehen *v* to deoxidize, to deoxygenate; Verbindung *f* mit ~, Oxyd *m* oxide; ~gas *n* oxygen gas; ~haltig *a* oxidized, containing oxygen; ~salze *n/pl* oxygen salts; ~säuren *f/pl* oxy-acids, oxacids; ~verbindung *f* oxidized compound.

Saug ... ~apparat, Exhaustor *m* suction apparatus, exhauster; ~hahn *m* suction cock; ~heber *m* siphon; ~kamm *m* (Elektrisir=Masch.) comb; ~pumpe *f* sucking pump; ~rohr *n* suction pipe, air pipe, (Chem.) suction tube, suction pipe; ~ventil *n*, ~klappe *f* sucking valve, suction valve.

Säule *f* column, pillar, (Phys.) pile; (Ständer) post; ~nbriefkasten *m* pillar box.

Säure *f* acid, (konzentrirte) concentrated acid, (verdünnte) dilute acid; der ~ widerstehend *a* antacid; ~haltig *a* acidiferous; ~messer *m* acidimeter, acetimeter.

schaben *v* to scrape; Metall ~, scheuern to scour.

Schaber *m* scraping tool, scraper.

Schablone *f* (Techn.) pattern, model, form, mould.

Schachtel *f* box, case; (Papp=schachtel) band box.

Schaden *m* loss, damage, (Nachtheil *m*) detriment; ~ leiden *v* to suffer a loss; mit ~ with a loss; ~ersatz *m* indemnification, compensation, ~ersatz leisten *v* to render damages, to give compensation, s. auch Entschädigung.

schadhaft *a* spoiled, damaged.

schadlos halten *v* to indemnify.

Schaffner *m* (bei der Eisenb.) guard, railway guard; (bei der Post) mail guard, s. auch Post . . .; ~bahnpost *f* (*P*) travelling post office in charge of a mail guard.

Schaft, Griff *m* (Techn.) handle, shank, stock.

Schale, Hülse *f* shell, husk, peel; (Napf *m*) cup, bowl; (Becken *n*) basin; (einer Waage) scale of a balance.

schälen *v* to peel, to pare; Bäume ~ to bark, to decorticate trees.

Schall ... ~boden, Resonanzboden *m* sound board, sounding board; ~lehre *f* acoustics; ~messer, Tonmesser *m* (Phys.) echo-meter; ~telegraph *m* acoustic telegraph; ~trichter *m* bell; ~welle *f* undulation, sound wave.

schalten *v* (*T*) to connect up, to join up; die Rollen, die Batterie hintereinander ~ to join up the coils (the battery) in series; nebeneinander ~ to join up in quantity; die Leitung direkt ~ (*T*) to establish direct communication, s. auch einschalten.

Schalter *m* (*P*) window, counter, office window; am ~ etwas aufgeben to hand letters etc. over the counter; ~beamter *m* counter clerk, clerk at the counter, counterman (counterwoman); ~dienst *m* duty at the counter, window duty; ~kasse *f* official till; ~raum, ~vorflur *m* anteroom of the P. O., waiting room; ~verkehr *m* business at the counter; Briefannahme=~ s. Brief . . .; Packetannahme=~ s. Packet . . .; Postanweisungs=~ s. Post . . .; Zeitungs=~ s. Zeitungs . . .

Schaltung *f* (des *T* Apparats) putting in circuit; (der Drähte) connection, s. auch Einschaltung.

scharf *a* sharp, acute; (geschliffen) sharp, keen, edged, whetted, ground; (ätzend [Chem.]) corrosive; ~gängige Schraube *f* screw with a triangular thread; ~kantig *a* sharp-edged, diesquare.

Schärfapparat *m* sharpener.

Schärfe *f* edge, sharpness of the edge; (einer Säure u. s. w.) acidity; (eines Keils [Mech.]) acies (edge) of a wedge.

schärfen *v* to sharpen, to edge, to point, to whet, to grind; einem Pferde die Eisen ~ to rough-shoe a horse; die Zähne einer Säge ~ to sharpen the teeth of a saw.

Scharnier, Charnier, Gewerbe *n* hinge, joint, hinge joint, turning joint; ~band *n* joint hinge; ~stift *m* joint wire, joint pin; ~ventil, Klappenventil *n* flap valve, leaf valve, hanging valve.

Scharte *f* (Techn.) notch (in an edge tool).

schartig *a* notchy, full of notches.

Schatz ... ~amt *n* exchequer, public treasury office, Treasury; ~meister *m* treasurer.

schätzen *v* (taxiren) to apprize, to value, to estimate.

Schätzung *f* valuation, taxation, tax, estimation.

Schaufel *f* shovel, scoop; ~hacke *f* scuffle hoe.

Scheer ... ~festigkeit *f* (Mech.) shearing strength; ~spannung *f* (Mech.) shearing stress, shearing strain.

Scheere *f* scissors, shears, (gekröpfte) curved (bent) scissors; Metall-~ plate shears.

Scheibe *f* (Techn.) disc, disk, sheave, coil; (Glasscheibe) pane, square of glass; (Rolle *f* [Mech.]) sheave; (Unterlagscheibe) washer, collar; (Leier *f* einer Drahtziehbank) drum, drawing plate; (farbige ~ einer Signallaterne auf der Eisenb.) colored glass; (feste ~ [Mech.]) fast (fixed) pulley.

Scheiben ... ~elektrisirmaschine *f* plate-electrical machine; ~kupfer, Rosettenkupfer *n* rose copper, rosette copper, refined copper in round crusts or disks; ~signal *n* (Eisenb.) disk signal; ~umschalter s. Umschalter.

Scheide *f*, **Futteral** *n* (Techn.) case, sheath.

Scheide ... ~münze *f* billon, change; ~wand *f* diaphragm, poröse ~wand porous partition (diaphragm); ~wasser *n*, Salpetersäure *f* aqua regia, nitro-muriatic acid; ~weg *m* forked way, cross way.

scheiden *v* (Chem.) to part; (analysiren) to analyze, to decompose, to separate.

Schein *m* (Bescheinigung *f*) certificate, attestation; (Schuldschein) bond, bill; (Quittung *f*) receipt, acquittance; (Kassenschein, Bankschein) bank bill, note, paper money; (Zusammensetzungen s. unter den betr. Buchst.); ~horizont *m* (beim Vermessen) visual horizon, apparent level.

Scheit *n* (Techn.) block, log, billet of wood; ~ Holz, Holzscheit, Holzkloben *m* billet (log) of wood; ~holz, Holz *n* in Scheiten oder Kloben billet wood, log wood; ~maaß *n* log-wood measure; ~recht *a* vertical.

Scheitel *m* (Math.) vertex; (eines Dammes) crown of a dike; (eines Winkels) angular point, apex, summit, vertex of an angle.

Scheitel ... ~linie *f* vertical line; ~punkt *m* vertex, angular (vertical) point; ~recht *a* vertical; ~winkel *m/pl* vertical (opposite) angles.

Schellack *m* shell lac, shellac; ~firniß *m* shellac varnish.

Schelle *f*, **Schelleisen** *n* (Techn.) collar band.

Schema *n* model, pattern, design, specimen.

schematisch *a* according to a pattern; ~e Darstellung *f* der Apparatverbindungen (T) diagrams of the instrument connections.

Schenkel *m* leg, limb (of a horseshoe magnet).

Scheuerbock *m* (T Bau) fender, protecting pales (to prevent cattle from injuring t. poles by rubbing against them).

scheuern *v* (reinigen) to scour; (vom Kabel auf felsigem Untergrund) to chafe (against rocks).

Schicht *f* couch, bed; (Erd-~ layer of earth; Feuchtigkeits-~ film of moisture; Guttapercha-~ layer of guttapercha; Luft-~ stratum of air; ~ machen *v* to cease working.

schichten *v* to dispose into layers or rows, to pile up (wood).

schicken *v* to send (a current, letters etc.).

Schieb ... ~karre *f*, ~karren, Schubkarren *m* wheel barrow, hand barrow; ~lade *f* (am Tisch) drawer,

8*

till; ~leere, Schubleere *f* (Techn.) slide gauge.

schieben *v* (Techn.) to push, to slide.

Schieber *m* (Techn.) slide; ~ventil *n* slide valve, sliding valve.

Schiebering *m* (*HA*) slide.

Schieds... ~gericht *n* court of arbiters (of arbitration); ~richter *m* arbiter; ~richterliche Entscheidung *f* arbitration, award.

schief *a* (Techn.) oblique, inclined; (quer) aslant, oblique; (geneigt) sloping, inclined; ~e Ebene *f* inclined plane, slope, gradient; ~er Winkel *m* oblique angle.

Schiefer *m* slate; (im Eisen) flaw; ~stift *m* slate pencil.

schieferig *a* slate-like; (vom Eisen) scaly, flaky.

Schiene *f* (dünne Leiste) slat; (Eisenb.) rail, railway plate; (an der Taste, am Umschalter ꝛc. [*T*]) bar, strap, brass strap; (Typen-~ des ersten *MA*) port rule, composing stick; aus den ~n kommen *v* to get off the rails.

Schienen...~oberkante *f* (Eisenb.) head of the rails; ~strang *m* (Eisenb.) track, set of tracks; ~umschalter *m* (*T*) Swiss commutator, bar commutator; ~weite, Spurweite *f* (Eisenb.) gauge of way.

Schiff *n* ship, vessel; ~brücke *f* pontoon bridge, boat bridge; ~fahrt *f* navigation; ~fahrtsnachrichten *f/pl* shipping intelligence; ~sbrief *m* ship letter; ~smeldung *f* (telegraphische) telegraphic shipping intelligence, telegraphic advice of ships arriving; ~spost *f* ship letter office; ~stelegramm *n* telegram for a ship.

Schiffchen *n* (Techn.) shuttle.

Schild *n* (Aushängeschild) sign, sign board.

Schippe, Schaufel *f* shovel, scoop.

Schlacke *f* slag; (am Draht) cinders; ~nbildung *f* scorification; ~neisen *n* cinder iron; ~nwolle *f* slag wool, slag hair, silicate cotton.

schlackig *a* slaggy.

Schlaf... ~abteilung *f* (Eisenb.) sleeping compartment; ~wagen *m* (Eisenb.) sleeping carriage, (*Am*) sleeping car.

schlaff *a* slack; ~er Anker *m* (*T*) slack tie; ~ hängende Drähte *m/pl* (*T*) slack wires.

Schlag *m* (Techn.) blow, shock; ~baum *m* toll bar, toll gate; ~eisen *n* beater; ~loth, Hardloth *n* hard solder; ~weite *f* (Elektriz.) striking distance.

schlagende Wetter *n/pl*, **Grubengas** *n* fire damp, mine gas (explosive mixture of marsh gas & atmospheric air).

Schleif... ~kontakt *m* (*T*) sliding (rubbing) contact, glide contact; ~stein *m* grindstone.

Schleife *f* loop, knot, (*T*) loop; (Schlinge *f*) loop, knot, bend, eye of a rope.

schleifen *v* to drag, to draw along the ground, to trail; (poliren) to polish; (wetzen) to grind, to whet.

Schleifen... ~leitung *f* (*T*) looped wire; ~probe *f* bei der Kabelmessung (*T*) loop test.

Schlepp *m*. ~dampfer *m* steam tug; ~schiff *n* tow boat, tug boat; ~schifffahrt *f* navigation with tow boats, towage; ~tau *n* tow rope, track rope, dragging cable, in's ~tau nehmen to take in tow; ~wagen *m* (Eisenb.) truck, car; ~zug *m* (Eisenb.) train of waggons.

Schleuse *f* lock, sluice, flood gate, water gate; ~nmeister, ~nwärter *m* sluice master, lock keeper; ~nthor *n* sluice gate, lock gate; ~nwehr *n* lock weir; ~nzoll *m* lock dues, lockage.

schließen *v* to close, to lock; (den Dienst [*TP*]) to close the office; (eine Rechnung) to close an account; (den Stromkreis) to close the circuit.

Schließung *f* closing (of a circuit, of an account, of the office), vergl. schließen; ~sbogen *m* (Elektr., *T*) closing arc; ~sdraht *m* (*T*) closing wire; ~sinduktionsstrom *m* induced current on making contact (at closing).

Schlinge *f* (Techn.) sling, knot, loop; (im Drahte) kink, knot.

Schlingern *n* (der Lokomotive) irregular oscillating motion; (eines Schiffes) pitching.

Schlitten *m* (HA) chariot, (P) sledge, (Masch.) sliding carriage; Körper *m* des ~s (HA) contact maker; ~achse *f* (HA) axle of the chariot.

Schlitz *m* slit, slot; (Kerbe) cutting, slit; ~brenner, Fledermausbrenner *m* (bei der Gasbeleuchtung) bat's wing burner, split burner.

schlitzen *v* to slit, to split.

Schloß *n* (zum Schließen) lock; ~nagel *m* (großer) dog nail, (kleiner) tack; ~rad *n* (Mech.) circular plate; ~riegel *m* bolt of a lock, lock bolt.

Schlucht *f* ravine, gorge, mountain pass, (Am) gulch, cannon.

Schluß *m* (Techn.) fastening; (P) closing (of the mails); (einer Rechnung) settlement of an account; ~abfertigung *f* (zollamtliche) discharge from custom house, all formalities having been duly performed; ~bilanz *f* final balance; ~ergebniß *n* final result; ~kurs *m* closing price; ~note *f* s. ~zettel; ~protokoll *n* final protocol; ~rechnung *f* final account; ~summe *f* result, sum total; ~zeichen *n* (T) the „end of message" signal, signal denoting the completion of a message; ~zeit *f* für Postsendungen latest time of posting letters etc.; ~zettel *m*, ~note *f* broker's note, broker's memorandum, broker's contract (closing the bargain).

Schlüssel (MA) s. Taste; Schraubenschlüssel s. Schraube; (einer Geheimschrift) key; ~, Keil *m* (Techn.) key; ~ des Schraubenstocks spanner; ~bart *m* bit of a key; ~ring *m* split ring; ~rohr *n*, ~schaft *m* key pipe, key shank; ~schild *n* escutcheon.

Schmalspur *f* (Eisenb.) narrow gauge; ~bahn *f* narrow gauge railway.

Schmelz *m*, **Schmelzglas** *n*, **Emaille** *f* enamel; emaillirtes Schild am Briefkasten enamelled plate; mit ~ belegen, überziehen *v* to enamel.

Schmelz ... ~arbeit *f* smelting, smelting of metal, smelting process; ~bar *a* (Phys.) fusibe, leicht ~bar easily fusible, easy to fuse, schwer ~bar, feuerbeständig refractory; ~barkeit *f* fusibility; ~glas *n*, ~glasur *f* enamel; ~grad *m* degree of melting; ~mittel *n* fusing agent, flux; ~punkt *m* point of fusion; ~stahl *m* rough (natural) steel; ~tiegel *m* crucible.

schmelzen *v* (Phys., Metall.) to fuse, to melt, to liquefy, to smelt, to bring down; (an der Luft zerfließen) to deliquesce; (zusammenschmelzen) to melt (to fuse) together.

Schmied ... u. **schmied** ... ~bar *a* forgeable; (hämmerbar) malleable, ductile; ~bares Eisen *n* malleable iron; ~barkeit *f* malleability; ~eisen *n* (geschmiedetes Eisen) wrought iron, forged iron, soft iron; ~eisen *n* (schmiedbares Eisen) ductile iron, malleable iron; fertiges ~eisen, Handelseisen finished iron; sehniges, faseriges, zähes ~eisen fibrous iron; ~eisern *a* wrought iron, made of wrought iron; ~eiserne, gezogene Röhre *f* wrought-iron tube; ~kohle, ~eskohle *f* smith coal, smithy coal, forge coal; ~maschine *f* forging machine, forge machine; ~nagel *m* wrought nail, forged nail; ~schlacke *f* slag, clinker, slack produced in a forge; ~stück, geschmiedetes Eisenstück *n* forging; ~zange *f* tongs, smith tongs, forge tongs, fire tongs.

Schmiede *f* forge, smithy, smithery, workshop of a smith; (Feldschmiede) field forge, travelling forge, portable forge.

schmieden *v* to forge, to tilt, to hammer.

schmiegen, sich *v* (Techn.) to bend, to ply.

schmiegsam *a* pliant, flexible.

Schmier ... ~gelder *n/pl* (+) extraordinary fee for greasing the extra mail coach; ~öl *n* (Masch.) lubricating oil, mineral sperm.

Schmiere *f* (Masch., Techn.) grease; ~ zur Verminderung der Reibung (Masch.) antifriction grease.

schmieren *v* (Masch., Techn.) to lubricate, to grease, to oil; Fuhrwerke ~ to grease; (schlecht schreiben) to slur.

Schmirgel, Schmergel *m* emery; ~papier *n* emery paper; ~pulver *n*, ~staub *m* powdered emery, emery powder, emery dust; ~stein *m* emery stone (clay burnt with emery).

schmirgeln *v* (Techn.) to rub (to polish) with emery; Metall ~ to grind; mit Glaspapier ~ to glaze.

Schnapp... ~feder *f* catch spring; ~schloß *n* German spring lock, snap lock; halbtouriges ~schloß half-turning spring lock.

Schnarrwecker s. Wecker.

Schnecke *f* (Rolle, Spirale) scroll; (Schraube ohne Ende) endless worm.

Schnecken... ~bohrer *m* screw auger, twisted bit, spiral drill, (doppelter) double lipped screw auger, (einfach gewundener) single lip screw auger; ~feder *f* spiral spring; ~förmig *a* spiral; ~gewinde *n* helix; ~linie *f* helix, spiral line; ~rad *n* balance wheel, snail, snail wheel; ~rad, Schraubenrad, Wurmrad *n* (Masch.) screw wheel, scroll wheel, scroll gear, spiral wheel; ~rad, Trommelrad, Tympanum *n* (Masch.) tympan, tympanum; ~radgetriebe *n* (Masch.) worm & wheel, screw & wheel, worm gear; ~windung *f* volution.

Schneewehe, Schneeverwehung *f* snow drift, accumulation of snow.

Schneid... ~instrument *n* edge tool; ~eisen, Schrauben~eisen *n* screw plate; ~kluppe, Schraubenkluppe *f* stocks, stocks & dies, screw stock; ~messer *n* cutting (chopping) knife; ~zange *f* cutting plyers.

schneiden *v* to cut; (zerschneiden) to carve; (Eisen) to slit, to split iron; (Holz) to saw, to cut up timber; (der Länge nach) to rive, to cleave; (in die Quere) to cross-cut; (schräg) to bevel.

Schneidenblitzableiter s. Blitzableiter.

Schnell... u. **schnell**... ~feder *f* spring; ~flüssiges, leichtflüssiges Metall *n*, ~flüssige Legirung *f* (Chem.) fusible metal; ~kraft, Elastizität *f* (Phys.) elasticity; ~kräftig *a* elastic; ~loth *n* fusible metal, (Weichloth) soft solder, tin solder; ~presse, Buchdruckmaschine *f* fly press, steam press; ~schiff *n* express boat; ~schreiber *m* (*T*) rapid automatic instrument, fast speed apparatus, Typen-~schreiber *m* (*T*) fast type transmitter; ~segler *m* fast-sailing ship; ~zug *m* (Eilzug) fast train, (Kurierzug) express train.

schnellen *v* (Techn.) to jerk.

Schnelligkeit *f* speed, velocity.

Schnitt *m* (Techn.) cut, incision; (Durchschnitt) section; ~brenner, Schlitzbrenner *m* split burner; ~fläche *f* cut; ~linie *f* (Durchschnitts~) line of section; ~schraube *f* screw with split head.

Schnur *f* cord, tape; (zum Vermessen) line, measuring tape; (Schlagleine) line, chalk line; (ohne Ende) endless band; nach der ~ messen to lay out, to trace by the line; ~gerade *a* straight, in a line; ~scheibe *f* (Rinnenscheibe, ~rad *n* [Masch.]) groove wheel, rigger.

schnüren *v* (abschnüren, bei Vermessungen) to line out, to lay out by the line.

Schonung *f* (Pflanzung) nursery of young trees.

Schotter... ~belag *m* gravelling, coating with broken stones; ~straße *f* gravelled road, gravel road.

schottern, beschottern *v* to ballast.

schraffiren *v* to hatch (to shade by lines).

schräg *a* oblique, diagonal, slanting; (schiefwinkelig) bevel, bevelled; ~e Neigung *f* inclination, slope.

Schräge *f* (Techn.) obliquity, oblique direction, slant; ~, Neigung *f* inclination, slope; ~ einer Mauer shelvingness, slopeness of a wall.

schrägen, abschrägen *v* to make oblique, to slant.

Schrägung *f* (Techn.) chamfer.

Schrank *m*, **Spind** *n* cupboard, press; feuerfester ~ fire-proof safe; kleiner ~ closet; ~ mit Schubladen cabinet.

Schranke *f* bar, barrier, screen; (Einfriedigung *f*) rail, fence;

Schraub — schuhen

(Grenze *f*) boundary; (in einem Gerichtshofe) bar.

Schraub ... ~bolzen *m* pin with screw head; ~haken *m* screw hook; ~hebel *m* wrench; ~kloben *m* vice; ~zwinge *f* screw clamp, cramp; ~stock *m* vice, screw vice; die Backen *f/pl* eines ~stocks chops of a screw vice.

Schraube *f* screw; ~ ohne Ende endless screw, perpetual screw; (Schraubenmutter *f*) inside screw, screw nut; Mikrometer=~ micrometrical screw; Holz=~ wood screw (eif. Schraube zum Schrauben in Holz).

Schrauben ... ~bohrer *m* screw tap, twisted drill; ~bolzen *m* screw bolt, set bolt, (mit Gewinde an beiden Enden) stud bolt, ~bolzen *m* mit Mutter bolt and nut; ~docke *f* screw chuck; ~förmig *a* spiral, helicoidal; ~gang *m*, ~gewinde *n* turn, thread, worm, fillet of a screw; ~kluppe *f* zum Schraubenschneiden stocks, screw stock, stocks and dies, die stock; ~kopf, ~knopf *m* screw head, screw knob; ~mutter *f* female screw; ~nagel *m*, Holzschraube *f* wood screw, screw nail, coach screw; ~schlüssel *m* screw key, screw plate, wrench, turn screw, englischer ~schlüssel, Universal=~schlüssel universal screw wrench, monkey spanner, shifting spanner, gabelförmiger ~schlüssel fork wrench; ~stütze *f* iron bracket with screw thread; ~verbindung, ~verkuppelung *f* an Röhren union joint; ~verschluß *m* screw cap; ~winde *f* screw jack; ~zieher *m* screw driver, turn screw.

schrauben *v* to screw.

Schreib ... ~apparat *m* (*T*) recording instrument, recording register, s. auch Apparat; ~bedürfnisse *n/pl* stationery; ~brett *n* (Bahnpost) writing accommodation; ~buch, ~heft *n* copy book; ~hebel *m* (*MA*) writing lever; ~hülfe *f* der Postverwalter auxiliary (help) engaged & paid by the postal manager; ~materialien *n/pl* writing materials, stationery; ~papier *n* writing paper; ~pult, Pult *n* desk, shelf desk (in the ante-room of the P. O. for the convenience of the public); ~rad *n*, ~scheibe *f* (*T*) inking disc, printing (marking) wheel; ~stift *m* (*MA*) style, steel pricker; ~tafel *f* (*MA*) transmitting plate; ~telegraph *m* f. ~apparat; ~zeug *n* writing case, ink stand.

schreiben *v* to write; (in's Reine) to copy (out) fair; (leserlich) to write a legible hand.

Schreiben *n* letter, (in Antwortsschreiben oft) favor z. B. your favor d. d. July 3^d has been duly received.

Schrift *f* (Handschrift *f*) writing, handwriting; (Schriftwerk *n*) book, paper, publication; (Letter, Type *f*) letter, print, type; ~führer f. secretary, protocollist; ~locher f. Locher; ~sprache *f* written language, book language; ~stück *n* writing, paper, packet; ~wechsel *m* der Postanstalten official correspondence, correspondence on postal business; ~zeichen *n* (als Unterschrift) mark.

schriftlich *a* written, in writing.

schroten *v* to roll.

Schrotleiter *f* pulling ladder, drayman's ladder.

Schub ... ~fach *m* drawer, sliding box; ~festigkeit, Scheerfestigkeit *f*, Widerstand *m* gegen Abdrücken oder Abscheeren (Mech.) shearing strength, strength of shearing; ~karre *f*, ~karren *m* wheel barrow; ~kasten *m*, ~lade *f* drawer; ~kraft, Zerbrechungskraft, Bruchkraft *m* (Phys., Mech.) transverse strain, breaking strain, lateral strain; ~kurbel *f* (Masch.) crank; ~lade *f* drawer; ~ladenschrank *m* cupboard with drawers, cabinet; ~leere, Schiebleere *f* slide gauge, sliding gauge; ~leere, Kalibermaaß *n* calliper scale; ~tisch *m* drawer table, table with drawers; ~ventil *n* (Masch.) slide valve, sliding valve; ~walze *f* roller, rolling pin, cylinder.

Schuh *m* (als Maaß) foot; (Eisenbeschlag) shoe, shoeing, ferrule; ~ eines Pfahls, einer Stange shoe, pile shoe.

schuhen, anschuhen *v* (einen Pfahl) to shoe a pile.

Schuld *f* debt, money owing; ~en eintreiben to collect debts; in ~ stellen to enter (an item) on the receipt side of an account.

Schuld . . . ~entilgungsfond *m* sinking fund; ~forderung *f* demand, claim, active debt; ~posten *m* article (item) of debt; ~schein *m* bond, bill of debt, note of hand, obligation; ~verschreibung *f* s. Schuldschein.

schuldig *a* (eines Versehens) guilty of a fault.

schulwissenschaftliche Vorbildung *f* education received at school.

Schuppen *m* (Bauwerk) shed, hut; (Wagen-~) coach house, cart house, (für Eisenbahnwagen) waggon house.

schuppig *a* (v. Mineralien) scaly, scaled; ~blättrig *a* scaly-foliated, in foliated scales; ~körnig *a* scaly-granular.

Schutt *m* rubbish; ~ablageplatz *m* place for depositing rubbish; ~haufen *m* heap of rubbish.

Schutz *m* protection, guard; ~apparat *m* protecting apparatus; ~blech *n* guard plate; ~dach *n* (bei der Einführung der Drähte ins TAmt) cover; ~decke *f* protective covering; ~drähte *m/pl* (des Kabels) sheathing wires, protecting wires; ~feder *f* (HA) protective spring; ~hülle *f* (des Kabels) armor, iron envelope; ~kasten *m* (des Apparats) protecting case; ~kreis *m* (eines Blitzableiters) circle of protection; ~marke *f*, Fabrikzeichen *n* manufacturer's mark (sign), trade mark; ~pfahl *m* fender; ~platte *f* guard ring, guard plate; ~streifen *m* strip of land along-side of the road; ~vorrichtung *f* protective contrivance; ~zoll *m* protective duty; ~zollsystem *n* protective system, prohibitory system.

schützen *v* to protect, to guard.

schwacher Leitungsdraht *m* (T) light wire.

Schwächungsanker *m* (HA) movable soft iron keeper (armature to weaken the magnetism of the cores).

Schwaden *m* (schlagende Wetter *n/pl*, Grubengas *n*) fire damp, mine gas (explosive mixture of marsh gas & atmospheric air).

schwanken *v* to fluctuate, to variate; to oscillate (said of the needle).

Schwanken *n*, **Schwankung** *f* fluctuation (of prices, of the electric current); variation (of the barometer); oscillation (of the needle); balancing (of the scales).

schwankend *a* fluctuating, fluctuant (said of prices).

Schwanz *m* tail; (eines Riegels) handle of a bolt; (einer Schraube) shank of a screw; ~stück *n* tail piece.

Schwarz . . . ~blech *n* black sheet iron, black iron plate; ~brüchig s. Eisen; ~fichte *f* black fir (pinus nigra); ~kiefer *f* Austrian fir (pinus austriaca); ~kohle s. Steinkohle.

Schwärz . . . ~apparat *m* inking apparatus; (Stempelapparat) stamping apparatus; ~rolle s. Farberolle.

schweben *v* (in der Luft) to float in the air.

schwebende Beträge *m/pl* floating (pendent) items; ~ Schuld *f* floating debt.

Schwefel *m* sulphur, brimstone; (gediegener, gewachsener) native sulphur; mit ~ imprägniren, vulkanisiren *v* to vulcanize (caoutchouc etc.); mit ~ verbunden *a* sulphuretted.

Schwefel . . . ~kohlenstoff *m* bisulphide of carbon; ~sauer *a* sulphuric; ~saures Kali *n* sulphate of potassium; ~saures Kupferoxyd *n*, Kupfervitriol *m* blue vitriol; ~saure Magnesia *f* sulphate of magnesium; ~saures Salz *n* sulphate; ~säure *f* sulphuric acid, hydric sulphate, oil of vitriol, vitriolic acid, gereinigte ~säure purified (rectified, distilled) sulphuric acid, verdünnte ~säure dilute (diluted) sulphuric acid.

schwefeln *v* to sulphurize, to sulphurate; Kautschuk ~, vulkanisiren to vulcanize caoutchouc.

Schwefelung *f* sulphuration, (des Kautschuks, Vulkanisirung *f*) vulcanization.

schweflige Säure f sulphurous acid; gasförmige ~ ~, Schweflich=säuregas n sulphurous-oxide gas, gaseous sulphurous acid, sulphurous anhydride; wässerige ~ ~, eigentliche ~ ~ sulphurous acid.

schwefligsaures Ammoniak n sulphite of ammonia; ~ Natron n sulphite of soda; ~ Salz n sulphite.

schweifen v (Techn.) to curve, to cut into a curved form, to channel, to flute.

Schweifung f curve, curving, rounding.

Schweiß ... ~arbeit f welding; ~bar a welding; nicht ~bar unweldable; ~eisen n weld-iron, welded iron; ~hitze f welding heat; ~ofen m reheating furnace, welding furnace; ~prozeß m welding, process of welding; ~stahl m weld-steel, welded steel; ~stelle f welding point; ~warm a welding hot.

schweißen, anschweißen v to weld; zusammen~ to weld together, to weld; Gußeisen ~ to mend cast iron.

Schweißung f weld, welding, process of welding, s. auch Schweißstelle.

Schwelle f (Eisenb.) sleeper, dormer; (der Doppelständer [T Bau]) tie, ground beam, sill; hölzerne Eisenb.=~ timber sleeper.

schwer a (v. Gewicht) heavy, weighty; (schwierig) difficult; ~er Leitungsdraht m (T) heavy wire; ~ löslich a hardly soluble; ~ schmelzbar, strengflüssig a (Chem.) difficult of fusion, melting at high temperature, hardly fusible; im Feuer ~ schmelzbar, feuerbeständig refractory.

Schwer ... u. **schwer** ... ~flüssig s. schwer; ~kraft f gravity, gravitation; ~punkt m centre of gravity.

Schwere f, **Gewicht** n weight; spezifische ~, Schwerkraft f specific gravity; (Gewichtseinheit f) simple weight; das Gesetz der ~ (Phys.) the law of gravity; spezifisches ~, spezifisches Gewicht n specific gravity.

schwingen v to swing; (oszillieren) to oscillate, to vibrate.

Schwingung f (Phys.) oscillation, vibration, swinging (of a needle); ~ des Schalles reverberation of sound; ~sachse f axis of oscillation; ~sbogen m arc of oscillation; ~s=weite f amplitude of oscillation; ~swelle f undulation; ~szeit, ~s=dauer f time of oscillation (vibration).

Schwung m (Mech.) swing, swinging, oscillation, vibration; in ~ bringen to cause to swing; ~=kraft f oscillating (vibrating) power, (Zentrifugalkraft) centrifugal power; ~rad n fly wheel, balancing wheel, crown wheel; ~radachse f axis of the fly wheel; ~radwelle f axle of the fly wheel.

See f sea, ocean; zur ~ gehörig a maritime; ~beförderung f (P) sea conveyance; ~dampfer m sea-going steamer, ocean steamer; ~handel m maritime commerce (trade); ~hand=lungsgesellschaft f sea trading company; ~kabel f. Kabel; ~karte f sea chart, hydrographical map; ~kom=paß m mariner's (sea) compass; ~=meile f (1855 m) sea mile, knot; ~porto n (P) sea postage; ~postbe=förderung f sea conveyance; ~post=beförderungsgebühr f fee for sea conveyance; ~postbüreau n ship letter office; ~postdienst m maritime postal service, mail packet service; ~postverbindung f mail packet service; ~telegramm n semaphoric message; ~telegraph m semaphore; ~telegraphen=anstalt f semaphore station; ~transit m sea transit; ~=transport m sea conveyance; ~warte f maritime observatory.

Seele s. Kabel.

Seh ... ~achse f axis of vision, visual axis; ~feld n, ~kreis, Gesichts=kreis, Horizont m horizon; ~weite f visual distance, distance of vision; ~winkel m angle of sight, visual angle.

Sehne f (Geom.) chord, subtense; (Metall) fibre; (des Eisens) fibre of iron, fibre on the fracture of iron.

Sehneisen, sehniges Eisen n fibrous iron.

sehnig a fibrous; ~er Bruch des

seide-umsponnener Draht — Sicherungsmittel

Eisens m fibrous fracture; ~e Textur f fibrous texture.

seide-umsponnener Draht m silk covered wire.

Seil n rope, cord, cable; ~ ohne Ende endless rope (cord); ~bahn f rope way, rope railway; ~bahn, Drahtseilbahn f wire way, wire-rope way, wire-rope tramway; ~betrieb m (Masch., Techn.) cable-working, rope-gearing, rope; ~förderung f, ~betrieb m auf geneigter Ebene transportation on inclined plane by means of a rope.

Seiler ... ~bahn f rope walk, rope yard; ~spule f spindle.

S-Eisen n S-iron.

Seite f side, face, flank; (eines Buches) page; (einer Gleichung) member, side of an equation; (eines Körpers [Geom.]) face of a solid; (schmale Kante f eines Brettes u. s. w.) edge, narrow side of a board.

Seiten ... ~ablagerung f von Erde (an der Straße) spoil, spoil bank; ~ansicht f, ~aufriß m side view, side elevation; ~befestigung f lateral fastening; ~entladung f lateral discharge; ~fläche f flat side, lateral face; ~kraft, Komponente f component force, component; ~linie (Eisenb.) siding line; ~verbindungen f/pl (PT) cross country connections; ~verstärkung f eines Balkens u. s. w. fish piece; mit ~verstärkung versehen v to fish; ~weg m by-way, banquette.

seitlich a lateral, collateral; ~er Drahtzug m (T) tensile strain.

Sektor m (HA) sector.

sekundär a secondary.

Sekundärstrom m, induzirter Strom, Polarisationsstrom m secondary current, secondary circuit.

Sekundawechsel m second of exchange.

selbst a (Bestellvermerk [P & T]) delivered into the addressee's own hands.

Selbst ... ~auslöser s. Apparat; ~auslösung f (der Papierführung des MA) self starting & stopping contrivance; ~betrieb m, ~verwaltung f self-management; ~bewegend a (Mech.) automatic; ~entzündung f spontaneous ignition; ~induktion f self-induction; ~kostenberechnung f cost account; ~kostenpreis m purchase price, prime cost; ~ständige Postanstalt f independent P. O.; ~thätig a (v. Masch.) self-acting, (v. Apparaten) automatic; ~thätigkeit f automatic action; ~übertrager m (T) automatic repeater; ~unterbrecher m (T) automatic circuit breaker, self-acting make & break; ~unterbrechung f self-acting making & breaking of the circuit; ~verbrennung f spontaneous combustion; ~wirkend a self-acting.

Semaphor, optischer Telegraph, Flügeltelegraph m semaphore.

senden v to send.

Sendung f (P) mail matter, mail article; (Handl.) consignment; (Schifff.) shipment.

Senk ... ~blei, Bleiloth n plumb line; ~kolben m (Masch.) countersink; ~leine f fathom line; ~rechte f perpendicular; ~waage f, Aräometer n (Phys., Chem.) areometer, hydrometer, hydrostatic level; ~waage, Bleiwaage f plummet level, levelling plummet.

senken v to let down, to lower; einen Schacht ~, abteufen to sink a shaft.

Senkung f (Neigung) pitching; (Phys.) inclination; (eingesunkene Stelle) sunken spot, hollow.

Serie f emission, series; in ~n arbeiten (T) s. Reihe.

setzen v (Techn.) to plant, to set; (in Betrieb) to set to work; (in Rechnung) to put to account; J—n in den Stand ~ to enable smb; J—n außer Stand ~ to disenable smb; sich ~ (Chem.) to settle down, to precipitate.

Shunt s. Nebenschluß.

Sicherheits ... ~modul m factor of safety; ~streifen m (an der Eisenb.) side space; ~stück n (an der elektr. Lichtleitung) safety plug; ~ventil n (Masch.) safety valve.

sichern v to secure, to make sure, to guaranty.

Sicherungsmittel n securing means, supporting (securing) piece.

Sicht *f* (Handl.) sight; auf ~ at sight; bei ~ on demand; kurze, lange ~ short, long sight; ~tage *m/pl* days of respite; ~wechsel *m* bill payable at sight, sight bill.

sichtbare Zeichen *n/pl* (*T*) visual signs; nicht ~ ~ (*T*) transient signs.

Sieb *n* sieve; ~artig *a* sieve-like (perforated copper plate).

siebenadriges Kabel *n* seven-wire cable, cable with seven conductors.

Siegel *n* seal; feiner ~lack letter sealing wax; Pack-~lack bag sealing wax; ~lackstange *f* stick of sealing wax; ~marke *f* paper seal.

siegeln *v* to seal.

Siel, Siehl *n*, **Deichschleuse** *f* dike lock, dike drain, sluice in a dike; (Kloake *f*) drain, sewer; ~deich *m* dike with a drain; ~graben *m* water course of a dike drain; ~wasser, Kloakenwasser *n* sewage.

Sielengeschirr, Sielgeschirr *n* breast harness.

Signal *n* signal; (zur Abfahrt [Eisenb.]) starting signal; ~ geben *v* to signal; ~apparat s. Apparat; ~glocke *f* signal bell; ~klappe *f* (*F*) annunciator; ~ordnung *f* (Eisenb.) signal code; ~pfeife, Dampfpfeife *f* whistle, steam whistle, (der Landbriefträger) whistle; ~scheibe *f* disc, signal disc.

Silber *n* silver; (gediegenes) native silber; (gemünztes) coined silver, silver coin; (geschlagenes ~, Blattsilber) beaten silver, leaf silver; ~barren *m* bar (ingot) of silver, silver ingot, silver bar, bullion; ~chlorid *n* (Chem.) chloride of silver, silver chloride; ~draht *m* silver wire; ~währung *f* silver standard.

Silizium *n* silicon, silicium; ~bronzedraht *m* silicious bronze wire.

sintern, zusammensintern *v* to sinter, to sinter together; (versteinern) to harden, to petrify.

Sinus *m* sine; ~-Bussole *f* sine galvanometer; ~-Tangenten-Bussole *f* sine-tangent galvanometer; ~versus *m* eines Bogens versed sine of an arc.

Siphon-Recorder *m* (*T*) siphon (syphon) recorder.

Situationsplan *m* site plan, plan of site.

Sitz ... ~kasten *m* (des Wagens) seat box; ~platz *m* seat, einen ~platz in der Postkutsche bestellen to secure a seat in the mail coach.

Skala *f* (Physi.) scale; (Gradeintheilung *f*) graduation; mit einer Gradeintheilung versehen *v* to graduate.

Skizze *f* (Zeichnung *f*) sketch, outline, rough draught; ~ nach dem Augenmaaß eye sketch.

skizziren *v* to sketch, to make a rough draught of.

Skonto *n* discount.

Sockel *m*, **Plinthe** *f* socle; (Mauerfuß) footing of a wall; (einer eisernen Telegr.-Stange) stone block, granite socket.

Soda *f*, **kohlensaures Natron, neutrales (einfaches) kohlensaures Natron** *n* soda, soda of commerce, carbonate of soda, neutral carbonate of soda (sodium); ~lauge *f* soda lye; ~salz, Natronsalz *n* soda salt.

sofort zurück (*P*) to be returned at once; ~ zum Protest (*P*) to be protested at once.

Sohle *f* (eines Tunnels, Flusses, Grabens) bottom.

Solawechsel *m* sola bill, single bill, sole bill of exchange; (eigener, trockener Wechsel) bill to order, promissory note, note of hand.

Soldaten ... ~brief *m*, eigene Angelegenheit des Empfängers (*P*) soldier's (seamen's, sailor's) letter, contents referring solely to the private affairs of the addressee; ~packet *n* soldier's (seamen's, sailor's) parcel, contents referring solely etc. wie vorstehend.

Solenoid *n* (elektro-dynamische Spirale *f*) solenoid.

solid *a* (fest) solid; (zuverlässig) creditable.

solidarisch *adv* each for the other, jointly & separately, solidarily.

Solidarität *f* joint liability, liability of each for the whole.

Soll, Sollen, Debet *n* debit; ~ und Haben debtor & creditor; ~ in einem Hauptbuch debit side of a ledger.

Sommerweg *m* summer road.

Sonde *f* (der Seeschiffe) plummet, sounding lead; (Kabelsonde) grapnel.

sondiren *v* to sound, to fathom.

Sonstiges *a* (Bezeichnung des Raumes auf Postanweisungen, Packetadressen u. s. w., auf welchem Privatmittheilungen gemacht werden dürfen) particulars, space for particulars.

Sortenzettel *m* (Zettel mit Spezifikation der abgelieferten Geldsorten) specification (itemizing) of the money remitted.

Sortir ... ~beamter *m* (*P*) sorter, sorting clerk; ~briefträger *m* sorting carrier; ~fach *n* sorting case; ~schrank *m*, ~spinde *n* cupboard; ~tisch *m* distributing table.

sortiren *v* (*P*) to sort; grob ~ (*P*) to sort for the first time; fein ~ (*P*) to re-sort.

Sortirer *m* sorter, s. auch Sortir ...

Spalt *m* (Techn.) rent, crevice, fissure, claw, (im Draht) split; ~feile *f* slitting file, blade file; ~keil *m* wedge for cleaving.

Spalte *f* (Riß, Sprung *m*) rent, chink, cleft; (Schlitz *m*) slot; (Kolumne *f*) column; (im Holz) chink, cleft, flaw; (in einer Mauer) crevice, flaw.

spalten *v* to cleave, to split; sich ~ to split, to chink, to chop, to chap; sich gabelweise ~ to fork.

Span *m* a thin piece of wood, chip; Späne *pl* shreds, chips, shavings.

Spann ... ~feder *f* tension spring; (Druckfeder) split spring; ~isolator *m* (*T*) shackle, terminal insulator, stretching insulator; ~kappe *f* (des Isolators) straining cap; ~kette *f* stretching chain; ~kraft, (Elastizität, Expansivkraft *f* elasticity, elastic force, (des Dampfes) expansibility of steam, (der Dämpfe) tension of vapours; ~riegel *m* straining beam, strutting piece; ~rolle *f* tension roller, expanding roller, tightening pulley; ~schraube *f* draw vice; ~stange, ~strebe *f* (*T*) strut; ~vorrichtung *f* wire stretcher; ~weite *f* (*T*) span, äußerste ~weite, die einem Drahte gegeben werden kann limiting span.

spannen *v* to stretch (the wire), to strain; (eine Feder) to bend a spring; (den Dampf) to increase by heat the expansive power of steam.

Spannung *f* tension, strain; freie ~ free tension; (des elektrischen Stromes) tension, potentiel; straffe ~ eines Drahtes tight strained wire, wire pulled up tight, zu lose ~ der Drähte slack wires; ~smesser *m* (Dampfmasch.) indicator; ~srolle *f*. Spannrolle.

Spar ... ~marke *f* savings stamp; ~karte *f* savings form; ~kasse *f* savings bank; ~kasseneinlage *f* deposit; Post-~kasse *f* postal savings bank; ~ u. Vorschuß-Verein *m* s. Post-Spar u. s. w.

spät *a* late; Zu-~kommen *n* in den Dienst (*PT*) late attendance.

Spaten *m* spade; ~stich *m*, den ersten ~ thun (Eisenb. u. s. w.) to turn the first sod.

Spätlings ... ~brief *m* (*P*) late letter; ~briefpacket *n* (*P*) late letter packet; ~gebühr *f* (*P*) late letter fee; ~kartenschluß *m* (*P*) late letter mail.

spediren *v* (*P*) to send, to forward, to transmit, to despatch; unrichtig ~ to mis-send.

Spediteur *m* forwarding agent, transmitter, consignee, carrying & forwarding agent.

Spedition *f* forwarding, transmission of goods; ~sbureau, ~setablissement *n* forwarding office, carrying establishment; ~sgeschäfte *n/pl* forwarding business, carrying traffic, carrying & forwarding; ~skosten *pl* forwarding charges, charges of transmission; ~sliste *f* (*P*) circulation list; ~srechnung *f* bill of conveyance.

Speiche *f* spoke, (ausgekehlte) curved, bevelled spoke, (hohle) tubular spoke, (am Zeiger-Apparat) arm; ~nloch *n* spoke mortise; ~nrad *n* spoke wheel; ~ntriebrad *n* (Masch.) spider; ~nzapfen *m* sharp end of a spoke.

Speicher *m* (Dachboden *m*) loft, garret; (Lagerhaus *n*) warehouse, store house, magazine.

speisen *v* to feed (an engine), to supply (a battery, the printing roller with ink).

Spektral=Analyse *f* spectral analysis, spectrum analysis.

Spektrum *n* spectrum; Sonnen-~ solar spectrum.

Spekulant *m* speculator, commercial adventurer.

Spekulation *f* speculation, adventure, venture, enterprise.

spekuliren *v* to speculate.

Sperr... ~feder *f* click spring; ~gut *n* bulky goods; ~hahn *m* stop cock; ~haken *m* click; ~kegel *m* detent; ~klinke *f* catch, click, ratchet; ~rad *n* rack wheel, ratchet wheel, click wheel; ~vorrichtung *f* click and ratchet wheel; ~zahn *m* tooth of a ratchet wheel, cog.

sperren *v* (Masch.) to stop, to catch; (einen Hafen) to shut up (to blockade) a port; (eine Straße) to stop a road; ~, bremsen (Mech., Masch.) to stop a movement, to catch, to lock.

sperrig, voluminös *a* bulky, pesterable; ~es Gut *n* bulky goods.

Sperrung *f* (Mech., Masch.) stop, catch, lock; (Hemmung *f*) escapement; (eines Hafens) blockade or a port; (des Handels) embargo.

Spesen *f/pl* charges, expenses; ~frei *a* free (clear, exempt) of charges; ~rechnung *f* account (note) of charges.

Spezial... ~karte *f* topographic map; ~rechnung *f* detailed account; ~vollmacht *f* special power of attorney, warrant of attorney.

spezifisch *a* specific; ~es Gewicht *n* specific gravity (weight); ~e Wärme *f* specific heat; ~er Widerstand *m* (*T*, Elektriz.) specific resistance.

Spiegel *m* looking glass, mirror; (Phys.) speculum, reflector; ebener ~, Planspiegel plane mirror; Hohl~ concave mirror; Konvex~ convex mirror.

Spiegel... ~ablesung *f* mirror reading; ~bild *n* reflected image; ~eisen s. Eisen; ~fläche *f* smooth surface; ~galvanometer *n* astatic mirror (reflecting) galvanometer; ~magnet *m* magnet carrying a mirror.

spiegeln *v* to reflect, to throw back.

Spiegelung *f* reflection; (Opt.) mirage.

Spiel *n* des Hebels nach unten, nach oben (*MA*) play of the lever downward, upward play of the lever; (einer Maschine) working of an engine.

Spielraum *m* range (of the lever between the limiting stops); ~ haben (Masch.) to work loose; den ~ zwischen zwei arbeitenden Theilen vergrößern (Masch., Mech.) to ease a working part.

spielen *v* (Masch., Mech.) to play, to work, to have play, to work easy.

Spind, Spinde *n*, **Schrank** *m* cupboard; Sortir=~ s. Sortir...

Spindel *f* (Techn.) spindle, pivot, axis; (dünne, stehende Welle *f* [Masch.]) spindle, arbor; ~ einer Achse (Masch.) axle spindle; ~ einer Presse, Preßspindel male screw of a press spindle; ~ einer Schraube, Schraubenspindel, auswendige Schraube *f* outside screw, male screw; ~baum *m* (Mech.) beam of a spindle; ~blitzableiter s. Blitzableiter; ~bohrer, Zentrumbohrer *m* centre bit; ~getriebe *n* (Masch.) skew-bevil wheel work; ~hemmung *f* (Mech.) crown escapement, verge escapement; ~presse, Schraubenpresse *f* (Mech.) screw press; ~zapfen *m* (Masch.) pivot.

Spiral... ~feder *f* spiral spring, main spring; ~förmig *a* spiral, helicoid, helical; ~linie s. Spirale.

Spirale, Spirallinie *f* (Geom.) spiral, spiral line, scroll, helical line, helix; (Drahtspule *f*) coil, wire coil; (Spule *f* [*T*]) coil, spiral, spiral coil s. auch Spule.

Spiritus *m* spirit; ~lampe *f* spirit lamp.

spitz *a* (Techn.) pointed, taper, tapering; ~er Winkel *m* (Geom.) acute angle; ~ machen *v* to point; ~winkelig *a* (Geom.) acute-angled, acute-angular, oxygonal; ~winkeliges Dreieck *n* oxygon, acute-angled triangle.

Spitz ... ~bohrer *m* centre bit; ~feile *f* taper file, sharp (pointed) file; ~hacke, ~haue *f* pick, pick axe; ~zange *f* nibbing pliers.

Spitze *f* (Techn.) point, top; (eines Dreiecks) vertex, summit; (Elektriz.) point; (Stachel, Dorn *m*) spur; (eines Kegels, einer Pyramide) vertex, apex; (des Löthrohrs) nozzle of the blow pipe; mit einer ~ versehen *a* (Geom.) cuspidal.

Spitzen ... ~schraube *f* steel pointed screw; ~wirkung *f* (Elektriz.) power of points; ~blitzableiter s. Blitzableiter.

Spleißen s. Splissen.

Splint *m* (Keil *m*) splint, peg, key; (Vorsteckagel *m*) forelock, fox; (eines Baumes) sap, sap wood, auber; ~bolzen *m* (Techn.) eyebolt & key, forelock bolt; ~bolzen eines Hebezeuges (Masch.) shackle bolt.

splissen *v* (zwei Hölzer) to scarf; (zwei Taue) to splice.

Splißstelle, Splissung *f* (zweier Taue) splice.

Splitter *m* (Techn.) splinter, splint; (im Draht) scale.

splittern *v* to splinter, to split into fragments.

Spornrädchen *n* (Techn.) spur rowel, rowel of a spur, mullet.

Sporteln, Nebeneinkünfte *pl* emoluments.

Sprache *f* language; Telegramm in chiffrirter ~ cypher telegram; Telegramm in gewöhnlicher ~ ordinary telegram; Telegramm in verabredeter ~ telegram in conventional language, code telegram.

sprachwidrige Verbindung *f* (sprachwidrige Zusammenziehung *f* von Wörtern [T]) words incorrectly joined together contrary to the usage of the language.

Sprech ... ~telegraph *m* telephone; ~strom *m* (T) working current.

Spreng ... ~arbeit, Bohr- und Schießarbeit *f* shooting & blasting; ~bohrer *m* borer for shooting & blasting, jumper; ~bohrloch *n* blast hole; ~ladung *f* bursting charge; ~mittel *n/pl* explosives, blasting agents; ~pulver *n* blasting powder; ~zünder *m* blasting fuse.

sprengen *v* (Gestein) to blast, to spring, to blow up, to shoot & blast (stones).

Sprengen *n*, **Sprengung** *f* (von Gestein) blasting, blowing up.

Spring ... ~feder *f* (Mech.) spiral metallic spring, elastic spring; ~federwaage, Federwaage *f* spring balance.

springen *v* (Techn.) to spring, to crack, to burst; eine Feder ~ lassen to relax a spring.

Spritzleder *n* (am Wagen) splattering leather, splash leather.

spröde *a* (Techn.) brittle (iron), short, rough.

Sprödigkeit *f* brittleness (said of iron), roughness.

Sprosse *f* (einer Leiter) step, round, rundle of a ladder; (einer Stangenleiter) peg of a peg ladder.

Sprung, Riß, Spalt *m* (Techn.) cleft, crack, flaw; (im Metall) crack; einen ~ bekommen to crack.

Spule *f* (T) bobbin, spool, coil; induzirende ~ inducing coil; induzirte ~ secondary coil; primäre ~ primary coil.

Spur *f* (Radspur *f*, Gleis *n*) track; (Spurweite *f*) width of track, width between the rails; (Rinne *f* einer Rolle, Seilscheibe *f*) groove of a rope pulley; keine ~ finden to find no trace (of a lost letter); auf die ~ kommen to trace; ~rad *n* spur wheel, straight wheel; ~weite *f* (Eisenb.) gauge of way, railway gauge, width of track, width between the rails; ~zapfen *m* (Masch.) pin, pivot, vertical pivot.

Staat *m* state; Vereinigte ~en von Amerika United States of America (U.S.); ~enbund *m* confederation, confederacy.

Staats ... ~amt *n* public office; ~anleihe *f* government loan; ~aufsicht *f* superintendence (control) of the state; ~bank *f* national bank; ~beamter *m* (civil) officer, government employee (official, agent); ~behörde *f* office of the state, authority, (*Am*) state board; ~dienst *m* public service; ~eigenthum *n* state property; ~eisenbahn *f* railway of the state; ~haushalt *m* finances (of a state); ~kasse *f* exchequer, public treasury; ~kosten *pl* public expenses; ~secretair *m* Secretary of state; ~telegramm *n* government telegram, telegram on government service (on official business); ~wirthschaft *f* political economy.

Stab *m* (Techn.) staff, stick, rod; (von Metall) bar; (Leiste) ledge; ~eisen *n* bar iron; ~magnet *m* bar magnet, straight magnet.

stabil *a* (Physs.) stable; ~es, sicheres Gleichgewicht *n* (Mech.) stable equilibrium.

Stabilität, Standfestigkeit *f* (Physs., Mech.) stability (of a t. line).

Stadt ... ~bestellbezirk *m* (*P*) town post delivery; ~briefträger *m* town letter carrier; ~fernsprecheinrichtung *f* telephone exchange; ~fernsprechvermittlungsamt *n* telephone office; ~gemeinde *f* township, municipality, community; ~linie *f* (*T*) town line; ~postanstalt *f* town (local) post office; ~postbote *m* messenger, (Briefkastenleerer) letter-box clearer; ~telegramm *n* city message.

Stafette *f* courier, express, estafete, estafette.

Stahl *m* steel; angelassener, geglühter ~ annealed, tempered steel; Cement-~ cementation steel; entharteter ~ softened steel; gehärteter ~ hardened steel; ~dorn *m* (zum Untersuchen der Telegr.-Stangen) steel spike; ~draht *m* steel wire; ~eisen *n* steely iron, natural steel; ~magnet *m* steel magnet; ~platte *f* steel plate; mit ~spitzen versehen *v* to edge (to point) with steel.

stählen *v* (in Stahl verwandeln) to convert into steel; (verstählen) to steel, to edge (to point) with steel.

stählern *a* of steel, steel.

Stamm *m* (Baumstamm) trunk, stem of a tree; ~ende *n* (einer Stange) butt end of a pole; ~faul *a* rotten in the heart (trunk); ~holz *n* trunk wood, stock wood; ~vermögen *n* (einer Handlungsfirma, Aktiengesellschaft u. s. w.) capital, stock, fund.

Stampfe *f* stamp, stamper; (Ramme *f*) ram.

stampfen *v* (Techn.) to stamp, to beat; die Erde fest~ to ram, to beat down the earth.

Stand *m* (Techn.) position; (Lebensstellung) situation; (Rang) rank; zu ~e kommen to be brought about, to be achieved; in gutem ~e sein to be in good order; in gutem ~e halten to maintain in good condition, to keep well in repair; im ~e sein to be able; in ~ setzen to repair (a t. line, a building etc.), to enable (*smb* to do *smt*).

Stand ... u. **stand** ... ~eslifte *f* descriptive list; ~fähig, ~fest *a* stable; ~festigkeit *f* stability; ~punkt *m* point of view, (einer Stange) s. **Stangen** ..; ~weite *f* distance, interval.

Ständer *m* (große, starke Stange) upright, post, wooden pillar; (Pfeiler *m* [Masch.]) standard.

Standesliste s. **Stand** ...

ständig *a* im Dienste verwendet (*PT*) employed continuously upon a service; nicht ~ not employed continuously, only temporarily employed.

Stange *f* pole, post, mast, support; (Doppelständer *m*) A pole; angeschuhte ~ fished (shoed) pole; gekuppelte ~n *pl* coupled poles; zubereitete ~ prepared (injected, impregnated) pole, (mit Kupfervitriol) pole prepared with sulphate of copper, boucherized pole, (mit Zinkchlorid) pole prepared with chloride of zink, (mit Quecksilbersublimat) pole prepared with bichloride of mercury, kyanized pole, (mit Kreosot) creosoted pole; verankerte ~

stayed pole, pole tied with a wire rope, trussed pole; verstrebte ~ strutted (propped) pole.

Stangen ... ~abstand *m* distance, interval, f. auch Abstand; ~blitzableiter f. Blitzableiter; ~bohrer *m* auger, f. auch Bohrer; ~eisen *n* rod iron, bar iron; ~kappe *f* (*T*) cap; ~linie *f* (*T*) line of poles; ~loch *n* (*T*) hole, excavation for a pole; ~register *n* specification of the poles; ~reihe *f* (*T*) line of poles; ~schuh *m* sleeper; ~stamm-Ende *n* f. Stamm; ~standpunkt *m* (*T*) position of a pole, site (marked for a pole); ~zubereitungsanstalt *f* (*T*) yard for injecting and preparing timber poles, impregnating establishment; ~regulator *m* (*HA*) governor.

Stanniol *n*, **Zinnfolie** *f* tin foil, stain; ~belegung *f* tin-foil coating; ~blätter *n/pl* tin-foil sheets.

Stanze, Stampfe *f* (für Metalle) die, stamp.

Stapel *m* pile, heap (of boards etc.); (Stapelplatz) staple, emporium, market, mart; ein Schiff vom ~ laufen lassen to launch a ship.

Stärke *f* (Techn., Mech.) strength, thickness.

stätig *a* steady (burning of the electric light).

Stätigkeit *f* steadiness, steady burning (of the electric light).

Statik, Gleichgewichtslehre *f* (Phys.) statics; ~ der festen Körper, Geostatik statics of rigid bodies; ~ der tropfbar flüssigen Körper hydrostatics; ~ der luftförmigen Körper aërostatics, statics of elastic fluids.

Station *f* station, office (f. auch Amt); ~sbatterie *f* local battery; ~seinführung *f* (*T*) joining of the wires outside to the wires inside the office; leading-in of the wires, (f. auch Einführung); ~seinführungsbrett *n* (*T*) connection board; ~seinrichtung *f* (*T*) fitting up of an office, arrangement of the wires in an office; ~sleitung, Eisenbahn-Telegr.-Leitung *f* line wire; (eines Beamten) location; ~sstellung *f* (*T*) the receiving instrument is in circuit; ~svorsteher *m* (Eisenb.) station master; ~swagen *m* (*MT*) telegraph wagon, office waggon, (*P*) office wagon.

statisch *a* (Phys.) static, statical.

Statistik *f* des Waarenverkehrs statistics of the goods-traffic.

stauen *v* (die Güter) to stow the goods; (das Wasser) to pen (to dam up) the water.

Stauung *f* (der Güter) stowing, stowage; (des Wassers) swelling, swell, retaining of water.

Steckbrief *m* warrant of caption.

Steg *m* (Techn.) cross piece; (einer Schiene, des doppelt T Eisens u. f. w.) web; (einer Brücke) small (wooden) bridge, plank; (Weg *m*) foot-path.

Stehpult *n* desk for persons who write in standing.

stehend, fest *a* stationary, fixed; ~e Maschine *f* vertical engine; ~e Welle *f* (Masch.) upright shaft, spindle.

steif *a* (Techn.) stiff, rigid.

Steife *f* (Stütze *f*) prop, strut, stay.

steifen *v* (absteifen) to prop, to support.

Steifheit, Steifigkeit *f* (eines Seils) rigidity.

Steig ... ~eisen *n/pl* (Klettereisen [*T*]) climbing irons, climbing spurs; ~rad, Hemmungsrad *n* escapement (escape) wheel, balance wheel; ~rohr *n*, ~röhre *f* (Masch.) rising pipe, lifting tube, column lift, raising pipe, (Standrohr *n*) column pipe, stand pipe.

steigen *v* to rise, to mount; (vom Preise) to advance, to get up.

Steigung *f* (Topogr., Eisenb.) ascent, gradient, rising; (Ganghöhe einer Schraube) pitch; schwache ~ easy gradient; starke, steile ~ heavy (steep) gradient; zunehmende ~ increasing pitch; ~sanzeiger *m* gradient post; ~sverhältnis *n* pitch.

steil *a* steep; ein ~es Ufer *n* an acclivous shore.

Stein *m* stone; ~arbeit *f* stone work; ~block *m* stone block; ~boden *m* stony ground; ~bohrer *m* stone borer, (Berg- od. Sprengbohrer) rock drill, jumper; ~böschung *f* stone batter; ~bruch *m* quarry; ~damm *m* causeway; ~deich *m* dike built of stones; ~eiche *f* red oak; ~eisen *n*, ~meißel *m* chisel for working in stone; ~gut *n* stone ware, glazed earthen ware; ~haue *f* stone pick; ~klammer *f* cramp for fastening stones together; ~kohle, Schwarzkohle *f* coal, common coal, mineral coal; ~kohlentheer *n* coal tar; ~kohlentheer-Asphalt *m* coal tar asphaltum; ~lager *n* stone bed; ~mauerwerk *n* masonry; ~meißel *m* ſ. ~eisen; ~öl *n* earth oil, rock oil, petroleum, naphtha; ~pappe *f* cartonpierre, statuary pasteboard; ~pfeiler *m* stone pier; ~pflaster *n* stone pavement; ~pickel *m* double-pointed pick; ~platte *f* slab, plate of stone; ~porzellan *n* hard porcelain; ~schicht *f* layer of stones; ~schraube *f* screw to fasten iron brackets in masonry.

steinern, steinen *a* stone, of stone.

steinig *a* stony.

Stell... u. stell... ~bar *a* adjustable; ~mutter *f* jam nut, check nut; ~rad *n* regulating wheel, regulator; ~schraube *f* set screw, adjusting screw; ~stift *m* set pin; ~vertreter *m* substitute (for a clerk), (Bevollmächtigter *m*) proxy; ~vertretung *f* substitution, in ~vertretung (bevollmächtigt *a*) by proxy; ~vertretungskosten *pl* pay for a substitute, extra payment to substitutes; ~wagen *m* stage coach.

Stelle *f* place, spot; (eines Buchs) passage; (Amt, Posten) situation, place; an ~ instead of.

Stemm... ~eisen *n* driving chisel, bur chisel; ~meißel *m* caulking chisel.

Stempel *m* stamp; (Waarenmarke) mark; ~abdruck *m* impression (of a stamp); ~apparat *m* stamping apparatus; ~bogen *m* stamped sheet (of paper); ~eisen *n* stamping iron; ~farbe *f* stamping ink; ~gebühr *f* stamp duty, stamp fee; ~kissen *n* stamping pad; ~marke *f* zur Entrichtung der statistischen Gebühr stamp by means of which the statistic fees are levied; ~maschine *f* stamping machine; ~papier *n* stamped paper; ~tisch *m* (*P*) stamping table; ~unterlage *f* cushion; ~verwendung *f* application of stamps; ~werthzeichen *n* stamped form; Datum-~, Tages-~ date stamp; ~ zur Entwerthung der Freimarken (*P*) obliterating (defacing) stamp.

stempeln *v* to stamp, to mark, (*Am*) to postmark.

Stempelung *f* application of the stamp.

Sterbe... ~jahr *n*, ~monat *m* year (month) of a person's death; ~urkunde *f* document of a person's decease, certificate of death.

Stern *m* star, (Buchdr.) asterisk; ~förmig *a* in the shape of a star; ~rad *n* star wheel; ~warte *f* observatory.

Steuer *f* tax, duty; (Grundsteuer *f*) ground rent; eine ~ ausschreiben to make an assessment; ~amt *n* board of taxes, custom office; ~stelle *f* branch custom office; ~übertretung *f* der Postillone defraudation of customs committed by a mail cart driver.

Stiefel *m* (Techn.) tube, case, shank.

Stift *m* little nail, tack; (Techn.) pin; eiserner ~ iron pin; Morse-~ style, pricker; ~büchse *f*, ~gehäuse *n* (*HA*) box (containing the contact pins); ~(Relief-)schreiber *m* (*T*) embosser.

Stilllager *n* (*P*) stop; während des ~s (*P*) when off travelling duty.

Stillstand *m*, **Stillstehen** *n* (Masch.) standing still, stop.

Stimme *f* voice; (bei Wahlen) vote.

stimmen *v* (Musik) to tune; (übereinstimmen) to agree, to tally; die Bilanz stimmt the balance squares; die Bücher ~ mit einander the books agree with each other; die Rechnung stimmt nicht the account does not agree.

stimmend *a* (von Rechnungen u. ſ. w.) true.

Stimmgabel *f* tuning fork.

Stipendium *n* allowance; Reise-

~ travelling allowance; Studien-~ allowance to students.

Stirnrad n spur wheel, cylindrical wheel; ~ und Getriebe n driver wheel and pinion; Stirnräder pl, Stirnräderwerk, Stirnrädergetriebe n spur gear, spur gearing.

stocken v to stop; (stillstehen) to grow dull, (vom Verkehr) to languish, to come to a standstill.

Stockung f stopping, stagnation, standstill (vgl. stocken).

Stoff m, **Materie** f (Phys.) substance, matter; (Chem.) body, substance; (gewebtes Zeug) stuff, cloth.

stopfen v (anfüllen) to stuff, to fill; (stoppen) to stop.

Stöpsel m (T) plug, peg, pin, stopper; ~fehler m (T) mistake made by inserting the plug in a wrong place; ~umschalter s. Umschalter; ~ einsetzen s. stöpseln; ~ entfernen (aus dem Rheostaten), Widerstand einschalten to unplug.

stöpseln v, **Stöpsel einsetzen** to plug, to stopper; (= Widerstandausschalten im Rheostaten) to plug.

Störung f (Fehler in der Leitung [T]) fault; (Unterbrechung der telegr. Verbindung) interruption; (Drahtberührung) contact; (Drahtverschlingung) cross, (in Folge schlechten Wetters) weather cross; (durch Gewitter verursachte) lightning contact; (durch magnetische Ströme) perturbation.

Störungs ... ~scheibe f (Eisenb.) signal in case of telegraphic interruption; ~tagebuch n (T) journal of faults; ~ursache f cause of interruption.

Stoß m (Zusammenstoß) collision; (einer Maschine) jerk; (Erdbeben) shock, concussion; (ein ~ Akten) a pile of papers.

Stößer m (HA) rejector, rejecting plate.

Straf ... ~bestimmungen f/pl für Post- und Portoübertretungen penalties for infringements of the postal laws; ~erlaß m remission of punishment (of a fine); ~fällig a subject (liable) to punishment; ~festsetzungen f/pl penal laws; ~gelder n/pl s. Geldstrafe; ~gesetzbuch n penal code; ~liste f (P) book (register) of fines; ~porto n (P) surcharge; ~verfahren n in Post- und Portoübertretungsfällen judiciary proceeding in case of infringement of the postal laws; ~versetzung f compulsory change of residence as a mode of punishment for some breach of discipline.

Strafe f punishment, (für Vergehen gegen die postgesetzlichen Bestimmungen) penalty, (im Disciplinarwege z. B. wegen Zuspätkommens verhängte Geldstrafe) fine; eine ~ (Geldstrafe) verhängen to inflict a fine, s. auch strafen.

strafen v to punish, to impose a penalty, to inflict a fine.

straff a stretched, straight, tight; ~er Anker m (T) tight strained tie; ~ spannen v to strain; ~ gespannte Leitung f (T) tight strained wire, wire pulled up tight.

Strahl m (des Lichts) ray, beam of light; (einer Flüssigkeit) vein, spout, stream, jet; (einfallender) incident ray; (gebrochener) refracted ray; (zurückgebrochener) reflected ray.

Strahlen ... u. **strahlen** ... ~brechend a (Phys.) refracting; ~brechung f refraction; ~bündel n, ~büschel m luminous pencil, luminous brush, (elektrisches) electric aigret (aigrette); ~förmig a radiated; ~kegel m (Phys.) cone of rays.

strahlen v (Phys.) to radiate; ~de Wärme f radiant heat.

Straße f street; (Landstraße, Heerstraße) road; (Gasse f) lane; (chaussirte) high road, main road; (gepflasterte) paved road; (mit festem Steingrund versehene, Schotterstraße) ballast road; eine ~ entwerfen to lay out a road.

Straßen ... ~aufschüttung f road embankment; ~-Aufseher m surveyor of roads; ~damm m causeway; ~graben m road ditch, road drain; ~kanal, Abzugskanal m sewer; ~rinne f gutter; ~seite f front side; der Isolator ist auf der ~seite angebracht the insulator is placed at the front of the pole; ~übergang m,

Kreuzung *f* crossing; ~verkehr *m* traffic.

Strebfestigkeit, Zerknickungs=festigkeit *f* compressional strength, composed strength of compression.

Strebe *f* (*T*) strut, prop; ~n=holz, ~nklotz *m* stop, brace, ground beam, block of wood, cross piece; ~nschraube *f* screw for fastening struts to poles.

Streck ... u. streck ... ~bar *a* ductile, malleable; ~barkeit, Zieh=barkeit *f* ductility; ~zange *f* stretching tongs.

Strecke *f* (Baustrecke *f*, Bauab=schnitt *m*) section, length of line; horizontale ~ (Eisenb.) horizontal track, level; ~nwagen *m* truck (to carry materials), (Bahnmeisterwagen) trolly.

strecken *v* to stretch, to strain (the wire).

Streicheisen, Glätteisen *n* tooling iron.

streichen *v* (magnetisiren) to touch with a magnet f. Strich; (ausstreichen) to blot out, to strike, to erase; aus einer Liste ~ to strike off a list.

Streichung *f* (gestrichene Stelle) a word, a sentence stricken out.

Streifband *n* (*P*) wrapper.

Streifen *m* (Morse=, Hughes=~) slip, tape, Morse slip, Hughes slip; (Land) strip of land; (Messing) strip of brass; leuchtende ~ *pl* des elektr. Lichtes luminous striae of electric light.

Strich *m* (der Morseschrift) dash; (beim Magnetisiren) touch, einfacher ~ single touch, doppelter ~ double touch, getrennter ~ separate touch.

Strom *m* (galv.) current; (großer Fluß) stream; ~ableitung *f*, Neben=schluß *m* (*T*) derivation, escape, leakage; ~bahn *f* (*T*) course (path) of the current, passage of the current; ~brecher *m* (*T*) circuit breaker, (automatischer) self acting make & break; ~einheit *f* unit of current; ~fähigkeit *f* continuity, electrical continuity, conductivity; ~gebung *f* (*T*) sending of a current; ~intensi=tät *f* intensity of current; ~impuls *m* short current, pulsation; ~kreis *m* (*T*) circuit, (Linienkreis) main (primary) circuit, (der Lokalbatterie) local (secondary) circuit; ~kurve *f* curve of the current; ~lauf *m* (als Bild) connections; ~los *a* (*T*) without current; ~richtung *f* (*T*) direction of the current; ~schema *n* (*T*) circuit arrangements, diagrams; ~schwankung *f* (*T*) fluctuation of the current; ~stärke *f* (*T*) strength (power) of current; ~theilung *f* (*T*) division of the current; ~umkehrung *f* (*T*) reversal of current; ~unter=brechung *f* (*T*) interruption, breaking the current; ~verlust *m* (*T*) loss of current; ~verstärkung *f* (*T*) current strengthening, strengthening of the current; ~verzögerung *f* (*T*) retardation of current; ~weg f. ~bahn; ~wender *m* (*T*) commutator, permutator, switch, gyrotrope; den ~ öffnen, schließen (*T*) to open (to close) the circuit; abgehender ~ (*T*) out-going current; ankommender ~ (*T*) in-coming current; Ausglei=chungs=~ (*T*) equating current; elektrischer ~ electric current; ent=gegengesetzter ~ inverse (opposite) current; (Entladungs=~ discharging current; Erd=~ earth current; Erd=platten~ earth plate current; galva=nischer ~ galvanic current; Gegen~ counter current; Hülfs~ auxiliary current; Induktions=, induzirender ~ inducing (primary) current, indu=zirter ~ induced (secondary) current; intermittirender ~ intermittent current; Ladungs=~ charging current; pulsatorischer ~ pulsatory current; Rück~ return current; umge=kehrter ~ reversed current; Undula=tions=~ undulatory current; Wechsel=ströme *m/pl* alternating currents.

Struktur *f* structure, texture; blätterige ~ lamellar structure; faserige ~ fibrous structure; körnige ~ granular structure; sehnige ~ fibrous structure.

Stück *n* piece; ein ~ Arbeit a piece of work; Arbeit auf's ~ piece work; auf's ~ arbeiten to do piecework; ~weise Beförderung von Post=stücken despatch in open mails.

Stufen ... u. **stufen** ... ~folge *f* gradation; ~förmig *a* in the form of steps; ~rad *n* wheel in steps.

stumpf *a* blunt; ~er Winkel *m* obtuse angle; ~ machen *v* to blunt, to dull; ~kantig *a* blunt-edged; ~winkelig, ~eckig *a* obtuse-angled.

Stunden ... ~plan *m* time table; ~platte *f* (am Briefkasten) notice plate, small movable (enamelled, iron, brass) plate with the hours (of clearing) painted or engraved thereon; ~type *f* (P) type of the date stamp denoting the hour of posting; ~zettel *m* (P) time bill; ~zetteltasche *f* (P) pouch.

Stundung *f* des Portos crediting of the postage due; ~sbuch *n* (P) account of postage credited; ein ~sbuch unterhalten to keep a deposit account; ~sgebühr *f* fee for crediting the postage due.

Sturm *m* storm, tempest; ~beobachtungen *f/pl* observations of storms; ~warnungssignal *n* storm warning signal.

Stütze *f* support, prop, stay; gerade Isolator=~ stalk, pin, bolt; hakenförmige Schrauben~ mit Gewinde curved iron bracket with screw thread.

Stützpunkt *m* support, point of support; (Unterstützungspunkt) fulcrum, bearing, s. auch Drehpunkt; ~snachweis *m* specification of the supports.

Subvention *f* (Beihülfe *f*) subsidy.

subventioniren *v* to subsidize.

Submission *f* tender; in ~ vergeben to give out in tender; ~sbedingungen *f/pl* specification; ~s-Einladung *f* tenders wanted, inviting tenders.

Süd ... ~licht *n* southern light; ~magnetismus *m* south magnetism; ~pol *m* south pole.

summarisch *a* u. *adv* in bulk; ~es Verzeichniß *n* summary list.

Summe *f*, **Betrag** *m* sum, amount.

Summen *n* der Leitungsdrähte (T) humming of the wires.

summiren *v* to sum up, to cast up (an account).

Sumpf *m* marsh, bog, swamp; ~boden *m* marshy ground; ~eisen *n*, ~eisenstein, Raseneisenstein *m*, ~erz *n* bog iron ore, moor ore.

sumpfig *a* marshy, boggy.

Superrevision, Nachprüfung *f* re-examination.

suspendiren von Dienstgeschäften s. entheben.

Suspension s. Enthebung.

synchron, gleichzeitig *a* synchronous, isochronous.

Synchronismus *m* synchronism, isochronism.

T.

T-Eisen *n* T-iron; doppeltes ~ ~ double T-iron, H-iron.

tabellarische Uebersicht *f*, **tabellarischer Bericht** *m* tabular statement.

Tabelle *f*, **Register** *n* table, register.

Tachometer, Tachymeter *n*, **Tachygraph, Geschwindigkeitsmesser** *m* (Mech.) tachometer, timing apparatus.

Tackbolzen, Bolzen mit Widerhaken *m* rag bolt, barb bolt.

Tafel, Platte *f* table, plate, slab; (Blechtafel) sheet; (Kupfertafel) copper sheet; (Schiefertafel) slate; (Tabelle *f* [Math.]) table; (Aushang *m*) notice; ~förmig *a* tabular, lamellar; ~land, Hochplateau *n* table land; ~waage, Brückenwaage *f* weighing machine, weigh-bridge; ~zinn, sächsisches Zinn *n* Saxon tin.

Täfelung *f*, **Täfelwerk** *n* (am Fußboden) parquetry; (Wandverkleidung *f*) wainscot, wainscotting, panelling.

Taffet, Taft *m* taffeta, taffetas; ~papier *n* satin paper.

Tage ... ~arbeit *f* day work; ~arbeiter s. ~löhner; ~buch *n* journal, day book, account of daily transactions; ~geld n, ~gelder n/pl daily allowance; ~lohn m daily wages; ~löhner m day laborer, journeyman; ~weise a by the day; ~werk n day work, day's task.

Tages ... ~arbeit f work of a day; ~bericht m daily account; ~dienst m (TP) the office is open for transaction of business during the day, day-duty; ~kurs m course of a day; ~ordnung f order of the day; ~platte f am Briefkasten (P) notice plate with the day (of clearing) printed or engraved thereon; ~stempel m (P) date stamp (f. auch Aufgabestempel); ~type f der Stempel (P) type of the date stamp indicating the day of posting; ~zeit f day time, time of the day.

takeln, antakeln, auftakeln v (ein Schiff) to tackle, to rig a ship.

Takelung f, **Takelwerk** n, **Takelage** f (das gesammte Tauwerk eines Schiffes) tackling, rigging, rig, set of rigging; ein Schiff mit ~ versehen to rig a ship.

Talg m tallow; mit ~ einschmieren to tallow, to grease; ~licht n tallow candle; ~stein, Speckstein m common talc.

Talk m (Talkstein m) talc; (gemeiner ~, Speckstein m) common talc, steatite; (schreibender ~, Kreide f) chalk; ~erde, Bittererde, Magnesia f, Magnesiumoxyd n (Chem.) magnesia, oxide of magnesium; ~spath, Bitterspath, Magnesit, Magnesitspath m, ~erde f magnesite, native carbonate of magnesia, neutral anhydrous carbonate of magnesium.

Talon m (eines Werthpapiers) dividend-warrant, talon.

Tangente f tangent; ~n-Bussole f tangent compass; ~n-Galvanometer n tangent galvanometer; ~n-Vieleck n (Geom.) polygon circumscribed round a circle.

Tangential ... ~beschleunigung f (Mech.) tangential acceleration; ~ebene, Berührungsebene f (Phys.) tangential plane, tangent plane; ~kraft f (Phys.) tangential force; ~punkt, Berührungspunkt m point of contact.

Tanne f, **Tannenbaum** m fir, fir tree (pinus abies).

tannen a of fir, fir ..., of deal wood; ~e Bohle f fir plank.

Tannen ... ~brett n fir board; ~holz n deal wood, fir wood; ~planke f thick fir board.

Tantième f, **Antheil** m extra allowance (for the transmission of telegrams); (Antheil am Gewinn) percentage, share in profits, portion.

Tara f tare (deficiency in the weight or quantity of commodities by reason of the weight of the package containing them); angenommene ~ computed tare; durchschnittliche ~ average tare; in der Faktura bemerkte ~ invoice tare; reine ~, durch Gewicht gefundene ~ real tare; ~rechnung f tare account.

Tarif m tariff, rates, table of rates; nach dem ~ as per tariff; ~Angelegenheit f tariff matter; ~bestimmung, ~festsetzung f establishment of the tariff; ~erhöhung f raising the rates; ~ermäßigung f reduction of rates; ~tabelle f table of rates.

tarifiren v to tariff.

tariren v (vgl. Tara) to tare; tarirte Waaren f/pl tared goods.

Tasche f (für Briefträger, Telegr.-Boten u. s. w.) pouch; verschlossene ~ für Abholer (P) private bag; Inhaber einer verschlossenen Post-~ private bag holder.

Taschen ... ~ausgabe f pocket edition; ~buch n pocket book, memorandum book, (beim Telegr.-Bau u. dergl.) field book; ~format n pocket size; ~galvanometer n pocket galvanometer, detector.

Tastatur f, **Tastenbrett** n, **Klaviatur** f (HA) key board.

Taste f, **Taster, Schlüssel** m (HA) key; (MA) key, manipulator; (des Schriftlochers) key, plunger; (des Einnadel-Telegr.) drop handle; Doppel-~ des Einnadel-Telegraphen pedal, tapper; Blank-~ f. Blank; Buchstaben-~ f. Buchstaben ..; Doppel∙~ f. Doppel..; Zahlen∙~ f. Zahlen..

Taften... ~druck m (M & HA) depression of the key; ~hebel m key lever, lever of the key; ~sender m (des Einnadel-Telegr.) pedal, tapper.

Tau, Seil n cord, rope; (Kabel) cable; (Ankertau) cable; ~ring m (aufgeschossenes Tau) coil; ~schifffahrt, Seilschifffahrt f, Schleppen n der Schiffe mittels Taues oder Kabels (Drahtseils) cable towage, cable towing, wire-rope towing, rope towing; ~werk n cordage, (eines Schiffes) rigging.

Tauchbatterie f (T, Elektriz.) plunging battery.

Tausch m exchange; in ~ in exchange; ~exemplar n einer Zeitung copy of a newspaper given in exchange for another one; ~handel m barter, exchange; ~handel treiben v to barter.

Tax... ~betrag m (P) amount of postage; ~einheit f unity of tax; ~frei a exempt of tax; ~holz, Taxusholz, Eibenbaumholz n yew wood; ~marke, Portomarke f (P) postage due stamp; ~werth m estimated value, appraisement; ~wort n tax word; ~quadrat n (P) tax-square.

Taxe f tax, rate, charge; internationale ~ (PT) foreign rate (of postage); interne ~ (PT) inland rate.

taxiren v to tax, to estimate, to value; zu hoch ~ to overrate, to overvalue.

Taxirung f tariff, taxation, estimate; ~sverfahren n (europäisches, außereuropäisches) european (extra-european) tariff.

Taxus m f. Taxholz.

Teakholz, Thekabaumholz n teak wood.

Technik f technics.

Techniker m technician.

technisch a technical, technic; ~er Dienst m (PT) practical service; ~e Hochschule f polytechnic school; ~e Sprache f, ~e Ausdrücke m/pl technical language, technical phrases (terms).

Technologie, Gewerbkunde f technology.

Telegramm n telegram, telegraphic message, telegraphic despatch (oft nur message od. despatch); Ankunfts-~ received telegram; ~-Annahmebuch f. Annahme...; ~beförderung f. Beförderung; ~ mit bezahlter Antwort telegram to which a reply has been prepaid; besonderes ~ special telegram; chiffrirtes ~ cypher telegram; Dienst-~ service telegram; dringendes ~ urgent telegram; Durchgangs-~ transmitted telegram; ~-Formular f. Formular; ~formularheft n pad; ~gebühr f charge for a telegram; gebührenfreies Eisenbahndienst-~ telegram franked by a railway pass; ~ in Geheimschrift code telegram; gewöhnliches ~ ordinary telegram; internes ~ inland telegram; internationales ~ foreign telegram; ~material n telegrams & paper slip; nachzusendendes ~ telegram re-directed to a second address, telegram to follow; offen zu bestellendes ~ telegram to be delivered open; ~ in offener Sprache telegram in plain (English, German) language; semaphorisches ~ semaphoric telegram, telegram for a person on board ship; unbestellbares ~ telegram which cannot be delivered; Ursprungs-~ forwarded telegram; ~ in verabredeter Sprache telegram written according to a preconcerted code; verglichenes ~ telegram to be repeated; zu vervielfältigendes ~ telegram with multiple addresses; Zeitungs-~ press telegram, news telegram.

Telegraph m telegraph, telegraphic instrument, instrument, telegraphic apparatus; akustischer ~ sounder, telegraph sounder; elektrischer ~ electric telegraph, magneto-electric telegraph, electro-magnetic telegraph; elektro-chemischer ~ electro-chemical telegraph; optischer ~ optical telegraph, signal-telegraph, semaphore; tragbarer ~ portable telegraph; unterseeischer ~ submarine telegraph; Vorposten-~ (MT) outpost-telegraph.

Telegraphen ... ~alphabet *n* telegraph alphabet; ~amt *n*, ~anstalt *f* telegraph office, telegraph station; ~anlage *f* telegraph plant; ~anstalt *f* mit Fernsprechbetrieb telephone office; ~=Apparat *m* telegraphic apparatus, register, instrument, (f. auch Apparat); ~=Apparatwerkstatt workshop of the telegraph department; ~bote *m* telegraph messenger; ~=Bau *m* telegraph construction, construction of telegraph lines, Rother's ~=Bau (als Buchtitel) A Manual of Telegraph Construction by Rother; ~=Bau-Abtheilung *f* telegraph engineering department; ~bauarbeiter *m* laborer, workman, line man, line repairer; ~bauaufseher *m* supervisor of telegraph constructions; ~baubüreau *n* office of telegraph construction; ~bauführer *m* telegraph engineer; ~bauholz *n* timber; ~baumaterialien *n/pl* telegraph materials, stores; ~bau=Ordnung *f* guide for the construction of telegraph lines; ~baupersonal *n* engineering staff; ~baustrecke *f* section; ~bauunternehmer *m* contractor for the construction of telegraph lines; ~beamter *m* telegrapher, telegraphist, operator, employee, telegraph official, instrument clerk, (im Büreaudienst) clerk; ~betrieb, ~betriebsdienst *m* technical (practical) telegraph service; ~draht *m* telegraph wire; ~=freimarke *f* telegraph stamp; ~geheimniß *n* secrecy of telegrams; ~gehülfin *f* female telegraphist, female operator; ~gesetz *n*, ~gesetzgebung *f* telegraph law, telegraph laws; ~gestänge *n* telegraph structure (poles & wires); ~=Ingenieur *m* electrician of the Telegraph Department; ~inspektor *m* inspector, surveyor of telegraphs, kaiserlicher ~ inspektor government telegraph inspector, inspector of the Imperial German Telegraph Department; ~=kabel *n* telegraph cable; ~leitung *f* wire, (zu Fernsprechbetrieb) telephone wire; ~leitungs=Revisor *m* supervising officer; ~linie *f* telegraph line, (an Straßen entlang geführte) road line, (über Häuser geführte) over house line, (mit vielen Drähten belastete) heavily wired line, (zu Fernsprechbetrieb) telephone line; ~linienbau f. ~bau; ~netz *n* network of telegraph lines; ~neubauten *pl* new telegraph lines, construction of new telegraph lines; ~schreibapparat *m* writing apparatus, recording instrument (f. auch Apparat); ~stange f. Stange; ~station *f* telegraph station; ~stations=Einrichtung f. Stations ... u. Amts ...; ~vorarbeiter *m* foreman; ~verein *m* Telegraph Union, Allgemeiner Welt=~ verein Universal Telegraph Union.

Telegraphir ... ~batterie f. Batterie; ~fehler *m* error in the transmission of telegrams; ~kontakt *m*, ~=Schiene *f* transmission contact; ~krampf *m* telegraph cramp, (Am) operator's paralysis; ~strom *m* transmitting current.

telegraphiren *v* to telegraph, to transmit by telegraph, to send telegraphic despatches, to announce by telegraph, to wire.

Telegraphirung *f* transmission (of telegrams); ~ mit der Hand transmission by hand; selbstthätige ~ automatic transmission.

telegraphisch *a* telegraphic, by telegraph; ~e Depesche f. Telegramm; ~e Postanweisung telegraphic money order, money order by telegraph.

Telegraphist f. Telegraphen-Beamter.

Telephon f. Fernsprecher.

Teleskop, Fernrohr *n* telescope; (kleines) field glass; ~goniometer *n* telescope goniometer; ~libelle *f* dumpy-level.

tellurischer Strom, Erdstrom *m* earth current.

Temperatur *f* temperature; mittlere ~ mean temperature; ~=Erhöhung *f* elevation of temperature; ~=Erniedrigung *f* lowering (depression) of temperature; ~schwankung *f* variation of temperature.

Tempo, Zeitmaaß *n* time, measure; (des Telegr.=Apparats) movement; das ~ des Apparats beschleunigen (*T*) to accelerate (to increase) the speed of the instrument; ~ halten to keep time; das ~ des Apparats verlangsamen (T) to retardate the movement of the instrument.

Termin m term; (Zahlungstermin) instalment, term of payment; in drei ~en zahlen to pay in three instalments; nach Ablauf des ~es at the expiration of the term; ~ haben to have to appear in court on a day appointed; einen ~ setzen to fix a day; ~weise adv by terms, by instalments.

Terminal... ~gebühr f (T) terminal tax; ~geschwindigkeit, Endgeschwindigkeit f (Phys.) terminal velocity.

Terminologie, chemische ~, chemische Nomenklatur f chemical terminology, chemical nomenclature, spoken language of chemistry.

Terpentin m turpentine; ~geist m essence of turpentine; ~öl n turpentine oil, benzene; ~spiritus m turpentine spirit.

Terrain n, **Gegend, Bodenbeschaffenheit** f country, ground; durchschnittenes ~ intersected country, broken ground; freies ~ open country; gebirgiges ~ mountainous country; das ~ rekognosziren (Telegr.-Bau) to survey the ground; ~abschnitt m section of country; ~skizze f sketch of the country.

Text m text; ~wort n text word.

Thal n valley, vale; ~fahrt f passage down a stream; ~weg m channel of a river, current.

That f act, action; ~bestand m statement.

Thäter m author, perpetrator; (der Schuldige) guilty person.

Thätigkeit f activity; in ~ setzen (einen Apparat) to put an instrument in action (motion); außer ~ setzen (einen Beamten) to suspend an officer from duty.

Theer m tar; (Holztheer) wood tar; (Steinkohlentheer, Gastheer) coal tar, gas tar; ~anstrich m coat of tar; ~öl n tar oil, (kreosothaltiges) creosote oil; ~werg m tarred tow.

theeren v to tar.
theerig a tarry.

Theil m part, portion.

Theil... u. theil... ~bar a divisible; ~barkeit f divisibility; ~ haben v to have a part (a share) in .., to share smt; ~haber m sharer, partner, (Abonnent, Subskribent m) subscriber; ~kreis m graduated circle; ~nahme f participation, interest; ~nehmen v to take part in, to partake of, to participate in (of) smt; ~nehmer m sharer, partaker, (Subskribent, Abonnent m) subscriber; ~strecke f (beim Telegr.-Bau) section.

Theilchen n particle; (Phys.) molecule.

theilen v to divide; (trennen) to separate.

theilend a, gleich ~ (Math.) aliquant; ungleich ~ aliquot.

Theiler m (Arithm.) divisor.

Theilung f division; ~ in zwei Theile bipartition; ~ in zwei gleiche Theile bisection; ~szahl f dividend; ~szeichen n hyphen.

thermoelektrisch a thermo-electric.

Thermoelektrizität f thermo-electricity.

Thermokette f thermo-electrical battery.

Thermometer n thermometer; ~kugel f thermometer bulb; ~röhre f thermometer tube; ~skala f thermometer scale.

Thon m clay, argil; ~becher, ~zylinder m porous cup; ~erde f argilaceous earth; ~röhre f earthenware tube (pipe); ~zylinder m (porous) clay cylinder.

thönern a earthen, of clay.

tiefer setzen v eine Stange (T) to replace the decayed butt end of a pole by a sound piece (thereby planting the pole deeper into the ground).

Tiefseekabel n deep sea cable.

Tinte f ink, writing ink, (farbige) colored ink, (rothe) red ink, (schwarze) black (writing) ink, (unauslöschliche) indelible ink; ~nfaß n inkstand; ~nlöscher m blotter.

Tisch m table; ~auszug m table drawer; ~blatt n table slab, leaf of a table; ~gestell *n* frame of a table; ~klemme *f* (*T*) binding screw; ~klemmleiste *f* (*T*) ledge (batten) for the binding screws; ~leitung, ~verbindung *f* (*T*) table connection; ~platte *f*. ~blatt; ~zarge *f* frame under the table board.

Tischler, Schreiner *m* joiner, cabinet maker; ~arbeit *f* joiner's work, joinery, cabinet work; ~hobel *m* plane; ~leim *m* glue, joiner's glue.

Titel *m* (eines Buchs, eines Beamten) title, (des Kostenanschlags) head; ~schild *n* zu Geldfahrpostsäcken, Beuteln u. s. w. (*P*) label, sliding label.

Titularrath *m* counselor by title only.

tobt *a* dead; ~e Erde *f*, richtiger: tödtende Erde, vollständige Erdverbindung einer gerissenen Leitung (*T*) dead earth; ~er Gang *m* (Mech.) back-lash, end-play, loss of time; ~es Gewicht *n* (Mech.) dead weight; ~er Punkt *m* einer Kurbel u. s. w. (Mech., Masch.) dead point, dead centre.

Ton *m* tone; ~art *f* key; ~dämpfer *m* damper; ~höhe *f* pitch; ~umfang *m* compass of tone.

tönen *v* (von Leitungsdrähten) to hum.

Tonne *f* tun, large cask; (als Gewichtseinheit) ton (die deutsche Tonne wiegt 1000 kg, die englische 20 hundred-weights = 1016,048 kg); ~, Boje buoy.

Tonnen . . . ~boje *f* cask buoy, tun buoy; ~brücke *f* cask bridge; ~fracht *f* freight by the tun; ~gehalt *m* registered tonnage; ~weise *adv* by tuns, by barrels.

Töpfer *m* potter; ~geschirr, ~gut *n* f. ~waaren; ~thon, plastischer Thon *m*, ~erde *f* potter's clay, plastic clay, argillaceous earth, argil; ~waare *f*, ~gut, irdenes Geschirr *n*, irdene Waare *f* potter's ware, earthen ware.

Torsion *f*, **Drehen, Winden** *n*, **Drehung** *f* torsion, wrenching.

Torsions . . . ~elastizität, Drehungselastizität *f* elasticity of torsion; ~galvanometer *n* torsion galvanometer; ~festigkeit *f* strength of torsion, resistance to torsional strain, torsional strength; ~kraft *f* force of torsion; ~vorrichtung *f* wrenching contrivance; ~winkel *m* angle of torsion.

total *a* total, entire, complete.

Total . . . ~arbeit *f* (Masch.) gross-work; ~bilanz *f* final balance; ~effekt *m*, ~leistung *f* einer Maschine whole effect; ~summe *f*, ~betrag *m* total amount, sum total, f. auch Gesammt . . .

Tour *f*, **Umgang** *m* (einer Maschine) revolution; ~anzahl *f* number of revolutions; ~zähler *m* counter of revolutions.

Trace *f* alignment.

Tracirpfahl *m* (zum Abstecken) tracing picket.

traciren *v* (abstecken) to trace, to lay out (a t. line).

Tracirung *f* tracing, laying out.

Trag . . . ~bahre *f* (Techn.) hand barrow; ~balken *m* beam, transom; ~band *n* (Techn.) carrying girth, strap, (Strebeband *n*) brace, strut; ~eisen *n* (Hängeeisen *n*) tie band; ~fähig *a* capable of bearing; ~fähigkeit, ~kraft *f*, ~vermögen *n* bearing capacity, capacity of bearing strength, supporting power, strength, working load; ~leiste *f* (*T*) wire board inside a station (for carrying the wires); ~modul *m* proof load, ultimate load; ~pfeiler *m* supporting pillar; ~riemen *m* strap.

Träger *m* (als Stütze) support; (des Pendel-Isolators) carrier, iron carrier.

Trägheit *f*, **Beharrungsvermögen** *n* inertia, vis inertiae; ~moment, Massenmoment, Beharrungsmoment *n* (Mech.) moment of inertia.

Trajekt *m* (Ueberfahrt *f*) traject, ferry; ~boot, ~schiff *n* railway ferry, traject boat.

Trakt *m* (Strecke, Straßenlinie *f*) tract.

Traktus *m* (Absteckungslinie *f*) traced line, alignment; Feststellung des ~ tracing, laying out.

tränken s. zubereiten.
transatlantisch *a* transatlantic.
Transformator *m* (Elektriz.) transformer.
Transit *m* transit, passage; ~ in geschlossenen Briefpacketen (P) closed mail; Einzel-~ (P) open mail; Freiheit des ~s right of transit; Land-~ land transit; See-~ sea transit.
Transit... ~beförderung *f* s. ~verkehr; ~gebühr *f* s. ~satz; ~handel *m* transit trade; ~kosten *pl* expenses of transit; ~linie *f* line of transit; ~porto *n* transit postage; ~satz *m* transit rate; ~verkehr *m* transit exchange of correspondence; ~verwaltung *f* (P) intermediate office; ~waaren *f/pl*, ~güter *n/pl* transit goods; ~zoll *m*, ~abgabe *f* transit duty.
Translation s. Uebertragung.
Transmission, Uebertragung *f* (Masch.) transmission, gearing, shafting; ~welle *f* (Masch.) connecting-shaft, counter-shaft.
Transport *m* transport, carriage, conveyance; (Uebertragung *f* einer Summe in Rechnungen) invoice continued, amount carried over, carried forward; ~kosten *pl* charges of transport (conveyance); ~skizze *f* (Telegr.-Bau) sketch of the materials to be deposited along a line; ~tabelle *f* (Telegr.-Bau) specification of the mode of conveyance of the materials or stores; ~wagen *m* (Eisenb.) transport car, truck, goods van; ~wesen *n* matters concerning the conveyance of goods.
transportabel *a* transportable.
transportiren, fortschaffen *v* to transport, to convey; (übertragen in Büchern) to carry forward, to carry over.
transversal *a* transversal.
Transversal... ~linie *f* transverse; ~schwingungen *f/pl* (Phys.) transversal vibrations.
Trapez, ungleichseitiges Viereck *n* (Geom.) trapezium.
Trassant *m* (Aussteller *m* eines Wechsels) drawer of a bill.
Trassat *m* (der Bezogene) drawee.
trassiren *v* (einen Wechsel) to draw a bill of exchange; (in Blanko) to draw in blank; (per Saldo) to draw per appoint; (über den Betrag des Saldo) to overdraw.
Tratte *f* (Wechsel) draft, bill of exchange.
Trauf... ~dach, Abdach *n* coping, caping, brow, descent of a gutter; ~rinne *f* eaves, gutter; ~röhre *f* gutter spout.
Traufe, Dachtraufe, Dachrinne *f* eaves, gutter.
Treib... ~achse, Triebachse *f* driving axle, motive axle; ~eis *n* floating ice, drifting ice; ~holz, Flößholz *n* drift wood; ~kraft s. Triebkraft; ~rad, Triebrad *n* driving wheel, main wheel; ~riemen *m* band, endless strap; ~sand, Triebsand *m* shifting sand, drifting sand; ~welle, Triebwelle, Kurbelwelle *f* driving shaft, main shaft.
treiben *v* (forttreiben) to propel; (eintreiben) to drive; (ein Gewerbe) to carry on, to exercise (a trade, a business); (eine Maschine) to work a machine, to set a machine going; (Waaren in die Höhe) to drive up goods.
Treiber *m* (Masch., Rohrpost) driver.
Treidel... ~leine *f* tow line, tracking rope; ~weg *m* towing path.
Trenn... ~amt *n* (T) way station with two wires & two instruments; ~schicht *f* separator.
trennen *v* (zwei Leitungen [T]) to disconnect; (die Leitung von der Batterie) to sever the connection of the line wire with the battery; (Chem.) to separate, to resolve, to part; (auftrennen) to rip up a seam.
Trennungszeichen *n* (T) break signal.
Trense *f* (Hanfumwickelung *f* im Kabel) serving of hemp, hemp packing.
treppenartig gegrabenes Stangenloch *n* hole dug in steps.
Treffenstreifen *m* (golden) lace (worn on the left sleeve).
Trieb *m* driving, drift; (bewegende Kraft) force of motion, mechanism; (im Getriebe) whirl.

Trieb ... ~achse f. Treibachse; ~feder f spring, moving spring; ~kraft f moving force, motive power; ~rad n (Masch.) spring wheel, spur wheel; ~sand f. Treibsand; ~welle f. Treibwelle; ~werk n (Getriebe n) gearing, machinery, (Motor m) motor, moving apparatus, (~werk in Bewegung) running gear, (~werk mit eckigen Rädern) angular gearing, (konisches ~werk) bevel gear.

Trinkgeld n s. Postillons-Trinkgeld.

Tritt m (HA) foot board; ~stange f rod of the foot board; ~umschalter s. Umschalter.

Trockenschreiber, Trockenschreibapparat m dass. wie Reliefschreiber.

Trog ... ~apparat, galvanischer ~apparat m trough battery, galvanic trough.

Trommel f (Rolle f an Hebevorrichtungen) drum, pulley; (Leier beim Drahtziehen) wire drum; (Gehäuse n) barrel, case; ~feder f barrel spring.

tropf ... ~bar a (Chem.) liquid, fluid; ~barflüssig liquid; ~barflüssiger Körper m liquid; Mechanik f der ~barflüssigen Körper, Hydromechanik, Hydraulik f hydraulics.

Trottoir n, Bürgersteig m foot path, side walk, banquet.

Troygewicht, Goldgewicht n troy, troy weight.

Tünche f, Anstrich m white-wash, lime-wash, rough-casting, plaster; (mit Lehm) loaming.

tünchen v to white-wash, to lime-wash, to rough-cast, to plaster; (mit Lehm) to loam.

Tunnel m tunnel; ~brücke f tunnel bridge; ~Einführungskasten, ~Ueberführungskasten m tunnel box; ~kasten wooden boxing in tunnels, trough.

Tülle f socket.

Type f, Buchstabe m type, printing letter.

Typen ... ~druckapparat m (T) type printing apparatus; ~rad n, ~scheibe f (HA) type wheel; ~schnellschreiber m type-printing fast speed apparatus, tachygraph.

U.

U-förmige Stütze f (T) bracket of U-shaped iron.

Ueberbau m (eines Bauwerks) superstructure.

überbieten v (bei Auktionen) to outbid, to overbid.

Ueberbleibsel, Residuum n (Chem.) residuum, remainder, residue.

überbringen v to deliver; (eine Nachricht) to report.

Ueberbringer m deliverer, bearer.

überbrücken v (einen Fluß) to throw a bridge across a river, to bridge over.

Ueberbrückung f bridging, bridge-communication; (eines Flusses) throwing a bridge across a river.

Ueberchlor ... ~säure f (Chem.) perchloric acid; ~sauer a perchloric; ~saures Salz n perchlorate.

Ueberdach, Wetterdach n jutty, shed, pent-house.

überdachen v to roof, to cover with a roof.

Ueberdachung f (eines Bauwerks) cope.

überdecken, übereinandergreifen v (Techn.) to lap over, to project.

Ueberdeckung f (von Bauwerken) covering; (Ueberwölbung f) overarching; (Uebereinandergreifen n) lapping-over, projecting.

Ueberdruck m (Dampfmasch.) pressure above the atmospheric pressure, pressure of air etc. above the atmosphere, additional pressure.

übereinander adv one upon another; ~greifen v to lap, to lap over; ~legen v to place one upon another (over one another).

übereinkommen v to agree, to come to an agreement.

Uebereinkommen n, Ueberein=

kunft *f* agreement, arrangement, convention, contract, settlement.

übereinstimmen *v* (gleichmäßig gehen [*HA*]) to keep time, to run with equal speed.

Uebereinstimmung *f* (*HA*) accord (between the type wheels & the chariot), unison, synchronism, analogy, coincidence.

überfahren *v* (hinüber fahren) to pass over, to convey over.

Ueberfahrt *f* passage, crossing (ſ. auch Trajekt); ~sgeld *n* (zu Schiff) passage money, fare.

überfälliger Wechſel *m* past due (overdue) bill.

überfirniſſen *v* to varnish over.

überfließen *v* to overflow.

überfordern *v* to overcharge.

Ueberfracht *f* over-freight, over-weight; (Ueberfrachtkoſten *pl*) overcharge (for freight); ~-Porto *n* (für das Gepäck der Poſtreiſenden) charge for the excess of the allowed weight of small luggage.

überfrachten *v* to over-freight.

überführen *v* to convert; (Chem.) to reduce; Güter ~ to transport (to carry, to convey) goods; Drähte über die Straße, über die Eiſenbahn ~ (*T*) to cross a road (the railway) with wires.

Ueberführung *f* (*T*) (der Drahtleitungen über das Straßenplanum) road crossing, (über die Eiſenb.) railway crossing.

Ueberführungs... ~kaſten *m* (*T*) junction box; ~ſäule *f* (*T*) hollow column for the joining of overhead wires with cable wires; ~ſtange *f* (*T*) junction post.

Uebergabe *f* delivery, transfer; ~ der Ladung bei den Poſten und im Eiſenbahnbetriebe (*P*) service of transferring mails to & from the vehicles of the P. O. oder from one travelling P. O. to another; ~-Bogen *m* (*P*) transfer sheet; ~-Geſchäft *n* (*P*) transfer, service of transferring.

Uebergang *m* (Weg *m*) passage, crossing; ~ von Buchstaben zu Zahlen (*HA*) change from letters to figures & signs (& vice versâ); ~ von Ka-beln zur Luftleitung (*T*) junction of cables with land lines; ~sbeſtimmungen *f/pl* transitory regulations; ~szettel *m* (*P*) bag list.

übergar, überfein *a* (v. Metallen) over-refined; ~es Kupfer *n* dry copper; ~es Eiſen *n* black pig-iron; ~er Stahl *m* burnt steel.

übergeben *v* to deliver, to transfer; (den Dienſt) to turn over the duty; (dem Verkehre [*TP*]) to open to the public, to open to traffic.

Uebergebot *n* outbidding.

übergehen *v* (übergeleitet werden) to pass over.

Uebergewicht *n* over-weight; das ~ haben to be overweight.

übergipſen *v* to plaster with gypsum.

übergreifend *a* lapping over, over-lapping; ~er Theil eines Brettes u. ſ. w. lag.

überhängen *v* (aus dem Loth treten) to overhang.

Ueberhängen *n*, **Ausladung** *f*, **Ueberhang** *m* (von Bauwerken) projecting, projecture.

überhäufen *v* to overload, to overstock; überhäuft *a* mit Telegrammen (a line is) crowded with telegrams; überhäuft mit Geſchäften pressed with business.

Ueberhäufung *f* (*PT*) pressure (of business); accumulation (of telegrams).

überheben *v* (mehr erheben) to overcharge; überhobener Betrag, zu viel eingezogenes Franko ob. Porto overcharge; einen Antrag auf Erſtattung eines überhobenen Betrages ſtellen to make an application with a view to the return of the overcharge.

überkitten *v* to cement over.

überkleben *v* to paste over, to glue over.

Ueberkunft *f* arrival.

überladen *v* to overload, to overcharge, to overfreight, to overlade (a vessel).

Ueberladung *f* overloading.

Ueberlager *n* (der Bahnpoſtbeamten) stop, stopping, (der Poſtpferde) halt; ~-Vergütung *f* (der

Bahnpoſtbeamten) allowance for stopping.

Ueberlandpoſt *f* overland mail.

überlaſſen *v* (abtreten) to make over, to cede, to abandon.

überlaſten *v* to overload; (ein Schiff) to overfreight a vessel.

überleiten, übergehen laſſen *v* to pass over.

überliefern, übergeben *v* to deliver, to transmit.

übermitteln *v* to transmit.

Uebermittelung *f* transmission, forwarding.

Uebernahme *f* der Poſt duty of receiving the mail, checking of the mail; (Beſitznahme *f*) taking possession; (Annahme *f*) acceptance; ~ eines Geſchäfts mit allen Aktiven u. Paſſiven taking possession of a business with all stock & debts.

übernehmen *v* (die Poſt) to receive (to check) the mail; (ein Geſchäft) to engage in a business; (die Verantwortlichkeit) to take the responsibility; (an Bord nehmen) to get on board; (eine Schuld) to charge one's self with a debt, to assume a debt; (in Beſitz treten von etw.) to take possession of *smt*.

Uebernehmer *m* receiver; (Unternehmer *m*) contractor.

Ueberproduktion *f* surplus production.

überquer *adv* crossways, across.

überrechnen *v* (nachrechnen) to count over, to calculate, to reckon over, to examine.

überreichen *v* to hand over; (behändigen) to deliver.

Ueberreſt *m* (in Rechnungen) remainder, balance; (Chem.) remainder, residue; (Reſte, Abfälle *m/pl*) scraps, chips, remains (said of metals), waste (said of paper etc.).

überſättigen *v* to supersaturate.

Ueberſättigung *f* supersaturation.

überſchicken *v* to send; (Geld ~, remittiren) to remit.

überſchießende Summe *f* surplus money.

Ueberſchlag *m* estimate; einen ~ (Koſtenanſchlag) machen to frame an estimate; reiner ~ fair estimate; ungefährer ~ rough estimate.

überſchlagen *v* (einen Ueberſchlag machen) to make a rough calculation, to compute, to estimate.

überſchlägliche Ermittelung *f* rough estimate.

überſchreiben *v* to superscribe; (Briefe) to address; (auf der Rückſeite den Namen des Schreibers vermerken) to back a letter.

Ueberſchreibung *f* superscription.

überſchreiten *v* (einen Fluß) to pass over; (einen Anſchlag) to exceed the estimated cost of the work.

Ueberſchreitung *f* excess; ~ des Etats ſ. Etats ...

Ueberſchrift *f* title, inscription; (Adreſſe *f*) direction, address; (eines Wechſels) head (heading) of a bill.

Ueberſchuß *m* net rest, excess, surplus (of the postal revenue); reiner ~ net rest; Abführung *f* der Ueberſchüſſe an die Ober-Poſtkaſſe remittance of the surplus cash to the chief postal cash office.

überſchwemmen *v* to overflow, to inundate, to flood.

Ueberſchwemmung *f* inundation, flood; ~sgebiet *n* land liable to inundation (to be flooded); ~szeit *f* season of inundation.

überſeeiſch *a* transmarine, transatlantic.

überſenden *v* to send, to transmit.

Ueberſendung *f* sending, transmitting (of mail articles); remittance (of cash).

überſetzen *v* (über einen Fluß) to cross over, to ship over; (ſprachlich) to translate; (die Morſeſchrift vom Streifen) to transcribe.

Ueberſetzung *f* translation, transcription.

Ueberſicht *f* return, returns; (amtliche Statiſtik *f*) returns, statistics; ~zeichnung, Totalzeichnung *f* general drawing.

überſpannen *v* (mit Drähten [*T*]) to span (the ocean), to clear (buildings).

überspinnen v to spin over, to cover; mit Seide übersponnener Draht silk covered wire.

überspringen v to pass, to strike across (said of the electric spark).

übersteigen v to exceed.

überstempeln v (die Freimarken) to deface, to obliterate the postage stamps, (Am) to postmark.

überstreichen v to paint over.

Ueberstunden f/pl hours in excess, extra attendance.

übertheuern v to overcharge.

Uebertrag m (Transport m) transfer, amount brought over; ~ der Faktura invoice continued.

übertragbar a transferable.

Uebertragbarkeit f quality of being transferable, transferability.

übertragen v to transport, to transfer; (Buchführung) to carry forward, to set over; (eine Bewegung) to transmit a motion; (einen Wechsel) to transfer a bill by endorsement.

Uebertrager m (Telegr.-Apparat) translator, repeater; (eines Wechsels) endorser.

Uebertragung, Translation f (durch Apparate [T]) translation; ~ zwischen Arbeits- u. Ruhestrom translation between a wire worked on the closed current system and one worked on the open system; auf ~ stellen to connect up for translating purposes; ~ einer Bewegung, Transmission f transmission of force; ~ einer Schuldforderung delegation; ~ eines Wechsels endorsement.

Uebertragungs ... ~amt n (T) translating office und dem entsprechend die übrigen Zusammensetzungen; ~relais n (T) repeating relay; ~stellung f (T) connection for translating purposes.

übertreten v (Gesetze) to transgress, to break, to infringe laws.

Uebertretung f infringement, transgression.

überwachen v to watch, to supervise.

Ueberwachung f supervision (of t. lines), watching.

Ueberweg m road crossing; ~stange f (T) post at a crossing.

Ueberweisung f der Zeitungen (+) assigning the delivery of a news-paper to another post office; ~skarte f für Postagenturen (+) bill containing all mail articles to be accounted for; ~spostanstalt f head post office; ~stelegramm n für telegr. Postanweisungen telegraphic advice, advice by telegram.

überwinden v (Mech., Masch.) to overcome.

überzählig a supernumerary, surplus; ~es Packet n one parcel more; es sind zwei Packete ~ there are two parcels more.

überziehen v (mit Ueberzug versehen) to cover, to coat; (inkrustiren) to incrustate; (belegen, plattiren) to plate; (galvanisch) to electro-coat.

Ueberzug m (von Guttapercha) guttapercha covering, guttapercha coating; schützender ~ protecting cover; (Plattirung f) plating; (galvanischer) electro-coating, coating (film) of electro-deposited metal.

Uebung f, ein zu militärischer ~ eingezogener Beamte an official (belonging to the Army Reserve) called out for training; ~sapparat m (T) learner's telegraph instrument; ~szimmer n (T) learner's room.

Uhrenzeichen n time signal.

Uhrwerk n clock work, clock movement.

umarbeiten v (die Post) to forward the mail; umzuarbeitende Korrespondenz (P) forward letters (s. a. T. forward); (Techn.) to work anew, to remodel.

Umbau m re-building.

umbiegen v to bend, to turn down (up).

umbrechen v to break; (Stangen sind umgebrochen t. poles are broken down).

Umbruch m break down (of t. poles).

umdrehen, sich umdrehen v to turn round, to rotate, to revolve.

Umdrehung f (rotirende Bewegung f) rotation (of the chariot [HA]); (Tour f, Umgang m einer

Maschine) revolution (of a machine), rotary motion; x ~en in der Minute (Masch., Techn.) x turns a minute.

Umdrehungs... ~achse f (Mech.) axis of rotation (revolution); ~anzeiger m revolution indicator; ~ebene f plane of rotation; ~geschwindigkeit f speed of rotation; ~punkt m (Mech.) centre of rotation.

Umdruck m, Umdrucksverfahren n (Metallographie f) reproduction from the original by the metallograph.

Umfahrt f (bei der Packetbestellung) trip, round; (Umweg m) circuitous route, round-about way.

Umfang m (Peripherie f) circumference, periphery, perimeter; (Volumen n) bulk; (der Geschäfte) extent (of business); ~sgeschwindigkeit f (Mech.) circumferential velocity; ~swinkel m des Kreises (Geom.) angle of circumference.

Umfassung, Umfriedigung, Einfriedigung f enclosure; ~smauer f enclosure wall, outer wall.

Umgang (Tour) s. Umdrehung.

umgehen v (einen Ort) to go round a place, to take a circuitous route.

umgehend a, mit ~er Post by return of post.

umgekehrt a inverted, inverse, reverse; ~es Verhältniß n inverse ratio (proportion).

umgießen v to refound, to cast anew, to re-cast (zinc cylinders).

Umgießung f, Umguß m recasting, refounding.

umgraben v to dig, to dig up, to turn up.

umhüllen v to wrap in, to cover, to protect.

Umhüllung f envelope; ~ der Kabel mit Schutzdrähten protecting armor, sheathing.

umkehrbar a reversible.

Umkehrbarkeit f reversibility.

umkehren v to reverse (a current, the polarity); (Math.) to invert.

Umkehrung f reversal (of the poles etc.); (Math.) inversion.

Umkreis m (Geom.) circumference, periphery; 5 Meilen im ~e 5 miles round.

umladen v to reload; (ein Schiff) to shift the cargo, to transship.

Umladung f reloading, reshipment.

Umlauf m (Umdrehung f) revolution, rotation; in ~ bringen, setzen v to put into circulation, to issue; im ~ sein v to circulate; (des Geldes) currency of money; ~buch n (PT) circular order-book; ~schreiben n circular letter; ~szeit f (Masch.) time of revolution.

umlaufen v (rundgehen) to revolve; (von Waaren, Schreiben u. s. w.) to circulate.

umlegen v (einhüllen) to cover, to lay round, to surround; (Stellung verändern) to turn, to transpose.

umleiten v (PT) to re-transmit the mail (telegrams); (auf einem andern Wege als dem gewöhnlichen befördern) to despatch the mail (telegrams) by another than the ordinary route.

Umleitung f (PT) re-transmission; ~s- (Umspeditions-) Postanstalt f re-transmitting office, forward office (s. a. T. forward).

ummauern v to surround with masonry.

umpacken v (anders verpacken) to repack; (einhüllen) to pack (all) round.

Umpackung f repacking, packing (all) round (vgl. das Vorst.).

umpfählen v to enclose with pales, to palisade.

umpflastern v to repave.

umrechnen v to reduce.

Umrechnung f reduction; ~stabelle f table of reduction, table of conversion.

Umriß, Entwurf m, Skizze f outline, sketch; den ~ zeichnen to draw (to trace) the outline.

Umsatz m (von Waaren) sale.

umschalten v (T) to switch, to change the current, to reverse the current.

Umschalter m (T) commutator, switch, circuit changer, current re-

verser, gyrotrope, rheotrope; ~ mit federndem Kontakt spring commutator; **~ mit gleitendem Kontakt** commutator with sliding contact; **~ für Zwischenstationen** intermediate station commutator; **(Ausschalter)** commutator for breaking contact, cut-out; **(Einschalter)** circuit closer, commutator for making contact; **Kurbel-~** lever switch, button switch; **Linien-~ (Linienwechsel** *m*) universal switch, Swiss commutator; **Scheiben-~** plate commutator; **Stöpsel-~** peg commutator, plug commutator; **Tritt-~** pedal commutator.

Umschaltung *f* commutation, changing the current, changing the instrument.

Umschlag *m* cover, covering; **(eines Briefes)** envelope, cover; **(Umsatz** *m*) sale, return; **~geld** *n* (+) charge for putting a newspaper to be sent to a foreign country under cover; **~papier** *n* wrapping paper.

umschlagen *v* to reverse (said of the needle of the galvanometer); **(umwickeln)** to wrap up.

umschmelzen *v* to refound, to recast, to remelt.

umschmieden *v* to reforge.

umschnüren *v* to tie with string.

umschreiben *v* to rewrite; **(einen Wechsel)** to endorse a bill.

Umschrift *f* einer Münze legend.

umsetzen *v* (auf eine andre Stelle) to transpose; **(Waaren)** to sell.

umsonst, unentgeltlich *adv* free of charge, gratis.

Umspann *m* (P) relay (of horses), putting fresh horses to; **~-Aufseher** *m* helper of the mail contractor with the special duty to superintend the putting of fresh horses to the mail coach; **~-Ort** *m* (P) way station where a change of horses takes place.

umspinnen s. überspinnen.

umstellbar *a* (Techn.) reversible.

umstellen *v* (Techn.) to reverse.

umwandeln *v* (Chem. u. s. w.) to convert, to transform; **(Eisen in Stahl ~)** to convert iron into steel.

Umwandlung *f* transformation (of postal establishments); conversion (of iron into steel).

umwechseln *v* (Geldsorten) to change, to exchange.

Umweg *m* circuitous way, side way, by-way.

umwenden *v* to invert, to turn upside down, to turn inside out (a mail bag).

umwickeln *v* to wrap, to wrap round; **(Draht)** to coat wire, to cover wire (with strands of hemp); **(einen Elektromagneten rechts, links ~)** to wind an electromagnet with the right handed (left handed) helix.

Umwickelung, Umwindung *f* **(Hülle)** sheath, covering (of a cable); **(des Elektromagneten)** coil, helix.

umtelegraphiren *v* to re-transmit telegrams; **(auf einen andern Weg leiten)** to despatch telegrams by another than the ordinary route.

Umzugskosten *pl* removal expenses.

unabgeholt *a* (von Postsendungen) not called for, not claimed, unclaimed.

Unabkömmlichkeitsbescheinigung *f* certificate showing that an officer (called out for military duty) cannot be spared.

unabsetzbar *a* irremovable, undeposable (said of officials who cannot be removed from office except by a disciplinary or judiciary proceeding).

unanbringlich *a* (PT) which cannot be delivered.

unauffindbar *a* not to be found.

unausgefüllt *a* left in blank.

unbefördert *a* not transmitted, not forwarded; **~ liegen gebliebenes Telegramm** telegram through some irregularity (error) not transmitted.

unbekannt *a* unknown, not known; **~e Größe** *f* (Math.) unknown quantity.

unbesetzte Linie *f* (T) unoccupied circuit.

unbestellbar *a* which cannot be delivered, undeliverable, undelivered; **~e Briefe** *m/pl* dead letters; **Ausschuß zur Eröffnung ~er Sen-**

dungen (*P*) Returned Letter Office, (*Am*) Dead Letter Office.

Unbestellbarkeit *f* non-delivery; ~smeldung *f* (*T*) telegram signalling the non-delivery.

unbeweglich, fest *a* immovable, immoveable, fixed; ~e Güter *n/pl* immovables.

unbezahlt *a* (von Rechnungen) unpaid, unsettled; (von Wechseln) dishonored.

unbiegsam *a* inflexible, unpliable.

unbrauchbar, abgängig *a* waste.

undehnbar *a* not extensible; (von Metallen) inductile, immalleable.

undicht *a* not tight, unretentive; ~ sein *v* (Dampfmasch.) to lose (said of valves); ~ sein, ~ werden, lecken to leak.

Undichtheit *f*, **Undichtwerden** *n* an den Verbindungsstellen od. Flantschen der Röhren (Techn.) leakage at the joints of pipes.

Undulation, wellenförmige Bewegung *f* undulation; ~stheorie *f* undulatory theory.

undulatorisch *a* undulatory.

undurchdringlich *a* impenetrable; (undurchlässig) impermeable, impervious.

Undurchdringlichkeit *f* impenetrability, impermeability, imperviousness.

uneben *a* (Topogr.) uneven, rugged; ~es Terrain *n* uneven ground, rugged country.

unecht *a* (falsch) false; (nachgemacht) counterfeit, fictitious; ~e Bronze *f* varnished bronze; ~e Folie, Kupferfolie *f* copper foil, German foil; ~e, verfälschte Waaren *f/pl* adulterated goods.

uneinbringlich, uneintreiblich *a* (von Forderungen) uncollectible.

uneinträglich, unergiebig *a* unprofitable, unproductive.

uneinziehbar *a* not to be obtained, uncollectible (said of postage, fees etc.).

unelektrisch *a* non-electric.

Unentbehrlichkeitsbescheinigung *f* certificate showing that the services of an officer cannot be dispensed with.

unentgeltlich *a* free of charge, gratis.

unerlaubt *v* illicit.

unerwartete Revision *f* unexpected visit of a superior officer, unexpected inspection.

unfahrbar, ungangbar *a* impracticable, impassable.

Unfall *m* (Eisenb.) accident; ~meldestelle *f* (*T*) arrangement to establish (by means of an alarum) telegraphic intercourse in case of accidents of any kind after the regular hours of business; ~versicherung *f* insurance against accidents; ~versicherungsgesetz *n* law regulating the insurances against accidents.

unförmlich *a* awkward (said of the shape of postal parcels).

unfrankirt *a* unpaid.

unganz *a* (vom Eisen) flawed, starred, not sound; ~e Stelle *f* im Eisen flaw, blister.

ungedeckt *a* (unbezahlt) uncovered, unpaid.

ungehärtet *a* untempered, unhardened (said of steel).

ungelöscht *a* unslaked, unslacked (said of lime); ~er Kalk, gebrannter Kalk, lebendiger Kalk *m* Kalziumoxyd *n* unslaked lime, burnt lime, quick lime, anhydrous lime, oxide of calcium.

ungenügend frankirt *a* insufficiently paid; ungenügende Frankirung *f* insufficient (short) payment.

ungerade *a* uneven, not straight; ~ Zahl *f* (Arithm.) odd number.

ungleich *a* unequal; uneven, not level.

Ungleich ... u. ungleich ... ~artig, ~förmig *a* dissimular, incongruous; ~heit *f* unequality, unevenness; ~namig *a* dissimilar s. auch Pol; ~schenkelig *a* (Geom.) not isosceles; ~seitig *a* (Geom.) scalene, scalenous; ~seitiges Dreieck *n* scalene.

ungültig *a* void; (Münzen) not current, out of circulation; für ~ erklären to cancel (a telegram), to annul, to make void.

unhämmerbar *a* immalleable (said of metals).

unipolar *a* unipolar.
Unipolarität *f* unipolarity.
Universal ... ~galvanometer *n* universal galvanometer; ~gelenk, Drehgelenk *n* swivel joint; ~schraubenschlüssel *m* universal screw-wrench, shifting key, monkey wrench, monkey spanner, shifting spanner.
Unkosten *pl* charges, expenses; ab an ~ charges to be deducted; ~rechnung *f* bill of charges.
unlösbar, unlöslich *a* (Chem.) insoluble.
unmeßbar *a* (Geom.) incommensurable.
unmittelbar *a* immediate; *adv* immediately.
Unregelmäßigkeit *f* im Dienst (*PT*) irregularity (in the treatment of registered letters, in regard to mail bags etc.).
Unreinigkeit, Verunreinigung *f*, **verunreinigender Bestandtheil** *m* (Chem.) impurity.
unrichtig zugegangene Postsendung *f* mail arrived out of course.
Unruhe *f* (Techn.) balance (of a watch); ~feile *f* balance web; ~kloben *m* balance vice; ~scheibe *f* balance ring; ~spindel *f* balance verge.
unsauber *a* dirty, unclean, impure.
unscheidbar *a* (Chem.) inseparable.
unschmelzbar *a* infusible, apyrous, refractory.
unschweißbar *a* not to be welded.
unstät, unstätig *a* (Mech.) unsteady; ~e Größe *f* (Math.) variable quantity.
unstreckbar *a* not ductile, not extensible.
untauglich *a* unfit, unqualified.
unter *prp* under, below; ~ Ihrer Adresse under your direction; ~ Kaution under bond; ~ dem Kurse below the day's rate; ~ Pari below par; ~ dem Preise below the price; ~ Segel under sail; ~ Zollverschluß under bond.
Unterabtheilung *f* subdivision, branch.

Unterbau *m* (von Bauwerken) substructure, foundation, ground work; (Eisenb.) ground work, earth works, foundation, road bed.
Unterbeamte *m* subaltern officer, subaltern official; ~nstelle *f* situation of a subaltern officer.
unterbieten *a* to beat down a person, to underbid *smb*.
unterbrechen *v* (*T*) to interrupt, to stop (the sending clerk); eine Verbindung ~ (*T*) to disconnect, to cut off.
Unterbrecher *m* (*T*) interruptor, circuit breaker; Selbst~ s. Selbst.
Unterbrechung *f* (*T*) interruption (in a message, of communication); disconnection (said when the wire is broken); ~srad *n* (*T*) interrupting wheel; ~szeichen *n* (*T*) the „break" signal.
unterdrücken *v* to intercept (a telegram), to purloin, to suppress (letters, money etc.).
Unterdrückung *f* interception, purloining, suppression.
unterfertigen *v* to sign, to subscribe; „der Unterfertigte" the undersigned.
Unterfläche *f* base, basis, bottom.
Unterführung *f* dass. wie Ueberführung.
Untergebene *m* subordinate officer.
Untergestell *n* frame, trestle.
Untergewicht *n* under-weight.
unterhaken *v* (anstreichen) to check, to check off.
unterhalten *v* to maintain, to keep in repair.
Unterhaltung *f* maintenance (of t. lines), care (of a battery), upkeep (of fittings); ~sbestände *m/pl* (*T*) stores of maintenance; ~skosten *pl* cost of maintenance, expenses of maintenance, repairs.
unterhandeln *v* to negotiate, to transact.
unterirdisch *a* underground, subterranean, subterraneous; ~e Eisenbahn *f* underground railway; ~e Führung *f* der Telegraphendrähte subterranean (underground) line; ~er Weg *m* subway.

Unterlag...~platte *f* bed plate, ground plate; ~scheibe *f*, Nietblech *n* (Masch., Techn.) rivet plate; ~scheibe, Dichtungsscheibe *f*, Schraubenmutterblech, Bolzenblech *n* (Masch.) washer, collar; ~scheibe von Blei lead washer; ~scheibe von Gummi India-rubber washer.
Unterlage *f* (Stütze) support, stay, bearer; (Bock) trestle, tiller; (Platte) ground plate; ~ eines Balkens prop, bearer of a beam; ~ eines Hebels (Mech.) fulcrum, hypomochlion.
Unterlegscheibe s. Unterlagscheibe.
unterliegen *v* to be subject to; einer Gebühr, der Taxirung ~ to be subject to taxation.
unternehmen *v* (eine Lieferung, Leistung u. dergl.) to contract.
Unternehmer *m* (einer Leistung), (Lieferant *m*); contractor.
Unternehmung *f*, **Unternehmen** *n* enterprise, venture; speculation.
Unterordnung *f* subordination.
unterpflügen *v* (beim Kabellegen) to plow under.
unterrichten *v* to instruct, to inform *smb*.
Unterrichtskursus *m* course of instruction.
Untersalpetersäure *f*, Stickstofftetroxyd, Salpetrig-Salpetersäureanhydrid *n* hyponitric acid, tetroxide of nitrogen, nitrosonitric anhydride, peroxide of nitrogen.
Untersatz *m* stand, stay; (Bock) trestle; (einer Säule) socle; ~kasten *m* stand box, (*MA*) drawer.
Unterschaffner *m* (Eisenb.) railway under guard, under guard.
Unterschied *m*, **Differenz** *f* (Math.) difference; ~ zwischen dem wirklichen und dem geometrischen Radius (Techn.) addendum.
unterschlagen *v* to intercept (a telegram), to embezzle, to purloin (money), to suppress (letters).
Unterschlagung *f* interception, embezzlement, embezzling, suppression.
Unterschleif *m* embezzlement, embezzling, defraudation, peculation.

unterschreiben *v* s. unterfertigen.
Unterschrift *f* signature.
unterschwefelsauer *a* hyposulphuric; ~es Salz *n* hyposulphate.
Unterschwefelsäure *f* dithionic acid, hyposulphuric acid; ~-Verbindung *f* hyposulphate, dithionate.
unterseeisch *a* submarine; ~e Telegr.-Linie *f* submarine t. line; ~es Kabel *n* submarine cable, ocean cable.
Unterstellung *f* (der Postwagen) keeping the P. O. vehicles in sheds.
unterstreichen *v* to underline (a word).
Unterstreichung *f* (*T*) underline; ~szeichen *n* (*T*) the underline signal.
unterstützen *v* to support, to sustain, to prop; (mit Geld) to relieve.
Unterstützung *f* support (of the wire), prop, stay (of a pole); (durch Geld) relief; fortlaufende Geld-~ relief afforded continuously; einmalige Geld-~ relief afforded once for all; ~skasse *f* fund for support, relief fund, endowment fund; ~spunkt *m* (Techn.) point of support, (*T*) support, (Mech.) fulcrum, hypomochlion.
untersuchen *v* to inspect, to investigate; (*T*, Chem.) to test; (Rechnungen) to examine, to audit accounts; (sondiren) to sound.
Untersuchung *f* (*T*) examination (of instrument connections), testing (of t. lines); (Sondirung) sounding; (Chem.) chemical analysis, chemical test; (einer betrügerischen Handlung) investigation of a fraud committed; (gerichtliche) judicial enquiry, official examination, trial; eine ~ anstellen (wegen einer verloren gegangenen Postsendung u. s. w.) to cause enquiry to be made on . . .; in gerichtliche ~ ziehen to investigate (to examine) judicially, to try; in ~ stehen to be under examination; ~sbatterie *f* (*T*) testing battery; ~sbrunnen *m* (*T*) testing box (made of cast iron or a pit lined with masonry); ~sführender Beamte *m* (*P*) officer charged with the investigation (examination) of a fraud committed; ~sgalvanoskop *n* (*T*) de-

tector; ~kommission *f* visiting committee; ~8richter *m* inquisitor; ~8stange *f* (*T*) control pole, testing pole; ~8station *f* (*T*) testing station.

Untertheil *m* (Techn.) lower part, bottom.

unterwaschen *v* to underwash, to undermine, to wash away.

Unterwasser=Linie *f* (*T*) subaquatic (subaqueous) line.

Unterwegsort *m* (*P*) way station.

unterwerfen *v* to subject to (charges, taxation etc.).

unterzeichnen *v* to subscribe, to sign, to subsign.

Unterzeichnung *f* (Unterschrift *f*) signature, signing, subscription.

untheilbar *a* indivisible.

untief *a* shallow.

Untiefe *f* shoal, shallow, shallow water.

ununterbrochen *a* unintermitted, unintermitting, continual, uninterrupted; *adv* unintermittingly, uninterruptedly; ~er Dienst, Tag- u. Nachtdienst s. Dienst.

unveränderlich *a* invariable.

Unveränderliche, Invariante *f* (Math.) invariant.

unverändert *a* unchanged, unaltered.

unverarbeitet *a* (Techn.) unwrought, not worked up; ~e Telegramme telegrams left, telegrams not cleared off.

unveräußerlich *a* inalienable.

unverbrennbar, unverbrennlich *a* incombustible; ~er Stoff *m*, ~es, feuersicheres Gewebe *n* incombustible stuff, fire-proof tissue.

Unverbrennbarkeit *f* incombustibility.

Unverletzlichkeit *f* eines Siegels u. s. w. inviolability, sanctity (of a seal etc.).

unverpackt, lose *a* in bulk.

unversehrt, unbeschädigt *a* uninjured, safe.

unversichert *a* uninsured.

unverständlich *a* unintelligible (combination of letters in telegrams).

unversteuert, unverzollt *a* unentered, uncustomed, duty unpaid; ~ im Zollhaus in bond.

unverzinslich *a* paying no interest.

unvorhergesehen *a* unforeseen, unlooked for, unexpected; ~e Ausgaben *f/pl* (die im Etat nicht vorgesehen sind) not accounted for charges.

unvortheilhaft *a* unprofitable, disadvantageous.

unvollständig *a* incomplete, insufficient (address), imperfect.

unwägbar *a* (Chem., Phys.) imponderable; ~e Substanzen *f/pl* imponderables.

Unwägbarkeit *f* imponderability.

Unze *f* ounce.

unzerbrechlich *a* infrangible.

unzerlegbar, unzersetzbar *a* (Chem.) indecomposable; ~er Körper, einfacher Körper *m*, Element *n* (Chem.) element, elementary body, simple substance.

unziehbar *a* inductile, not ductile, immalleable (said of metals).

unzulässig *a* inadmissible (telegrams indecently worded will *not be transmitted*).

Unzulässigkeit *f* non-admissibility.

unzureichend *a* insufficient; ~ frankirter Brief *m* insufficiently paid letter.

Urbestandtheile, Urstoffe, einfache Stoffe *m/pl*, **Elemente** *n/pl* (Chem., Phys.) ultimate elements, elements, simple substances, elementary bodies.

Urkunde *f* document, instrument, deed.

Urlaub *m* leave of absence, absence on leave; ~ geben to grant leave of absence to *smb*; ~ nehmen to apply for leave of absence to ...; ~sgesuch *n* application for leave of absence.

Urschrift *f* (*T*) original text.

Ursprung *m* origin; ~sland *n* country of origin; ~stelegramm *n* forwarded telegram.

Urtheil *n* judgment, sentence; ~ fällen to pass sentence (jugdment); ein ~ abgeben to give one's judgment upon *smt*.

Utensilien *n/pl* utensils, tools; (Geräthschaften *pl*) working stock.

V.

Vacat s. Vakat.
Vache f leather case in coaches; ~=Decke f (auf Güterpostwagen) leather cover, tilt.
Vacuum n, luftleerer Raum m, Luftverdünnung, Luftleere f vacuum; barometrisches ~, Toricelli'sche Leere f (Phys.) barometric vacuum, Toricellian vacuum (vacuous space above the barometric mercurial column); ~Manometer m (Dampfmasch.) vacuum gauge; ~pumpe f vacuum pump, air pump for the vacuum pan.
Vakanz f vacancy; ~enliste f list of vacancies, vacancy report.
Vakat n (leerer Raum) blank; ~liste f blank return (in England mit „nil" = nihil ausgefüllt).
Valuta f value, standard; (beständige) regular standard, (unbeständige) fluctuating standard.
Varia, vermischte Ausgaben pl sundry expenses.
variabel, veränderlich a (Math.) variable.
Variation f variation, change; ~srechnung f (Math.) calculus of variation.
Ventil n valve; (zum Absperren) stop valve; Sicherheits=~ safety valve.
Ventilation f ventilation, the act of ventilating; ~srohr n ventilating pipe.
Ventilator, Windfang m ventilator, pneumatic machine; (Zentrifugalventilator m, Wettertrommel f, Luftsauger, Windbläser m) centrifugal ventilator, centrifugal fan, rotary fan, fan blast.
ventiliren, lüften v to ventilate.
verabfolgen v to deliver, to surrender, to hand over.
verabreden v to agree upon; verabredete Adresse (T) abbreviated address; verabredete Sprache (T) preconcerted language; verabredetes Zeichen (T) preconcerted (conventional) signal; Telegramm in verabredeter Sprache mit Wörtern aus den 8 zugelassenen Sprachen code telegram; Telegramm in verabredeter Sprache mit Wörtern, die in keinem Wörterbuch zu finden sind cypher telegram.
veränderlich a variable (state), fluctuating (traffic), s. auch variabel.
verändern v to alter, to change, to vary.
Veränderung f variation, change, alteration.
verankern v to stay (a t. pole with wire ropes), to truss a pole; (ein Gewölbe) to bind a vault.
veranlassen v etw., to occasion, to cause; ~ daß etw. gethan wird to cause smt to be done.
Veranlassung f occasion, cause, motive; bei jener ~ on that occasion; eine Sache zur weiteren ~ übersenden to transmit a matter *for further action*; ich beehre mich, Ihnen die Angelegenheit zur weiteren ~ vorzulegen I have the honor to submit the matter to you that you may *take the necessary measures* oder *to take such measures as you may deem proper*.
veranschlagen v to estimate, to tax, to rate; die Einnahme auf x Mark ~ to estimate the receipts at x Mark.
Veranschlagung f estimate; (der Materialien, Transporte u. s. w.) specification.
verantworten, sich v to justify one's self in writing.
verantwortlich a answerable, accountable, responsible for.
Verantwortlichkeit f responsibility.
verarbeiten v (Techn.) to work; es sind 200 Telegramme in der Stunde verarbeitet worden 200 telegrams have been despatched (forwarded, transmitted) in an hour.
verästeln, sich v to ramify (said of the electric current, of t. lines etc.).
Verästelung f ramification.
verauktioniren v to sell by auction, to sell at public sale, to sell to the highest bidder.

Verauktionirung *f* auction, public sale.

verausgaben *v* to pay out, to pay away.

Verband *m* (Techn.) bond; (Mauerverband) bond in masonry, bond in brickwork; (Zimmerverband) bond, joining; (Verzahnung *f*) joggling, dovetailing; ~bolzen *m* holding bolt; ~mauerwerk *n* masonry in bond; ~stück, ~holz *n* framing piece, framing timber, scantlings.

verbiegen *v* to give a wrong bend, to strain.

verbinden *v* (Techn.) to join, to join together; (Chem.) to combine with, to unite, to unite with; (T) to connect, to put (two stations) in communication, to join (two wires).

verbindlich *a* liable; sich ~ machen to engage.

Verbindlichkeit *f* engagement, liability; eine ~ auf sich nehmen to assume a liability; seinen ~en nachkommen to meet one's engagements.

Verbindung *f* (Techn.) joining, fastening, junction; (Chem.) combination, compound; (der Telegraphen-Leitungen) connection, joining, (der oberirdischen mit den unterirdischen Leitungen) junction of the overhead line with the cable; (Zimmer- ~en [*T*]) leads, office wires; in ~ stehen to be in communication with *smb*.

Verbindungs... ~klemme *f* (*T*) binding clamp, splicing pliers; ~linie *f* (*T*) line of communication; ~röhre *f* joint pipe, connecting tube; ~schraube *f* coupling screw; ~zeichen *n* hyphen.

verbleien, einbleien *v* to lead; (eif. Stangen in den Sockeln) to fix the poles in the socles by means of melted lead; (Waaren) to seal with lead, to lead goods.

verblenden, verkleiden *v* (Bauwerk) to face, to line.

Verblendungsstein *m* facing stone.

verbolzen *v* to bolt, to fasten with bolts.

verborgen *a* (Phys.) latent.

Verbrauch *m* consumption (the act of consuming by use, waste etc.); use (the act of employing anything); ~snachweisung *f* (Telegr.-Bau) specification of the stores & materials used.

verbrauchen *v* to consume, to use, to use up, to employ, vgl. das Vorstehende.

verbreitern *v* (Techn.) to enlarge, to expand, to widen.

verbrennen *v* to burn; (Holz) to burn up wood; (Holz zu Kohlen) to burn wood into charcoal.

verbrennlich, brennbar *a* combustible, inflammable.

Verbrennlichkeit, Verbrennbarkeit *f* combustibility, inflammability.

verbunden *a* (Masch. u. s. w.) connected, joined; (Chem.) combined, united.

verbürgen *v* to bail, to answer for; sich ~ to become bail, to give (to stand) security.

verdampfen *v* (Phys.) to evaporate, to vaporize.

Verdampfung *f* evaporation, vaporization; ~sfähigkeit *f* evaporative power; ~spunkt *m* point of evaporation.

Verdeck *n* (Decke) covering, awning, tilt; (Dach des Wagens) head, (zum Niederschlagen) folding head; ~laderaum f. Deckladeraum; ~leder *n* (des Wagens) head leather, coach leather.

verdichten *v* (Phys. u. s. w.) to condense, to condensate, to solidify; (zusammendrücken) to compress.

Verdichtung *f* (Phys.) condensation, concentration, solidification; (Zusammendrückung *f*) compression.

Verdichtungsapparat *m* condenser, condensator.

Verdienstgelder *n/pl* des Posthalters (+) profit (receipts) accruing to the mail contractor out of the extra mail coach service.

Verding *m*, etwas in ~ oder auf ~ übernehmen to undertake *smt* by the job (by contract); ~arbeit, Akkordarbeit *f* work by contract, work upon terms.

verdingen *v* to let out for hire, to hire out; (eine Arbeit) to give a work by the job, to let out a work in contract; freihändig ~ f. freihändig.
Verdingung *f*, **Gedinge** *n* bargain; ~ einer Arbeit letting out a work in contract; f. auch Gedinge; freihändige ~ f. freihändig.
verdrehen *v* (Techn.) to wrench, to distort, to twist out of its regular shape, to force.
Verdrehung *f* torsion.
verdübeln *v* to peg, to dowel.
Verdübelung *f* pegging, dowelling.
verdünnen *v* to thin; eine Farbe ~ to temper (to wash) a color; eine Flüssigkeit, eine Säure ~ (Chem.) to dilute a liquid, an acid; die Luft mittels der Luftpumpe ~ to rarefy the air by the air pump.
verdünnt *a* (Chem.) dilute, diluted (said of liquids), (Phys.) rare, rarefied (said of air & gases), thinned (said of axes, rods etc.).
Verdünnung *f* (Chem.) dilution, (Phys.) rarefaction, (Mech.) thinning; ~, Ausdehnung *f* (Phys., Mech.) expansion, dilatation; ~öl *n* diluting oil.
verdunsten *v* (Phys.) to evaporate, to exhalate.
Verdunstung *f* (Phys.) evaporation, vaporization, exhalation.
vereiden *v* to administer an oath to *smb*, to swear *smb* in, to bind by oath.
vereidigt *a* sworn, sworn-in, bound by oath.
Vereidung *f* swearing-in, binding by oath.
Verein *m*, Post=~, Telegraphen=~ f. Post ... u. Telegraphen ...; ~taxe *f* (P) Union rate.
vereinbaren *v* to agree upon *smt*.
Vereinbarung *f* agreement.
vereinigen *v* (Chem., Mech.) to combine; sich ~ (Chem., Phys.) to unite, to combine, to unite together; sich wieder ~ to recombine.
Vereinigung *f* combination; (chemische) chemical union, chemical combination; (in einem Mittelpunkte) concentration.

vereinnahmen *v* to receive, to book as received, to enter on the receipt side of an account.
Vereinnahmung *f* (eines Betrages) receiving, receipt.
verengen *v* to narrow, to make narrow.
Verengerung *f* narrowing.
verengt *a* narrowed; ~er Theil *m* (Techn.) waist.
verfahren *v* (handeln) to proceed, to act; (Materialien ~) to transport stores, to export goods.
Verfahren *n* proceeding, method, manipulation, mode; diszplinarisches ~ (+) disciplinary proceeding; gerichtliches ~ proceedings at law, judicial procedure; ~ der Eisengewinnung mode of extracting iron from ore.
Verfall *m* (eines Bauwerks) ruin, decay, dilapidation; (eines Rechtes) loss, forfeiture; (eines Wechsels) falling due, maturity; bei ~ when due, at maturity; in ~ gerathen *v* to decay, to go to ruin, to go out of repair.
Verfall ... ~tag *m* (eines Wechsels u. f. w.) day of expiration, day when a bill becomes due; ~zeit *f* time when a bill becomes due, time of payment, expiration, maturity; bis zur ~zeit till due; vor der ~zeit before due; zur ~zeit when due.
verfallen *v* (Bau) to decay, to go to ruin; (ablaufen) to expire, to elapse; (fällig werden) to fall due, to become due.
verfallen *a* fallen to decay, ruinous, dilapidated; (von Wechseln) due; ~es Pfand *n* forfeited pledge, forfeited mortgage; ~e Postanweisung *f* lapsed money order.
verfassen *v* to write, to draw up (a report).
verfaulen *v* to rot, to get rotten, to putrefy, to become putrid.
verfehlen *v* to miss (the train); (seinen Zweck) to be disappointed, to fail; ich werde nicht ~ zu erscheinen I shall not fail to appear; verfehlte Anschlüsse *m/pl* (P, Eisenb.) missed connections.

verflüchtigen, sich *v* (Chem.) to volatilize, to subtilize.
Verflüchtigung *f* (Chem.) volatilization, subtilization.
Verfolg *m*, in ~ meines Schreibens in pursuance of my letter.
verfolgen *v* to pursue (a criminal), to trace (a letter); (sein Recht) to prosecute; (gerichtlich) to prosecute at law; (seinen Weg) to continue.
verfrachten *v* to send (to carry) goods, to pay the freight, to charter a ship.
verfügen *v* (anordnen) to order; (über etw.) to dispose of *smt.*
Verfügung *f* disposition, disposal, order, ordinance; ~ treffen *v* to dispose, to order; die nöthigen ~en treffen to give the necessary orders; weitere ~en abwarten to wait for further directions; zur ~ stehen to be at the disposal, eine zweite Leitg. steht zur ~ (*T*) a second wire is available; zur ~ stellen to place at the disposal (of the public).
verganten *v* to sell by auction, to expose to public sale; Einen ~ to declare a person a bankrupt.
Vergantung, Gant *f* public sale, auction.
vergeben *v* (Arbeit) to give out work, to strike a bargain for any work to be done; Arbeit freihändig ~ to give out work without any preliminary formalities.
Vergebung *f* einzelner Arbeiten ob. Lieferungen giving out work; (freihändige) giving out work without inviting tenders (without any preliminary formalities).
vergießen *v* (Techn.) to stop up, to cast in (with molten lead etc.); (durch Guß verlöthen) to burn together, to burn over.
vergipsen, gipsen *v* to plaster.
vergittern *v* to grate, to grate up, to lattice; (umgittern) to enclose with a lattice (trellis).
Vergleich *m* agreement, accord, treaty, arrangement, contract; einen ~ schließen to enter into a contract, to make an arrangement.
vergleichen *v* (Schriftstücke) to collate; (Telegramme) to collate (to repeat) telegrams; sich ~ to make an arrangement, to come to terms.
Vergleichung *f* (der Telegramme) collation, repetition; ~ bezahlt (*T*) repetition paid.
verglühen *v* (Techn.) to cool down; (das Porzellan) to give the biscuit-baking to porcelain.
Vergnügungs... ~fahrt *f* (zu Schiff) pleasure trip; ~zug *m* (Eisenb.) excursion train, pleasure train.
vergriffen *a* out of print (said of books).
vergüten *v* to make up, to make amends for, to compensate; (in Geld ersetzen) to reimburse (to re-imburse); Einem etw. ~ to make one compensation for; eine Summe in Rechnung ~ to allow a sum in account, to place to one's credit.
Vergütung *f* compensation, amends; (in Geld) reimbursement; (für zerbrochene Waaren) breakage; (außergewöhnliche) reward, gratuity; ~ auf Amtskosten (*PT*) allowance for office expenses; ~ für die Beförderung der Post auf Eisenbahnen allowance for the conveyance of mails by railway; ~ssatz *m* rate of compensation.
verhaften *v* to arrest, to capture, to apprehend, to put in prison.
Verhaftsbefehl *m* warrant of capture.
Verhaftung *f* arrest, capture, imprisonment.
verhalten, sich ~ *v* (Math. u.s.w.) to bear a proportion, to be of a ratio; die Telegraphen-Apparate ~ sich gut the telegraph instruments work well (keep in good order).
Verhältniß *n* (Math.) ratio, rate, (Chem.) proportion; im direkten ~ stehen to be directly in proportion to, to be in direct proportion; im umgekehrten ~ stehen to be inversely in proportion, to be in inverse proportion; ~ des Werthes, Werthverhältniß *n* (Techn.) rate.
Verhältniß... u. verhältniß... ~mäßig *a* proportional; ~mäßiger Antheil, ~mäßiger Werth *m*, Werthverhältniß *n* rate; ~regel *f* (Arithm.)

rule of proportion; ~zahl *f* (Math.) number of the proportion.

Verhaltungsregel *f* directive rule.

verhandeln *v* (Verhandlungen führen) to transact, to negotiate, to treat.

Verhandlung *f* transaction, negotiation, treaty, protocol, verbal process; eine ~ aufnehmen to draw up a protocol (a verbal process); ~sschrift *f* protocol, the minutes, the particulars of a case.

verjähren *v* (alt werden) to become superannuated, to grow out of date; (verfallen) to fall under the right of prescription, to become null & void; to outlaw; eine Forderung ist verjährt a claim is outlawed.

verjährt *a* (alt geworden) antiquated, superannuated; ein ~es Recht a prescriptive right (title); eine ~e Schuld a debt beyond the statute (of limitations); an outlawed debt.

Verjährung *f* superannuation, antiquation, prescription; die ~ (einer Schuld) geltend machen to plead prescription; ~sfrist *f* term of prescription.

verjüngen *v* (Techn.) to reduce to a small scale, to diminish, to lessen, to taper; sich ~, spitz zulaufen to taper.

verjüngter Maaßstab *m* (Techn.) reduced (tapering) scale.

Verjüngung *f* (Techn.) diminution, reduction, taper.

Verkauf *m* sale; (fingirter) colorable sale; (gerichtlicher) judicial sale; (aus freier Hand) sale by private contract; ~sanweisung *f* bill of sale; ~sbedingungen *f/pl* terms (conditions) of sale; ~sbuch *n* sales book, day book; ~skontrakt *m* sales contract; ~slokal *n* sale room, auction room; shop, store; ~spreis *m* selling price, sale price; ~srechnung *f* account of sales, account sale; ~sstelle *f* von Postwerthzeichen (P) s. Postwerthzeichen.

verkaufen *v* to sell; (im Kleinen) to retail, to sell by retail; (im Großen) to sell in gross, to do wholesale business.

Verkehr *m*, **Verkehrswesen** *n* trade, commercial intercourse, (P, T, Eisenb.) traffic; (schriftlicher) correspondence; (telegraphischer) telegraphic correspondence (intercourse); dem ~ übergeben to open to traffic.

Verkehrs... ~anstalt *f* office (open to postal, telegraphic & railroad business); ~bericht *m* traffic return; ~chef, Betriebsinspektor *m* (Eisenb.) traffic manager; ~entwickelung *f* development of traffic; ~mittel *n* means of traffic; ~steigerung *f* increase in the traffic (on the t. lines, in the postal exchange); ~straße *f* high road; ~umfang *m* amount of business transacted.

verkehrt *a* (Techn.) turned the wrong way, inverted, reversed.

verkeilen *v* (Techn.) to fasten (to steady) with wedges.

verkitten *v* to cement, to seal, to putty, to stop up with cement.

verklammern *v* (Techn.) to fasten with cramp irons, to cramp.

verkleiden *v* (ein Bauwerk) to line, to face, to case; (mit Brettern) to wainscot, to board, to line with boards.

Verkleidung *f* lining, facing, casing; ~, Bekleidung *f* (Masch., Techn.) cleading, jacket; ~smauer *f* lining wall, revetment wall.

verkleinern *v* to diminish (the battery power, the resistance).

verkohlen, in Kohle verwandeln *v* to carbonize, to convert into carbon; (ankohlen) to char; (karbonisiren [Chem., Techn.]) to carbonize; Holz ~, Kohlen brennen to carbonize wood, to burn (to make) charcoal; Steinkohlen ~, verkoken to coke coal, to char coal, to coke.

Verkohlung *f* carbonization; (Ankohlung) charring.

verkröpfen *v* (Techn.) to bend at right angles.

verkröpft *a* (Techn.) bent, carried round.

verkupfern *v* (Techn.) to copper,

to copper over, to coat (to sheathe) with copper.

verkuppeln *v* f. kuppeln.

Verkuppelung *f* f. Kuppelung.

Verkürzung *f* des Drahtes durch Kälte contraction of the wire by cold.

verladen *v* (mittels Fuhrwerks) to load, to lade, to transport; (zu Schiff) to ship, wieder ~ to reship.

Verladung *f* loading, shipping, shipment, vgl. das Vorstehende; ~s-kosten *pl* shipping charges, shipping expenses; ~spreis *m* shipping price; ~sschein *m* bill of lading.

Verlag *m* (Buchhandel) publication, publishing of books; right of publication; in ~ nehmen *v* to undertake the publishing of; im ~ von ... published by ...; ~s-buchhändler *m* publisher, publishing bookseller, copy purchaser; ~s-buchhandlung *f* publishing house (firm); ~shandel *m* publishing business; ~skosten *pl* publishing expenses; ~spostanstalt *f* (+) post office of publication; ~srecht *n* copy right; ~swerk *n* publication.

verlangen *v* to request, to desire, to ask.

Verlangen *n*, auf ~ des Empfängers upon the request of the receiver.

verlängern *v* (Techn.) to lengthen, to extend in length; (den Zahlungstermin) to prolongate the term of payment.

Verlängerung *f* (Techn.) lengthening, extension; (des Zahlungstermins) prolongation of the term of payment.

verlangsamen *v* to retard (the movement of a t. instrument).

Verlangsamung *f* retardation.

verlegen *v* to transport, to transplace; (ein Buch) to publish a book; (ein Kabel) to lay (to submerge) a cable; (eine Station) to change, to remove an office from one building into another one; (eine Telegr.-Linie) to alter an established telegraph route, to remove a telegraph line.

Verlegung *f* einer Telegraphenlinie removal.

verletzen *v* to damage, to injure.

verlöthen *v* to solder, to solder up, to close by soldering.

Verlust *m* loss; einen ~ erleiden to incur a loss; ~-Konto *n*, ~-Rechnung *f* loss account.

vermauern *v* to lay bricks; to wall in, to wall up, to fill up with brick-work; to enclose with walls.

vermehren *v* to increase.

Vermehrung *f* increase.

Vermerk *m* remark; dienstlicher ~ im Kopfe des Telegramms service instructions.

vermerken *v* to mark, to note down.

vermessen *v* to measure, to survey; (mit der Kette) to survey by the chain, to chain.

Vermessung *f* measuring, measurement, survey; (mit der Kette) chain-measuring.

vermiethen *v* to let, to let out, to hire out.

Vermiethung *f* letting, leasing, hiring-out.

vermindern f. verkleinern, verminderter Maaßstab f. verjüngter Maaßstab.

Verminderung f. Verkleinerung.

Vermittlung *f* mediation; durch ~ der Post by the mediation of the P. O.; ~samt *n* (*PT*) exchange office.

Vermögen *n* property, means; (bewegliches) movable goods, personal estate; (unbewegliches) real estate, real property, immovables; ~smasse *f* assets of a bankrupt.

vernichten *v* to destroy.

vernickeln *v* to plate (to coat) with nickel, to nickel; galvanisch ~ to electro-nickel.

Vernickelung *f* nickel plating, nickel coating, nickeling; (galvanische) electro-nickeling.

vernieten *v* to rivet, to clinch, to clench.

Vernietung *f* riveting, rivet joint, rivet work; (doppelte) double riveting; (einfache) single riveting; (versenkte) countersunk riveting, flush riveting.

verordnen *v* to order.

Verordnung *f* order, ordinance; ~sblatt *n* (*P*) post office circular; ~sbüreau *n* (*P*) office of instructions.

verpachten *v* to let out on lease, to lease.

verpacken *v* to pack up, to embale (goods); ~, libern (Masch.) to pack, to leather.

Verpackung *f* packing up; ~skosten *pl* packing charges; ~sstoff *m* wrapping, packing cloth.

verpfänden *v* to pawn, to pledge, to mortgage, to give in mortgage, to hypothecate.

Verpfändung *f* pawning, pledging, mortgaging, giving in mortgage.

verpflichten *v* to bind *smb*.

verquicken, amalgamiren *v* to amalgamate.

Verquickung, Amalgamirung *f* amalgamation.

verrechnen *v* to account for, to place to account, to put down; das Franko muß in Freimarken verrechnet werden prepayment of postage *can be effected* only by means of postage stamps; sich ~ to mis-reckon, to make an error in reckoning.

Verrechnung *f* der Einnahmen u. Ausgaben keeping the accounts of the receipts & expenditures; gegenseitige ~ der Zuschüsse u. Ueberschüsse settlement of surplus funds remitted to the General Postal Treasury.

verrosten *v* to rust.

verrostet *a* rusted, rusty.

versagen *v* to fail (said of instruments, of batteries etc.).

versanden *v* to get choked up with sand (said of a river etc.).

Versandt, Export *m* exportation (of goods).

Versäumnißstrafe *f* penalty (fine, forfeiture) in case of non-fulfillment of stipulations.

verschicken *v* to forward, to send, to transmit.

verschiebbar *a* sliding (contact, weight).

verschiffen *v* to ship, to send by water, to convey by ship.

Verschiffung *f* shipping, shipment; ~sspesen *pl* shipping charges, shipping expenses.

Verschlag *m* partition, wooden partition; (abgeschlagener Raum *m*) box; ~nagel, Brettnagel *m* plank nail.

verschlagen *v* (von Schiffen) to cast away, to drive away; (abschlagen) to partition off; (mit Brettern) to board, to board up.

Verschleiß *m* consumption; (Verkauf *m*) sale.

verschleppen *v* to carry to the wrong place; **verschleppt** *a* miscarried, lost.

verschließen *v* to lock, to close up; verschlossener Brief closed letter.

verschlingen *v* to twist, to twine together; (von Drähten) to be in contact.

Verschlingung *f* der Drähte (*T*) contact, cross, (in Folge Unwetters) weather cross.

Verschluß *m* locking, locking-up, place (receptacle) locked up; ~vorrichtung *f* shutting, closure.

verschmelzen *v* to melt, to smelt, to fuse; (zusammenschmelzen) to melt together; (die Post u. Telegraphie) to fuse the Departments of Posts & Telegraphs.

Verschmelzung *f* melting, smelting, fusion; (zweier Verwaltungen) fusion.

verschränken *v*, zwei Balken ~, verzahnen to joggle two beams.

verschrauben *v* to screw up, to screw on, to fix by screws.

Verschraubung *f* (durch Schrauben befestigtes Verbindungsstück *n* an Röhren u. s. w.) screw joint.

verschreiben *v* (J—m etw. schriftlich übermachen) to make over, to transfer, to assign *smt* to *smb* in writing; sich ~ to bind one's self by writing, to enter into a bond, to give one's bond; sich für J—n ~ to pass one's bond in security for *smb*.

Verschreibung *f* bond, obligation; eine ~ ausstellen to give a bond (an obligation).

Verschwiegenheit f (im Amte) official secrecy.

versehen v (Techn.) to furnish, to provide, to supply; sich ~ to err, to fail, to commit an error, to make a mistake.

Versehen n (Fehler m) blunder, fault, error, mistake, inadvertence; ein bei der Beförderung eines Telegramms vorgekommenes ~ error in the transmission of a telegram; aus ~ inadvertently, by mistake.

verseilen v to twist into a single strand, s. strand a. T.; verseilte Eisendrähte m/pl iron strands.

Verseilungsmaschine f (bei der Telegr.-Kabelfabrikation) stranding machine.

versenden v to forward, to transmit, to despatch, to convey; (einen elektrischen Strom) to send an electric current; (zu Schiffe) to ship.

Versendung f transmission, conveyance; ~sgebühr f charges for forwarding goods; ~skosten pl (Frachtlohn m) freightage, cartage; ~schein m (Frachtbrief m) bill of carriage, letter of conveyance, way bill.

Versenk ... ~bohrer m countersink, countersink drill, countersunk bit; ~bolzen m driving bolt, driver.

versenken v to sink, to countersink; (ein Kabel) to submerge a cable.

versenkt a sunk, countersunk; ~e Leitung (T) sunk wire, s. auch unterirdisch; ~es Niet n flush rivet.

versetzen v (Techn.) to displace, to remove; (einen Beamten) to subject an officer to a change of residence, to transfer an official to another place.

Versetzung f change of residence to which officials are subjected; (auf Wunsch des Beamten) change of residence in consequence of private convenience; die zwangsweise ~ in den Ruhestand herbeiführen to enforce the retirement of an officer; ~skosten pl allowance, s. auch Pensionirung.

versichern v to insure, to assure, to give security for; (verwahren) to secure.

Versicherung f insurance, assurance; ~sanstalt f insurance office; ~sgebühr f insurance fee; ~sgesellschaft f insurance company; ~skapital n stock of insurance; ~schein m written obligation, bond, (Police) policy of insurance; ~ssumme f capital insured; die ~ssumme f bestimmen to assess the damages.

versiegeln v to seal, to seal up.

verspäten, sich v to delay, to be late, to be too late; verspäteter Protest m retarded protest.

Verspätung f delay (in the transmission or delivery of mail matter); die Post hat ~ the mail is late; (eines Eisenbahnzuges) delay; ~ haben v (Eisenb.) to be behind time.

verspleißen v s. spleißen.

verstählen v (in Stahl verwandeln) to convert into steel; (anstählen) to steel, to overlay with steel, to point (to edge) with steel (an iron tool); galvanisch ~ to coat with electro-deposited iron.

Verstählung f von Werkzeugen steeling.

Verstanden-Zeichen n (T) the „Understand" signal.

verständigen, sich ~ v (T) zwei Aemter ~ sich gut two offices work well together.

Verständigung f in der Leitung ist gut (T) the line is in good working order; gute, schlechte ~ haben (T) to work well (badly).

verständlich a intelligible (ordinary telegrams are composed of words, letters etc. conveying an *intelligible* meaning).

verstärken v to strengthen (a t. line), to increase (the battery power), to fish (a piece of timber); (armiren) to truss.

verstärkter Träger m (Techn.) trussed girder, truss girder.

Verstärkung f strengthening, increase, vgl. das Vorstehende.

Verstärkungs ... ~batterie f (T)

auxiliary battery; ~mittel *n/pl* strengthenings (of a t. structure); ~stück *n* strengthening piece.

Versteckbeutel *m* (P) enclosure bag, (*Am*) inner sack.

versteifen *v* (mit Steifen versehen) to prop, to strut, to brace.

versteigern *v* to sell by auction (by public sale), to auctioneer.

Versteigerung *f* auction, public sale.

versteinen *v* (eine Straße) to ballast (to gravel) a road, to cover a road with stones; (mit Meilensteinen versehen) to provide a road with mile stones (kilometer stones).

verstellbar *a* adjustable.

verstemmen *v* (Mech.) to calk (riveted plates); eine Fuge ~ to stem a joint (with tow, lute etc.).

versteuern *v* to pay the taxes, duties or imposts upon an article.

Versteuerung *f* der vom Auslande kommenden Postpackete paying the taxes (duties) for foreign parcels.

verstopfen *v* to stop up, to plug up, to close; to choke, to obstruct; sich ~ (Masch.) to get dirty, to get greasy.

verstreben *v* to strut, to prop (t. poles).

verstümmeln *v* to mutilate, to alter (a telegram).

Verstümmelung *f* mutilation, alteration (of a telegram).

Versuch *m* (Techn.) essay, trial, experiment.

Versuchs... ~station *f* establishment for experimental working; ~weise *adv* by way of trial, experimentally.

vertagen *v* to adjourn, to put off; (auf unbestimmte Zeit) to adjourn *sine die*.

vertheilen *v* to distribute, to divide; sich ~ to be distributed, to divide itself.

Vertheiler, Vertheilungsapparat *m* distributor.

Vertheilung *f* distribution; (der Dienstgeschäfte) allotment of the work; (der Baumaterialien in die Lager auf der Baustrecke) distribution of the materials (stores) to the depots along the route selected for the t. line; (Elektrisirung durch ~ electrisation by statical induction.

Vertheilungs... ~apparat, Vertheiler *m* distributor; ~liste *f* (der Baumaterialien) list of distribution, s. Vertheilung; ~plan *m* (Telegr. Bau) plan of distribution, s. Vertheilung; ~vermögen *n* inductive capacity.

vertiefen *v* to deepen.

Vertiefung *f* deepening, cavity, excavation; (in einer Mauer) break.

Vertikal... ~ebene *f* vertical plane; ~galvanometer *n* vertical galvanometer; ~projektion *f*, Aufriß, Standriß *m* upright projection, orthographic projection, orthograph, elevation; ~schnitt *m* vertical section.

Vertrag, Kontrakt *m* contract, agreement, bargain, treaty, settlement, accord, stipulation; (mit dem Auslande) international convention; einen ~ schließen to make a contract, to conclude a treaty, to strike (to close) a bargain.

Vertrags... ~artikel *m* article (term) of agreement; ~gemäß *adv* according to the terms of agreement; ~widrig *adv* contrary to the terms of agreement.

vertragschließende Theile *m/pl* contracting parties.

Vertrauensarzt *m* der Postverwaltung Medical Officer of the P. O. Department.

vertraulich *a* (als Mittheilung) confidential.

vertreten *v* to act as substitute; (verantwortlich sein) to be responsible.

Vertreter *m* substitute, acting for...; (eines Handlungshauses) deputy, proxy, representative.

Vertretung *f*, in ~ (als Unterschrift) acting as substitute; ~skosten *pl* payment to a substitute; ~sverbindlichkeit *f* der Beamten responsibility of the officials, obligation of an official to make up for a loss incurred through his negligence.

Vertrieb *m* sale (of postage stamps, of news papers etc.).

vervielfältigen v to multiply; (ein Telegramm) to make out copies of a telegram to be delivered to several persons in the same place; ein zu ~des Telegramm a multiple telegram.

Vervielfältigung f reproduction from the original by any mechanical process; (von Telegrammen) making out copies of a (multiple) telegram & delivering them to several persons in the same place; ~sgebühr f (T) charge for multiple copies.

verwahren v to guard, to keep, to preserve, to keep in deposit.

Verwahrgut n deposit, property deposited; baares ~ der Ober-Postkasse cash deposit (deposit in cash) of the chief postal cash office.

Verwahrung f keeping, guard, preservation; in ~ geben to commit to one's keeping; in ~ nehmen to take in one's keeping.

verwalten v to manage, to superintend.

Verwalter m administrator, manager, inspector, overseer.

Verwaltung f administration, management.

Verwaltungs... ~bericht m Annual Report; ~sbezirk m department; ~einrichtung f organization; ~kosten pl expenses of management; ~rath m council of administration; ~zweig m branch (department) of administration.

verweigern v die Annahme to refuse the acceptance (of a letter, telegram etc.); „Annahme verweigert" (dienstl. Vermerk) Refused.

Verweis m reprimand; J—m einen ~ ertheilen to reprimand smb.

verwenden v to employ; Schnellschreiber ~ (T) to use fast speed instruments.

verwerfen v (ausschießen) to cast off, to condemn; (Material bei der Abnahme) to reject (not to receive) materials.

verwerthen v to convert into money, to dispose of, to realize, to turn to account.

verwirken v to forfeit; (die gestellte Kaution) to forfeit one's bail.

verwittern v (Chem.) to effloresce; (v. Mineralien) to weather, to be decomposed (disintegrated) by the atmosphere.

Verwitterung f (Chem.) efflorescence; (von Mineralien) weathering, decomposition (disintegration) by the influence of the atmosphere.

verzählen, sich v to miscount, to misreckon.

verzahnen v (Masch.) to tooth, to cog; (Bretter, Balken u. s. w.) to joggle, to indent, to scarf with indents.

Verzahnung f (Masch.) toothed wheel work; (Bauwerk) tooth work, toothing; (Zimmerei) indent, joggling; (Uhrwerk) catch; innere ~ (Masch.) inside-toothed wheel work, internal gear, inside-gearing; konische ~ (Masch.) conical (bevel, angular) gearing.

verzeichnen v to note down, to record.

Verzeichniß n register, list (of the daily receipts & expenditures), inventory, specification; Einzel~ detailed list; summarisches ~ summary list; ~ von Waarenproben price current.

Verzicht m renunciation, renouncement, act of disclaiming; auf etwas ~ leisten to renounce, to resign smt, to withdraw all claim to smt.

verzinken v to coat with zinc, to galvanize; verzinkter Draht m zinc coated wire, galvanized wire.

verzinnen v to tin; (Kupfer) to blanch copper; verzinntes Eisenblech, Weißblech n tin plate, tinned sheet-iron, sheet-iron coated with tin.

verzinsen v to pay interest on (for); sich ~ to yield interest.

verzinslich a paying (bearing) interest; eine ~e Anleihe f a loan bearing interest; Geld ~ austhun, ausleihen v to put money out at interest.

Verzinsung f paying interest on (for).

verzögern v to retard; (v. Telegrammen) to delay; (verlangsamen)

Verzögerung — Volta'scher Bogen

to loose speed (said of the movement of a telegraph apparatus); verzögerte Bewegung f (Mech.) retarded motion, retarded velocity. **Verzögerung** f (Mech.) retardation; (v. Telegrammen) delay.

verzollen v to pay duty on (for). **Verzollung** f paying duty; ~ von Packeten u. s. w. s. Versteuerung; ~sgebühr f customs duty.

Verzug m delay; ~sstrafe f penalty (fine, forfeiture) in case of non-fulfillment of stipulations; ~szinsen pl interest for delay, extra interest on account of dilatory payment.

verzweigen, sich v to ramify, to fork; (in zwei Zweige) to bifurcate. **Verzweigung** f ramification; (in zwei Zweige) bifurcation.

Bezirkschloß n (Kombinationsschloß) combination lock, permutation lock, puzzle lock.

Viadukt m viaduct.
Vicinal... s. Vizinal.
Viel... ~eck, ~seit, Polygon n (Geom.) polygon; ~eckig, ~kantig a polygonal, multangular; ~fach a multiple; ~fach=Telegraphie f multiplex telegraphy; ~gängige, mehrgängige Schraube f multiplex-threaded screw, screw with multiplex thread.

Vier... ~eck, ~seit n (Geom.) quadrangle, quadrilateral figure, rectangle; reguläres ~eck, Quadrat n square, quadrate; ~eckig, ~seitig a square, quadrangular, quadrilateral; ~eckiger Kopf m (einer Schraube u. s. w.) square head; Schraube f mit ~eckigem Kopf square-headed screw; ~eckig gewundene Schraube, Schraube mit rechteckigem Gewinde square-threaded screw; ~eckig machen v to square; ~gespann n team of four horses; ~kantig a square; ~kantige Feile f square file; ~stöckiges Haus n four-storied house.

Viertel n fourth part, quarter; (Stadtviertel n) quarter, ward; ~jahr, Quartal n quarter; ~jährlicher Rechnungsabschluß m quarterly account; ~jahrsweise a u. adv quarterly.

Vignette f vignette, flourish, printer's flower.
Visir... ~linie f (beim Vermessen) line of direction, line of sight; ~scheibe f sliding vane; ~stab m, ~latte f ranging pole.
visiren v (beim Bau) to take the level; (aichen) to gauge, to gage.
Vitriol m (Chem.) vitriol, copperas; blauer ~, Kupfervitriol m blue vitriol, blue copperas, copper vitriol, sulphate of copper; grüner ~, Eisenvitriol m green vitriol, green copperas, iron vitriol, sulphate of protoxide of iron, ferrous sulphate; ~öl n, rauchende Schwefelsäure, Vitriolsäure f fuming sulphuric acid, fuming oil of vitriol, vitriolic acid.
Vizinal... ~bahn f (Eisenb.) secondary railway; ~weg m parish road.

Volkszählung f census.
voll a full; (vollständig) whole, complete, entire.
Voll... ~druck m (Dampfmasch.) full pressure; ~gewicht n full weight; ~macht, Prokura f power of attorney, full power, procuration, proxy, (notarielle ~macht) power of attorney, (Einem ~macht geben to empower, (to authorize) one, schriftliche ~macht letter of attorney; ~machtgeber m constituent; ~machthaber m attorney; ~wichtig a (von Münzen) of weight, of full weight; nicht ~wichtig deficient in weight, short.

Vollstreckung f eines Beitreibungsbeschlusses (+) execution of a decree to collect forcibly fees or any other amount due to the Treasury, execution of an executory decree.

vollziehen v, ein Schriftstück ~ to sign a paper, to put one's signature to a paper.
Volt n (Phys.) volt (unit of electro-motive force).
Voltameter n voltameter.
Volta'scher Bogen, galvanischer Lichtbogen m voltaic arc; ~sche Berührungstheorie f Volta's contact theory; ~sches Element n voltaic couple, voltaic element; ~sche

Säule *f* voltaic pile, voltaic battery.

Volum, Volumen *n*, **Rauminhalt, geometrischer Inhalt** *m* eines Körpers (Phys.) volume, solid content, cubical content, bulk; **spezifisches ~, Molekularvolum, Atomvolum** (Phys., Chem.) specific volume, molecular volume, atomic volume; **das ~ berechnen** *v* to cube; **~gewicht** *n* weight of volume; **~messer** *m*, **~meter, Volumenometer, Stereometer** *n* (Phys.) volumenometer, stereometer; **~verhältniß** *n* proportion of volume; **~vermehrung** *f* increase of bulk.

volumetrisch, das Raummaaß betreffend *a* volumetric; **~e Analyse, Maaßanalyse** *f* volumetric analysis, analysis by measure or by volume.

voluminös *a* voluminous, bulky.

Voranschlag *m* rough estimate, previous calculation.

Vorarbeiten *f/pl* preliminary work, preparatory work, preparation for work.

vorarbeiten *v* (Techn.) to prepare the work, to do the preliminary works.

Vorarbeiter *m* foreman, headman.

Voraus... u. voraus... **~bestellung** *f* einer Extrapost securing (taking) an extra mail coach in advance; **~bezahlen** *v* to pay in advance, to anticipate payment; **~transport** *m* (P) s. Vortransport.

Vorbehalt *m*, **Klausel** *f* reservation, reserve, proviso; **ohne ~** without restriction (reserve).

vorbehalten, sich *v* to reserve to one's self, to stipulate for ...

vorbehältlich *adv* under restriction, except, with the exception of.

Vorbereitungsarbeiten *f/pl* preliminary works, preparation for work.

Vorbescheid *m* preliminary answer, acknowledgment of receipt.

Vorder... ~ansicht *f* front view, front elevation; **~fläche, Außenfläche** *f* (Techn.) face, surface, superficies; **~gebäude, ~haus** *n* fore-part of a building, front building; **~glied** *n* eines Verhältnisses (Arithm.) antecedent of a ratio; **~hof** *m* forecourt, fore-yard; **~laderaum** (P) receptacle for parcels in the fore-part of the mail coach; **~seite, Frontseite** *f* façade, frontal side, frontispiece, front; **~sitz** *m* (eines Wagens) front seat.

voreilen *v* to run ahead (said of synchronous t. instruments).

Vorgelege, Rädervorgelege, Zwischengeschirr *n* (Masch.) communicator, connecting gearing, intermediate gearing, purchase, pinion; **inneres ~** wheelwork with internal pinion; **~ mit Stirnrädern** spurgear.

vorgeschrieben *a* (amtlich) official; *adv* officially.

Vorgesetzter *m* superior, superior officer.

Vorhalle, Thürhalle *f* porch, entry hall, lobby; (**Vorplatz** *m*, **Vorzimmer** *n*) anteroom, antechamber, vestibule.

Vorhängeschloß, Vorlegeschloß *n* padlock; **mit einem ~ verschließen** to padlock.

Vorhof *m* fore-court, front-court, fore-yard.

Vorkauf *m* preemption.

vorläufig *a* preliminary, (provisorisch) temporary; **~e Wiederherstellungsarbeiten** *f/pl* (Telegr.-Bau) temporary repairs.

Vorlege... ~scheibe, Unterlegscheibe, Dichtungsscheibe *f* (Techn.) washer; **~schloß, Vorhängeschloß** *n* padlock.

vorliegende Zeichen *n/pl* kommen an (HA) the letters or signs of the receiving instrument advance in the alphabet.

vormerken *v* to note down, to take note of.

Vormerkung *f* note.

Vorposten-Telegraph *m* telegraph connecting the out-stations (out-posts, pickets etc.) with the rear body.

Vorprüfung *f* previous examination.

Vorrang *m*, **Priorität** *f* priority, precedence; **~ in der Beförderung**

haben (*T*) to take precedence of other telegrams (Government telegrams are entitled to priority over other telegrams).

Vorrath *m* stock, store, supply; im ~ haben *v* to have in store (on hand).

vorräthig *a* in store, on hand, stored up, ready.

Vorraths... ~leitung *f* (*T*) spare wire; ~stück *n* spare piece; ~wagen *m* (Eisenb.) tender.

Vorreiber *m* (am Fenster) turn buckle, button.

Vorrichtung *f* (Mech.) contrivance, mechanism, apparatus; (Masch.) gear; ~ zum Ausrücken, Ausrückvorrichtung, Ausrückung, Ein- und Ausrück-~, Ausrücker *m*, Ein- und Ausrückzeug *n* (Masch.) disengaging gear, engaging & disengaging gear, engaging & disengaging coupling; selbstthätige ~ self-acting apparatus.

vorschieben *v* to feed (the paper slip of telegraph instruments).

Vorschlag *m* proposition; in ~ bringen to make a proposition, to propose.

vorschreiben *v* (das via eines Telegramms) to specify a particular route for the transmission of a telegram.

Vorschrift *f* instruction, regulation, order; ~smäßig *a* u. *adv* according to the instructions given.

Vorschuß *m* (*P*) s. Nachnahme; advanced money, advance, cash-advance; (Darlehn *n*) loan; J—m Vorschüsse *pl* leisten *v* to accommodate *smb*; ~postanweisung dasj. wie Nachnahme - Postanweisung; ~verein s. Post-, Spar- und ~verein; ~weise *adv* in advance.

Vorsichtsmaaßregeln *f/pl* treffen to take precautionary measures.

Vorspann *m* relay, fresh set of horses.

vorspringen *v* (Bauw.) to jut out, to project, to be salient.

vorspringend *a* (Techn.) projecting, prominent, salient; ~er, ausspringender Winkel *m* salient angle.

Vorsprung *m* (Techn.) projecting part, projection; (eines Hafendammes) jetty, jettee, jutty.

Vorstadt *f* suburb.

Vorsteck... ~nagel *m* key, wedge, pin; ~schild *n* (*P*) slide label, reversible label; ~stift *m* forelock, key, cottrel, pin.

Vorstecker, Vorstecknagel, Vorsteckstift *m* (Techn.) forelock, key, cottrel, pin; (am Wagen) linch pin, axle pin; (hölzerner) peg, pin.

Vorsteher *m* (Techn.) inspector, chief; (eines großen Post-Amts) head postmaster; (einer Post - Agentur) subpostmaster; (einer Anstalt nur für Briefannahme) receiver; (eines Telegr.-Amts in England) controller, (einer vereinigten Verkehrsanstalt in England) postmaster engaged in telegraph business; (in einer Werkstatt) foreman, head-man.

Vortrag *m* (in Rechnungen) balance carried forward.

Vortransport *m* (*P*) preliminary mail; einen Kartenschluß als ~ anfertigen to despatch a preliminary mail.

Voruntersuchung *f* preliminary examination, preliminary trial.

vorzeigen *v* to show (forth), to produce; (einen Wechsel zum Akzept, zur Zahlung) to present a bill.

Vorzeiger *m* dieses bearer of this; ~ eines Wechsels bearer of a bill of exchange.

Vorzug *m* s. Vorrang.

vulkanisiren *v* to vulcanize, to sulphurize, to cure India rubber or gutta percha.

vulkanisirt... ~e Guttapercha *f* vulcanised gutta percha; hornisirte ~e Guttapercha, Ebonit *n* hardened gutta percha, ebonite; hornisirter ~er Kautschuk, Hartgummi *m* hardened caoutchouc, ebonite, vulcanite.

W.

Waage *f* balance, pair of scales; (Federwaage) spring balance; (Schnellwaage mit verschiebbarem Gewicht) Roman balance, steel yard; (Gleichgewicht n) equilibrium, equipoise; (Waageamt, Waagehaus n) office where goods are weighed.

Waage ... ~amt n office where goods are weighed; ~arm m balance arm; ~balken m (Mech.) beam, lever of a balance, board of a balance; ~bescheinigung f certificate of the weight; ~brett n wood scale, board of a balance; ~gebühren f/pl, Wägegeld n weighage, duty paid for weighing, pesage; ~haus n s. Waage; ~kette f scale chain; ~meister m keeper of the public balance; ~recht a horizontal, level; ~schein, ~zettel m weigh-bill.

Waare f ware, article, article of commerce, merchandise, goods.

Waaren ... ~artikel m article of merchandise; ~lager n store, stock, assortment of goods; ~muster n sample & pattern of merchandise; ~muster mit Kaufwerth (P) sample (pattern) containing merchandise of intrinsic value; ~muster-Post f Pattern Post, Sample Post; ~marke f, ~stempel m mark; ~niederlage f store house, entrepot; ~partie f allotment; ~preis m price of goods; ~probe f, ~muster n pattern, sample of merchandise; ~rechnung f invoice, bill of parcels; ~sendung f consignment (shipment) of goods; ~verzeichniß n list (catalogue) of goods, invoice, bill of lading; ~zeichen n mark made on goods, brand; ~zoll m custom-house duty.

Wachs n wax; ~draht m (T) wax-covered wire; ~kerze f, ~licht n wax candle, wax taper, taper; ~leinwand f oil cloth; ~tuch n oil cloth, wax cloth.

Wächter m (Techn.) watchman, keeper; ~-Kontroluhr f tell-tale, time detector; elektromagnetische ~-Kontroluhr electro-magnetic watch-clock.

Wachtzimmer n watchroom.

Wadelzeit f season for felling timber (when the circulation of the sap is least active).

Wag ... s. Waag ...

Wage u. **wage** ... s. Waage.

Wagen m carriage, car, cart, wagon (waggon); (des HA) chariot, contact maker; (Eisenbahn-~ wagon (waggon), railway carriage, (Am) car, railway car; Post~ s. Post; Haupt~, Bei~, Schlaf~ s. die betr. Buchstaben.

Wagen ... ~bau m building (manufacturing) of carriages; ~decke f cover, awning of a carriage, waggon tilt, (von gefirnißtem Zeuge) tarpaulin, tarpauling; ~geld n bei Extraposten (+) amount charged against the passengers by extra mail coach for the use of the coach; ~gestell n frame (body) of a carriage; ~kasten m body (trunk, chest) of a waggon; ~klasse f (Eisenb.) class; ~ladung f cart load, waggon load; ~schmiere f grease; ~schuppen m coach house, carriage shed, (Eisenb.) waggon house; ~spur f track; ~stücke, bloßgehende Packete n/pl (P) separate parcels; ~wärter m (Eisenb.) porter; ~wascher s. Post-wagenwascher.

wägen, wiegen v to weigh, to poise, to balance.

Wägung f, **Wägen** n weighing.

Waggon m (Eisenb.) wagon, waggon, railway wagon, railway carriage, (Am) railway car.

Wagner, Stellmacher m cartwright, wheelwright.

Wahr ... u. **wahr** ... ~nehmbar a perceptible; ~scheinlichkeitsrechnung f (Math.) calculus of probability, rule of probabilities; ~zeichen n mark.

Währung f value, standard, standard currency of money.

Waisen ... ~geld n pension to children of deceased officers; ~haus n orphan asylum, orphan house.

Walz ... ~blech, gewalztes Blech n rolled plate, rolled metal; ~blei n rolled lead, milled lead, sheet lead; ~draht m rolled wire; ~eisen, gewalztes Eisen n rolled iron, drawn iron, drawn-out iron; ~werk (Draht-~werk) n rolling mill, (zum Verkleinern der Erze, Erz-~werk) ore-crushing mill, crushing mill; ~zinn, gewalztes Zinn *n* rolled tin, laminated tin, sheet tin.

Walze *f* (Geom.) cylinder; (Masch.) roll, roller, drum, cylinder; (Rolle *f*, Haspel *m*) reel; (für die Papierführung der Telegr.-Apparate) roller; (Chaussee-~) street roller.

walzen, auswalzen *v* to roll, to mill, to pass between rollers; (Eisen) to draw iron into bars.

Walzenumschalter *m* (*T*) cylindrical contact.

Wand *f* (Techn.) side cheek; (in Bauwerken) wall, partition; ~leiste *f* (*T*) ledge, batten, wooden batten (fixed to the wall); ~dicke, ~stärke *f* thickness (of metal, of the guttapercha etc.); ~träger *m* bracket; ~uhr *f* house clock, time piece.

Wange *f* (Techn.) cheek, side, side piece; (Gestellplatte, vordere, hintere des *HA*) front plate, hind plate.

Wappenschild *n* (Reichswappen *n*) an den Bahnpostwagen coat of arms of the German Empire, arms of the Empire.

Warm... u. **warm**... ~brüchig *a* red-short, brittle when hot (said of metals); ~laufen *n* (Masch.) heating; ~luftheizung *f* heating by hot air, hotair heating; ~luftheizungsapparat *m* apparatus for hot-air heating; ~wasserheizung *f* hot-water heating.

Wärme *f* (Phys.) heat; ~ ausstrahlen *v* to radiate heat; ~ binden *v* to absorb heat, to consume heat; ~ durchlassen *v* to transmit heat; nutzbare ~ available heat.

Wärme... ~einheit *f* (W. E.) unit of heat, calory, (French) calorie, heat unit, thermal unit (quantity of heat required to raise 1 kilogramme of water 1° C.); ~entwickelung *f* disengagement, development of heat, evolution of heat; ~grad *m* degree of heat; ~leitend *a* heat conducting; ~leiter *m* conductor of heat, (guter, schlechter) good (bad) conductor of heat; ~messer *m* calorimeter, thermometer; ~wirkung *f* thermal (heating, calorific) effect.

Warnungstafel *f* warning-table; (an einem Niveauübergang der Eisenbahn) notice, notice board.

Warte... ~geld *n* der Beamten (in Deutschland) three quarters of the annual salary but not less than 450 marks, (in England) half pay; ~geld bei Extraposten (+) charge for keeping the extra mail coach waiting; ~saal *m*, ~zimmer *n* (*P*, Eisenb.) waiting room; ~zeichen *n* (*T*) the „wait" signal.

Wärter *m* (Bahnwärter) watchman; ~bude *f*, ~haus *n* watchman's house.

Wasser... ~abgabe, ~taxe *f* water rate; ~ablauf, ~abzug *m* letting off of water, drainage, draining; ~abzug *m* einer Stadt town-drainage, sewer, sewage; ~bau *m* hydraulic work; ~bau-Ingenieur *m* hydraulic engineer; ~baumeister *m* hydraulic architect; ~behälter *m* reservoir, cistern, tank; ~dampf *m* steam; ~dampfheizung *f* steam heating; ~dicht *a* water proof, water tight, impervious to water, impermeable; ~graben *m* water ditch, water trench; ~haltige Erdschicht *f* aqueous earth; ~höhe, ~linie *f*, ~stand *m* height of water, water level; ~kraft *f* hydraulic power, water power; ~lauf *m* water course, drain; ~leitung *f* aqueduct, conduit of water, water conduit; ~leitungsrohr *n* conduit pipe, water-main; ~messer *m* hydrometer, water-meter; ~rinne *f* water channel, gully; ~rohr *n*, ~röhre *f* s. ~leitungsrohr; ~spiegel *m* surface of water; ~stand *m* s. ~höhe; ~standsmeldung *f* river report (in order to give notice of the conditions of the rivers affecting navigation & floods); ~standsmesser *m* tide gauge, water gauge; ~stoff *m* hydrogen; ~stoffhaltig *a* hydrogenous; ~stoffgas *n* hydrogen gas; ~transport *m* conveyance by water, water conveyance; ~waage *f* (Libelle *f*) level, water level, (Aräometer *m*) areometer, hydrostatic level.

wässerige Lösung f aqueous solution, solution in water.

Wechsel m (Veränderung f) change, alteration; (Geldanweisung f) bill of exchange, bill; (Umschalter) s. Umschalter; (am HA) s. Uebergang; einen ~ akzeptiren to accept a bill; einen ~ ausstellen to issue a bill; einen ~ begeben to negotiate a bill; einen ~ bezahlen to pay a bill; einen ~ diskontiren to discount a bill; einen ~ honoriren to honor a bill; einen ~ indossiren to endorse a bill; einen ~ präsentiren to present a bill; einen ~ remittiren to remit a bill; einen ~ trassiren to draw a bill; domizilirter ~ domiciliated bill; eigener, trockener ~, Sola~ promissory note, note of hand; an den Inhaber zahlbarer ~ bill payable to the bearer; kurzer, langer ~ short dated (long dated) bill; Prima~, Secunda~ first (second) bill of exchange; ~ auf Sicht bill at sight (on demand); Sola~ sole bill of exchange; überfälliger ~ bill overdue; fälliger ~ bill due.

Wechsel... ~geld n money of exchange, exchange money, small change; ~hebel m (HA) figure changing lever; ~klemme f alteration screw; ~kurs m course (rate) of exchange; ~protest m protest of a bill, Erhebung von ~protest durch die Post-Verwaltung protesting of unpaid bills of exchange by the postal administration; ~recht n law on bills of exchange; ~schlüssel m (T) reversing key; die Isolatoren ~ständig gruppiren (T) to place the insulators in vertical distance alternately on opposite sides of the poles; ~stempel m bill stamp; ~stempelmarke f bill stamp; Vertrieb der Reichs-~stempelmarken business connected with the sale of Imperial bill stamps; ~stempelsteuer f bill-stamp tax; ~strom m (T) alternating current; ~strom-Apparat s. Apparat; ~strom-Maschine f (Elektrz.) alternate current dynamo, alternating current machine; ~stromtaste f (T) double current key, pole changing key; ~verkehr m (Geldanweisungs-Verkehr) circulation of bills of exchange, (Austausch von Postsendungen) exchange of correspondence; ~vordruckblatt n (gestempeltes) stamped form for bills of exchange; ~wirkung f (Mech.) reciprocal action.

wechseln v to change (money, poles of the battery etc.); (verändern) to alter; (auswechseln) to exchange (correspondence), to replace (old stores).

Wecker m (T, F, Eisenb.) bell, alarm, alarum, call bell; ~ mit einem Schlage single stroke bell; ~ mit zwei Schlägen double stroke bell; (Schnarrwecker, ~ mit Selbstunterbrechung) trembling bell; ~batterie f ringing battery, battery for ringing; ~knopf m call button, s. auch ~taste; ~linie f bell line; ~taste, Läutetaste f bell push, bell key, electric plunger, (mit Rücksignal) repeating bell push; ~schaltung, ~stellung f (T, F) joining up a bell in circuit; ~uhr f alarm clock; auf ~ stellen (T, F) to put a bell in circuit.

Weg m way, road, passage, walk, course, track; gebahnter ~ beaten way; ~ mit festem Steingrund metalled road.

Wege... ~aufseher m inspector of roads; ~aufsichtsbehörde f (local) road authority; ~bau m construction of roads; ~bauamt n office for road building and inspection, road office; ~geld n wheelage, (Chausseegeld) turn-pike toll, gate toll; ~überführung f (Eisenb.) upper bridge; ~übergang m im Niveau level crossing, railway crossing; ~unterführung f (Eisenb.) under bridge; ~weiser m sign post, s. auch Straße.

Weich... u. **weich** a soft; ~eisen n soft iron; ~loth, Schnellloth n soft solder, tin solder.

Weichbild n (einer Stadt) precincts, boundary of a town.

Weiche f (Eisenb.) switch, turn-out place, shunt.

Weichen... ~apparat m (Eisenb.)

switch stand; ~ſteller, ~wärter m switch man, points-man, guard watching the switches; ~ſtellerhaus n points-man's house.

Weiß... ~blech, verzinntes Eiſenblech n tin plate, tinned plate, tinned sheet-iron; ~brüchig a ſ. Eiſen; ~glühen n, ~gluth f white heat, incandescence; ~glühend a white hot, incandescent; ~tanne, Edeltanne f white pine.

Weite f distance, width; (im Lichten) intermediate open space, inside width, width in the clear; lichte ~ (eines Bogens) span, chord; (magnetiſche) magnetical amplitude; ~ des Schraubenganges, Ganghöhe f (Techn.) pitch of a screw.

weiterbefördern v to retransmit (by telegraph), to redirect (letters, parcels etc.); to transmit, to send (by train, by post); an eine zweite Adreſſe ~ (P) to redirect, (T) to retransmit.

Weiterbeförderung f retransmission (by telegraph), redirection, transmission (by post, by train).

Well... ~baum m arbor, shaft, beam, axle tree, spindle, revolving beam; ~baum, Federſtift m arbor, spring arbor; ~daumen, Hebedaumen, Mitnehmer m cog, arm; ~rad n arbor wheel, axle tree wheel; ~zapfen m pin, pivot, axle end, gudgeon.

Welle f (Achſe f, Wellbaum m) ſ. Wellbaum; (Walze f) cylinder, roller, roll; (Daumenwelle f) cam shaft, tumbling shaft; (ſich drehende) revolving shaft; (horizontale) horizontal (lying) shaft; (vertikale) vertical (upright) shaft; (Stromwelle im Kabel) wave.

Wellen... ~bewegung f undulation, undulating motion; ~förmig a undulatory; ſich ~förmig bewegen v to undulate; ~lager, Achſen-, Zapfenlager n bearing, journal bearing, journal box.

Weltpoſt... ~verein m Universal Postal Union; Länder, die dem ~verein angehören Union countries, Länder, die dem ~verein nicht angehören non-Union countries; ~vertrag m convention of the Universal Postal Union.

Wende... ~eiſen n, ~haken m cant hook; ~punkt m einer Kurve (Geom., Mech.) turning point, point of inflection.

Weniger s u. **weniger** a (bei den Kaſſenabſchlüſſen der Poſtanſtalten) cash short, deficiency.

Wenigſtfordernde m the lowest contractor.

werfen v to cast; Anker ~ to cast anchor, to drop anchor; ſich ~ to warp, to cast (said of wood).

Werft n wharf, quay; dock, dockyard, shipyard; ~dock n, ~docke f dry dock; ~geld n wharfage.

Werg n oakum; (getheertes) black oakum; (ungetheertes) white oakum.

Werk n (Techn., Mech.) work, mechanism; ~ſtatt f workshop, workroom, (Chem.) laboratory; ~zeug n tool, instrument; ~zeugkaſten m tool box, chest of tools.

Werth m value; den ~ beſtimmen v to tax, to appraise, to apprize; ~ empfangen value received; ~angabe f declaration of value; Packete ohne ~angabe (P) uninsured parcels; ~beſtimmung f apprizement, valuation, estimation; ~brief m letter with value declared, letter containing money, letter of value; ~briefannahme (+) counter where business with letters of value is transacted; ~brieftarif m table of postage for letters with value declared; ~gelaß n strong box; ~packet n parcel with value declared; ~papiere n/pl papers of value, value papers, securities; ~ſendung f (P) article of value; ~verminderung f depreciation, deterioration.

Wetter n weather; (Unwetter) tempest, storm; ſchlagende ~ pl, Grubengas n mine gas; ~büreau n meteorological office; ~leuchten n sheet lightning, heat lightning; ~ſeite f weather side, west side; ~telegramm n weather telegram, obs. telegram; ~vorherbeſtimmung, ~prognoſe f forecast of the weather, (Am) weather probability; ~zeiger *m* (Phyſ.) weather gage, (Windzeiger *m*) anemoscope.

Wheatſtone'ſche Brücke *f* (*T*, Elektriz.) Wheatstone's bridge, bridge of balance; ~'ſcher Selbſtübertrager *m* Wheatstone's automatic transmitter.

Wickel ... ~draht *m* ſ. Bindedraht; ~löthſtelle *f* ſ. Löthſtelle.

wickeln, ſpulen *v* to wind, to spool.

Widerſtand *m* (widerſtehende Kraft *f* [Mech.]) resistance, resisting force; (elektriſcher) electric resistance, resistance to the passage of an electric current; (Leitungswiderſtand) resistance to conduction, conductivity resistance; ~ der Nebenſchließungen, Iſolations-~ (*T*) insulation resistance; (Normal-~) standard resistance, ohm; (äußerer) external, (exterior) resistance; (innerer) internal resistance; (künſtlicher) artificial resistance; (reduzirter) reduced resistance; (ſpezifiſcher) specific resistance; ~ einſchalten to insert resistance.

Widerſtands ... ~einheit *f* unit of resistance; ~fähigkeit *f* resisting force, stability (of t. lines); ~kaſten *m* box containing resistance coils; ~kraft *f* ſ. ~fähigkeit; ~meſſung *f* measurement of resistance; ~rolle *f* resistance coil, rheostat (Poggendorff's) rheocord; ~ſäule *f* standard resistance coil.

Wieder ... u. **wieder** ... ~aufnehmen *v* (das Verfahren) to take up the proceeding again (anew); die Arbeit ~ aufnehmen to resume, to take up (the work); ~ eröffnen *v* to reopen; ~eröffnung *f* reopening (of an office); ~erſtatten *v* to restore, to repay; ~herſtellen *v* to reconstruct, to repair; ~herſtellungsarbeiten *f/pl* repairs, work of reconstruction; ~holen *v* to repeat (a telegram); ~holung *f* repetition (of a telegram).

Wiege ... ~meiſter *m* weighmaster; ~vorrichtung *f* contrivance for weighing.

wiegen *v* to weigh.

Wieger *m* weigher.

Wind ... ~bruch *m* windfall, windfallen wood, rolled timber; ~brüchiges Holz *n* cracked wood, chinky timber; ~druck *m* pressure of the wind; ~fang *m*, Flügelrad *n* fly, expanding fly (contrivance for catching the wind or the air); ~ofen *m* wind furnace, draught furnace; ~ſeite *f* direction from which the wind may be expected.

Winde *f* (Mech.) winding engine; (zum Heben v. Laſten) lifting jack, hoisting jack; (zum Drahtziehen) pulley; ~eiſen *n* wrench, draw tongs; ~haken *m* tackle hook, claw of a jack, sling dog.

Windung *f* coil, helix, *pl* helices (of an electro-magnet), convolution; (Gewinde einer Schraube) thread, worm of a screw.

Winkel *m* (Geom.) angle; (Techn.) corner, nook; ſpitzer ~ acute angle; ſtumpfer ~ obtuse angle; vorſpringender ~ projecting angle.

Winkel ... ~band *n* corner band, angular iron band, (von Holz) brace, strut, strutting piece; ~eiſen ſ. Eiſen; ~förmig *a* angular; ~haken *m* (zur Beſtimmung des Draht-Durchhangs) angle iron; ~hebel *m* (*MA*) elbow lever; ~maaß *n* square, rule; ~meßinſtrument *n* angular instrument; ~meſſung *f* goniometry; ~punkt *m* corner; ~ſchiene *f* angle iron; ~ſtange *f* (*T*) angle pole; ~ſtütze *f* stalk, bracket of angle iron.

winkelig *a* angular, angulous.

Wippe *f* (Umſchalter [*T*]) Wippe, tumbler switch.

wirken *v* (Mech.) to act on *smt*; ~d *a* acting; automatiſch ~d self acting; doppelt ~d double acting; einfach ~d single acting; ~de Kraft *f* acting force, force, actuation.

wirklich *a* real.

wirkſam *a* (Mech.) effective.

Wirkung *f* (einer Kraft, Maſchine u. ſ. w.) action, effect of a force (of a machine); äußere ~ external action; direkte ~ direct action; einfache ~ single action; ~ in die Ferne (Phyſ.) action at a distance; innere ~ (Phyſ.) internal action; gegenſeitige ~ zweier Körper auf einander, ~ u. Rückwirkung ob. Gegenwirkung, dynamische Einwirkung f (Phys., Mech.) mutual action between two bodies, action & reaction, stress.

Wirkungskreis m der Post sphere of business falling to the share of the postal service.

Wismuth, Bismuth n u. m bismuth; ~loth, Schnellloth n fusible metal (2 bism., 1 tin, 1 lead); ~oxyd n oxide of bismuth.

Witterung f weather, temperature; ~sbeobachtungen f/pl meteorologic observations; ~skunde, ~slehre f meteorology; ~stelegramm n s. Wettertelegramm.

Wittwen ... ~geld n pension to the widows of deceased officers; ~kasse f widow's relief fund; ~ u. Waisengeld-Beitrag m contribution towards the widow's & orphan's pensions.

Wochen ... ~bericht m weekly report; ~blatt n weekly paper of publication; ~lohn m weekly pay, weekly wages; ~rechnung f weekly bill (account); ~tag m week day, day of the week; ~täglich a & adv on a week day, on week days.

Wohlfahrtseinrichtung f provident institution.

Wohn ... ~haus n dwelling house; ~ort m place of residence; ~stube, ~zimmer n sitting room, parlor.

Wohnungs ... ~anzeiger m city directory; ~geldzuschuß m lodging allowance; ~veränderung f change of residence, removal.

Wort ... ~bildung, ~fassung f formation of words; ~kürzung f abbreviation; ~laut m wording, text; ~tarif m word tariff; ~taxe f word rate, word tax; ~verbindung f joining of words; ~zahl f number of words; ~zählung f (T) counting of words & signs; ~zusammenziehung f combination of words.

wörtlich a literal, verbal, word for word; 100 Mark, ~ Einhundert Mark 100 marks, expressed in words: One hundred marks.

Wucht ... ~baum, ~klotz m, Unterlage f eines Hebels fulcrum, prop, rest of a lever, lever prop; (Mech.) colstaff; ~kette f chain for lifting up (raising).

wuchten v to lift up, to raise by means of a lever.

Wulst f u. m pad, padding; (Masch.) collar.

Würfel m die (pl dice); ~förmig a cubic; ~inhalt, Kubikinhalt m cubic contents; ~kohlen f/pl cubical coal, round coal, lumps.

Würge ... ~band, Ziehband n clip, iron clip; ~löthstelle s. Löthstelle.

Wurzel f (Arithm.) root; die ~ ausziehen to extract the root of a quantity; die vierte ~ biquadratic root; ~größe f radical quantity; ~zeichen n radical sign.

X.

Xylograph, Holzschneider m xylographer, engraver on wood.

Xylographie f xylography, wood engraving.

xylographiren v to engrave on wood, to print from wooden types.

xylographisch a xylographic, xylographical.

Y.

Yacht f, **Luftschiff** n yacht.

Yard f yard (a measure of three English feet).

Z.

zäh *a* tenaceous, tough; ~es Eisen, Zäheisen *n*, Stahlguß *m* toughened cast-iron.

Zäh... ~festigkeit *f* der Metalle tenacity, resistance to rupture by traction; ~flüssig, strengflüssig *a* hardly fusible, difficult of fusion, difficult to be melted, refractory.

Zähigkeit *f* tenacity (of the wire); (Zugfestigkeit, absolute Festigkeit *f*) tensile strength, strength of extension; (Biegungsfestigkeit, relative Festigkeit *f*) strength of flexure; ~smodul *m* modulus of tenacity.

Zahl *f* (Math., Arithm.) number; (Zahlzeichen *n*) figure, cipher; (ganze) integer number, integer; (gebrochene) fractionary number, fraction; (gerade) even number; (ungerade) odd number.

Zahl... u. **zahl**... ~bar *a* payable, to be paid, due, einen Wechsel ~bar machen to domiciliate a bill of exchange, ~bar werden to fall due; ~tag *m* day of payment, pay day; ~tisch, Zähltisch *m* counter, counting board; ~zeichen *n* (Arithm.) figure, cipher, numeral character.

Zähl... ~apparat, Zähler *m* (Masch.) counter, operameter, telltale; ~brett *n*, ~tisch *m* counting board, counter; ~rad *n* tell-tale.

zählen *v* to count, to reckon (the words & signs in telegrams); als ein Wort ~ (T) to count as one word; fünf Zahlenzeichen werden als ein Wort gezählt five figures are counted as one word or are reckoned one word.

Zahlen... ~=Blanktaste (HA) cipher blank key; ~gruppe *f* (T) group of figures; ~verhältniß *n* arithmetical proportion, ratio.

Zähler *m* (Masch.) counter; (eines Bruches) numerator.

Zahlung *f* payment, pay; gegen baare ~ for cash; seine ~en einstellen *v* to suspend (to stop) payment; ~ leisten *v* to pay; Mangel an ~ non-payment; ~ pro rata, Ratenzahlung ratable payment, payment pro rata.

Zahlungs... ~anweisung *f* order of payment; ~einstellung *f* suspension (stopping) of payment; ~fähig *a* solvent; ~fähigkeit *f* solvency; ~frist *f*, ~termin *m* term of payment, term fixed for the payment; gesetzliches ~mittel *n* legal tender; ~unfähig *a* insolvent; ~unfähigkeit *f* insolvency, inability to pay; ~werth *m* einer Münze numerical value of a coin.

Zahn *m* (Techn.) tooth; (Masch.) tooth, cog (of a toothed wheel); (Einschnitt) indent, joggle; ~getriebe *n* (Masch.) rack work; ~lücke *f* (eines Zahnrades, Zwischenraum zwischen zwei Zähnen) space between two teeth, interdental; ~rad, gezähntes Rad *n* toothed wheel, cogged wheel, gear wheel, gear, pinion, (mit eingesetzten Zähnen, Kammrad) cog wheel; konisches ~rad bevelled wheel, bevel gear; ~rad u. Sperrklinke *f* ratchet & pawl; ~radbahn *f* inclined-plane road; ~räder *pl*, ~räderwerk, ~getriebe *n* toothed gear, toothed wheelwork, engaged wheels; ~stange, gezahnte Stange *f* rack; ~stangengetriebe *n* rack gear; ~trieb *m* (HA) tooth spring; ~werk *n* tooth work, indented work.

Zange *f* pincers, tongs; Beiß~, Draht~, Zieh~ pliers, plyers, nippers; Flach~ flat pliers, flat nose pliers; Rund~ round pliers, round nosed pliers.

Zapfen *m* (Techn.) pin, peg, pivot, plug; (Schraubspitze *f* [HA]) pivot; (Dreh~, Well~) trunnion, journal, axle, pivot; (Dübel *m*) peg, plug; ~lager *n* bearing, journal box; (einer horizontalen od. liegenden Welle) pillow, collar, socket; ~loch *n* mortice, peg hole, pivot hole; ~mutter *f* sole, socket of a pivot; ~schraube *f* pivot screw.

Zarge *f*, **Rand** *m* (Techn.) border, raised border, edge, rim, brim; Thür~ wooden door case; Tisch~ frame under a table board.

Zaun *m*, **Einfriedigung**, **Ein**=

Zäunung f fence, fencing, paling; (Hecke f) hedge; (lebendiger) hedge, quickset hedge; (todter) fence; ~draht m fence wire; ~draht mit Stahlstacheln steel-barbed fencing wire; ~pfahl m fence pale, fence stake, pale, fence rail.

Zehrgeld n allowance, subsistence allowance.

Zeichen n (T) signal, sign, character; (Techn.) sign, mark, stamp; (Eisenb.) signal; chemische ~ pl chemical symbols; (Merkzeichen) sign, mark; (eingebranntes) brand; (konventionelles [T]) conventional sign, signal of a conventional character; ~ geben v (auf dem Telegr.-Apparat) to signal.

Zeichen ... ~geber m (T) indicator, communicator, (Eisenb.) signal man; ~scheibe m (Zifferblatt n) dial; ~tafel f table (of the Morse alphabet); ~wechsel m (HA) change from letters to figures & signs (od. umgekehrt).

zeichnen v to draw, to delineate, to design; (unterzeichnen) to sign; (bezeichnen) to mark; per Prokura ~ to sign per procuration.

Zeichnen n drawing; (aus freier Hand) free-hand drawing.

Zeichnung f drawing, draught, design, delineation; (in Bleistift) lead-pencil drawing; (geometrische Figur f) diagram; (getuschte) drawing in Indian ink; (kolorirte, ausgemalte) illuminated drawing; (perspektivische) perspective drawing; (schraffirte) hatched drawing.

Zeiger m (des Telegr.-Apparats) needle, pointer, index, hand; (einer Waage) pointer of a pair of scales; ~achse f axis of the pointer; ~apparat m (T) pointer instrument s. auch Apparat; ~telegraph m dial instrument, pointer telegraph, (mit Selbstunterbrechung) step-by-step motion telegraph, pointer telegraph with self acting make & break.

Zeit f time; mittlere ~ mean (equated) time; seiner ~ adv in time, in due time; ~abschnitt m period; ~angabe f date, time, (Uhrenzeichen) time signal; ~ball m time ball; ~bestimmung f determination of the time; ~dauer f duration (space) of time; ~einheit f unit of time; ~folge f succession of time, chronological order; ~gleiche, ~gleichung f equation of time, equated (mean) time; ~messer m chronometer; ~messung, ~meßkunst f chronometry; ~punkt m moment; ~raubend a requiring much time; ~raum m space of time, period; ~signal n time signal; ~schrift f periodical paper, periodical publication; ~verlust m loss of time.

Zeitungs ... ~abonnement n f. ~bestellung; ~abgangszettel m (+) list of bags containing newspapers; ~amt, Post-~amt n Postal Newspaper Office, P. O. mediating the subscription to newspapers; ~ausgabe f (+) delivery of newspapers; ~ausgabestelle (+) counter (place) for the delivery of newspapers; ~ausschnitt m cut from a newspaper; ~beilage f supplement; ~bestellung f, ~bestelldienst m durch die Post (+) subscription to newspapers through the mediation of the P. O., newspaper subscription service; ~bestellungsbuch n (+) subscription list; ~bestellungs- u. Vertheilungsbuch n am Absatzorte list of subscription & distribution of newspapers; ~bezug m durch die Post subscription to, & transmission of, newspapers through the medium of the P. O.; ~eingangsbuch n (+) book of arrival of newspapers; ~gebühr f commission; ~gebührenrechnung f (+) quarterly account of the commission on newspapers; ~gelder n/pl subscription price, subscription; ~kontobuch n (+) account kept by the office of publication of newspapers sold; ~kursbeutel m newspaper bag; ~packet n packet of newspapers; ~preisliste f price list of newspapers; ~schalter m (+) counter for the delivery of newspapers; ~telegramm n news message, press telegram; ~überweisung f (+) assigning to another post office the delivery of a newspaper; ~überweisungsgebühr *f* (+) fee (charge) for assigning to another post office the delivery of a newspaper; ~verkehr *m*, ~wesen *n* exchange of newspapers.

Zelle *f* (*T*, Elektriz.) cell; (mit nur einer Flüssigkeit) single-fluid cell; poröse ~ porous cell.

Zelt, Lötherzelt *n* (*T*) tent for the jointers; ein ~ aufschlagen to pitch a tent.

Zement s. Cement.

Zentesimal ... ~=Eintheilung *f* centigrade scale; ~waage *f* centesimal balance.

Zenti ... ~gramm *n* centigram; ~meter *n* centimetre.

Zentner *m* (englischer) hundredweight; (ausländischer) quintal (gewöhnlich 50 kg).

zentral *a* central.

Zentral ... ~organ *n* central organ; ~station *f* central station.

Zentralisation *f* centralization.

zentralisiren *v* to centralise.

zentrifugal *a* centrifugal.

Zentrifugalkraft *f* centrifugal force.

zentripetal *a* centripetal.

Zentripetalkraft *f* centripetal force.

zentriren, den Mittelpunkt bestimmen *v* to centre, to set off the centre.

Zentrum *n* centre, center; ~bohrer *m* centre bit.

zerbrechen *v* to break, to break down, to fracture.

zerbrechlich *a* brittle, fragile.

Zerbrechlichkeit *f* brittleness, fragility.

Zerbrechungs ... ~festigkeit, Bruchfestigkeit, relative Festigkeit *f* (Mech.) resistance to breaking strain, transverse strength, strength of flexure; ~kraft *f* (Mech.) breaking strain, transverse strain.

zerbröckeln *v* to crumb, to fall into small pieces.

Zerdrückungsfestigkeit, Druckfestigkeit, rückwirkende Festigkeit *f* (Mech.) strength of compression, resistance to compressive strain, crushing strength, resistance to crushing strain.

zerfressen *v* (Chem.) to corrode.

Zerfressen *n*, Zerfressung *f* (Chem.) corrosion, corrosive action.

Zerknickungsfestigkeit, Strebefestigkeit *f* (Mech.) compressional strength, composed strength of compression.

zerlassen *v* to melt, to dissolve, to liquefy.

zerlegbar *a* (Chem.) decomposable.

zerlegen *v* (Chem.) to analyse; (zersetzen) to decompose, to resolve; (auseinander nehmen [Masch. rc.]) to take asunder, to take to pieces; sich ~ (Chem.) to decompose.

Zerlegung *f* (Chem.) analysis; (Zersetzung) decomposition, resolution; (der Kräfte) decomposition of forces.

zerreißen *v* to rend asunder, to tear; (vom Drahte) to break.

zersetzbar *a* (Chem.) decomposable.

zersetzen *v* (Chem.) to decompose, to split up; sich ~ to be decomposed, to split up.

Zersetzung *f* (Chem.) decomposition; (elektrochemische) decomposition by electricity; ~sfähigkeit *f* (Chem.) decomposing power; ~smittel *n* means of decomposition.

zersprengen *v* to burst asunder, to split asunder.

zerspringen *v* to burst, to crack; (explodiren) to burst, to explode, to fly into pieces (said of steam engines).

Zerstörung *f* (eines Drahtes, einer Stange durch atmosphärische Einflüsse rc.) decay.

zerstreuen *v* (Opt.) to diffuse, to disperse.

Zerstreuung *f* (Opt.) divergence, dispersion, dispersive power; ~sglas *n*, ~slinse *f* diverging glass, dispersing lens; ~skreis *m* circle of divergence; ~spunkt, virtueller Brennpunkt *m* point of diversion, virtual focus; ~sspiegel *m* convex mirror; ~svermögen *n* dispersive power.

Zettel *m* billet, scrip, note; (Etikette) ticket, label; (Anschlagzettel) placard, notice; (auf das Gepäck geklebter [Eisenb.]) baggage check; (vom Postboten in der Wohnung zurückgelassener Benachrichtigungszettel) notice; ~bank f bank of issue (of circulation).

Zeug n (Techn.) materials for the fabrication of anything; (Geräth n) tools; (Stoff m, Gewebe n) stuff, cloth; (Mörtel m) mortar; ~ zur Papierfabrikation, Papierzeug n pulp, paper pulp; (Geschirr n) harness, trappings.

Zeug ... ~kammer f (Techn.) tool room; ~kasten, Werkzeugkasten m tool box, tool chest; ~schmied m tool smith, artillery smith; ~schmiede f tool-smith's forge.

Zeugniß n certificate, attestation; (ärztliches) medical certificate; ein ~ ausstellen to certify.

Ziegel m (Mauer~) brick; (Dach~) tile; eine Schicht ~steine a layer of bricks; ~brenner m brick maker, tile burner; ~brennerei, Ziegelei f brick kiln, tile kiln, tilery, brick works; ~dach n tile roof; ~erde f, ~lehm m brick clay, brick earth, loam.

Zieh ... ~band n clip, iron clip, band; ~bank, Platte f mit Ziehlöchern zum Drahtziehen drawing bench, wire drawer's bench; ~brücke f s. Zugbrücke; ~eisen n (zum Drahtziehen) wire plate, draw plate, wire-drawing iron; ~kraft s. Zugkraft; ~pfad, Leinpfad m tow path, towing path; ~rolle f (Mech.) pulley; ~schraube, Stellschraube f (Techn., Mech.) adjusting screw; ~waage, Schnellwaage f steel yard; ~zange f (beim Drahtziehen) plyers, nippers.

ziehbar a ductile.

Ziehbarkeit f (der Metalle) ductility (property of permanently extending by traction).

ziehen v to draw; (die Bilanz) to strike the balance; (Draht) to draw wire; (Draht) recken) to stretch the wire; (einen Wechsel) to draw a bill of exchange; sich ~, sich ausdehnen (Techn.) to stretch, (sich werfen) to warp, to cast (said of wood), to distort (said of steel).

Ziehung f (Tratte f) draft; (bei der Lotterie) drawing; ~sliste f (bei der Lotterie) list of the prizes drawn; ~stag f day appointed for the drawing of a lottery.

Ziel n (Termin m) term; drei Monate ~ at three months' credit; ~tag m term day, quarter day, day when payment is due.

Ziffer f, **Zahlzeichen** n character, figure, cipher; arabische ~n pl Arabian or English characters or figures; römische ~n Roman characters or figures; ~blatt n dial, dial plate, (des Zeigertelegraphen) telegraph dial; ~taste f (HA) cipher blank key.

Zimmer ... ~leitung f (T) office wires; ~verbindungen f/pl (T) office connections; ~leitungsdraht m (T) office wire; ~leitungskabel n (T) office cable.

Zink n zinc; ~blech n sheet zinc, zinc plate; ~chlorid n chloride of zinc; ~erz n zinc ore; ~folie f zinc foil; ~oxyd n oxide of zinc; ~pol m zinc pole; ~stab m zinc rod; ~strom m zinc current; ~überzug m zinc coat; ~vitriol n zinc vitriol, sulphate of zinc; mit ~ überzogener Draht zinc coated wire, galvanized wire.

Zinn n tin, stannum; ~folie f, Stanniol n tin foil, stain.

Zins m rent, ground rent; (Zinsen pl) interest; ~abschnitt, Coupon m coupon; ~fuß m rate of interest, percentage; ~rechnung f account (calculation) of interest or rent, interest account; ~schein m coupon, dividend warrant; ~tabelle f table of interest, interest table.

Zinsen pl interest; auf ~ ausleihen v to put out upon interest; ~ tragen to bear interest; ~rechnung, ~berechnung f interest account.

Zinseszins m compound interest, interest upon interest.

Zirkular ... ~schreiben n circular letter; ~stellung f (T) the apparatus at the way station is included in circuit.

zirkuliren v to circulate; ~lassen to put in circulation.

Zisterne f cistern, tank.

zivilversorgungsberechtigter Militäranwärter m military (naval) pensioner (one who is entitled to an appointment in the civil service in consideration of his military [naval] services).

Zivilversorgungsschein m certificate of capacity issued by the military authority for holding an appointment in the civil service.

Zoll m (Maaß) inch = 2,540 cm; (Abgabe) custom, duty, tax; ~abfertigung f going through the custom's formalities; ~abgaben f/pl duties, custom house charges; ~amt n custom house, custom office, board of customs; ~amtliche Behandlung f der Poststücke fulfilment of the custom's formalities the parcels forwarded in the mails are subjected to; ~amtliche Schlußabfertigung f final discharge from custom house, all formalities having been duly performed; ~ bezahlt a duty paid; ~formalitäten f/pl = ~vorschriften; ~frei a free from duty; ~gebühren f/pl s. ~abgaben; ~inhalts-Erklärung f customs declaration; ~pfund n pound of the German Zollverein, half a kilogram; ~pflichtig a dutiable, liable to pay customs (duties); ~übertretungen f/pl der Postillone offences against the custom laws committed by mail drivers; unter ~verschluß in bond, under bond; ~vorschriften f/pl custom's formalities; Deutscher ~verein m German customs-union, Zollverein.

Zone f (Math., Geom.) zone; ~ntarif m tariff of distances.

Zopf m eines Baumes top of a tree; ~ende n (einer Stange, eines Baumes) top end, small end of a pole (of a tree); ~stärke f diameter of the top end; ~trockner Baum m tree withered at the top.

Zotten f/pl (Kupfertheilchen, die sich im Zink-Kupfer-Element auf der Zinkplatte niederschlagen) threads, filament.

Zubehör n appurtenance, accessory; (Masch.) set of machinery & tools, lumber.

zubereiten, imprägniren, tränken v to impregnate, to preserve, to prepare, s. auch Zubereitung.

Zubereitung, Imprägnirung, Imprägnation f impregnation, saturation, preparation; ~ der Stangen mit Kupfervitriol boucherizing, injecting of poles with sulphate of copper; ~ mit Quecksilbersublimat, Kyanisirung f impregnation with bichloride of mercury, kyanizing; ~ mit Steinkohlentheeröl creosoting; ~ mit Zinkchlorid impregnation with chloride of zinc; ~ im luftleeren Raume preparation in the vacuum pan; Kessel-~ s. Kessel; ~sanstalt f (T) yard (establishment) for preparing telegraph poles.

Zubuße f supply, contribution.

zucken v to flicker (said of the electric light).

Zufluß m flux, afflux, affluent; ~rohr n supply pipe.

Zufuhr f (von Waaren) import, importation; (Techn.) supply; (Anfuhr) carting (of telegraph stores from the depot).

Zuführung f (Masch.) feed; (Techn.) supply; ~leitung f (T) lead; ~rad n (Masch.) feed-wheel.

Zug m drawing, pulling, pull; (Drahtzug) wire drawing, implements for wire drawing; (Eisenb.) train; (Gespann) team, set; (Rolle mit Seil) pulley; ~ um ~ hand-to-hand; ~anker m (Telegr.-Bau) iron tie, iron rod; ~anschluß m junction, ~anschluß haben to join, to time with another train, to make connection; ~brücke f draw-bridge; ~elastizität u. Festigkeit, absolute Elastizität und Festigkeit f (Mech.) elasticity & strength of extension; ~festigkeit f (Mech.) tensile strength, strength of extension, resistance to tensile strain; ~führer m (Eisenb.) train conductor; ~kontakt m pull contact; ~kraft f traction, tractive power (force); ~personal n (Eisenb.) men attending a railway train; ~pferd n draught horse, cart horse; ~rolle f (Mech.) pulley; ~spannung f (Mech.) tensile strain, pulling strain; ~stange f (HA) drawing rod; ~telegraph m (der Telegr.=Draht, welcher zum Signalisiren der Eisen.=Züge dient) train wire; ~winde f (Mech.) pulley, draw-beam.

Zulage f increment, increase, augmentation (of salary, wages).

zulassen v to admit.

zulässig a admissible, permissible, allowed.

Zulässigkeit f admissibility.

zuleimen v to glue up.

Zuleitung f (T) lead, connection; ~klemme f (T) binding screw, terminal.

Zünder, Anzündeapparat m (beim Sprengen) fuze, fuse, fusee.

Zündschnur f fuze.

zunehmen v to increase.

Zunge f (Techn.) tongue; (einer Waage) tongue, cock, index, needle of a balance; ~npfeife f (F) mouth piece, trumpet mouth.

Zurück ... u. zurück ... ~erstatten v to refund, to repay; ~erstattung f refunding, repayment, reimbursement; ~fordern v to reclaim, to redemand, to recall; ~forderung f von Postsendungen demand that mail articles be returned to the sender; ~geben v to give back, to return; ~gelegte Akten s. ~legen; ~gleiten v (v. Draht) to run; ~legen v to lay aside, to put aside; ~gelegte Akten pl official papers put aside; ~nahme f (eines Briefes) taking back, withdrawal of a letter, (einer Beschwerde) withdrawal of a complaint, (einer Beleidigung) recantation; ~nehmen v (vgl. ~nahme) to take back, to withdraw, to recant; ~prallen v to strike back, to spring back, to rebound (said of a spring), (Opt.) to be reflected; ~schicken v to return; ~schnellen v to spring back, to fly back; ~schrauben v to slacken a screw, to unscrew; ~telegraphiren v to reply by telegraph; ~weisen v (einen Anspruch :c.) to reject (a claim etc.); ~weisung f rejection; ~werfen, ~strahlen v to reflect, to be reflected; ~zahlbar a repayable; ~zahlen v to pay back, to repay, to reimburse; ~ziehen (ein Telegramm) to withdraw, to cancel (a telegram); ~ziehung f withdrawal, cancelling (of a telegram).

Zusammen ... u. zusammen ... ~binden v to bind, to tie together; ~brechen v to break down (said of t. lines); ~bruch m breakdown, (pl breaks down u. breakdowns); ~drängen v to press, to crowd together; ~drehen v to twist; ~drücken v to compress; ~fassen v to recapitulate; ~fließen v to flow together, to join; ~fluß m confluence (of water courses), conflux (of people), coincidence (of circumstances); ~fügen v to join together, to unite; ~hang m coherence, connexion, (Phys.) cohesion, coherency; ~hängend a coherent; ~laufen v (Phys.) to converge, (von Morsezeichen) to run together, (in einem Mittelpunkte) to meet in a common centre; ~löthen v to solder, to joint; ~passen v to fit, to match; ~pressen v to compress; ~rechnen v to add together, to sum up; ~schmelzen v to melt together; ~schrauben v to screw together; ~schweißen v to weld; ~setzen v to combine, to join, to compose, sich ~setzen (Chem.) to combine, eine Batterie ~setzen to set up a battery, Wörter ~setzen to combine words, ~gesetzte Wörter n/pl compound words; ~spleißen v to splice together; ~sprechen v (T) to be in contact, the currents pass from one wire into another; ~stellung f list; ~treffen v to meet, (von Umständen) to coincide; ~wirken v to act together; ~wirken n der elektrischen u. mechanischen Kraft (HA) mutual action of the electrical & mechanical power; ~ziehen v to combine (words); ~ziehung f combination (of words).

Zusatz m addition; (Chem.) admixture, ingredient; ~akte f additional act; ~artikel m additional clause.

Zuschlag m (bei Versteigerungen)

adjudication, the act of knocking down to a bidder at a public sale; den ~ erhalten v to receive the contract bargained for, to be the lowest bidder at a public sale; den ~ ertheilen v to knock down a thing to a bidder at a public sale; ~, Flußmittel n flux; ~, Geldbetrag m supply to make up a (certain) sum, (P) additional payment, additional charge; ~ zu den Futterkosten (für die Pferde des Posthalters) supplementary allowance for food (forage).

zuschlagen v s. Zuschlag ertheilen.

Zuschlags ... ~gebühr f (P) supplementary rate, additional charge; ~porto n (P) surcharge; mit ~porto belegen v to surcharge; ~taxe f (P) supplementary rate, surcharge.

Zuschuß m addition, supply; ~ beantragen v (seitens der Postämter) to apply for funds.

zuspitzen v (Techn.) to point, to sharpen to a point, to taper; sich ~ (Techn.) to taper, to end in a point.

Zustand m state; (betriebsfähiger) working condition; (veränderlicher) variable state.

zustehender Betrag m amount due.

zustellen v s. bestellen.
Zustellung f s. Bestellung.
Zutaxe f s. Zuschlagstaxe.
zutaxiren v to surcharge; zutaxirtes Porto n surcharge, s. auch Zuschlags ...

Zwei ... u. zwei ... ~monatlich a & adv once in two months; ~spänner m carriage & pair, coach & pair; ~spännig a two horse (coach); ~theilig a in two parts; ~werthig a (Chem.) bivalent, divalent, diatomic.

Zweig m branch; ~bahn f (Eisenb.) branch line; ~kasse f branch cash-office; ~leitung f (T) derived wire, branch wire; ~strom m (T) branch current, derived current.

Zwischen ... ~amt n, ~station f (P, T) intermediate station, way station; ~apparat m (T) intermediate apparatus, instrument at the way station; ~ort m (P) way station; ~quittung f provisory receipt; ~raum m interval, distance, (zwischen den Elementen der Morseschrift) space, gehörigen, gleichmäßigen ~raum beim Abtelegraphiren der Morseschrift machen to space; ~stück n intermediate piece; ~wand f partition, (Phys., Chem.) diaphragm.

Zweiter Theil:

Englisch-deutsch.

———

A.

A form *s* (*T*) das Telegramm-Aufgabeformular.

A poles *s/pl* (*T*) der Doppelständer, Bock.

A.B.C. circuit s (*T*) die Telegraphenlinie, auf welcher mit dem A.B.C. Apparat gearbeitet wird; ~ instrument, Wheatstone's ~ instrument (*T*) der elektro-magnetische Zeiger-Apparat von Wheatstone.

to abandon *v* a telegraph office ein Telegraphen-Amt aufheben.

abandonment *s* die Aufhebung (eines Amtes); die Abtretung, Verzichtleistung.

to abate *v* herabsetzen (Preise), nachlassen; to ~ a house ein Haus niederreißen.

abatement *s* der Abzug, Nachlaß, Rabatt.

to abbreviate *v* abkürzen; (Arithm.) auf das kleinsten Nenner bringen; abbreviated address (*T*) die abgekürzte Telegramm-Adresse.

abbreviation *s* die Abkürzung.

aberration *s* (Opt.) die Abweichung oder Brechung der Lichtstrahlen.

abroad *adv* draußen (in der Welt), nach außen; the mail ~ (*P*) die Post nach dem Auslande.

absence *s* from illness die Abwesenheit vom Dienst in Folge von Krankheit; ~ on leave die (erfolgte) Beurlaubung, s. auch leave; ~ under medical certificate (M. C.) die Abwesenheit vom Dienst auf Grund eines ärztlichen Zeugnisses.

absentee *s* from duty der vom Dienst Abwesende.

to absorb *v* einsaugen; (Chem.) absorbiren.

absorption, absorbition *s* (Chem.) die Einsaugung, Absorption.

absorptive *a* (Chem.) einsaugungsfähig, absorptionsfähig.

abstract *a* abgezogen, allgemein; ~ book *s* (England) das Telegramm-Annahmebuch; ~ mathematics *s/pl* die reine Mathematik.

to accelerate *v* beschleunigen.

accelerating force, accelerative force *s* (Mech.) die beschleunigende Kraft.

acceleration *s* (Phys.) die Beschleunigung; normal ~ die Normalbeschleunigung; tangential ~ effect die Tangentialbeschleunigung.

accelerative, acceleratory *a* beschleunigend.

to accept *v* annehmen, akzeptiren.

acceptance, acceptation *s* die Annahme, der Akzept (of a bill of exchange, eines Wechsels); ~ in blank der Blanko-, ungedeckte Akzept; general ~ der reine Akzept; ~ upon protest der Interventionsakzept; qualified ~ der bedingte Akzept.

accepter, acceptor *s* der Akzeptant, Annehmer eines Wechsels.

access *s* der Zugang, Zutritt; ~ to letters (*P*) die Befassung mit Briefen.

accessory *a* hinzukommend, nebensächlich; ~ work *s* die Nebenarbeit.

accident *s* der Unfall (auf Eisenbahnen); ~ while on duty der im Dienst erlittene Unfall.

acclivity *s* die Steigung, Böschung.

acclivous *a* steigend, steil.

accomodation bill *s* der Gefälligkeitswechsel, Kellerwechsel.

to account *v* berechnen, tariren, schätzen; rechnen, Rechnung ablegen; to ~ for verrechnen.

account *s* die Rechnung, das Konto, der Betrag; ~ of charges die Unkostenrechnung; ~ current die laufende Rechnung, das Konto; ~ of exchange das Wechselkonto; joint ~ die gemeinschaftliche Rechnung; open ~ die offene, laufende Rechnung; ~ of settlement die Schlußrechnung; to balance an ~ eine Rechnung saldiren, bezahlen; to bring in one's ~ seine Rechnung einreichen; to carry to ~ in Rechnung stellen; to cast, to prepare an ~ eine Rechnung aufstellen; to check, to examine, to verify an ~ eine Rechnung prüfen; to have an ~ with *smb* in Rechnung stehen mit J—d.; to keep ~s Bücher führen; on ~ auf Rechnung, auf Abschlag; to pay on ~ auf Abschlag bezahlen; to settle, to close an ~ eine Rechnung bezahlen; to turn to ~ verwerthen, Gewinn bringen; ~ agreed upon, close of ~s, balance of ~s der Rechnungsabschluß; rendering of ~s die Rechnungslegung; statement of ~s der Rechnungsabschluß.

accountant *s* der Rechnungsführer, Rechnungsrevisor, Buchhalter, Buchführer; ~'s office (*Am*) das Rechnungsbureau.

to accouple, to couple *v* paaren, paarweise verbinden.

to accumulate *v* anhäufen.

accumulation *s* die Anhäufung (der Elektrizität); ~ test die Methode, einen kleinen Fehler in einem sonst guten Kabel während der Fabrikation einzugrenzen.

accumulator *s* der Akkumulator, Ansammlungs-Apparat, Sammler für mechanische Arbeit.

accuracy *s* (of a telegram) die Richtigkeit.

acid *s* (Chem.) die Säure.

acid *a* (Chem.) sauer.

acidiferous *a* (Chem.) eine Säure enthaltend, säurehaltig.

acidifiable *a* (Chem.) der Verwandlung in eine Säure fähig.

acidification *s* (Chem.) die Säuerung.

to acidify *v* (Chem.) in eine Säure verwandeln.

to acidulate *v* (Chem.) ansäuern; ~d water mit Kohlensäure gesättigtes, gesäuertes Wasser.

to acknowledge *v* anerkennen, anzeigen; to ~ the receipt den Empfang anzeigen, bestätigen.

acknowledgment *s* of receipt die Anerkennung, der Empfangsschein, die Empfangsanzeige; (*T*) die Quittung am Apparat.

acoustic *a* (Phys.) akustisch, den Schall, das Gehör betreffend; ~ reading (*T*) die Aufnahme nach dem Gehör; ~ telegraph der akustische (Schall-)Telegraph.

acoustics *s/pl* die Lehre vom Schall, Akustik.

to acquit *v* abtragen, quittiren (a debt eine Schuld).

acquittance *s* die Quittung (über geleistete Zahlung).

to act *v* on *smt* (Mech.) auf etwas wirken.

acting *a* (Mech.) wirkend; double ~ doppelt wirkend; self ~ selbst wirkend, automatisch wirkend; single ~ einfach wirkend.

action *s* der Prozeß; (Mech.) die Wirkung, Kraft; direct ~ of a machine die direkte Wirkung einer Maschine; ~ & reaction Wirkung und Gegenwirkung; uniform ~ der regelmäßige Gang einer Maschine.

active *a* thätig, wirksam; ~ commerce der Aktivhandel, Ausfuhrhandel; ~ property, ~ capital das Aktivvermögen, die Aktiva.

to actuate *v* in Bewegung setzen, antreiben; to ~ an apparatus (*T*) einen Apparat zum Ansprechen bringen.

actuation *s* (Mech.) die wirkende Kraft.

acumen *s* die scharfe Spitze.

acute *a* scharf, spitz; ~ angle (Geom.) der spitze Winkel; ~-angled spitzwinklig.

to adapt *v* anpassen.

adaptation, adaption *s* die Anpassung.

to add *v* abbiren, hinzufügen; to ~ the interest to the capital die Zinsen zum Kapital schlagen.
adding machine *s* die Rechenmaschine.
additional *a* hinzugefügt, vermehrt; ~ act die Zusatzakte; ~ charges *pl* die Nebenkosten, Nebenspesen.
to address *v* adressiren (a letter); konsigniren, absenden (goods).
address *s* die Adresse; ~ in case of need die Nothadresse (nur im Handel gebräuchlich); ~ label *s* (*P*) der Packetaufgabezettel.
addressee *s* der Adressat.
to adhere *v* anhangen, anhaften, ankleben; the armature ~s to the cores der Anker klebt an den Kernen; to ~ to a convention einem Vertrage beitreten.
adherence *s* das Anhangen, Ankleben.
adherent *a* anhaftend, anklebend.
adhesion *s* (Phys.) die Adhäsion, die gegenseitige Anziehung von Körpern in großer Nähe; to declare one's ~ to a treaty seinen Beitritt zu einem Vertrage erklären.
adhesive *a* anhangend, anklebend; ~ postage stamp das gummirte Postwerthzeichen.
adjacent *a* anstoßend, naheliegend.
adjoining *a* anstoßend; Neben...; ~ room das Nebenzimmer.
to adjudicate *v* (gerichtlich) zuerkennen, zuschlagen (bei Auktionen).
adjudication *s* die (gerichtliche) Zuerkennung; public ~ der Zuschlag (bei Auktionen).
to adjust *v* (Techn., Mech.) anpassen, einrichten; to ~ an instrument (*T*) einen Apparat einstellen, reguliren; to ~ accounts Rechnungen berichtigen; ~ing screw die Stellschraube; zero ~ing lever (*HA*) der Einstellhebel.
adjustable *a* verstellbar, regulirbar; ~ stop (*T*) der Regulir-Kontakt.
admeasurement *s* (Techn.) die Abmessung, die Aichung.

admiralty *s* die Admiralität; high court of ~ das Admiralitätsgericht; the Lords Commissioners of the ~ das Admiralitätskollegium.
admissible *a* zulässig; ~ articles (*P*) zur Beförderung durch die Post zugelassene Gegenstände.
to admit *v* zulassen; to ~ the steam (Masch.) den Dampf zulassen, eintreten lassen.
admixture *s* die Beimischung, der Zusatz.
adobe *s* der Lehmstein, ungebrannte Ziegelstein.
adrift *adv* flott, treibend, den Wellen preisgegeben.
aduncous *a* hakenförmig, gekrümmt.
to advance *v* (money) Geld vorschießen, vorausbezahlen; to ~ the price of goods den Preis der Waaren erhöhen.
advance *s* der Geld-Vorschuß, to make ~s Vorschüsse leisten; an ~ of money eine baare Entschädigung, Vergütung; to pay in ~ im Voraus bezahlen; prices are on the ~ die Preise steigen; ~ money der Vorschuß.
advertisement *s* die Anzeige, Bekanntmachung, das Inserat.
advice *s* der Bericht, Avis, s. auch money order advice; as per ~ laut Bericht; letter of ~ der Avisbrief; for want of ~ wegen Mangels an Bericht; ~ boat das Avisschiff, die Postjacht; ~ slip (*P*) der Benachrichtigungszettel.
adze *s* die Krummart, das Zimmerbeil; ~, hollow ~ der Dächsel, das Dachsbeil.
aërial *a* (Phys.) luftig, ätherisch; ~ cable (*T*) das Luftkabel; ~ line, ~ wire (*T*) die Luftleitung, oberirdische Leitung; ~ meteor der Meteorstein, Aërolith; ~ navigation die Luftschifffahrt.
aëriform *a* (Phys.) luftförmig, gasförmig.
aëro-dynamics *s/pl* (Phys.) die Aërodynamik, Dynamik elastisch-flüssiger Körper, Luftkraftlehre.
aëronaut *s* der Luftschiffer.
aëronautics *s/pl* die Luftschifffahrtskunde.

aërostat *s* der Luftballon, das Luftschiff.
aërostatics *s/pl* (Phys.) die Äroftatif, Luftgleichgewichtslehre.
aëro-steam-engine *s* (Masch.) die Dampfmaschine mit Benutzung komprimirter Luft.
aeruginous *a* (Chem.) roftig, mit Grünspan bedeckt.
aerugo *s* (Chem.) der Rost, Grünspan.
affidavit *s* die eidliche Erklärung, das eidliche Zeugniß; to make an ~ eidlich erhärten.
affinity *s*, chemical ~ (Chem.), die chemische Verwandtschaft.
affirmative quantities *pl* (Math.) positive Größen.
to affix *v* anheften, anschlagen, ankleben, aufkleben; to ~ a postage stamp to a letter eine Freimarke auf einen Brief aufkleben.
afflux *s* (Elektriz.) die Einströmung (Bewegung des elektr. Fluidums nach einem Punkte); point of ~ der Einströmungspunkt.
agate *s* der Achat, Achatstein; ~ stud das Achathütchen.
agency *s* die Agentur, Vermittlung, Kommission; ~ business das Agenturgeschäft, Kommissionsgeschäft.
agent *s* der Agent, Kommissionär, Geschäftsträger; (Phys., Chem.) die wirkende Kraft, das Agens; ~ for fusion das Schmelzmittel; government ~ der Regierungsbeamte.
to agglomerate *v* (Phys., Chem.) zusammenballen, agglomeriren, sich zusammenballen, gerinnen.
to agglutinate *v* zusammenleimen.
agglutination *s* das Zusammenleimen.
aggregate *s* das Aggregat, das Angehäufte; ~ length die Gesammtlänge.
agreement *s* der Vertrag, Kontrakt, die Uebereinkunft; to come to an ~ zu einem Einverständniß kommen; ~ by piece das Affordauf's Stück.
ahead *adv* vorwärts, voran; to run ~ (von synchronen Telegraphen-Apparaten) voreilen.

aid *s* die Hilfe; der Helfer, Aushelfer.
aigret, aigrette *s* die Aigrette, der Büschel; electric ~ der elektrische Strahlenbüschel.
to air *v* lüften, ventiliren.
air *s* die Luft; compressed ~ verdichte Luft; rarefied ~ verdünnte Luft; ~ balloon der Luftballon, das Luftschiff; ~ blast, ~ blower das Gebläse; ~ bubble die Luftblase; ~ chamber (Masch.) der Windkasten, Windkessel; ~ compressor, ~ condenser der Luftverdichter, die Luftkompressionsmaschine; ~ drain das Luftloch, der Windfang; ~ dried *a* an der Luft getrocknet; ~ escape die Vorrichtung zur Entfernung der Luft; ~ exhauster (Mech.) der Luftabzug, Windfang, die Luftpumpe; ~ flue der Luftkanal; ~ furnace der Windofen, Flammofen; ~ heater der Luftheizungsofen; ~ heating apparatus der Luftheizungsapparat; ~ pipe (Masch.) die Luftröhre, Saugröhre, der Luftschlauch; ~ pressure machine die Luftdruckmaschine; ~ pump die Luftpumpe; ~ spring die pneumatische Feder; ~ tight *a* luftdicht, hermetisch; ~ trap der Luftabzug, Windfang; ~ tube die Luftröhre, der Luftschlauch, die pneumatische Röhre.
alarm, alarum *s* (T) der Wecker, der Alarmapparat; ~ bell, alarum bell die Weckerglocke; ~ clock die Weckeruhr; ~ lock das Sicherheitsschloß; ~ telegraph der Läutetelegraph.
alburn, alburnum *s* der Splint (von Bäumen), das Splintholz.
alhedada, alhidade, alidade *s* das Diopterlineal, Visir (am Meßtisch).
alignment *s* das Abmessen nach der Schnur, die Schnurrichte, die Trace.
aliquot *a* (Arithm.) gleich theilend, aliquot; ~ part einer von mehreren gleichen Theilen, ein aliquoter Theil.
all-wire circuit *s* (T) die Telegraphenleitung mit metallischer Rückleitung (unter Ausschluß der Erde).
alley *s* das Gäßchen, die Allee.

allotment s der Antheil.
allowance s die Zulassung, Erlaubniß, Genehmigung, Bewilligung, Vergünstigung, Vergütung; ~ during absence die Gehaltszahlung während einer Beurlaubung; ~ for drift die Zugabe eines Endes Kabel für den Abtrieb.
to alloy legiren, vermischen, versetzen.
alloy, alloyage, alloying s die Legirung, Vermischung; fusible ~ das Schnelllot.
alluvial, alluvious a angeschwemmt, angespült; ~ detritus Geschiebbänke, Sandbänke.
alluvium s die Anschwemmung, das angeschwemmte Land.
alteration s die Abwechslung, Aenderung; (of a telegram) die Verstümmelung.
alternate a abwechselnd, wechselständig; ~ angles pl (Geom.) die Wechselwinkel; ~ current machine die Wechselstrommaschine; ~ motion (Masch.) die hin- u. hergehende Bewegung.
alum, alumen s (Chem.) der Alaun; chlor ~ das Chloraluminium; natural ~, native ~, potash ~, potassium ~, common ~ der natürliche, gediegene Alaun, Kalialaun; ~ battery die Alaun-Batterie.
alumina, aluminium oxide, aluminic oxide, alum earth s die Alaunerde, Thonerde; silicate of ~, silicate die kieselsaure Thonerde, das Aluminiumsilikat.
aluminite s der Aluminit, die reine Thonerde.
aluminium, aluminum s das Aluminium; ~ bronze die Aluminiumbronze (Mischung von Aluminium u. Kupfer).
aluminous a alaunhaltig.
alumish, alumy a alaunartig.
amalgam, amalgama s das Amalgam, die Quecksilberlegirung.
to amalgamate, to amalgamize v amalgamiren, andere Metalle mit Quecksilber verbinden.
amalgamation s die Amalgamirung, die Verbindung eines Metalls mit Quecksilber.

to amass v häufen, anhäufen.
amber s der Bernstein.
ambient a umgebend; ~ air (Phys.) der Dunstkreis.
ambit, ambitus s der Umfang, Umkreis; der freie Raum um ein Gebäude.
ammonia s das Ammoniak; aqueous ~, caustic ~, liquor ~ der Salmiakgeist, die Ammoniakflüssigkeit; gaseous ~ das Ammoniakgas.
ammoniac s das Ammoniakgummi; sal ~ der Salmiak.
ammonium s das Ammonium; ~ chloride, sal ammoniac der Salmiak, das Chlorammonium.
amortization, amortizement s of a debt die Amortisation, Tilgung einer Schuld.
to amortize v a debt eine Schuld amortifiren, tilgen.
amount s der Betrag, die Summe; ~ of balance der Saldobetrag; gross ~ Roh-(Brutto-)Betrag; net (clear) ~ Rein-(Netto-)Betrag.
amplitude s der Umfang, die Weite; die Amplitude (Bezeichnung für Winkel u. s. w.); der größte Ausschlag (bei Schwingungsbewegungen); magnetical ~ die magnetische Weite; ~ of oscillation die Schwingungsweite.
analysis s (Math., Chem.) die Analyse, Zerlegung in die Grundbestandtheile, Scheidung, Auflösung; qualitative ~ die qualitative Analyse; quantitative ~ die quantitative Analyse, Gewichtsbestimmung.
to analyze v analysiren, auflösen, zerlegen.
analytic, analytical a analytisch, auflösend, zerlegend; ~ chemist der öffentliche Chemiker, Handelschemiker.
anchor s der Anker; dragging ~ der schleppende Anker; ~ flukes pl die Ankerflügel.
anchorage s das Ankergeld; ~, anchoring, anchor ground der Ankergrund, Ankerplatz.
ancient s die Nationalflagge (von Schiffen auf See).

anemographer s (Phys.) der Anemograph, Windzeiger.

anemometer, wind gauge s (Phys.) der Windmesser.

anemoscope s (Phys.) der Windzeiger, die Wetterfahne.

angle s der Winkel; acute ~ der spitze Winkel; adjacent ~ der Nebenwinkel; alternate ~ der Wechselwinkel; bevel ~ der schiefe Winkel; ~ of deviation, ~ of emergence (Opt.) der Abweichungswinkel; external ~ der Außenwinkel; ~ of incidence (Opt.) der Einfallswinkel; ~ of inclination der Neigungswinkel; internal ~ der innere Winkel; oblique ~ der schiefe Winkel; obtuse ~ der stumpfe Winkel; optic, visual ~ der Sehwinkel; ~ of reflection der Absprungs-, Ausfallwinkel; ~ of repose (Mech.) der Ruhewinkel; right ~ der rechte Winkel; salient ~ der vorspringende Winkel; vertical ~ der Scheitelwinkel.

angle bar, ~ iron s das Winkeleisen, Eckeisen, die Eckschiene; ~ bevel, bevel rule das stellbare Winkelmaß, der Stellwinkel; ~ brace, ~ bracket das Winkelband; lower ~ brace die Strebe; ~ meter der Winkelmesser; ~ pole (T) die Winkelstange, Stange im Winkelpunkte.

angled a (Geom.) winkelförmig, winkelig; acute - ~ spitzwinkelig; obtuse - ~ stumpfwinkelig; right - ~ rechtwinkelig.

angular a winkelig, eckig; ~ gearing (Masch.) das Triebwerk mit eckigen Rädern; ~ point (Geom.) der Scheitel eines Winkels; ~ thread das eckige Schraubengewinde; ~ velocity die Winkelgeschwindigkeit.

angulous a winkelig, eckig.

anhydride s (Chem.) das Anhydrid, die wasserfreie Säure.

anhydrite s der Anhydrit, wasserfreie Gips.

anhydrous a wasserfrei.

animal a animalisch, thierisch; ~ clutch das Fangeisen; ~ power die Pferdekraft.

anione s (Phys.) das Anion (das an der Anode ausgeschiedene Element).

to anneal v kühlen, brennen, ausglühen und langsam abkühlen; to ~, to temper, to soften (metals) anlassen, ausglühen; to ~ bricks Ziegel brennen.

annex s der Nebenbau, Anbau; die Anlage (eines Schreibens).

annual a jährlich; ~ account die Jahresrechnung; ~ balance die Jahresschlußbilanz; ~ rings pl die Jahresringe im Holze; ~ variations of the needle die Jahresschwankungen der Magnetnadel.

annuity s die jährliche Leibrente; government ~ die staatliche Jahresrentenversicherung; deferred ~ die Jahresrentenversicherung, die erst nach einer bestimmten Zeit in Kraft tritt; immediate ~ die Jahresrentenversicherung, die sofort in Kraft tritt.

to annul v amortisiren, widerrufen, abbestellen; to ~ a telegram ein Telegramm kassiren.

annular a ringförmig; ~ borer der Ring-, Kreisbohrer; ~ cylinder steam engine die Ringzylindermaschine.

annulary, annulate, annulated a ringförmig.

annulment s die Amortisation; der Widerruf.

annunciator s der Haustelegraph; (F) die Klappe am Klappenschrank.

anode s (Phys.) die Anode, der positive elektrische Pol.

to answer v antworten, beantworten; to ~ a bill of exchange einen Wechsel honoriren; to ~ a purpose einem Zweck entsprechen.

antagonistic a widerstrebend; ~ force die Gegenkraft; ~ spring (M & HA) die Abreißfeder.

anteroom s das Vorzimmer; (P) der Schaltervorflur.

to anticipate v vorausnehmen; to ~ payment vor der Zeit Zahlung leisten; ~d bill of exchange der vor der Verfallzeit eingelöste Wechsel.

anticlinal, anticlinical, anticlinic a sich in entgegengesetzter Richtung neigend; ~ axis, ~ line (Geom.) die Neigungslinie.

antifriction grease *s* (Masch.) die Schmiere zur Verminderung der Reibung; ~ metal (Masch.) das Zapfenlagermetall (zur Verminderung der Reibung); ~ wheel das Rad mit Vorrichtung zur Verminderung der Reibung.

antiseptic *a* Fäulniß verhindernd; ~s *pl* Fäulniß verhindernde Stoffe.

anvil *s* der Amboß; (*MA*) der Telegraphirkontakt.

aperture *s* die Oeffnung, das Loch, die Einwurfsöffnung am Briefkasten; ~ of an angle (Geom.) der Raum zwischen den Schenkeln eines Winkels, die Oeffnung eines Winkels; ~ of a bridge die Brückenöffnung.

apex *s* die Spitze, der Scheitel.

aphlogistic *a* aphlogistisch, flammlos.

apparatus *s* (*T*, Chem., Phys., Masch.) der Apparat, die Vorrichtung (gewöhnlich für mail bag apparatus gebraucht); mail bag ~ (*P*) die Briefbeutel - Fangvorrichtung, Vorrichtung um Briefbeutel einzunehmen und abzugeben, ohne daß der Zug hält.

appetence, appetency *s* (Phys.) das Streben, die Anziehung, Attraktion.

application *s* (Mech.) die Anwendung, das Ansetzen; ~ of a law die Anwendung eines Gesetzes; der Antrag; to make (to hand in) an ~ for leave of absence ein Urlaubsgesuch einreichen; ~ form das Antragsformular.

to apply *v* anwenden, ansetzen; beantragen; to ~ to *smb* for *smt* ein Gesuch an J—d richten.

appoint *s* (im Handel) der Appoint, Abschnitt; payment per ~ die Saldozahlung.

to appoint *v* bestimmen, festsetzen, ernennen (Beamte), bestellen; to ~ one to a place J—d zu einem Amte ernennen.

appointable *a* anstellbar (von Beamten).

appointee *s* der Angestellte.

appointment *s* die Bestellung, Ernennung von Beamten; die Verabredung; der Beschluß; die Besoldung, das Gehalt; ~ branch das Personalbüreau im englischen General-Postamt.

to appraise *v* taxiren, abschätzen.

appraisement *s* die Taxation, Schätzung; der taxirte Werth.

appraiser *s* der Taxator, Abschätzer.

appropriation *s* die Anwendung, Anpassung; (*Am*) die von der gesetzgebenden Körperschaft für irgend einen Zweck etatsmäßig bewilligte Summe, die Anleihe.

approximate *a* annähernd; ~ value der Näherungswerth.

approximation *s* (Math.) die Annäherung, Approximation.

apyrous *a* feuerfest, unschmelzbar.

aqueduct, aquaduct, conduit *s* die Wasserleitung, der Aquädukt.

aqueous *a* wasserartig, wasserhaltig, verdünnt; ~ vapor der Wasserdampf.

arbitrary *a* willkürlich; ~ address (*T*) die abgekürzte Adresse; ~ signs *pl* (*T*) verabredete Zeichen.

arbitration *s* das Schiedsgericht, die schiedsrichterliche Entscheidung; ~ of exchange die Wechselarbitrage, Vergleich der verschiedenen Wechselkurse.

arbor, arbour *s* (Mech.) der Baum, die Welle, Achse; ~ ring der Wellring; ~ wheel das Wellrad.

arc *s* (Geom.) der Bogen, Kreisbogen; graduated ~ der Gradbogen; ~ of oscillation der Schwingungsbogen; Voltaic ~ (Phys.) der Volta'sche (galvanische Licht-) Bogen.

arch *s* (Gewölb-)Bogen; ~ bridge die Bogenbrücke, Jochbrücke.

arched *a* gewölbt, bogenförmig.

arching *s* die Wölbung, das Gewölbe.

are *s* der Ar (ein französisches Flächenmaaß von 100 □ Meter).

area *s* (Geom.) die Fläche, der Flächenraum; ~ of cross section der Querschnitt; ~ of surface der Inhalt der Fläche, Flächeninhalt.

areometer, araeometer *s* der Aräometer, die Wasserwaage; scale

~ der Skalen-Aräometer; weight ~ der Gewichtsaräometer.

areometry *s* die Aräometrie (Bestimmung des spezifischen Gewichts von Flüssigkeiten).

argentan, German silver *s* das Neusilber, Weißkupfer, Nickelkupfer.

argillaceous, argillous *a* thonartig, thönern; ~ earth die Thonerde; ~ ironstone der Thoneisenstein; ~ sand der thonige Sand.

argilliferous *a* thonhaltig.

aright *a* aufrecht, gerade.

to arm *v* armiren, bewaffnen; to ~ a magnet einen Magnet armiren.

arm *s* (Mech.) der Arm; der (hölzerne) Querträger der Telegraphenstange; ~ badge der Armbinde.

armature *s* die Armatur; ~ of a condenser (Elektriz.) die Belegung eines Kondensators; ~ of a dynamo-electrical machine die Armatur einer Dynamo-Maschine; ~ of a magnet die Armatur, Einfassung eines Magneten; ring ~ die Ring-Armatur (das charakteristische Zeichen der Gramme'schen dynamo-elektrischen Maschine).

to armor *v* mit Schutzdrähten bekleiden (ein Kabel); ~ing machine die Drahtumspinnungsmaschine (bei der Kabelfabrikation).

armor, armour *s* die Schutzhülle, der Schutzdraht (eines versenkten Kabels).

to arrange *v* ordnen, richten; to ~ an office ein Amt einrichten; to ~ a set of books Bücher anlegen; to ~ an account eine Rechnung liquidiren.

arrangement *s* die Anordnung, die Uebereinkunft, der Vergleich; ~ of an office die Amtseinrichtung; to come to an ~ sich vergleichen.

arrear *s* die rückständige Summe; to be in ~s mit der Zahlung im Rückstande sein, (Am, P) mit dem Schatzamt noch nicht abgerechnet haben.

arrest *s* (Mech.) die Hemmung, der Einfall; die Beschlagnahme; to lay ~ on mit Beschlag belegen.

arrester *s* (T) der Elektrizitätsableiter, s. auch lightning arrester.

arrival *s* die Ankunft; on ~ nach Ankunft; ~ place der Landungsplatz (eines Schiffes), die Anfahrt.

arrow *s* der Pfeil.

article *s* der Artikel, Waarenartikel; ~s of agreement die Kontraktsbedingungen.

articled *a* kontraktlich verpflichtet.

articulate, articulated *a* gegliedert.

artificial *a* (Techn.) künstlich, nachgemacht; ~ resistance (T) der künstliche Widerstand.

ascent, gradient *s* das Aufsteigen, die Steigung, das Gefälle; die Auffahrt (eines Gebäudes), Rampe.

to asphalt *v* asphaltiren, mit Asphalt umgeben, mit Asphalt pflastern.

asphalte, asphaltum, bitumen, mineral pitch *s* der Asphalt, das Erdpech; ~ concrete der Asphaltbeton; ~ pavement das Asphaltpflaster.

assemblage *s* (Techn.) die Verbindung, der Verband.

to assemble *v* (Techn.) verbinden, zusammenfügen.

to assess *v* schätzen, (zur Steuer) einschätzen; to ~ the damages die Versicherungssumme bestimmen.

assessment *s* of post office property die Schätzung, der Werth des Eigenthums der Postverwaltung.

assets *s/pl* die Aktiva, der Vermögensbestand eines Falliten; ~ & liabilities das Aktiv- u. Passiv-Vermögen; no ~ kein Guthaben (die Tratte wird nicht akzeptirt).

to assign *v* anweisen, festsetzen; zediren.

assignee *s* der Bevollmächtigte, Kurator, Zessionar.

assigner *s* der Zedent; der Bevollmächtigende.

assignment *s* die Anweisung, Zessionsurkunde; der trassirte Wechsel, die Tratte.

assistant *s* helfend, Hülfs ...;

Assistent (in England Diener des Postmeisters, nicht der Post-Verwaltung); ~ engineer der Ingenieur-Assistent.

to assort v sortiren.

astatic a (Phys.) unstät, unberührt; ~ galvanometer das astatische, von der Wirkung des Erdmagnetismus unberührte Galvanometer; ~ needle die astatische Nadel.

atmosphere s die Atmosphäre, der Dunstkreis; der Druck, die Kraft der Atmosphäre.

atmospheric, atmospherical a atmosphärisch; ~ electricity die atmosphärische Elektrizität, Luftelektrizität; ~ pressure der Luftdruck, Druck der Atmosphäre.

atomic, atomical a (Phys.) atomistisch, Atom ...; ~ attraction die Anziehungskraft der Atome; ~ theory die atomistische Theorie; ~ volume das Atomvolumen; ~ weight das Atomgewicht, Mischungsgewicht.

atomicity s (Chem.) die Atomigkeit, atombindende Kraft, Werthigkeit der Elemente.

to attach v befestigen, anheften; mit Beschlag belegen.

to attend v to the public (P) das Publikum (am Schalter) abfertigen; to ~ the P. O. Telegraph School die Telegraphen-Schule besuchen.

attendance s (P, T) die Anwesenheit im Dienst, der Dienst; ~ book (P) Präsenzbuch, in welches die englischen Beamten und Unterbeamten, wenn sie in den Dienst kommen und wenn sie denselben verlassen, ihre Namen einschreiben müssen; es enthält die vorschriftsmäßigen sowie die wirklich geleisteten Dienststunden und u. A. den Betrag der Strafe für Zuspätkommen; hours of ~ (P, T) die Dienststunden; late ~ das Zuspätkommen in den Dienst.

to attenuate v verdünnen, schwächen.

attenuation s die Verdünnung, Schwächung.

attest, attestation s das Zeugniß; der Beweis durch Zeugen.

attorney s der Anwalt; letter of ~ die schriftliche Vollmacht; power of ~ die Vollmacht; full & absolute power of ~ die Generalvollmacht.

to attract v (Phys.) anziehen; the magnet ~s iron der Magnet zieht das Eisen an.

attraction, attraction power, attractive power s (Phys.) die Anziehung, Anziehungskraft; capillary ~ die Haarröhrchen-Anziehung; ~ of gravity (Phys.) die Schwerkraft.

attribute s das Merkmal, Sinnbild.

attrite a abgerieben, abgenutzt.

attrition s die Abnutzung, Abgenutztheit.

auber, sap wood s der Splint, das Splintholz.

auction, public sale s die Auktion, öffentliche Versteigerung; to sell by ~ versteigern.

to audit v an account eine Rechnung untersuchen.

audit s die Rechnungsuntersuchung.

auditor s der die Rechnungen prüfende Beamte, Rechnungsrevisor; ~ of public accounts (Am) der Auditeur, der höchste Rechnungsbeamte des Staats.

auger s der große Bohrer, Holzbohrer (zur Untersuchung v. Telegr.-Stangen); annular ~ der Ringbohrer; earth boring ~, ground ~ der Erdbohrer; rivet ~ der Nietbohrer; screw ~, twisted ~ der Schraubenbohrer, Schneckenbohrer; shell ~ der Hohlbohrer, Löffelbohrer; taper ~ der Spitzbohrer, konische Hohlbohrer; ~ bit die Bohrspitze.

aural a akustisch, das Gehör oder den Schall betreffend.

auricular tube s das Hörrohr.

aurora borealis, polar aurora s das Nordlicht.

authority s die Autorität; die Behörde; local ~ die Ortsbehörde; under the ~ of im Auftrage von.

automatic a automatisch, sich von selbst bewegend, selbstthätig.

auxiliary a zur Hülfe dienend, Hülfs..., Aushülfs...; (P) Hülfs-Unterbeamter; ~ sorter Hülfs-Sor-

tirer u. f. w.; ~ battery die Verstärkungs-Batterie.
available *a* zur Verfügung stehend.
average *s* die Havarie, der Verlust durch Beschädigung der Ladung, Seeschaden; die Entschädigung für den Verlust durch Beschädigung der Ladung; on an ~ im Durchschnitt.
average *a* durchschnittlich; the ~ amount der Durchschnitts-Betrag; ~ price der Durchschnittspreis; upon the ~ durchschnittlich.
avoir du poids *s* das schwere Handelsgewicht (1 Pfund = 16 Unzen).
ax, axe *s* die Art, das Beil; broad ~ die Zimmerart; felling ~ die Waldart; ~ head der Rücken der Art; ~ helve der Helm, Stiel einer Art.
to axe *v* mit der Art zurichten.
axis *s* (Geom.) die Achse, Mittellinie; ~ of incidence (Physi.) das Einfallsloth; optic (optical) ~ die Sehachse; ~ of revolution die Umdrehungsachse.
axle, axle tree *s* (Mech.) die Radachse; rank ~ die Kurbelachse; driving ~ die Treibachse; turning ~ die rotirende Achse; wheel & ~ (Masch.) Rad und Wellbaum zusammen; ~ arm der Achsenarm, Achszapfen; ~ bed das Achsenfutter; ~ end der Wellzapfen; ~ journal das Achsenlager; ~ tree, winch (Mech.) der Wellbaum, die Radwinde, das Rad an der Welle.
azimuth *s* der Azimuth, (astronomische) Scheitelkreis; magnetical ~ der Abweichungswinkel der Magnetnadel (Unterschied zwischen dem magnetischen und dem wirklichen Meridian eines Ortes).

B.

B form, transmitted message form (*T*) das Durchgangs-Telegramm-Formular.
back *s* der Rücken, die Rückseite (von Briefen, Postkarten u. s. w.); ~ armature das Verbindungsstück zwischen den beiden Elektromagnetkernen; ~ contact, ~ stop (*MA*) der hintere oder Ruhe-Kontakt; ~ freight die Rückfracht; ~ pressure (Masch.) der Gegendruck; ~ train (Eisenb.) der Rückzug.
badge *s* das Zeichen, Ordenszeichen; arm ~ (*P, T*) die Armbinde.
to bag *v* in einen Sack stecken.
bag *s* der Sack; collecting ~ (*P*) der Sammelsack der Briefkastenleerer; enclosure ~ (*P*) der Versteckbeutel; mail ~ (*P*) der Briefbeutel; private ~ (*P*) der Postbeutel von Privatpersonen; transit ~, enclosure ~ (*P*) der Versteckbeutel; a ~ out of course (*P*) ein fehlgeleiteter Briefbeutel; a ~ out of repair (*P*) ein schadhafter Briefbeutel; ~ duty book (*P*) das Verzeichniß der auf den Eisenbahnstationen des Kurses abzugebenden u. einzunehmenden Briefbeutel; ~ list (*P*) der Abgangszettel, Eingangszettel, Uebergangszettel.
baggage *s* das Gepäck, Reisegepäck; ~ bill (Eisenb.) der Gepäckschein; ~ check, ~ label der auf das Gepäck geklebte Zettel; ~ train, goods train der Güterzug; ~ waggon, goods van der Gepäckwagen, Packwagen.
bagging *s* die Packleinwand.
to bake *v* backen, härten, brennen; to ~ bricks Ziegel brennen; ~d earthenware die Töpferwaare, das Steingut, die gebrannte Thonwaare.
to balance *v* abwägen, vergleichen, saldiren; (Mech.) entlasten, in's Gleichgewicht bringen; to ~ an account eine Rechnung abschließen, saldiren; to ~ accounts with *smb* Abrechnung halten mit J—m; ~d in account durch Gegenrechnung saldirt; the expenses ~ the receipts die Ausgaben sind den Einnahmen gleich, sie saldiren sich.
balance *s* 1) die Waage, das Gleichgewicht; spring ~, elastic ~ die Federwaage; the tongue of a ~

die Waagezunge; ~ arm der Waagenarm; ~ wheel, escapement wheel das Steigrad; 2) die Bilanz, Schlußrechnung, der Saldo; ~ of accounts der Rechnungsabschluß; amount of ~ der Saldobetrag; to have a ~ in one's favor eine Summe gut haben; Postmaster's daily ~ der tägliche Kassenabschluß bei den Postämtern; daily ~ sheet das zum Kassenabschluß erforderliche Formular, der Bilanzbogen; net ~ die reine, Netto-Bilanz; rough ~ die rohe, Brutto-Bilanz; the ~ squares die Rechnung stimmt; to strike a ~ einen Saldo ziehen; ~ account das Bilanzkonto; ~ bill der Saldowechsel.

to bale v einballen, einballiren.

ball s der Ball, die Kugel; ~s pl (Elektriz.) die Korkkügelchen; ~ & socket (Mech.) der Kugelzapfen; ~ & socket joint das Kugelscharnier, Kugelgelenk.

to ballast v ballasten; to ~ a road eine Straße beschottern; to ~ a ship ein Schiff mit Ballast beladen.

balloon s der Luftballon, das Luftschiff; (Chem.) der Ballon, die kugelförmige Flasche; ~ element das Meidinger'sche Ballon-Element.

ballot s der Stimmzettel; das Abstimmen; ~ papers (P) die auf die Vornahme von Wahlen zum Parlament bezüglichen Papiere.

band s der Saum, die Leiste; endless ~ (Masch.) der Laufriemen, die Schnur ohne Ende; iron ~ die Bandschiene, das Eisenband; ~ of paper (T) der Papierstreifen.

band chain s die Bandkette, Gelenkkette; ~ pulley, ~ wheel (Masch.) die Riemenscheibe.

to bank v 1) dämmen, eindämmen, 2) in die Bank einlegen, bei der Bank deponiren.

bank s 1) das Ufer; die Bank, Sandbank; (Eisenb.) das Bankett, die Böschung; der Damm, Straßendamm; 2) die Bank, Wechselbank; ~ of circulation die Girobank; ~ of deposit die Depositenbank; ~ of discount die Diskontobank; ~ of issue die Zettelbank; ~ for loans die Leihbank; ~ for savings, savings ~ die Sparkasse; joint stock ~ die Aktienbank; payment in ~ die Bankzahlung.

bank account s das Bankkonto, ~ bill die Banknote; ~ holiday der Bankfeiertag (Bankfeiertage sind: Ostermontag, Pfingstmontag, der erste Montag im August und der zweite Weihnachtstag. An diesen Tagen werden in Folge Parlamentsbeschlusses die Banken geschlossen); ~ note die Banknote; ~ share, ~ stock die Bankaktie.

bankable a diskontirbar.

banker s der Bankier, Geldwechsler; ~'s commission die Bankiersprovision; ~'s note die Bankanweisung.

bankrupt s der Zahlungsunfähige, Insolvent; the active property (estate) of a ~ die Konkursmasse; ~'s letter der an eine im Konkurs befindliche Person gerichtete Brief; ~ office der Gerichtshof in Bankrottsachen.

bankruptcy s der Bankrott, das Fallissement, die Zahlungsunfähigkeit; court of ~ das Konkursgericht; declaration of ~ die Bankrotterklärung; fraudulent ~ der böswillige, betrügerische Bankrott.

banquet, banquette s das Bankett, der Fußweg, der Seitenweg an Straßen und Eisenbahnen.

to bar v (Mech.) verriegeln, versperren, vergittern.

bar s die Stange, der Stab, der Riegel, das Querholz; der Querstrich, der Bruchstrich; der gegossene oder geschmiedete Stab, die Schiene; ~ for crossings der Eisenbahnschlagbaum; ~ of a magnet der Magnetstab; ~, switch ~ (T) die Kurbel des Kurbelumschalters.

bar commutator s (T) der Linienumschalter, Linienwechsel; ~ copper das Stangenkupfer; ~ iron das Stabeisen, Stangeneisen, Handelseisen; flat ~ iron das Flacheisen, Flachstabeisen; round ~ iron das Rundstabeisen; forged ~ iron, hammered ~ iron das geschmiedete, gehämmerte Stabeisen; laminated ~

iron, laminated ~, rolled ~ das gewalzte Stabeisen; square ~ iron das Quadrateisen; ~ steel der gemeine Stahl, Stangenstahl; ~ wimble der Riegelbohrer, Zwickbohrer, Zapfenbohrer; ~ wood das Sandelholz.

barb *s* der Widerhaken; ~ bolt der Bolzen mit Widerhaken.

to bare *v* entblößen, bloßlegen.

baritel *s* (Mech.) der Göpel; ~ of horses der Pferdegöpel.

to bark *v* trees Bäume entrinden.

bark *s* die Rinde, Borke.

barometer *s* der (das) Barometer, das Wetterglas; cistern ~ der Gefäßbarometer; siphon ~, syphon ~ der Heberbarometer.

barometrical, barometric *a* barometrisch; ~ height die Barometerhöhe.

baroscope *s* (Phys.) das Baroskop (ein Instrument, um die Veränderungen des Luftdrucks aufzuzeichnen).

barrel *s* das Faß, die Tonne; der Zylinder, die Trommel; (MA) das Federgehäuse; ~ of a pulley der zylindrische Theil eines Flaschenzugs, die Rolle, Scheibe.

barrier *s* die Schranke, der Schlagbaum; ~ with a rod (Eisenb.) die Stangenbarrière an Straßenübergängen von Eisenbahnen.

barrow *s* die Bahre, Tragbahre; hand ~, wheel ~ der Schiebkarren, Schubkarren, die Rabbahre.

base *s* die Basis, Grundfläche; (Chem.) die Base; ~ of a column der Säulenfuß; ~ of logarithms (Algebra) die Grundzahl, Basis; ~ of a wedge (Mech.) der Kopf eines Keils.

base board *s* (MA) das Grundbrett; ~ line, datum line die Standlinie, Grundlinie (beim Vermessen); ~ plate of a machine die Grund-(Fundament-)Platte.

base *a* gering, niedrig; geringhaltig; ~ coin das geringhaltige Geld; ~ metals *pl* unedle Metalle; ~ tin das Halbzinn.

basement *s* die Grundmauer, das Fundament; ~ story das Erdgeschoß, Parterre; ~ das Souterrain, das Kellergeschoß.

basic *a* (Chem.) basisch, Grund...; ~ salt das basische Salz.

basification *s* (Chem.) die Basenbildung.

basket *s* der Korb; ~ for parcels (P) der Packetkorb.

batch *s* der Satz, die Serie; ~ working (T) das Arbeiten in Serien, das abwechselnde Arbeiten.

batten *s* die Latte, Leiste, s. auch timber.

battery *s* (Phys., Elektriz., T) die elektrische Batterie; to charge the ~ die Batterie laden; to set up a ~ eine Batterie ansetzen; to supply a ~ eine Batterie speisen; to take a ~ to pieces eine Batterie auseinander nehmen; auxiliary ~ die Hülfsbatterie, Verstärkungsbatterie; constant ~ die konstante Batterie; Daniell's ~ die Daniell'sche Batterie; electric ~ die elektrische Batterie; equating ~ die Ausgleichungsbatterie; field ~ (MT) die Feldbatterie; galvanic ~ die galvanische Batterie; graphite ~ die Graphitbatterie; line ~ die Linien-(Telegraphir-)Batterie; magnetic ~ die magnetische Batterie; main ~ dass. wie line ~ polarization ~ die Polarisationsbatterie; quantity ~ die nebeneinander geschaltete Batterie; sand ~ die Sandbatterie; secondary ~ die sekundäre Batterie; testing ~ die Untersuchungsbatterie; thermo-electric ~ die thermo-elektrische Batterie; translating ~ die Uebertragungsbatterie; trough ~ die Trogbatterie; ~ box der Batterieschrank; ~ jar das Batteriegefäß; ~ plate die Batterieplatte; ~ rack das Batteriegestell; ~ room das Batteriezimmer; ~ stand das Batteriegestell; ~ switch der Batterieumschalter.

to baulk *v* wood Holz an beiden Seiten behauen.

baulk *s* auch balk s. timber.

bay *s* die Bai, Bucht; das Fach; das Joch; ~ of a bridge das Brückenjoch; ~ of a window die Fensteröffnung, Fensternische; ~ window das Bogenfenster; ~ work das Fach-

werk (in Gebäuden), Bindwerk, die Fachwand.
bayonet catch, bayonet joint s der Bajonettverschluß.
beacon s die Bake, Boie; der Leuchtthurm, die Feuerwarte.
beak s die Schnauze, Nase, Röhre; der Gasbrenner mit runder Oeffnung.
beam s der Balken, Hauptbalken; ~ of a balance, of a pair of scales (Mech.) der Waagebalken, Hebel; horizontal ~ of a bridge der Brückenträger, Brückenbalken.
bearer s der Träger, der Ueberbringer; ~ of a bill der Inhaber, Vorzeiger eines Wechsels.
bearing s (Mech.) der Stützpunkt, Hebelpunkt, das Lager; die Richtung, Visirlinie; ~ of a shaft (Masch.) das Achslager, Wellenlager; ~ of a spindle (Masch.) das Fußlager, untere Zapfenlager.
bearing block s (Masch.) das Zapfenlager.
to beat v schlagen, stampfen; to ~ down the earth die Erde feststampfen, rammen; to ~ down a person J—d unterbieten; to ~ the iron das Eisen schmieden; to ~ out ausbauchen, aushämmern.
beat s 1) der Schlag, der Schwingungsschlag; 2) das Revier (des Briefträgers, des Polizisten u. s. w.).
beaten a geschlagen; ~ gold das Blattgold; a ~ road ein gebahnter Weg; ~ silver das Blattsilber.
beater s der Klopfer, Schlägel.
to bed v betten; to ~ an engine (Masch.) eine Maschine dem Fundament anpassen.
bed s das Bett, Lager; die Bettung, Lagerung; die Schicht; das Flußbett; ~ of bricks die Ziegelschicht; ~ of coal die Kohlenschicht; ~ of a river das Flußbett.
bed plate s of a machine die Bodenplatte des Gestells.
bedding s of hemp die Hanflage, die beim Kabel als Mittelglied zwischen der Guttapercha und den eisernen Schutzdrähten dient.
beech, beech tree s die Buche, der Buchenbaum.

to begrease v befetten, einölen.
bell s die Glocke; die Schelle, Klingel; ~, electric bell (T) der elektrische Glockenapparat, der Wecker; Bright's ~ der elektrische Glocken-, akustischer Telegraphen-Apparat von Bell; double ~ die Doppelglocke, der Doppelglocken-Isolator; single stroke ~ der Wecker mit einem Anschlag; vibrating ~, trembling ~ der Klingelwecker.
bell clapper, ~ hammer, ~ tongue s der Glockenklöppel; ~ crank der Glockenarm, Glockenwengel; ~ crank lever der Hebel, die Ziehklammer eines Klingelzugs; ~ lever (Mech.) ein rechtwinkliger Hebel; ~ mouth der Schalltrichter eines Sprachrohrs; ~ mouthed a trichterförmig; ~ pull der Schellenzug; ~ push (T, F) der Druckknopf (des Weckers); ~ shaped a glockenförmig.
belt s der Gürtel, Leibriemen; (Masch.) der Treibriemen; endless ~ der Riemen ohne Ende; ~ pulley, ~ sheave (Mech.) die Riemenscheibe; ~ saw die Bandsäge.
bench s die Bank; working ~ die Werkbank, der Arbeitstisch; ~ clamp die Kluppe, der Schraubstock an der Werkbank; ~ hook der Bankhaken, das Bankeisen; ~ vice with screw clamp der Schraubstock mit Schraubzwinge.
to bend v biegen, krümmen; to ~ at angles kröpfen; to ~ a spring eine Feder spannen.
bend s die Krümmung; der Knoten im Draht; ~ of a river die Flußkrümmung; ~ of a rope der Knoten.
bending s das Krümmen, Biegen; ~ machine die Biegemaschine; ~ test die Prüfung des Drahtes auf Zähigkeit.
beneficiary s (P) der Empfänger einer Postanweisung, überhaupt eines Geldbetrages.
bent a gekrümmt, gebogen; ~ at angles gekröpft; ~ lever (Mech.) der Winkelhebel; ~ lever balance die Zeigerwaage mit rechtwinklig gebogenem Hebelarm.
Bessemer process s der Besse-

merprozeß; ~ pig das Roheisen für Bessemerstahl; ~ steel der Bessemerstahl.

„best" Bezeichnung für den gewöhnlichen in England verwendeten Telegraphen-Draht; „best best" besserer Telegraphen-Draht als „best"; „extra best best" der beste Telegraphen-Draht.

best bidder s der Meistbietende (bei Auktionen); ~ kind die feinste Qualität, erste Sorte; ~ metal der Konzentrationsstein (vom Kupfererzschmelzen).

beton s der Beton, s. auch concrete.

to bevel v schräg abschneiden; to ~ an edge eine Kante abschrägen, abkanten.

bevel, bevil s der schiefe Winkel von zwei Flächen, die Abgratung, Schmiege.

bevel, bevelled a schräge, schief; a ~ angle ein schiefer Winkel; ~ gear, ~ gearing, bevelled gear die schiefe Gehrung, das konische Räderwerk.

bevel pinion s (Masch.) das Kegelradgetriebe; ~ square das Schrägmaaß; ~ wheel das konische Rad, Winkelrad.

biangular, biangulate, biangulous a (Geom.) mit zwei Winkeln oder Ecken, zweiwinklig, zweieckig.

to bias v auf die Seite neigen.

bias s die schiefe Seite, schiefe Richtung, der Ueberhang.

bias a schief.

bicarbonate s (Chem.) das zweifach kohlensaure, doppelkohlensaure Salz.

bichloride s (Chem.) das Doppelchlorid.

bichromate s das zweifach chromsaure Salz (im Handel gewöhnlich für doppelchromsaures Kali gebraucht); ~ of potash das zweifach chromsaure Kali.

to bid v bieten (bei Versteigerungen); to ~ up überbieten, in die Höhe treiben.

bidding s das Bieten, Gebot (bei Auktionen).

bidder s der Bieter; the best ~, the highest ~ der Meistbietende.

bifurcate, bifurcated, bifurcous a zweiastig, gabelförmig.

bifurcation s die Gabelung (eines Weges, des elektrischen Stromes u. s. w.).

bight s of a rope die Schleife eines Taues.

bill s der Zettel, Anschlagzettel, die Bekanntmachung; die Rechnung; der Wechsel, Schuldschein, die Tratte; to draw a ~ einen Wechsel ziehen, trassiren; to protest a ~ einen Wechsel protestiren; hand ~ der Handschein, Handschuldschein; letter ~ (P) die Briefkarte; short ~, shortsighted ~ der kurzsichtige Wechsel; sight ~, ~ at sight, ~ after sight der Sichtwechsel, der bei Präsentirung zahlbare Wechsel; single ~, sole ~, only ~ der Solawechsel; ~ receivable der zu empfangende Wechsel, die Rimesse; time ~ (P) der Stundenzettel; ~ at usance der Usowechsel, Usanzwechsel, der nach Handelsbrauch zahlbare Wechsel; way ~ (P) das Begleitpapier, Begleitverzeichniß, Begleitzettel, vgl. auch a. T. Begleit…; ~ of cost die Kostenrechnung; ~ of the course of exchange der Kurszettel; ~ of debt der Schuldschein; ~ of delivery der Lieferungsschein; ~ of exchange der Wechsel; ~ of freight, ~ of lading der Frachtbrief, Verladungsschein; ~ of law der Gesetzesvorschlag.

bill account s das Wechselkonto; ~ stamp der Wechselstempel.

billet s der Zettel, das Billet; ~ of wood das Scheit Holz.

bimensal, bimestrical a zweimonatlich.

to bind v binden; to ~ with iron mit Eisen beschlagen; to ~ the insulator (T) den Draht am Isolator festbinden.

binder s (T) 1) der Binder (Arbeiter, der den Draht an den Isolatoren festbindet); 2) die Bindung (am Isolator).

binding piece s der Spannriegel, das Gurtholz; ~ post,

screw (*T*) die Klemmschraube; ~ wire (*T*) der Bindedraht.

binoxyd, binoxide *s* (Chem.) das Bioxyd, Dioxyd.

bipartition *s* die Theilung in zwei Theile.

bipolar *a* (Phys.) zweipolig.

biquadrate *s* (Algebra) das Biquadrat, die vierte Potenz.

biquadratic *a* biquadratisch; ~ equation die biquadratische Gleichung; ~ root die vierte Wurzel.

Birmingham wire gauge *s* (B. W. G.) das Birminghamer Drahtmaaß, die Birminghamer Drahtleere.

to bisect *v* in zwei gleiche Theile theilen, halbiren.

bisection *s* die Theilung in zwei gleiche Theile, Halbirung.

bisulphate *s* of potash (Chem.) das zweifach schwefelsaure Kali.

bisulphite *s* of soda (Chem.) das zweifach schwefligsaure Natron.

bit *s* das kleine Stück; boring ~ (Mech.) das Bohreisen, die Bohrschneide; chamfering ~ der Abschrägbohrer; ~ for soldering, soldering ~ der Löthkolben; spoon ~ der Löffelbohrer; square ~ der Kreuzbohrer, vierschneidige Kernbohrer.

bitumed *a* mit Erdpech beschmiert.

bitumen, compact bitumen, bitume *s* der Asphalt, das Erdharz, Judenpech, Bitumen; earthy ~ der Erdasphalt; viscid ~ der flüssige Asphalt.

to bituminate, to bituminize *v* mit Asphalt belegen, mit Erdpech sättigen.

bituminous *a* asphaltartig; ~ coal die bituminöse Kohle; ~ mastic, ~ cement der Asphaltkitt, Asphaltmastik.

black, blacking *s* die Schwärze, schwarze Farbe; die Druckerschwärze; bone ~ die Knochenkohle; ivory ~ das Elfenbeinschwarz, die Knochenschwärze; jet ~ das Pechschwarz; lamp ~ das Lampenschwarz, Lampenruß, Kienruß; platinum ~ das Platinschwarz, der Platinmohr.

black *a* schwarz; ~ copper das Schwarzkupfer, Rohkupfer; ~ damp der Schwaden, die bösen Wetter; ~ iron das hämmerbare, streckbare Eisen; ~ mud der Batterie-Rückstand der Zink-Kupfer-Elemente; ~ oakum das getheerte Werg ~ oxide of copper die Kupferschwärze; ~ oxide of iron das Eisenoxyduloxyd.

blade *s* das Blatt, Sägeblatt; ~ of an axe das Artblatt.

to blanch *v* weiß machen; weiß sieden, abglühen; to ~ copper Kupfer verzinnen.

blanch *s* of ore ein mit anderen Massen vermischtes Erz.

blank *s* 1) der leere Raum; to make a ~ (HA) einen leeren (Zwischen-)Raum auf dem Streifen machen; 2) (Am) das Formular.

blank acceptance *s* das Blanko-Akzept; ~ day (*P*) ein Tag, an welchem (in England) der gewöhnliche Postdienst ganz oder theilweise ruht. Zunächst ist jeder Sonntag ein solcher Tag, ferner in England und Irland der Weihnachtstag und Charfreitag; endlich in Schottland jeder sog. *sacramental fast day* (s. dens.); ~ key (HA) die Blanktaste; ~ portion, ~ space of the type wheel (HA) das Blank des Typenrades; ~ power of attorney die Blanko-Vollmacht, unausgefüllte Vollmacht.

to blast *v* sprengen; to ~ rocks Felsen sprengen.

blast *s* die Gebläseluft, das Gebläse; cold ~ die kalte Gebläseluft; hot ~ das heiße Gebläse; steam ~ das Dampfgebläse; ~ air die Gebläseluft; ~ apparatus die Gebläsevorrichtung.

blasting *s* das Sprengen; shooting & ~ die Sprengarbeit.

blasting agents *s/pl* Sprengmittel; ~ fuze der Sprengzünder; ~ powder das Sprengpulver; ~ tools *pl* das Bohr- und Schießzeug.

blea *s* der Splint im Holze.

blind reader *s* (*P*) der Beamte, welcher die mangelhaften, unleserlichen u. s. w. Adressen entziffert.

blister *s* die Blase, blasige Stelle in Metallen; ~ copper das Blasenkupfer, Rohkupfer; ~ steel, blistered steel der Blasenstahl, Zementstahl.

to block *v* the line (Eisenb.) die Linie absperren; das Signal zum Stehenbleiben nach dem block system geben.

block *s* der Block; der Holzblock; der Kloben, die Rolle; ~ of carbon das Kohlenstück im Mikrophon; gin ~ (Mech.) der Kloben des Hebebocks; ~ & pulley, ~ & fall, ~ & tackle (Mech.) der Flaschenzug, das Zugwerk; ~ signal (Eisenb.) das Haltezeichen; ~ system, ~ telegraph system das Blocksystem; ~ work das Mauerwerk (um den Fußpunkt der Telegraphen-Stangen).

blotter *s* der Tintenlöscher.

blotting pad *s* die Schreibunterlage, ~ paper das Löschpapier.

to blow up *v* sprengen (Gestein).

blower *s* der Luftfang, Windfang.

blue stone, blue vitriol *s* der Kupfervitriol, das schwefelsaure Kupferoxyd.

blunt *a* stumpf, abgestumpft, ungeschärft; ~ cone (Geom.) der abgestumpfte Kegel; ~ file die Stumpffeile.

board *s* das Brett, f. auch timber; ~ of customs das Zollamt; ~ of directors die Direktion, das Direktorium; free on ~ (fob) frei an Bord geliefert; ~ of roads das Wegeamt; sounding ~ der Resonanzboden; ~ of surveyors der Aufsichtsrath; ~ of a table die Tischplatte; ~ of trade das Handelsministerium; ~ of works die Baukommission, das Bauamt.

boat *s* das Boot, der Kahn, Nachen, die Barke; advice ~ der Aviso; drag ~, dredging ~ das Baggerboot, der Prahm; ferry ~ die Fähre; life ~ das Rettungsboot; tug ~ das Bugsirschiff, Schleppschiff; ~ bridge die Schiffbrücke; ~'s crew die Bootsmannschaft.

bobbin *s* (*T*) die Spule, die Induktionsrolle.

body *s* der Körper, Haupttheil; das Gestell; fluid ~ (Phys.) der flüssige Körper; simple ~ der Grundstoff, das Element; solid ~ der feste Körper.

bog *s* das Moor; ~ earth die Moorerde; ~ iron ore, ~ ore das Sumpferz, der Raseneisenstein.

bogie *s* der kleine, niedrige Wagen zum Transport schwerer Lasten.

boiled *a* gekocht; entschält; ~ bar das Rohschieneneisen; ~ oil, ~ linseed oil das gekochte Leinöl, der Leinölfirniß; ~ plaster der gebrannte Gips.

boiler *s* die Pfanne, der Kessel; der Dampfkessel; ~ house das Kesselhaus.

boiling point *s* of a thermometer (Phys.) der Siedepunkt.

to bolt *v* (Techn.) verbolzen, mit Bolzen befestigen; verriegeln.

bolt *s* (Techn., Masch.) der Bolzen; (T) die gerade Stütze des Isolators; der Riegel; countersunkheaded ~ der Bolzen mit versenktem Kopfe; eye ~, eyed ~, key ~ der Splintbolzen, Vorstecker; fender ~ der Kopfbolzen; fish ~ der Laschenbolzen; flat-headed ~ der flachköpfige Bolzen; fore- ~ der Splintbolzen; French ~, dormant ~ der französische Riegel; hook ~, hooked ~ der Hakenbolzen; ~ & nut der Schraubenbolzen mit Mutter, Mutterbolzen; pointed ~ der Spitzbolzen; screw ~ der Schraubenbolzen; soldering ~ der Löthkolben; square-headed ~ der Bolzen mit viereckigem Kopfe; tie ~ der Ankerbolzen; ~ eye das Auge des Bolzens, die Oese; ~ head der Bolzenkopf; ~ screw die Bolzenschraube.

bond *s* 1) der Verband, die Verbindung; der Holzverband; der Mauerverband; 2) die Schuldverschreibung, Obligation, (P, T) die Kaution; to give ~ Kaution stellen; goods in ~ Waaren unter Zollverschluß; ~ of obligation der Schuldschein; under ~ gegen Kaution.

to book *v* in ein Buch eintragen, buchen; (Eisenb.) eine Fahrkarte ausgeben; ein Billet nehmen; to ~ the baggage das Gepäck bezet' mit Etikett nach einem bestim Ort versehen lassen.

book *s* das Buch; ~ of accounts das Rechnungs- (Konto-)Buch; ~ of cargo das Frachtbuch; ~ of charges das Unkostenbuch; ~ of complaints das Beschwerdebuch; copying ~ das Kopirbuch; ~ of postages das Portobuch (des Kaufmanns); receipts & expenditures ~ das Einnahme- und Ausgabebuch.

book keeper *s* der Buchführer; ~ keeping by double entry die doppelte Buchführung; ~ packet (*P*) das Bücherpacket (Sendung von Büchern durch die Bücherpost); ~ post (*P*) die Bücherpost.

booking office *s* (Eisenb.) das Billetausgabe-Büreau, die Fahrkarten-Ausgabe.

to bore *v* bohren, durchbohren, anbohren; to ~ a blast ein Bohrloch schlagen (um Gestein zu sprengen); to ~ a mortice ein Zapfenloch bohren.

bore *s* die Bohrung, das Bohrloch; ~ for blasting das Sprengbohrloch.

bore bit *s* das Bohreisen; ~ hole das Bohrloch.

borer *s* der Bohrer; ~ for blasting & shooting der Sprengbohrer, Steinbohrer; ~ with circular bit der Kreisbohrer; ~ for metal der Metallbohrer; pointed ~ der Spitzbohrer; ~ for wood der Holzbohrer.

borne (Partiz. von to bear); all charges ~ nach Abzug aller Kosten.

to boss *v* (Techn.) mit Buckeln beschlagen, mit erhabener Arbeit versehen; to ~ metal Metall treiben.

boss *s* (Techn.) der Buckel; (*Am*) der Arbeitgeber, der Meister.

bossed, bossy *a* (Techn.) mit Buckeln versehen.

bottom *s* der Grund, Boden; der Bodensatz; ~ of a ditch (trench) die Grabensohle; false ~ der Doppelboden.

boucherizing *s* die Zubereitung (s. auch a. T.) des Holzes mit Kupfervitriol nach der Methode des Dr. Boucherie.

boulder, bowlder *s* der rundliche Kiesel, Feldstein; das Geschiebe, Geröll, Gestein; ~ paving, ~ pavement das Feldsteinpflaster.

bound *a* (Techn.) gebunden, verbunden; ~ home, ~ in heimwärts bestimmt; ~ out auswärts bestimmt (von Schiffen gesagt).

bowlder *s* f. boulder.

box *s* die Büchse, Schachtel, der Kasten, das Futteral; ~ containing the contact pins (*HA*) das Stiftgehäuse; (*P*) das Abholungsfach, ~ rent (*P*) die Miethe für ein Fach; feed ~ (Masch.) die Speisebüchse, Speisevorrichtung; safe ~, strong ~ die Geldkiste.

box car *s* (Eisenb.) der bedeckte Güterwagen; ~ tree der Buchsbaum; ~ waggon (Eisenb.) der Gepäckwagen, Güterwagen; ~ wood das Buchsbaumholz.

boxen *a* von Buchsbaum.

boy clerk *s* der Angestellte im englischen Civildienst bis zum vollendeten 19. Lebensjahre; ~ sorter (*P*) der junge Sortirer; ~ telegraph messenger der Telegraphen-Bote, Knabe im Alter von 14—19 Jahren.

to brace *v* (Techn.) absteifen, mit Ankern oder Klammern verbinden.

brace *s* das Band; der Querriegel; ~ pin der Spannnagel.

bracket *s* der Träger, die Konsole, die Knagge; (*T*) die Isolatorstütze; bridge ~ (*T*) die Mauerkonsole; double bridge ~ die Mauerkonsole für zwei Isolatoren, single bridge ~ die Mauerkonsole für einen Isolator; pole ~ die Isolatorstütze, die direkt in die Stange eingeschraubt wird; wall ~ die Mauerkonsole.

brad *s* der Lattennagel, Fußbodennagel, Nagel ohne Kopf; ~ awl der Nagelbohrer.

to braid *v* flechten, mit Litzen besetzen; ~ed wire (*T*) der verseilte Draht.

braid *s* die Flechte, die Flechtschnur.

to brake *v* bremsen.

brake *s* die Bremse; Appold's ~ die Bremsvorrichtung beim Abrollen des Kabels vom Schiffe; friction ~ das Bremsdynamometer;

~ arrangement die Bremsvorrichtung; ~ lever der Bremshebel; ~ man der Bremser; ~ ring der Bremsring; ~ spring die Bremsfeder; ~ wheel das Bremsrad.

to branch v sich in Zweige theilen; to ~ off abzweigen.

branch s der Zweig, Arm; der Ast; der Geschäftszweig; ~ current (T) der Zweigstrom; ~ line (P, T, Eisenb.) die Zweiglinie, Nebenlinie; ~ post office die Zweigpostanstalt s. auch post; ~ road (Eisenb.) die Nebenlinie; die Nebenstraße; ~ wood das Astholz.

brand s das eingebrannte Zeichen; ~ iron, branding iron das Brenneisen.

branded a mit eingebranntem Zeichen versehen.

brass s das Mischmetall, Erz, die Bronze; das Messing; hard ~ der Hartguß; latin ~, latten ~, plate ~, sheet ~ das Messingblech; red ~ der Tombak.

brass bar s die Messingschiene; ~ clamp die Messingklammer; ~ frame der Messingrahmen; ~ lever der Messinghebel; ~ pillar der Messingständer; ~ pin der Messingstift; ~ plate das Messingblech, die Messingplatte; ~ rod die Messingstange; ~ side die Messingwange (des Apparats); ~ solder das Messingschlagloth; ~ standard der Messingständer; ~ strip das Messingstäbchen; ~ terminal die messingene Klemmschraube; ~ wire der Messingdraht; ~ work die Messingwaare; ~ works pl das Messingwerk, die Messingfabrik.

to braze v hart löthen, zusammenlöthen; in Erz arbeiten; to ~ over bronziren.

brazier, brasier s das Kohlenbecken; der Gelbgießer; der Klempner; ~'s work die Klempnerarbeit an einem Gebäude.

breadth s die Breite; ~ in the clear die Breite im Lichten eines Raumes.

to break v brechen; to ~ bulk anfangen, die Ladung eines Schiffes zu löschen; to ~ the electric current den elektrischen Strom unterbrechen; to ~ up (Techn.) abreißen, aufbrechen; to ~ up a pavement ein (Straßen-)Pflaster aufreißen.

break s der Bruch; (T) die Bruchstelle (im Draht); der Fehler (als Leitungsunterbrechung); die Unterbrechung (des Lokalstromkreises); der Apparat (Umschalter) zum Unterbrechen der Leitung.

break down s der Zusammenbruch (der telegraphischen Verbindungen); ~ signal das Trennungszeichen des Morse-Alphabets.

breakage s der Bruch, das Brechen; die Refaktie (Vergütung für ramponirte Waaren); free from ~ bruchfrei.

breaking strain s (Techn.) die Bruchkraft, Bruchspannung; ~ test die Probe auf Bruchkraft; ~ weight das Bruchgewicht.

brick s der Backstein, Mauerstein, Ziegel, Brennstein; Dutch ~, Flemish ~, clinker der Klinker; ~ works pl die Ziegelei, Ziegelbrennerei.

bridge s die Brücke; (Wheatstone's) ~ of balance die Wheatstone'sche Meßbrücke; ~ bracket s. bracket; (T) ~ method, ~ principle die Brückenmethode (beim Gegensprechen).

bridle s das Querholz; (Mech.) der Bügel.

to brief v (Am) ein Schriftstück mit Inhaltsangabe versehen.

brief s to counsel die von dem Anwalt (solicitor) dem plaidirenden Advokaten (barrister, counsel) ertheilte schriftliche Instruktion.

bright a glänzend, funkelnd, metallblank.

Bright's bell instrument s (T) der elektrische Glockentelegraph von Bell.

to brighten v poliren, metallblank machen.

brimstone s der Schwefel, Rohschwefel.

to bring v to account in Rechnung bringen; to ~ over, to ~ up auf die andere Seite übertragen (in Büchern, Rechnungen u. s. w.).

brittle *a* zerbrechlich, spröde, brüchig; ~ iron spröder Eisen; ~ metal der Rotguß, das Rotmessing.
brittleness *s* die Sprödigkeit, Zerbrechlichkeit.
broad axe *s* die Zimmeraxt, das Breitbeil.
broker *s* der Mäkler, Makler; exchange ~ der Wechselmakler; insurance ~ der Assekuranzmakler; ~'s memorandum, ~'s note der Schlußzettel des Maklers; ship ~ der Schiffsmakler; stock ~ der Makler in Staatspapieren; sworn ~ der vereidigte Makler, der Sensal.
bronze *s* die Bronze, das Erz; die kleine Kupfermünze; phosphor ~ die Phosphorbronze.
to bubble *v* aufwallen, Blasen aufwerfen.
bubble *s* die Blase (von aufsteigenden Gasen); ~ of air die Luftblase; ~ of a level die Weingeiströhre einer Libelle.
to buckle *v* sich werfen (von den Akkumulator-Platten gesagt, die ihre Gestalt verändern).
bulb *s* das Gefäß, die Kugel, Kapsel; ~ of a barometer die Kugel, das zylindrische Gefäß eines Barometers.
bulk *s* die Masse, der Klumpen; to break ~ s. to break; in the ~ im Ganzen; in ~ unverpackt, lose; purchase in the ~ der Kauf im Ganzen, der Bauschkauf.
bulky *a* voluminös, sperrig; ~ goods *pl* das Sperrgut.
bulled message *s* (*Am*) das verstümmelte Telegramm.
bulletin *s* of verification s. verification.

bullion *s* das ungemünzte Gold oder Silber.
bundle *s* das Bündel, Packet (Briefe), die Rolle; ~ of paper zwei Ries Papier; ~ iron das Bundeisen; ~ of wire das Drahtbündel.
buoy *s* die Boie, Baake; beacon ~ die Baakentonne; cask ~, ~ cask die Tonnenboie.
bur, burr *s* (Masch.) das Schraubenmutterblech, Nietblech; ~ chisel das Stemmeisen.
bureau *s* der Schreibtisch; ~, office die Schreibstube, das Geschäftszimmer.
burglar alarm *s* die Diebsschelle; ~ lock das Alarmschloß, Diebsschloß.
to burn *v* brennen, verbrennen; to ~ bricks Ziegel brennen; to ~ together vergießen, durch Guß löthen.
burner *s* der Brenner, Lampenbrenner; flat ~ der Flachbrenner; jet ~ der Strahlbrenner; round ~ der Rundbrenner.
burnetizing *s* die Zubereitung des Holzes mit Zinkchlorid.
busy circuit *s* (*T*) eine besetzte Telegraphen-Linie.
butt *s* die Butte, das große Faß; das dicke Ende einer Stange; ~ end of a tree das Stammende, Wurzelende eines Baumes.
button *s* (Techn.) der Knopf; ~ repeater (*T*) der Kurbelumschalter als Uebertrager; ~ switch (*T*) der Kurbelumschalter.
by- coach *s* der Beiwagen; ~ law das Ortsstatut, das besondere Gesetz einer Gesellschaft; ~ path der Nebenpfad; ~ road der Nebenweg, Umweg.

C.

C form, received message form *s* (*T*) das Formular für Ankunfts-Telegramme.
to cable *v* kabeln, ein Telegramm mittels Kabels versenden; a cabled river ein mittels Kabels überschrittener Fluß.
cable *s* das Kabel; deep sea ~ das Tiefseekabel; ocean ~, submarine ~ das unterseeische Kabel; river ~ das Flußkabel; underground ~ das versenkte Erdkabel; core of the ~ die Kabelseele; laying of a ~ die Kabellegung; path of the ~ das Kabellager; to coil the ~ das Kabel aufrollen (in Ringe legen); to land

a ~ ein Kabel landen; to lay a ~ ein Kabel legen; to pay out a ~ ein Kabel abrollen.

cable box *s* der Kabelkasten, Kabelschrank; ~ buoy die Kabelboie; ~ company die Kabelgesellschaft; ~ core die Kabelseele; ~ house das Kabelhaus; ~ rate die Kabeltaxe (Gebühr); ~ shed die Kabelschutzhütte; ~ ship das Kabelschiff; ~ test die Kabeluntersuchung (Messung); lift ~, elevator ~ das Kabel für Aufzugsvorrichtungen.

cablegram *s* das Kabeltelegramm.

cabotage *s* die Küstenschifffahrt; der Küstenhandel; die Küstenkenntniß.

„**caf**" d. h. cost, assurance, freight Kosten, Versicherung und Fracht (einbegriffen).

to calcinate, to calcine *v* glühen, erhitzen; (Metalle) brennen, rösten, verkalken.

calcium light *s* das Drummond'sche (Kalk-)Licht, Hydroxygen-Licht.

to calculate *v* berechnen, abmessen; to ~ the expenses die Unkosten berechnen.

calculation *s* die Rechnung, Berechnung; ~ of exchange die Wechselkursberechnung; at the lowest ~ nach dem niedrigsten Anschlage.

calculus *s* (Math.) das Rechnen, die Rechnungsart; die Infinitesimalrechnung (auch infinitesimal ~); differential ~ die Differenzialrechnung; integral ~ die Integralrechnung; literal ~ die Buchstabenrechnung.

calendar *s* der Kalender, Almanach; ~ month der Kalendermonat; ~ year das Kalenderjahr.

to calibrate *v* kalibriren.

caliper *s* s. calliper.

to calk, to caulk *v* (Mech.) verstemmen (zusammengenietete Bleche); kalfatern; die Fugen mit Werg verstopfen.

to call *v* a station (*T*) ein Amt anrufen; to ~ at a port anlegen (von Schiffen); to ~ for bestellen; to be ~ed for (auf Briefen) postlagernd, poste restante.

call *s* (*T*) der Anruf (eines Amtes); ~ bell (*T*) der Klingelwecker; ~ box (*F*) die Fernsprechzelle; ~ signal (*T*) das Rufzeichen; ~ station (*F*) die öffentliche Fernsprechstelle; station ~ (*T*) der Stationsruf.

calliper scale *s* der Kalibermaaßstab, die Schubleere; sliding square (Techn.) das verstellbare Winkelmaaß, die Schmiege.

callipers, calipers, calibers *s/pl* der Greifzirkel, Tastzirkel, Dickzirkel.

caloric *s* (Phys.) der Wärmestoff; conductor of ~ der Wärmeleiter.

caloric *a* kalorisch.

calorie, calory *s* die Kalorie, Wärmeeinheit (der Hitzegrad, der nöthig ist, um ein Gramm Wasser um 1^0 C. mehr zu erhitzen).

calorific *a* Wärme erzeugend; ~ effect, ~ intensity, ~ power die Heizkraft.

cam (Masch., HA) der Daumen, Hebedaumen; correcting ~ (HA) der Korrektionsdaumen; detent ~ (HA) der Daumen der Auslösung des Einstellhebels; paper-moving ~ (HA) der Daumen der Papierführung; ~ shaft (HA) die Daumenwelle; ~ wheel (HA) das Daumenrad.

camion *s* ein zweirädriger Wagen, Karren (meist in der Armee gebraucht).

to cancel *v* vernichten, annulliren; to ~ a telegram ein in der Beförderung begriffenes Telegramm zurückziehen; ~ling stamp (*P*, *T*) der Entwerthungsstempel.

candle *s* das Licht, die Kerze; dip ~, dipped ~ das gezogene Licht; mould ~, moulded ~ das gegossene Licht; ~ power die Lichtstärke, Kerzenstärke.

to canker *v* orydirt werden, angefressen werden, rosten.

to cant *v* (Techn.) umkanten, umkehren; abkanten, behauen; (im Handel) versteigern, an den Meistbietenden verkaufen.

cant *s* (Techn.) die Kante, ausspringende Ecke; bevelled ~, bevil ~ die abgefaßte Kante; full ~, sharp ~ die scharfe, volle Kante; ~ hook der Kanthaken, Kenterhaken.

canted, cant *a* eckig, kantig; behauen, zugerichtet; ~ wall die Mauer, die mit einer anderen einen Winkel bildet.

canting wheel *s* das Kronenrad.

caoutchouc, gum elastic, India rubber *s* das Kautschuk, Federharz, elastische Gummi; hardened ~, ebonite ~ das Ebonit, Hartgummi; sheet ~ das Kautschuk in Tafeln, Plattenkautschuk; vulcanised ~ das vulkanisirte (geschwefelte) Kautschuk; ~ band das Kautschukband (zum Umwickeln der Telegraphendrähte u. s. w.); ~ plate die Kautschukplatte; ~ tube, elastic tubing der Kautschukschlauch.

to cap *v* bedecken, bekleiden, bekappen; to ~ a magnet einen Magnet bewaffnen; to ~ a pole eine Telegr.-Stange mit einer Kappe versehen.

cap *s* die Kappe, Stangenkappe, Haube, der Aufsatz; ~ paper das graue Packpapier; eine Art Schreibpapier (blau liniirt heißt es foolscap, roth liniirt mit Rand ist es legal cap).

capacity *s* die Größe des Raumes; (Geom.) der kubische Inhalt; (Chem., Phys., T) die Ladungsfähigkeit, Aufnahmefähigkeit; ~ of weight das Tragvermögen; specific inductive ~ (Elektriz.) die spezifische Kapazität der Induktion; ~ test (T) die Prüfung der Kabel auf Ladungsfähigkeit.

capillarity, capillary attraction *s* die Haarröhrchenkraft, die Haarröhrchenanziehung.

capillary, capillaceous, capilliform *a* haarförmig, haarfein; ~ attraction s. capillarity; ~ tubes *pl* die Haarröhrchen.

capital *s* das Kapitäl, der Knauf; das Kapital; acting ~, circulating ~ umlaufendes Kapital; ~ paid in ~ put in das Einlagekapital; trading ~ das Betriebskapital; unproductive ~, dead ~ das todte Kapital.

capstan, capstern *s* (Masch.) die stehende Welle, die Erdwinde; moveable ~, field ~, crab ~ die Erdwinde.

car *s* der Karren, Wagen; (*Am*) der Eisenbahnwagen, Passagierwagen, (auch railway car, in England stets carriage oder waggon); baggage ~ der Gepäckwagen; flat ~, platform ~ der offene Güterwagen; freight ~ der verschlossene Güterwagen; passenger ~ (Eisenb. *Am*) der Personenwagen; sleeping ~ der Schlafwagen; street ~ der Wagen einer Straßeneisenbahn; tram ~ der Tramwagen, der Wagen einer Trambahn; ~ bed, ~ body der Wagenkasten; ~ brake die Wagenbremse; ~ seat der Wagensitz; ~ truck das Untergestell, Rahmwerk des Wagens mit den Rädern, die gesammte Bewegungsvorrichtung des Wagens.

carbon *s* (Chem.) der Kohlenstoff; ~ filament die Kohlenfaser; ~ light (Phys.) das Kohlenlicht; ~ protector, lightning bridge protector (T) der Varley'sche Blitzableiter mit Kohlenspitzen.

carbonic paper *s* das Durchdruckpapier (in England vielfach für postalische Zwecke, namentlich auch bei der Ausfertigung von Telegrammen gebraucht, um eine Kopie des Anfunfts-Telegramms für Dienstzwecke zurückzubehalten).

Carcel *s* Carcel (die französische Normallichtstärke = 9,6 engl. u. 8,7 deutschen Kerzen); ~ lamp die Carcel-Lampe.

card *s* die Karte, der Pappdeckel; die Handkratze (zum Reinigen der Zinkplatten des Daniell-Elements); post ~ die Postkarte; ~ board der Pappdeckel.

caret *s* das Anführungszeichen, Gänsefüßchen.

cargo *s* die Fracht, die Schiffsladung, das Gut; ~ in grains Stürzgüter; ~ in parcels Stückgüter.

cariole *s* ein kleiner Wagen; (*P*) das Kariol.

carriage *s* der Wagen, das Fuhrwerk; die Anfuhr, Verfrachtung, der Transport; der Fuhrlohn, die

Fracht; (Eisenb.) der Eisenbahnwagen, Personenwagen (auch railway carriage); post office ~ der Eisenbahnpostwagen; ~ by sea der Seetransport; sleeping ~ der Schlafwagen; ~ road die Landstraße, der Fahrweg; ~ shed der Wagenschuppen.

carrier s der Fuhrmann; der Beförderer; die Büchse (bei der Rohrpost); der Treiber; (Mech.) das intermittirende Rad; common ~ der Lohnfuhrmann; letter ~ der Briefträger; head letter ~ der Oberbriefträger; rural letter ~ der Landbriefträger; mounted (driving) rural letter ~ der fahrende Landbriefträger.

to carry v tragen; (Eisenb.) befördern; to ~ to account auf Rechnung setzen; to ~ out a sum eine Summe auswerfen; to ~ over (forward) übertragen, transportiren (in Rechnungen); carried forward, amount carried forward der Uebertrag; carried over der Vortrag.

carrying s das Fahren, das Fuhrwesen; ~ & forwarding das Speditionsgeschäft; ~ establishment das Speditionsetablissement; ~ trade das Geschäft des Rheders oder der Eisenbahngesellschaft; ~ traffic der Güterverkehr auf der Eisenbahn.

to cart v karren, mit einem Karren transportiren; to ~ the ground (the earth) die Erde abfahren (bei Erdarbeiten).

cart s die Karre, der Karren (meist zweirädrig); hand ~ die Handkarre; registered letter ~ (P) der Karren für den Transport eingeschriebener Briefe; ~ house der Wagenschuppen; ~ load die Wagenladung.

cartage s das Fahren; das Rollgeld, der Fuhrlohn, die Transportkosten.

carter s der Kärrner, Fuhrmann.

carting s die An- u. Abfuhr (beim Bau).

to case v in ein Futteral stecken; einfassen, einrahmen; bekleiden, verkleiden.

case s das Futteral, Gehäuse, Behältniß; das Fach; die Verkleidung (bei Maurerarbeiten); (Masch.) der Mantel, die Hülle, das Gehäuse; door ~ der Thürrahmen; ~ of a window der Fensterrahmen.

case-hardened a durch Einsatz gehärtet, glashart (von Metallen gesagt); ~hardening der Prozeß, durch den dem Eisen eine Stahloberfläche gegeben wird.

to cash v ablösen, einwechseln, zu Gelde machen; to ~ a bill einen Wechsel bezahlen, einkassiren; to ~ at a house sich eines Hauses als Bankier bedienen.

cash s das baare Geld; die Kasse; die Baarzahlung; the balance in ~ der Kassenbestand; for ~ gegen baar; in ~ in Kassa, per Kassa; the proceeds in ~ der Kassenertrag; ready ~ baares Geld; running ~ das zirkulirende Geld; ~ over der Ueberschuß; ~ short der Minderbetrag; ~ account das Kassakonto; ~ advance der Geldvorschuß; ~ book das Kassabuch; ~ box der Geldkasten, die Schatulle; ~ keeper der Kassirer; ~ note die Anweisung; ~ purchases pl die Baarkäufe.

cashier, cash keeper s der Kassirer.

casing s der Ueberzug, die Umhüllung; das Gehäuse; die Verkleidung (an Gebäuden); die Verpackung (Emballage); ~ of timber work die Holzbekleidung; the ~ of a wall die Bekleidung einer Mauer.

cask s das Faß, die Tonne; large ~ das Stückfaß; ~ bridge die Tonnenbrücke.

to cast v gießen, schmelzen; sich werfen, sich krümmen (vom Holze); to ~ a balance den Saldo ausziehen; to ~ up zusammenrechnen.

cast s der Guß, das Gießen; das Gußstück; ~ iron das Gußeisen; ~ steel der Gußstahl; ~ work die Gußarbeit.

to catch v (Masch.) hemmen, aufhalten; in einander greifen, anhaken; to ~ with the spring (Mech.) einspringen.

catch s (Masch.) der Mitnehmer, die Knagge, Nase; (Mech.) der Sperrhaken, Stützhaken; die Verzahnung, die Sperre; ~ of a wheel der Sperrkegel; ~ station (*P Am*) die Eisenbahnstation mit einem catcher, s. dens.

catcher, mail catcher *s* (*P Am*) der Fangapparat für Briefbeutel auf Eisenbahnstationen, an denen der Zug nicht hält; cow ~ der Kuhfänger (eine an amerikanischen Lokomotiven allgemein angebrachte Vorrichtung); ~ post office (*Am*) das Postamt, welches seine Postsachen durch den catcher erhält, bz. befördert; ~ pouch (*P Am*) der Briefbeutel aus Segeltuch, der an den catcher gehängt wird.

catching *s* (Mech.) das Ineinandergreifen gezahnter Räder; ~ hook der Fanghaken.

catenarian *a* kettenförmig; ~ curve die Kettenlinie.

catenary *s* (Math.) die Kettenlinie, s. auch chain curve.

cathode *s* (Elektriz.) die Kathode, der negative Pol.

cation, kation *s* (Phys.) der Kation (elektro-positiver Bestandtheil eines durch den elektrischen Strom zerlegten Körpers).

to caulk, caulking s. calk.

causeway, causey *s* der Hochweg, die Chaussee, Kunststraße.

caution *s* die Kaution; (als Eisenbahnsignal) „die Bahn ist nicht frei!"; ~ plate (*P*) die Platte am Briefkasten mit einer Warnung für das Publikum, daß Briefe mit Werthinhalt, die in den Briefkasten gelegt werden, der doppelten Einschreibegebühr unterliegen.

to cave in *v* nachstürzen, einstürzen (von den Seiten einer Erdvertiefung gesagt).

caveat *s* die Anmeldung eines einzuholenden Patents.

caveatee *s* der Einholer eines Patents.

cavity *s* die Vertiefung, Höhlung, Mulde.

cell *s* die Zelle; galvanic ~ (oft auch nur cell) das galvanische Element.

cellulose *s* die Zellulose, der Pflanzenzellenstoff.

to cement *v* zementiren; mit Zement befestigen; verkitten; Zementstahl verfertigen; to ~ well gut binden (vom Mörtel).

cement *s* der Zement, hydraulischer Mörtel, Mörtel, das Bindemittel im Allgemeinen; der Kitt, das Zementirpulver; ~ steel der Zementstahl.

cementation *s* das Zementiren, die Zementirung; ~ steel der Zementstahl.

census *s* die Volkszählung.

center, centre *s* der Mittelpunkt, s. auch centre; ~ bit der englische Zentrumbohrer.

centering *s* das Zentriren, das Aufsuchen des Mittelpunktes.

centesimal *a* hunderttheilig.

centigrade *a* hundertgradig, huntertheilig; ~ scale die Zentesimaleintheilung.

centigram *s* das Zentigramm.

centimeter *s* der Zentimeter.

centre *s* der Mittelpunkt; ~ of attraction der Punkt, der die größte Anziehung ausübt, der Attraktionsmittelpunkt; dead ~ der todte Punkt einer Kurve; ~ of gravity der Schwerpunkt; ~ of gyration, rotation, motion der Drehpunkt, Schwingungspunkt.

centrifugal *a* zentrifugal, vom Mittelpunkt wegstrebend; ~ force die Zentrifugalkraft.

centripetal *a* zentripetal, nach dem Mittelpunkte hinstrebend; ~ force die Zentripetalkraft.

certificate *s* das Zeugniß; ~ of the custom office die Zollquittung; ~ of posting (*P*) der Einlieferungsschein für Packete; medical ~ das ärztliche Zeugniß.

to chafe *v* schaben, reiben, scheuern, durch Reiben beschädigen.

to chain *v* mit der Kette messen.

chain *s* die Kette, die Meßkette (ein in 100 links getheiltes Längemaaß von 22 yards = 20,1166 m); endless ~ die endlose, geschlossene Kette; gearing ~ (Masch.) die Treibkette; surveyor's ~ die Meßkette; ~ curve die Kettenlinie (des Telegr.-Drahts) mit folgenden Ausdrücken: vertex *s* der tiefste Punkt des

Drahtes zwischen zwei Stangen; modulus, parameter *s* die Linie, die vom vertex bis zu der zwischen den zwei Isolatoren gedachten Horizontalen gezogen wird; dip *s* der Unterschied der Entfernung von irgend einem Punkte der Kettenlinie bis zu der durch den vertex gelegten Horizontalen; der Durchhang; ~ wheel das Kettenrad.

chair *s* (Techn.) die Fußplatte für Telegraphenstangen auf Hausdächern.

to chalk *v* mit Kreide abzeichnen; to ~ a line abschnüren (beim Bau).

chalk *s* die Kreide, der kohlensaure Kalk; red ~ der Rothstift; ~ line die Schlagleine, Schlagschnur (beim Bau).

chalking *s* die Kreidebezeichnung, der Entwurf.

chamber *s* die Kammer, das Gemach; ~ of commerce die Handelskammer; ~ battery die in England gebräuchliche Art der Daniell'schen Zink-Kupfer-Batterie.

to chamfer, to chamfret *v* auskehlen, kanneliren, abschrägen, abkanten.

chamfer *s* die Auskehlung, Hohlrinne an einer Säule, Schrägfläche.

chamfered, chamfretted *a* abgeschrägt; ~ edge die abgestoßene Ecke.

chamfering bit *s* der konische Bohrer.

to change *v* ändern, wechseln; to ~ money Geld wechseln.

change *s* die Veränderung, der Wechsel; das kleine Geld; die Börse; der Wechselkurs; das Agio, Aufgeld; ~ of carriage (Eisenb.) der Wagenwechsel; ~ from letters to figures & signs (HA) Figurenwechsel.

changing place *s* (Eisenb.) der Ausweicheplatz, die Weichestelle.

to channel *v* auskehlen, kanneliren; to ~ on edge ausbogen, bogenartig ausschweifen.

channel *s* der Kanal; das Strombett, der Thalweg; die Auskehlung, Kannelirung, Hohlrinne; die Nuth.

channeled *a* gefurcht, gerippt.
chapiter *s* das Kapitäl, s. auch capital.

to char *v* verkohlen, ankohlen, (Holz) zu Kohlen brennen.

charcoal *s* die Holzkohle; ~ iron das mit Holzkohlen gefrischte Eisen; ~ wire der aus mit Holzkohle gefrischtem Eisen hergestellte Draht.

to charge *v* laden (das Kabel); anschreiben; fordern.

charge *s* die Last, Ladung; die Gebühr, der Preis; die Aufforderung zur Zahlung, (Elektriz.) die Ladung; additional ~ die Zuschlaggebühr.

charges *s/pl* die Belastung, die Kosten, Spesen, Unkosten; additional ~, extraordinary ~ die unvorhergesehenen, außerordentlichen Kosten; book of ~ das Unkostenbuch; ~ to be deducted ab an Unkosten; packing ~ die Verpackungskosten; petty ~ kleine Unkosten.

chariot *s* der Wagen, Karren, (HA) der Schlitten.

charring *s* das Verkohlen; (Telegr.-Bau) das Ankohlen (der Stangen-Stammenden).

chart *s* die Tabelle; ~, sea chart, hydrographic ~ die Seekarte.

to charter *v* a ship ein Schiff verfrachten; ein Schiff befrachten, heuern, miethen.

charter *s* der Vertrag, Kontrakt; die Schiffsmiethe, Befrachtung.

charterer *s* der Verfrachter; der Befrachter eines Schiffes.

to check *v* Gegenrechnung führen; (Mech.) einhalten, aufhalten, hemmen; to ~ the mail (P) die Post übernehmen; to ~ tickets die Fahrkarten kontroliren.

check *s* die Kontrolle; Gegenrechnung; das Billet; der Bankschein, die Bankanweisung; ~ account das Gegenregister, die Kontrolle; ~ book das Kontrollbuch, s. auch cheque book; ~ hook (Masch.) der Sperrhaken, Absperrhaken (durch den die Bewegung abgesperrt wird); ~ nut (Mech.) die Stellmutter, Gegenmutter.

cheek *s* (Techn.) der Backen, die Wange, der Seitentheil, die

Seitenwand; ~ of a block der Backen eines Blocks; (Mech.) das Gehäuse, der Kloben, die Flasche eines Flaschenzugs; ~ of a crank axle (Masch.) der Arm einer Kurbelachse; ~ of a door der Thürpfosten; ~ of a ladder der Leiterbaum; ~s *pl* of a vice die Backen, das Maul eines Schraubstocks.

chemical *a* chemisch; ~ marking system das System der Schriftgebung des chemischen Telegraphen; ~ telepraph, ~ printing telegraph der chemische Telegraph, chemische Drucktelegraph.

chemist *s* der Chemiker.

chemistry *s* die Chemie.

cheque *s* der Cheque, die Bankanweisung; ~ book das Chequebuch, Buch mit eingehefteten Bankanweisungen zum Ausreißen.

chief *s* das Haupt, der Haupttheil; ~ clerk der erste Kommis, (P & T) der Aufsichtsbeamte, der Büreauvorsteher; ~ engineer der Oberingenieur; ~ examiner (P) der Ober-Rechnungsbeamte; ~ Office, Metropolitan Office (P) das Hauptpostamt (es giebt deren drei in Großbritannien: in London, Edinburgh und Dublin).

china *s* das Porzellan; ~ clay die Porzellanerde, der Kaolin; ~ ware das Porzellan; English ~ ware das englische feine Steingut.

chinese reed *s* der Jute-Hanf, ostindische Hanf.

to chink *v* spalten, aufreißen; Geld durch den Klang untersuchen.

chinked *a* rissig.

to chisel *v* mit dem Meißel bearbeiten, meißeln; to ~ off abmeißeln, abstemmen, behauen.

chisel *s* der Meißel; das Stemmeisen; der Steinmeißel.

chloric acid *s* die Chlorsäure.

chloride *s* das Chlorid, die Chlorverbindung; ~ of calcium, calcium ~ das Chlorcalcium.

chlorine, chlorine gas *s* das Chlor, Chlorgas; ~ water das Chlorwasser.

chromate *s* das chromsaure Salz; ~ of iron der Chromeisenstein, das Chromeisenerz; ~ of lead das chromsaure Bleioryd, Chromgelb; ~ of potassa, potassium ~ das chromsaure Kali; ~ of zinc das chromsaure Zinkoryd, Zinkgelb.

chrome, bichrome, bichromate *s* das doppelt chromsaure Kali.

chrome alum *s* der Chromalaun.

chromic *a* Chrom . . . ., chromsauer; ~ acid die Chromsäure; ~ oxide das Chromoryd, Chromgrün; ~ salts *pl* die Chromorydsalze.

chromium *s* das Chrom.

chronometer *s* der Chronometer, die Seeuhr, Längenuhr; ~ escapement (Techn.) die freie Hemmung.

Chubb's lock *s* das Chubb'sche Sicherheitsschloß (nach dem Erfinder benannt).

churn jumper, churn drill *s* ein langer Bohrer, an beiden Enden mit Meißeln versehen.

cinders *s/pl* ausgebrannte Kohlen, Schlacken (am Draht).

cinnabar *s* der Zinnober, das Schwefelquecksilber.

cipher, cypher *s* (Math.) die Ziffer, das Zahlzeichen, die Null; das geheime, verabredete Zeichen; ~ key der Schlüssel zu einer Geheimschrift; ~ system das Zeichensystem; ~ telegram das Telegramm in Geheimschrift.

circle *s* der Kreis, der Zirkel; ~ divided into 60 degrees der Gradbogen; graduated ~ der Theilkreis.

circuit *s* die Kreisbewegung; der Kreislauf, der Umfang; (T) der Stromkreis, die Leitung; to place in ~ einschalten; two stations in ~ zwei Stationen in demselben Stromkreise, in einer Leitung; all wire ~ s. metallic ~; busy ~ eine besetzte Linie; circular ~ der Kreisstrom; closed ~ der geschlossene Stromkreis, Ruhestrom; combined ~ zwei oder mehrere Stromkreise so verbunden, daß die telegr. Zeichen aus dem einen in den andern übergehen; conductive ~ Draht, Apparate u. s. w., kurz Alles, was den Stromweg bildet; local ~ der Lokalkreis; die durch Relais geschlossene Ortsbatterie eines Amtes; metallic ~, all wire ~ Stromkreis aus Drähten gebildet (unter Ausschluß der Erde); Morse's ~ die Morseleitung; open ~ der geöffnete Stromkreis, Arbeitsstrom; a line worked on the open ~ plan eine mit Arbeitsstrom betriebene Leitung; telegraphic ~ die direkte Verbindung (ohne Uebertragung).

circuit breaker *s* (T) der Stromunterbrecher; der Schlüssel; automatic ~ der Selbstunterbrecher; ~ closer *s* der Stromschließer, der Schlüssel.

circular *s* das Rundschreiben; official ~ of the Post Office das Post-Amtsblatt; ~ of inquiry (P & T) der Laufzettel.

circular *a* rundförmig, kreisförmig; ~ letter das Rundschreiben; ~ saw die Kreissäge; ~ sector (Geom.) der Kreistheil.

to circulate *v* kreisförmig bewegen; umlaufen lassen, umlaufen, zirkuliren; to ~ bills Wechsel giriren.

circulation *s* der Umlauf; (P) die Spedition; bank of ~ die Girobank; ~ of bills, notes der Wechselverkehr; to put in ~ in Umlauf setzen; ~ list (P) die alphabetisch eingerichtete Leitübersicht, Speditionsliste; general ~ book (P) die allgemeine Abfertigungs-Uebersicht.

circumambient air *s* die uns umgebende Luft.

circumference *s* der Kreisumfang, die Peripherie.

claim *s* der Anspruch, die Forderung; der Antrag; acknowledged ~ die anerkannte, liquide Forderung; ~ on mortgage die hypothekarische Forderung; ~s *pl* die Rechte, Ansprüche; die Zölle; ~ form (P) das Antragsformular.

clamp *s* die Klampe, Klammer, Klemmvorrichtung; die Kluppe; ~ vice die Kluppe; ~ screw die Preßschraube.

clasp *s* der Haken; die Klammer; ~ nut die Schraubenmutter.

claw *s* die Klaue; der Haken; die Drahtzange; devil's ~ die Steinklaue, das Kropfeisen; ~ bar das Brecheisen; ~ hammer der Hammer mit gespaltener Klaue, mit Zangenvorrichtung; ~ wrench der Nagelzieher, die Beißzange.

clay *s* der Thon; china ~, chinese ~, porcelain ~ die Porzellanerde, der Kaolin; layer of ~ das Thonlager; potter's ~ der Töpferthon ~ cylinder der Thoncylinder.

clayey *a* thonig, lehmig; ~ ground (soil) der Lehmboden.

to clean *v* reinigen, säubern; to ~ a battery throughout (T) eine Batterie auseinandernehmen.

to clear *v* hell machen; reinigen; to ~ the letter boxes (P) die Briefe aus den Briefkasten abholen; to ~ the balance den Saldo in's Reine bringen; to ~ a port aus einem Hafen auslaufen; to ~ the tract (Telegr.-Bau) die Baustrecke aufräumen; ~ed out & all right (T) „Alles in Ordnung".

clear *a* hell, klar; to be ~ of frei sein von . . .; to keep ~ from frei halten von . . .

clearance *s* der Spielraum; die Zollquittung, Ausklarirung vom Zollamt (von Paketen gesagt, die verzollbare Gegenstände enthalten); (Masch.) die Freimachung; (P) die Abholung der Briefe aus den Briefkasten.

cleared out & all right s. to clear.

clearing house *s* das Abrechnungsbüreau (für den gegenseitigen Verkehr von Eisenbahnen u. s. w.).

cleat *s* (Techn.) die Klampe.

cleft *s* der Sprung, Spalt, die Ritze.

to clench *v* vernieten, verklinken, s. to clinch.

clench bolt *s* (Mech.) der Klinkbolzen; ~ nail der Schraubennagel.

clerk *s* der Kommis, der Handlungsgehülfe; government ~ der Staatsbeamte, Regierungsbeamte, Beamte im engl. Zivildienst; ~ (1st class, 2nd class) der Zivilbedienstete im Alter von 18 bis 24 Jahren; ~ at the counter (P) der Annahmebeamte; issuing ~ der Beamte, der Rechnungen anweist; managing ~, head ~ der Büreauchef; paying ~ der auszahlende (Kassen=)Beamte; town ~ der Stadtschreiber, Stadtsyndikus; ~ of the ruler der Kanzlist.

to click *v* anschlagen (vom Morsehebel gesagt).

click *s* der Schlag, der Anschlag (des Morsehebels); (Mech.) der Sperrhaken, Sperrkegel, die Sperrklinke; ~ & ratchet-wheel (Masch.) die Sperrvorrichtung, das Gesperre; ~ with ratchet-crown-wheel die Sperrklinke mit Kronensperrrad; ~ spring die Sperrfeder; ~ wheel (Masch.) das Sperrrad; ~ wire, ~ steel der Sperrkegeldraht.

to climb *v* klettern.

climbers *s/pl* die mit Sporen versehenen Kletterstiefel zum Ersteigen der Telegr.=Stangen.

climbing appliance, ~ contrivance *s* die Klettervorrichtung; ~ iron das Klettereisen, Steigeisen.

to clinch *v* (Mech.) vernieten, verklinken; to ~ a bolt einen Bolzen umnieten, verklinken.

clinch *s* (Mech.) die Klinke, Klinkung; ~ bolt der Klinkbolzen, vernietete Bolzen; ~ ring der offene Ring mit übereinander tretenden Enden.

clinched *a* vernietet; ~ & riveted niet= u. nagelfest.

clincher *s* die Klammer; (T) der Kabelhalter; ~ nail der Schraubennagel.

to clink *v* reißen, bersten (von Metallen).

clinker *s* der Klinker (hartgebrannte Backstein); die Schlacke, Steinkohlenschlacke (beim Metall=Schmelzprozeß).

to clip *v* beschneiden, abschneiden, stutzen; clipped signals (T) abgerissene Zeichen.

clip *s* das Band, Eisenband, der eiserne Reifen; das Ziehband.

clipper *s* die Schneidzange, Scheere; der Klipper (schnellsegelndes Schiff).

clippers *s/pl* die Scheere.

clock *s* die Schlaguhr, Penduluhr (jede nicht zum Tragen eingerichtete Uhr); sidereal ~ die Uhr nach Sternzeit; standard ~ (T) die Normaluhr; watchmaker's ~ der Regulator; weight-moved ~ die Gewichtuhr; ~ alarm der Wecker; ~ movement das Uhrwerk; ~ work das Uhrwerk; ~ work lamp die Carcellampe, Lampe mit Pumpenwerk.

to close *v* schließen, verschließen; to ~ an account eine Rechnung abschließen; to ~ the circuit, the current (T) die Leitung, den elektr. Strom schließen.

close *s* der Schluß; ~ of an account der Rechnungsabschluß.

closing hour *s* (P & T) die Schlußstunde; ~ time die Schlußzeit; self ~ *a* mit Selbstverschluß versehen.

clout nail *s* der Bandnagel, Plattnagel.

clutch *s* (Mech.) der Griff, die Klaue, die Kuppelung.

coach *s* die Kutsche, der geschlossene Wagen zur Personenbeförderung; express ~ der Schnellwagen; ~ & four die vierspännige Kutsche; mail ~ die Post= u. Personenkutsche; ~ & pair die zweispännige Kutsche; stage ~ der Eilwagen; ~ box der Kutscherbock; ~ fare das Fuhrlohn; ~ hire das Fahrgeld, der Kutscherlohn; ~ shed die Wagenremise; ~ office das Passagier=Einschreibebüreau; ~ stand die Haltestelle für Wagen; ~ wrench (Mech.) der Universalschraubenschlüssel.

to coagulate *v* (Chem.) gerinnen, zusammenlaufen.

to coak *v* mit Zapfen aneinanderfügen, verzapfen.

coak *s* der Zapfen; sunk ~ das Zapfenloch.

coal *s* die Kohle; gas ~ die zu Leuchtgas taugliche Kohle, Gaskohle; ~ oil das Theeröl; ~ tar der Steinkohlentheer.

coarse *a* grob, (von Metallen) roh; ~ grained grobkörnig (von Metallen).

to coat *v* überziehen, überstreichen, bekleiden, belegen (mit Guttapercha u. s. w.).

coat s (Techn.) die Schicht, der Ueberzug; ~ of arms das Wappen; ~ of paint der Anstrich; first ~ der erste Bewurf (eines Gebäudes); ~ of plaster der Gipsbewurf; ~ of tin die Verzinnung.

coating s der Ueberzug, die Schicht, die Bedeckung; ~ of a Leyden jar die Belegung einer Leydener Flasche.

cock s (Techn.) der Hahn, Krahn; ~ of a balance das Züngelein einer Waage; blow-off ~ (Dampfmasch.) der Abblasehahn; delivery ~ der Ablaßhahn; feed ~ der Speisehahn; stop ~ der Hahn, Krahn; ~s & valves pl (Masch.) das Klappenwerk.

cocoon s der Kokon, das Gehäuse des Seidenwurms; ~ fibre devoid of torsion der Kokonfaden ohne Torsion.

code s (T) das konventionelle Alphabet od. Phrasenbuch für Telegraphie, die Geheimschrift; ~, ~ letters (T) das Rufzeichen; ~ of commerce das Handelsgesetzbuch; Morse ~ das Morse-Alphabet; single needle ~ das Alphabet des Einnadel-Telegraphen; ~ book (T) das amtliche Verzeichniß der Telegraphen-Anstalten; ~ name (T) die abgekürzte Bezeichnung einer Telegraphen-Anstalt; ~ of signals das Flaggensignalsystem (zwischen Schiffen auf See); ~ time (T) die Aufgabezeit. Sie wird in England in Buchstaben ausgedrückt: A = 1 Uhr (Vorm. od. Nachm), B = 2 Uhr, C = 3 Uhr u. s. w. Diese Buchstaben bedeuten gleichzeitig Zeitabschnitte von je 5 Min.; z. B. AA = 1 Uhr 5 Min., AB = 1 Uhr 10 Min. u. s. w. — Die 4 Zwischenminuten jedes Abschnittes werden durch die Buchst. R, S, W, X bezeichnet, z. B. AR = 1 Uhr 1 Min., BS = 2 Uhr 2 Min., MAR = 12 Uhr 6 Min. u. s. w. Diese Zeitangabe wird im Kopf des Telegramms gleich nach dem *prefix* (s. denf.) gegeben.

coefficient s der Koeffizient, Mitwirker; ~ of linear expansion der lineare Ausdehnungs-Koeffizient.

coercive force s (Phys.) die Koerzitivkraft.

coffer dam s der Fangdamm (beim Wasserbau).

to cog v a wheel ein Rad mit Zähnen versehen.

cog (Masch.) der Wellbaumen, Hebedaumen; ~ of a wheel der Zahn eines Rades, der Kamm; ~ shaft (Masch.) die Daumenwelle; ~ tooth (Masch.) der Kamm, eingesetzte Zahn; ~ wheel das Kammrad, Zahnrad mit eingesetzten Zähnen; ~ wheel of a crane das Krahnenrad.

cogged a (Masch.) gezahnt.

cohesion, coherency s die Kohäsion, die Anziehungskraft der Theile eines Körpers; absolute ~ die Zähigkeit.

to coil v (Techn.) spiralförmig aufwinden; to ~ a cable ein Kabel aufschießen, in einen Ring zusammenlegen; to ~ the wire (T) den Draht aufrollen.

coil s die Spirale; die Drahtrolle; die Spule; der Ring; die Rolle; (T) die Umwindungen, die Spule, Induktionsrolle; directing ~ die richtende, induzirende Spule; induction ~ (T, Elektriz.) die Induktionsrolle; primary ~ die Hauptspule, primäre Spule; secondary ~ die Nebenspule, sekundäre Spule; ~ of wire der Drahtring, die Drahtrolle.

coin s die Münze (Geldstück); ~, quoin der Keil, die keilförmige Unterlage; base ~, light ~ die geringhaltige, schlechte Münze; counterfeit ~ die falsche Münze; worn-out ~ blindes Geld; ~ assorting machine die Münzensortirmaschine; ~ counter der mechanische Apparat zum Geldzählen; ~ letter (P) der Brief mit Geldinhalt, der eingeschrieben werden muß (nicht Geldbrief im deutschen Sinne).

to coincide v übereinstimmen.

coincidence s die Uebereinstimmung.

coke s, **cokes** s/pl der Koke, die Kokes, ausgeglühte, entschwefelte Steinkohle; gas ~ der Gaskoke.

cold-short iron s das kaltbrüchige Eisen.

collar s (Techn.) der Ring, Reifen, die Hülse; das Querstück, der Querbalken; (Masch.) die Pfanne, das Zapfenlager; der Kragen, die Flantsche; ~, washer das Schraubenmutterblech, die Unterlagscheibe; ~ of an arbor or beam das Zapfenlager; ~ of a bolt der Hals eines Bolzens; leather ~ die Manschette zwischen Kolben und Zylinder einer hydraulischen Presse; ~ band das Schelleisen, Eisenband; ~ beam der Querbalken; ~ pin (Masch.) der Vorstecknagel; ~ plate das Schraubenfutter.

to collate v (T) kollationiren, vergleichen, berichtigen.

collation s (T) die Kollationirung, Vergleichung.

to collect v sammeln, einkassiren; to ~ letters (P) die Briefe aus den Briefkasten sammeln; to ~ gases (Chem.) Gase auffangen; to ~ in debts ausstehende Schulden einfordern.

collecting plate s (Elektriz.) die untere Platte eines elektrischen Kondensators.

collection s die Sammlung; ~ of letters, ~ from the letter boxes (P) die Briefeinsammlung aus den Briefkasten.

collector s der Sammler; (T) der Einsammler der angekommenen und abtelegraphirten Telegramme an den Apparaten; (Phys., Elektriz.) der Sammelapparat; ~ of taxes der Zolleinnehmer.

collet s (Techn.) die Buchse, Hülse.

colliery s die Kohlenzeche, Kohlengrube, das Kohlenbergwerk; ~ explosion der Grubenbrand, die schlagenden Wetter in Kohlenbergwerken.

colored glass s (Eisenb.) die farbige (Signal-) Scheibe, das Licht.

colstaff s der Hebebaum, Wuchtbaum.

column s die Säule; (Phys.) die Luftsäule, Wassersäule; mercurial ~ die Quecksilbersäule.

comb s (Elektriz.) der Saugkamm an der Elektrisirmaschine; ~ protector (T) der Spitzenblitzableiter.

combination s die Verbindung, Vereinigung, Mischung; chemical ~ die chemische Verbindung; ~ instrument (T) der amerikanische Typendruck-Apparat von Phelps; ~ of words (T) die Wortzusammenziehung.

combining a (Chem.) kombinirend; ~ capacity, ~ atomicity die Atombindekraft der Elemente; ~ proportion (Chem.) das Mischungsverhältniß.

combustible a brennbar, entzündlich; readily ~ leicht brennbar; slow ~ schwer brennbar.

commander s (Mech.) die Handramme.

commerical s (P) der weiße Briefumschlag mit eingedrucktem Werthzeichen, der von der englischen Post verkauft wird.

commercial a geschäftlich; ~ papers s/pl (P) Geschäftspapiere; the ~ union of Germany der deutsche Zollverein.

commission s die Kommission, der Auftrag, die Provision; das Patent, Ernennungsdekret; ~ on money order transactions (P) die Gebühr für angenommene Postanweisungsgelder.

committee s das Komité; visiting ~ die Untersuchungskommission.

commotion s (Elektriz.) die Erschütterung.

communication s die Mittheilung; die Verbindung, der Zusammenhang; ~ of motion (Mech.) die Fortpflanzung der Bewegung.

communicator s (T) der Zeichengeber; (Masch.) die Zwischenmaschine, das Vorgelege-, Zwischengeschirr.

commutator s (T) der Stromumkehrer, Umschalter; (Synonyma sind:) break, contact breaker, circuit changer, switch, rheotrope.

companion hatch s der Ueberbau über der Kajütentreppe auf Kauffahrteischiffen; ~ ladder, ~ way, ~ stairs pl die in die Kajüte führende Treppe, Kajütentreppe.

company s die Kompagnie, Han-

belsgesellschaft; joint-stock ~ die Aktienkompagnie.
compartment s die Abtheilung, das Feld; das Fach, der Behälter; (Eisenb.) das Coupé.
compass s der Kompaß, die Bussole; der Umfang; declination ~ die Deklinationsbussole, das Deklinatorium; inclination ~ die Inklinationsbussole; tangent ~ die Tangentenbussole.
compasses s/pl, a pair of ~ der Zirkel.
compassing s das Abmessen, Abzirkeln.
to compensate v kompensiren, ausgleichen, entschädigen.
compensation s die Ausgleichung; der Schadenersatz, to give, to grant ~ to the amount of ... Schadenersatz bis zur Höhe von ... leisten; die Gegenrechnung, Abrechnung; (Mech.) die Ausgleichung; ~ current (T) der Ausgleichungsstrom; ~ method (Phys.) die Kompensationsmethode.
competent a kompetent, zuständig; ~ judge, ~ party der Sachverständige.
component s der Bestandtheil, die Komponente; ~ forces, components s/pl (Phys.) die Seitenkräfte; ~ parts pl die Bestandtheile.
composing stick s (T) der Setzerstock (für das Alphabet des ersten MA).
composite a zusammengesetzt; ~ wire (T) Draht mit verschiedenem Durchmesser, in der Mitte zwischen 2 Stützpunkten dünner, nach den Stützpunkten zu dicker.
composition s die Zusammensetzung, Mischung; ~ by volume (by weight) die Zusammensetzung nach dem Volumen (nach dem Gewicht).
compound s die (chemische) Verbindung; ~ cell (Phys., T) das durch Nebeneinanderschaltung mehrerer Elemente gebildete (Element); ~ engine, ~ steam engine die Woulf'sche Dampfmaschine, Hoch- u. Niederdruckmaschine; ~ interest Zinseszinsen; ~ magnet das Magnetstabbündel (mehrere Magnete mit ihren gleichen Polen zusammengefügt); ~ metal die Legierung; ~ wire Draht bestehend aus Gußeisen-Draht, der dann verzinnt und mit einem Kupferüberzug versehen, schließlich durch ein Bad von geschmolzenem Zinn gezogen ist; ~ words pl zusammengesetzte Worte, Zusammenziehungen von zwei od. mehreren Worten.
compressed a komprimirt; ~ air die komprimirte Luft; ~ airengine die mit komprimirter Luft getriebene Maschine.
compresser s, wooden ~ der Kabelhalter (beim Löthen).
compressible a zusammendrückbar.
compressibility s die Zusammendrückbarkeit.
compressing engine, ~ machine s die Druckmaschine; ~ strain, ~ strength s. compressive.
compression s der Druck, die Zusammendrückung, Zusammenpressung.
compressive a zusammendrückend, pressend; ~ strain die Druckspannung; ~ strength die rückwirkende Festigkeit, Druckfestigkeit.
compressor s die Presse; air ~ der Luftkompressor, die Luftkompressionsmaschine (der Apparat zur Erzeugung komprimirter Luft).
to comprise v within the same account in ein u. dieselbe Rechnung bringen.
comptroller, controller s der Kontroleur; der Rechnungsrevisor; ~ of the Money Order Office der Vorsteher des Postanweisungsamts; ~ of Postal Telegraphs der Vorsteher eines Haupt-Telegraphen-Amts.
computation s die Berechnung; der Kostenüberschlag.
to compute v berechnen, den Preis bestimmen; ~d tax die Durchschnittstaxe.
concatenation s eine Reihe von Kettengliedern.
concave a konkav, ausgehöhlt; ~ lens (Opt.) die Hohllinse; ~ mirror der Hohlspiegel, Brennspiegel.
to concentrate v in einen Mittelpunkt vereinigen; verdichten,

konzentriren; verdicken; ~d acid (Chem.) die konzentrirte Säure.

concentric, concentrical *a* konzentrisch, einen gemeinsamen Mittelpunkt habend; ~ engine, ~ steam engine die rotirende Dampfmaschine.

concession *s* die Konzession, Bewilligung.

concessionary *s* der Konzessionar.

concrete *s* der Beton, der Guß- od. Steinmörtel (mit Wassermörtel angemacht); asphalte ~ der Asphaltbeton.

concrete *a* verdichtet, fest; to become ~ sich verdichten, gerinnen.

to condensate, to condense *v* kondensiren, verdichten; condensed water das destillirte Wasser.

condensator *s* (Phys.) der Elektrizitäts-, Wärmesammler; der Kondensator.

condenser *s* (Phys., Techn., Elektriz.) der Kondensator, der Ansammlungsapparat; outer (inner) coating of the ~ (Elektriz.) die äußere (innere) Bekleidung des Kondensators; sliding ~ der Schleifkondensator; spark ~ der Funkensammler.

condensing lens *s* die Sammellinse; ~ plate (Elektriz.) die obere Kondensatorplatte.

to conduct *v* leiten, führen; (Phys., Elektriz.) leiten.

conductibility *s* (Phys.) die Leitungsfähigkeit.

conducting power, ~ property *s* (Phys.) das Leitungsvermögen; ~ wire (*T*) der Leitungsdraht.

conduction *s* (*T*) die Leitung (die Führung des Stroms von einem Punkt zum andern); surface ~ (Elektriz.) die Oberflächenleitung.

conductivity *s* (Phys.) das Leitungsvermögen; specific ~ das spezifische Leitungsvermögen; ~ resistance der Leitungswiderstand.

conductor *s* der Leiter eines Baues, Bauführer; (Phys., Elektriz.) der Leiter, der Leitungsdraht; lightning ~ der Blitzableiter; non-~ der Nichtleiter; railway ~ der Eisenbahnschaffner.

conduit *s* (Techn.) die Leitung, die Wasserleitung durch Röhren, der Abzug; ~ of gas die Gasleitung; ~ of pipes die Röhrenleitung; ~ water ~ die Wasserleitung; ~ pipe das Leitungsrohr.

cone *s* (Geom., Techn.) der Kegel; blunt ~ der stumpfe Kegel; ~ gear (Masch.) die konische Verzahnung, Winkelverzahnung; ~ wheel, conical wheel das Kegelrad, konisches Rad.

coned *a* (Masch.) zugespitzt, spitz, kegelförmig.

conic, conical *a* (Geom.) kegelförmig, konisch; conical gearing (Masch.) die konische Verzahnung, Winkelverzahnung; conical wheel (Masch.) das konische Rad, Kegelrad.

to connect *v* verbinden, zusammenfügen; (Masch.) koppeln; to ~ up (*T*) einschalten.

connection *s* die Verbindung; ~ in trade die Handelsverbindung.

connector *s* der verbindende Theil; (*T*) die Klemmschraube; (Eisenb.) die Wagenkuppelung.

consignatary *s* der Depositar.

consignee *s* derjenige, an welchen Waaren konsignirt sind, der Empfänger; der Faktor.

consigner, consignor *s* der Waareneinsender, der Deponent.

consignment, consignation *s* die Uebersendung, Konsignation; ~ of goods die Waarensendung; ~ of (in) specie die Baarsendung.

constant *a* beständig, konstant; ~ battery (Elektriz.) die konstante Batterie; ~ current (Elektriz.) der konstante Strom; ~ quantities *pl* (Math.) konstante Größen.

constant *s* die Konstante (einer Batterie, eines Galvanometers, d. h. der konstante Leitungswiderstand).

constituent *a* wesentlich, hauptsächlich; ~ parts *pl* die Bestandtheile; ~ party der Auftraggeber.

to construct *v* bauen, erbauen; (Math.) konstruiren.

construction *s* die Konstruktion, die Erbauung, der Bau; cost of ~ die Baukosten; railway ~ der Eisenbahnbau; ~ way die für die Dauer eines Eisenbahnbaues zu Bauzwecken angelegte Baulinie.

contact *s* der Kontakt, die Berührung; (*T*) der Kontaktfehler (der Strom der einen Leitung geht auf eine andere über); full ~, metallic ~ die metallische Berührung (die Drähte hängen zusammen, od. sind durch einen dritten Draht verbunden); intermittent ~ zeitweiser Kontakt (durch Wind u. f. w.); metallic ~ dasf. wie full ~; partial ~ zeitweiser Kontakt (fremde, schlechtleitende Körper hängen in den Drähten; schlechte Erde, mangelhafte Isolation); weather ~ Berührung der Leitungen (in Folge schlechten Wetters, Nebels, Regens u. f. w.); ~ breaker (Elektriz., *T*) der Kontaktbrecher; ~ maker (*HA*) der Körper des Schlittens; ~ piece das Kontaktstück; ~ pin der Kontaktstift; ~ point der Berührungspunkt; ~ theory die (Volta'sche) Berührungstheorie, Theorie des Galvanismus.

content *s*, **contents** *s/pl* der Inhalt, Gehalt; cubic ~ der Kubikinhalt; solid ~ der körperliche Inhalt; das Volumen; superficial ~ der Flächeninhalt.

contingencies *s/pl* die Kosten für unvorhergesehene Fälle, Nebenausgaben.

continuation *s* die Fortsetzung; (Techn.) der Fortsatz, Ansatz.

continuity *s* der metallische Zusammenhang (einer Telegraphen-Leitung), die Stromfähigkeit.

contract *s* der Kontrakt, Vertrag; in ~ in Akkord, in Entreprise; to give, to let out a work in ~ eine Arbeit in Akkord geben.

contractor *s* der Kontrahent; der Lieferant; der Unternehmer.

contrate wheel *s* (Mech.) das Kronrad, das Steigrad.

controller *s* f. auch comptroller; ~ of stores (*T & P*) der Materialienverwalter.

conventional *a* signs (*T*) verabredete Zeichen.

conversion *s* die Umwandlung; die Geldumrechnung; ~ of coins die Umschmelzung der Münzen; ~ of steel die Zementirung des Stahls; table of ~ die Umrechnungstabelle.

to convert *v* umwandeln; to ~ steel Stahl zementiren; to ~ timber into slabs Stammholz zurichten, bearbeiten; ~ed steel der Brennstahl.

converting *s* das Zementiren, Zementirverfahren; ~ chest der Zementirkasten; ~ process das Bessemerverfahren.

to convey *v* befördern, transportiren, versenden.

conveyance *s* die Beförderung, der Transport; das Fuhrwerk; die Fahrgelegenheit; charges of ~ die Transportkosten; ~ of despatch die Eilfracht, das Eilgut; letter of ~ der Frachtbrief; means of ~ das Transportmittel; mode of ~ die Versendungsart; sea ~ der Seetransport.

cope *s* die Bekrönung, Ueberdachung (eines Gebäudes); die Mauerkappe.

copper *s* das Kupfer; bar ~ das Stangenkupfer; sheet ~ das Kupferblech; sulphate of ~ das schwefelsaure Kupferoxyd, der Kupfervitriol; ~ bit der Löthkolben; ~ plate die Kupferplatte; das Kupferblech; der Kupferstich; ~ solder das Kupferloth; ~ strand der Kupferstrang, der aus verseilten Drähten bestehende Leiter im Kabel.

to copy *v* kopiren, abschreiben; to ~ fair ins Reine schreiben.

copy *s* die Kopie, Abschrift; das Exemplar.

copying clerk *s* der Kopist; ~ ink die Kopirtinte; ~ telegraph der Kopirtelegraph.

core *s* das Innerste einer Sache, der Kern; ~ of a cable der metallische Kern eines Kabels, die Kabelseele; ~ of soft iron der weiche Eisenkern; ~s *pl* of an electromagnet die Kerne eines Elektromagneten; ~ joint die Kabellöthstelle.

corner band *s* das Winkelband; ~ pole, ~ post (*T*) die Winkelstange; ~ stone der Eckstein.

to correct *v* berichtigen; ~ing cam (*HA*) der Korrektionsbaumen;

~ing telegram das Berichtigungstelegramm; ~ing wheel (HA) das Korrektionsrad.
correspondence s die Korrespondenz; der Briefwechsel; (P) die Briefpostsendungen.
to corrode v (Chem.) zerfressen, anfressen.
corrosion s das Anfressen, Einfressen.
corrugated a gerippt, gewellt; ~ bearing das Ringlager einer Schraubenwelle; ~ plate, ~ iron das geriefte, gewellte Eisenblech.
cost s der Preis, die Kosten; ~s pl die Kosten, Ausgaben; ~ of construction die Baukosten; first ~, prime ~ die Anschaffungskosten, der Selbstkostenpreis; ~ of maintenance die Unterhaltungskosten; ~ for repair die Reparatur- (Wiederherstellungs-)kosten; ~ account, ~ sheet die Selbstkostenberechnung; ~ price der Selbstkostenpreis.
cottar, cotter, cottrel s der Keil, Splint; (Masch.) der Vorstecker, Keil.
cottrel s. cottar.
couch s (Techn.) die Schicht, die Lage.
counter s (P & T) der Tisch in engl. Postämtern, über den weg mit dem Publikum verhandelt wird; der Schalter; ~ clerk (P & T) der Schalterbeamte; ~ current (T) der Gegenstrom; ~ foil s. counterfoil; ~ list (P) das Packetannahmebuch; ~ man dasj. wie ~ clerk; ~ nut die Stellmutter; ~ proof die Gegenprobe, Nachprobe; ~ woman der weibliche Annahmebeamte.
counterfoil form s (P) das Formular für die im Wege des Durchdrucks hergestellte Annahmeliste einzelner nachzuweisender Gegenstände, mittels welcher die Überweisung dieser Gegenstände an eine andere Dienststelle erfolgt; ~ list (P) die im Wege des Durchdrucks hergestellte Annahmeliste einzelner nachzuweisender Gegenstände u. s. w. wie vorstehend.
to countersink v a rivet ein Niet einlassen, versenken.

countersunk a versenkt; ~ bit der Rundstahl; ~ nail der Nagel mit versenktem Kopf.
to couple v kuppeln, koppeln; paarweise verbinden.
coupled a poles pl (T) gekuppelte Stangen.
coupling s die Kuppelung, Verkuppelung; ~ chain die Kuppelungskette; ~ screw die Kuppelschraube, Verbindungsschraube.
course s der Lauf; (Techn.) die Reihe, der Kurs eines Schiffes; der Wechselkurs; ~ of the current (T) der Stromweg.
to cover v decken, bedecken; überziehen, belegen; to ~ money into the Treasury (Am) Geld in die Kasse des Schatzamts abliefern.
cover s die Decke, der Deckel, der Ueberzug; der Briefumschlag.
covering s das Bedachen, Decken; die Decke, das Futteral; der Ueberzug; guttapercha ~ machine die Guttapercha (Kabel-) Umspinnungsmaschine; ~ plate, ~ slab die Deckplatte.
crab s (Masch.) die vertikale, stehende Winde, die Erdwinde; ~ with communicator der Haspel mit Vorgelege, die Vorgelegewinde; ~ bar der Hebebaum; ~ engine der Dampfgöpel.
to crack v reißen, aufreißen; rissig werden, bersten.
crack s der Riß, Sprung, der Bruch; (Elektriz.) das Knistern; ~ in steel der Härteriß; ~ in wood der Windriß.
cracked a rissig.
to cramp v mit Klammern befestigen, einklammern; mit Krampen befestigen; anhaspen.
cramp s die Klammer; der Holzhaken; die Zarge; die eiserne Klammer; iron ~ das Zugband, der Klammerhaken; ~ iron die eiserne Klammer, Krampe; ~ iron for wood die Holzkrampe, das Bankeisen.
crane s (Techn.) der Krahn, Aufzug; post ~ der Ständerkrahn; steam ~ der Dampfkrahn.
crank (Mech.) die Kurbel; der Bügel; ~ of a bell, bell ~

14

der Glockenschwengel, Glockenarm; double ~ der doppelte Krummzapfen, die gekröpfte Welle; ~ axle die Kurbelachse; ~ handle die Handhabe einer Kurbel; ~ lever der Kurbelhebel, Kurbelarm, die Ziehklammer; ~ shaft die Kurbelwelle, gekröpfte Welle; ~ of a wheel der Schwengel an einem Rade.

to craze v (Techn.) rissig werden, Risse bekommen.

to credit v kreditiren; gutschreiben; to ~ in account in Rechnung kreditiren; to ~ by balance per Saldo kreditiren.

creosote s das Kreosot; coal-tar ~ das Steinkohlentheerkreosot; wood-tar ~ das Holztheer= (eigentliche) Kreosot.

creosoting s die Zubereitung (von Holz) mit Steinkohlentheeröl.

to cross v kreuzen, kreuzweise legen; sich kreuzen; to ~ wires (T) an Stelle einer gestörten Leitungsstrecke das Stück einer Parallelleitung einschalten.

to cross-cut v in die Quere schneiden, durchschneiden.

cross s das Kreuz; (T) die Kreuzung, Berührung von Drähten; weather ~ (T) Berührung der Drähte in Folge schlechten Wetters; die Verschlingung; ~ arm (T) der Querträger; ~ bar (T) die Querschiene; der Querriegel; ~ country circuit (T) die Querverbindung (telegr. Verbindung auf Umwegen); ~ feet pl das Kreuzholz (Befestigung der Strebe im Erdboden); ~ grain die Hirnseite des Holzes; ~ grained timber quer durchgeschnittenes Holz; ~ leakage (T) die Nebenschließung in Folge einer Drahtverschlingung; ~ piece das Querstück, das Verbindungsstück der Eisenkerne des Elektromagneten; ~ post bag (P) der direkte Briefbeutel zwischen zwei Provinzial-Postämtern ohne Vermittlung eines Haupt=Postamts; ~ post letters pl Briefe, die in einem ~ post bag (s. dens.) befördert werden; ~ road der Kreuzweg, die Wegscheide; ~ stay die Kreuzstrebe; ~ way der Kreuzweg; ~ way of timber das Hirnholz.

crossing s (Eisenb.) die Kreuzung; der Kreuzweg, Straßenübergang; level ~ (Eisenb.) der Uebergang in gleicher Höhe, Niveauübergang.

crotch, crotchet s der Haken.

crow bar s (Techn.) das Brecheisen, die Brechstange.

crowded a line (T) die mit vielen Leitungen belastete Linie.

crown wheel s das Kronrad.

crucible s der Tiegel, Schmelztiegel.

cryptograph, cryptography s die Geheimschrift; ~ code das Buch, das die Geheimschrift enthält; ~ key der Schlüssel zur Geheimschrift.

cubic foot s der Kubikfuß.

culvert s die Abzugsrinne, der überwölbte Wasserabzug, Durchlaß; wooden ~ die Holzrinne.

cuneate, cuneiform a keilförmig.

cup s der Becher; porous ~ der poröse Becher (im Batterie=Element).

cupboard s der Schrank; (T) der Batterieschrank.

cupric a oxide das Kupferoxyd; ~ sulphate das schwefelsaure Kupfer, Kupfersulphat.

cuprous a oxide das Kupferoxydul.

curb sender s der Curb sender (von W. Thompson u. Fl. Jenkin erfundener Apparat, der bei der Kabel-Korrespondenz die Entladung des Kabels beschleunigt); ~ stone der Prellstein, Bordstein.

currency s der Kurs, die Währung, der Geldumlauf.

current s der Strom, die Strömung; battery ~ der Batteriestrom; ~, electric ~ der elektrische Strom; constant ~ (Elektriz.) der konstante Strom; continuous ~ (Elektriz.) der ununterbrochene Strom; derived ~ (Elektriz.) der Zweigstrom; double ~ working (T) das Arbeiten mit Wechselströmen; double ~ key (T) die Wechselstromtaste; induced ~ der induzirte Strom; inducing ~, inductive ~ der induzirende Strom, Induktionsstrom; inverse ~ der entgegengesetzte Strom; line ~ der Linienstrom; line ~ working die mit Arbeitsstrom (ohne Relais) betriebene Leitung; local ~ der Lokalstrom; magneto-electric ~ der magneto-elektrische Strom; marking ~ der positive Strom des Wheatstone-Senders, der die Punkte auf dem Streifen macht; open ~ der durchgehende Strom; primary ~ der primäre Strom; secondary ~ der sekundäre Strom, Polarisationsstrom; spacing ~ der Strom, der die Zwischenräume auf dem Streifen bewirkt; terrestrial ~ der Erdstrom.

curvature s die Biegung, Krümmung; radius of ~ der Krümmungsradius.

curve s die krumme Linie, Krümmung, Kurve; radius of ~ der Kurvenradius; ~ line die Kurve.

cushion s das Polster, das Kissen; (P) die Stempelunterlage; ~ of an electrical machine das Reibkissen, Reibzeug.

custom s der Zoll; ~s duty die Verzollungsgebühr, Zollgebühr; ~s inwards pl der Eingangszoll; ~s outwards pl der Ausfuhrzoll; ~ free a zollfrei; ~ house das Zollamt, Zollhaus; ~ house charges pl die Zollspesen; ~ house officer der Zollbeamte.

to cut v schneiden, zerschneiden, behauen; to ~ off the connection (T) die Verbindung abschneiden; to ~ out (T) ausschalten.

cut s der Schnitt, (Terrain-) Einschnitt; der Graben; der Brückendurchlaß; ~-out (T) der Umschalter, Ausschalter.

cutting s (Eisenb.) der Einschnitt, Durchstich; ~ nippers, ~ pliers pl die Drahtzange, Schneidzange.

cylinder s der Zylinder, die Walze; die Rolle; die Welle.

cylindric, cylindrical a zylindrisch, walzenförmig.

cypher s. cipher.

D.

dam s der Deich, Damm, Wasserbrecher; ~ of a harbor der Hafendamm; coffer ~ der Fangdamm (beim Wasserbau).

to damage v beschädigen, verderben.

damage s die Beschädigung; ~ of a parcel (P) die Beschädigung eines Postpacketes; ~s pl der Schadenersatz; ~ certificate das Schadenattest.

damper s (Techn.) die Klappe; das Register, die Zugklappe; der Dämpfer, Tondämpfer.

danger! (Eisenb.) Gefahr! Halt! (Signal für den Lokomotivführer).

danger light s (Eisenb.) die Signallaterne; ~ signal, ~ whistle das Nothsignal.

dash s der Strich, Gedankenstrich; (T) der Strich des Morse-Alphabets.

date s das Datum; die Verfallzeit eines Wechsels; bearing ~ de dato; long ~ die lange Sicht; ~ stamp, dated stamp (P) der Tagesstempel.

datum s das Gegebene, der Stoff, Vorwurf, die Grundlage; ~ line die Basis, Vergleichungslinie, Grundlinie (beim Vermessen).

day book s das Journal; ~ work das Tagewerk, die Arbeit im Tagelohn; ~ worker der Tagelöhner.

dead a todt; matt, glanzlos; ~ angle der todte Winkel; ~ beat, ~ beat escapement die ruhende Hemmung; ~ beat galvanometer das aperiodische Galvanometer (eine Art Spiegelgalvanometer von Thomson); ~ centre der todte Punkt; ~ earth (T) die tödtende Erde (ein Ende der gerissenen Leitung liegt im Wasser oder auf feuchter Erde); ~ letter (P) der unbestellbare Brief; ~ letter office (P) das Büreau für unbestellbare (Rück-)Briefe; ~ level die ebene, horizontale Fläche; ~ load s. load; ~ weight die Last; das todte

deal — to deliver

Kapital; die Schiffslast; (Eisenb.) die todte Last; ~ wood das faule Holz.

deal *s* das Fichtenholz, Föhrenholz, Rothtannenholz; das Brett, die Diele; fir ~ das Fichtenholz; das Fichtenbrett; red ~ das Kiefernholz; white ~ das Tannenholz.

to debate *v* an account eine Rechnung anfechten.

debenture *s* der Schuldschein, die Obligation; (am Zollamt) die Obligation über den Rückzoll, der Rückzollschein.

debentured *a* goods *pl* Waaren, auf welche der Rückzoll bezahlt worden ist.

to debit *v* debitiren, belasten.

debit *s* das Debet, das Soll; die Debetseite des Hauptbuchs.

debt *s* die Schuld; prescribed ~ die verjährte Schuld; receivable ~ die Schuldforderung.

debtor *s* der Schuldner; the ~'s side die linke Seite des Hauptbuchs.

to decarbonize, to decarburate, to decarburize *v* den Kohlenstoff entziehen, entkohlen.

to decay *v* verfallen (von Bauten); verfaulen; (Chem.) verwesen.

decay *s* der Verfall; die Zerstörung (einer Stange, eines Drahtes durch atmosphärische Einflüsse); (Chem.) die Verwesung.

declaration *s* of contents die Zollinhaltsangabe; ~ of duty die Zollangabe; ~ of urgency die Dringlichkeitserklärung; ~ of the value die Werthangabe.

declination *s* die Deklination, astronomische Abweichung; ~ of the needle, magnetic ~ die Abweichung der Magnetnadel; ~ compass die Deklinationsbussole.

declivity *s* die Abdachung, der Abfall, das Gefälle.

to decompose *v* (Chem.) zersetzen; (Phys., Mech.) zerlegen.

decomposition *s* (Chem.) die Zersetzung; (Phys., Mech.) die Zerlegung; ~ by electricity die elektrochemische Zersetzung, Elektrolyse; ~ of forces (Mech.) die Zerlegung der Kräfte.

to decorticate *v* abrinden, die Rinde abschälen.

to decrease *v* abnehmen.

decrease *s* die Abnahme.

decrement *s* die Abnahme.

deep sea cable *s* das Tiefseekabel; ~ sea sounding die Tiefseelothung.

to deepen *v* a trench einen Graben ausschachten.

deepening *s* die Vertiefung, Aushöhlung, der Tiefschnitt; die Austiefung, Abteufung.

defacing stamp *s* (P) der Entwerthungsstempel.

deferred *a* verschoben; ~ annuity *f.* annuity; ~ telegram ein gelegentlich zu beförderndes Telegramm.

deficiency *s* der Ausfall, das Defizit, der Fehlbetrag.

to deflect *v* ablenken (von der Magnetnadel).

deflection *s* die Abweichung, Biegung; die Ablenkung (der Magnetnadel).

to defray *v* the charges die Kosten bestreiten.

degree *s* (Math., Geom., Phys.) der Grad; ~ of heat der Wärmegrad; ~ below zero der Kältegrad; ~s *pl* of oxydation (Chem.) die Oxydationsstufen.

delay *s* die Verzögerung; (Eisenb.) die Verspätung; ~ of payment die Fristung, Stundung.

del credere, to stand ~ ~ Bürgschaft leisten.

delicacy *s* die Empfindlichkeit (eines Apparats, einer Waage u. s. w.).

delicate *a* empfindlich, fein (von Apparaten u. s. w.).

deliquescent *a* zerfließlich; ~ salt ein Salz, das fähig ist, aus der Luft Feuchtigkeit aufzunehmen und dadurch flüssig wird.

delinquency *s* die Dienstvernachlässigung.

to deliver *v* abliefern, einhändigen, bestellen; ~ing officer (P) der die Post übergebende Beamte; ~ing post office die Bestell-Postanstalt.

delivery s die (Brief-, Telegramm-)Bestellung; der Bestellbezirk; to live within the ~ of a post office innerhalb des Bestellbezirks einer Postanstalt wohnen (wobei es gleichgiltig ist, ob die Bestellung gebührenfrei oder gebührenpflichtig erfolgt); free ~ gebührenfreie Bestellung; private box ~ die Fachabholung; town postal ~ der Ortsbestellbezirk; window ~, ~ at the window die Abholung, Briefausgabe am Schalter; bill of ~ der Lieferungsschein; contract of ~ der Lieferungsvertrag; ~ department (T) die Lokalexpedition; ~ docket (P) das Bestellbuch, (in England) das Bestell- und Abrechnungsbuch des bestellenden Boten; ~ sheet (P) der Uebergabe- und Abrechnungsbogen.

to demagnetize v entmagnetisiren.

demand s der Bedarf (z. B. ~ for poles der Stangenbedarf); die Einforderung; der Anspruch; die Nachfrage; der Auftrag; in ~ begehrt; on ~ bei Sicht.

demurrage s (P) die Liegezeit, das Liegegeld; ~ account die Verrechnung der Lagergebühr; ~ charge die Lagergebühr.

demy s ein großes Zeichen-, Druck- u. Pack-Papierformat.

density s die Dichtigkeit, Dichte; das spezifische Gewicht; electric ~ (T) die elektrische Dichte; maximum of ~ die größte Dichtigkeit.

dent s (Techn.) der Zahn, die Zahnung, die Einzahnung; ~s pl die Verzahnung.

dentated a gezahnt.

dented a gekerbt, ausgezackt, gezahnt; ~ wheel das Zahnrad.

to deoxidate, to deoxidize, to deoxygenate v einem Stoffe den Sauerstoffgehalt entziehen, desoxydiren.

to depart v abweichen; abreisen; (Eisenb.) abgehen.

departure station s (Eisenb.) die Abgangs-, (Absende-)Station.

deponent s der Hinterleger, Deponent.

to deposit v niederlegen, hinterlegen, eine Einlage machen; (Chem.) sich abscheiden, sich absetzen; (Phys.) sich niederschlagen; to ~ the amount den Betrag deponiren; to ~ the earth die Erde aufschütten.

deposit s das Depositum, die hinterlegte Geldsumme, die Einlage; (Chem., Phys.) der Niederschlag; to keep a ~ account ein Stundungsbuch haben; ~ receipt die Quittung über ein Depositum.

depositary s der Depositar.

depot s das Depot, die Niederlage; (T) das Materialienlager; (Eisenb.) die Güterexpedition; (Am) der Eisenbahnhof, das Bahnhofsgebäude.

depredation s of a mail die Beraubung einer Post, der Postdiebstahl; ~ case der Fall einer Postberaubung.

to depress v the key (T) den Schlüssel drücken.

deputy s der Bevollmächtigte, Stellvertreter; ~ Postmaster General der Stellvertreter des General-Postmeisters.

derailment s (Eisenb.) die Entgleisung.

derangement s of the line (T) die Linien-, Leitungsstörung.

derivation s die Ableitung (des Wassers, des galvanischen Stromes u. s. w.).

to derive v ableiten; ~d current (T) der Zweigstrom.

derrick s (Mech.) der bewegliche Ausleger eines Krahns oder einer Hebevorrichtung; ~ crane der Krahn mit Ausleger.

descent s (Techn.) der Hang, Fall, die Neigung, das Gefälle; ~ of bodies (Phys.) der Fall der Körper.

desiccation s das Trocknen, die Austrocknung (von Holz).

to design v zeichnen; (Masch.) erdenken, erfinden, Pläne entwerfen; to ~ a telegraph line eine Telegr.-Linie entwerfen.

design s die Zeichnung; der Aufriß; der Entwurf.

desk s das Pult; reading ~ das Lesepult; writing ~ das Schreibpult.

despatch, dispatch s die Depesche, das Telegramm; ~ book (P) das Abfertigungsbuch; ~ good (Eisenb.) das Eilgut; ~ table (P) der Abfertigungstisch; table of ~ (P) die Leitübersicht; ~ tube, pneumatic ~ die pneumatische Einrichtung, Rohrposteinrichtung.

to despatch v absenden, befördern; ~ing office (P & T) Absendungspostanstalt; ~ing officer, ~ing clerk (P) der Abfertigungsbeamte; ~ing table (P) der Abfertigungstisch.

destination s die Bestimmung; office of ~ die Bestimmungs-Postanstalt.

to destroy v the magnetism den Magnetismus aufheben; to ~ materials altes Zeug vernichten.

to detach v ablösen; ~ed a abgelöst, lose, abgesondert.

detector, current detector s (T) das Untersuchungs-Galvanoskop; der Alarmapparat bei der Dampfmaschine (der das Sinken des Wasserstandes anzeigt; auch low water detector); ~ lock das Sicherheitsschloß.

detent s der Kegel, Springkegel; (Masch.) die Sperrklinke, der Sperrkegel; ~ lever (HA) der Auslösehebel; ~ pin der Anhaltstift, Sperrkegel.

deviation s die Abweichung, Ablenkung; angle of ~ der Ablenkungswinkel; ~ of the compass die Abweichung der Magnetnadel.

device die Vorrichtung, der Apparat; der Entwurf, Plan; die Erfindung; ~ of building costs der Kostenanschlag, Bauanschlag.

devil s (Mech.) die Kluppe, Schraubenschneidekluppe; ~'s claw die Teufelsklaue, das Kropfeisen, die Froschklemme.

diagonal a diagonal, schräg, querlaufend; ~ brace die Kreuzstrebe.

diagram s die Darstellung, die schematische Zeichnung, die Illustration in wissenschaftlichen Werken; (Masch.) der Riß.

dial das Zifferblatt; telegraph ~ (T) das Zifferblatt eines Zeigertelegraphen; ~ instrument dass. wie ~ telegraph; ~ plate das Zifferblatt; ~ telegraph der Zeigertelegraph.

diamagnetic a diamagnetisch.

diamagnetism s der Diamagnetismus.

diameter s der Durchmesser; ~ inside der Durchmesser im Lichten; ~ outside der äußere Durchmesser.

diaphragm s die Scheidewand, das Diaphragma.

die s (pl dice) der Würfel; (pl dies) die Stanze; das Loch des (Draht-)Zieheisens; ~s pl of the die stock die Backen der Kluppe; ~ stock die Kluppe.

dielectric s der dielektrische Körper, der Nichtleiter.

differential block, ~ pulley s der Differenzialflaschenzug oder Kloben; ~ calculus (Math.) die Differenzialrechnung; ~ galvanometer das Differenzialgalvanometer; ~ principle (T) die Differenzialmethode (des Gegensprechens).

diffusion s of liquids der Austausch der Flüssigkeiten (in der galvanischen Batterie).

to dig v graben; to ~ earth Erde ausheben.

to dilate v ausdehnen, ausweiten, strecken; sich ausdehnen.

dilation s (Phys.) die Ausdehnung (Expansion).

to dilute v (Chem.) verdünnen.

dilute, diluted a verdünnt.

dilution s die Verdünnung.

dimension s die Dimension, die Abmessung; figured ~ die Maaßbezeichnung, das eingeschriebene Maaß; linear ~ die Längenausdehnung; superficial ~ die Flächenausdehnung.

diode working s (T) die gleichzeitige Versendung von 2 Telegrammen mit dem Delany'schen Apparat.

dip s der Fall, die Neigung, die magnetische Inklination; der Durchhang (des Drahtes) s. auch chain curve; ~ candle das gezogene Licht.

direct a direkt; ~ inkwriter (T) der Farbschreiber, der direkt (ohne Relais) arbeitet.

to direct v a letter einen Brief adressiren.

direction *s* die Richtung; die Anweisung; die Verfügung; die Adresse; ~ in case of need die Nothadresse (bei kaufmännischem Versandt).

directive *a* Richtung gebend, richtend.

directory, city directory *s* das Adreßbuch.

„Dis" (*P Am*, Abkürzung von *dis*tribution, Vertheilung) Bezeichnung eines Kartenschlusses, dessen Inhalt unterwegs vertheilt werden soll.

disc f. disk.

to discharge *v* entlassen; entladen; leeren; to ~ a bill eine Rechnung bezahlen; to ~ a contract eine Verbindlichkeit aufheben; to ~ a ship ein Schiff löschen.

discharge *s* das Ausladen; die Bezahlung; (Elektriz.) die Entladung; lightning ~ der Blitzschlag; silent ~ die Entladung ohne Funken.

discharger *s* (Phys.) der Entlader; lightning ~ der Blitzableiter; plate ~ der Plattenblitzableiter; point ~ der Spitzenblitzableiter.

discharging key *s* (*T*) der Entladungsschlüssel.

to disconnect *v* (Mech.) loskuppeln, entkuppeln; (*T*) eine Verbindung aufheben.

disconnection *s* (*T*) die Leitungsunterbrechung (der Draht ist gebrochen); intermittent ~, partial ~ die zeitweise, stellenweise Unterbrechung (in Folge mangelhafter Löthstellen u. dergl.); total ~ die gänzliche Unterbrechung (der Draht ist gerissen).

discontinuance *s* die Aufhebung (einer Postanstalt).

to discontinue *v* a post office eine Postanstalt aufheben.

discontinuous *a* intermittirend, unterbrochen.

to discount *v* diskontiren, Diskontwechsel nehmen.

discount *s* der Skonto, der Rabatt; ~ bank die Diskontobank.

discrepancy *s* in the number of words (*T*) die Verschiedenheit in der Wortzahl.

to disengage *v* (Phys.) entbinden; to ~ the gear (Masch.) die Maschinerie ausrücken, außer Gang setzen.

disengaging gear *s* (Masch.) die Vorrichtung zum Ausrücken.

to dish *v* aushöhlen, austiefen.

dished, dished out *a* (Techn.) ausgehöhlt, vertieft.

dishonored *a* nicht honorirt, protestirt (von einem Wechsel).

disintegration *s* die Auflösung; die Verwitterung.

disk, disc *s* die Scheibe; (Eisenb.) die Signalscheibe; receiving ~ of the telephone die Empfangsscheibe; sending ~ of the telephone die Absendescheibe; ~ signal (Eisenb.) die Signalscheibe.

to dismantle *v* niederreißen, abbrechen.

dispatch *s* f. despatch.

to dissolve *v* auflösen, schmelzen; auslaugen; sich lösen; to ~ a contract einen Vertrag auflösen.

distance *s* die Entfernung, der Abstand, Zwischenraum; long ~ telephony das Fernsprechen auf weite Entfernungen; ~ testing (*T*) die Bestimmung der Entfernung eines Fehlers durch Messung.

distant signal *s* (Eisenb.) das Distanzsignal, Einfahrtsignal.

distributor *s* (*T*) der Vertheiler, die Vertheilerscheibe; der Unterbeamte, welcher die abzutelegraphirenden Telegramme an die Apparate legt.

district *s* der Bezirk; ~, postal ~ (*P*) der Bestellbezirk.

ditch *s* der Graben; draining ~ der Entwässerungsgraben.

diurnal *a* täglich; ~ variations *pl* die täglichen Schwankungen der Magnetnadel.

to divide *v* theilen, zertheilen, eintheilen; sich theilen; Dividende geben.

dividend *s* die Dividende, der Gewinnantheil; ~ warrant der Coupon, Zinsschein.

divisibility *s* (Phys.) die Theilbarkeit.

division *s* die Theilung, Vertheilung; die Scheidewand; die Abtheilung, der Bezirk; (P) der Postbezirk, Kartirungsbezirk; railway ~ (P) der Kartirungsbezirk (nach den von London ausgehenden Eisenbahnkursen); ~ sorting list (P) die Vertheilungsliste nach den von London ausgehenden Eisenbahnkursen; ~ superintendent (Eisenb.) der Abtheilungschef.

to docket v summarisch verzeichnen, ausziehen, einen Auszug machen; in ein Verzeichniß eintragen; auf dem Rücken von Schriften den Inhalt derselben verzeichnen; (im Handel) mit Adreßzetteln versehen.

docket s die kurze Inhaltsangabe; das Verzeichniß; ~ of goods der Waarenadreßzettel; delivery ~ s. delivery.

dog s (Techn.) der Bock, das Gestell; die Klammer; ~ bolt der Bolzen mit viereckigem Kopf; ~ nail der große Schloßnagel.

domestic a inländisch.

domiciled a bill der domizilirte, auf ein Zahlungsdomizil angewiesene Wechsel.

door case s die Thüreinfassung, Thürzarge.

dormant, dormer s der Grundbalken, die Bodenschwelle; ~ tree der Tragbalken, Hauptbalken.

to dot v a line eine Linie punktiren (beim Zeichnen).

dot s (T) der Punkt; ~s & dashes pl die Punkte und Striche des Morse-Alphabets.

double s das Doppelte, Zweifache; die Kopie.

double acting a (Masch.) doppelt wirkend; ~ angle iron das Doppelwinkeleisen, U-Eisen; ~ bell, ~ cup insulator die Doppelglocke; ~ entry, book keeping by ~ entry die doppelte Buchführung; ~ key (T) der Doppeltaster; ~ line (Eisenb.) das Doppelgeleise; ~ needle telegraph der Doppelnadeltelegraph; ~ plate sounder (T) der Doppelklopfer (ein Telegr.-Apparat, welcher mit zwei verschiedene Klänge erzeugenden Klopfvorrichtungen versehen ist. Der Strom geht vom Relais zu der einen Klopfvorrichtung, deren Schläge die Punkte, oder zur andern, deren Schläge die Striche der Morsezeichen darstellen); ~ thread of a screw das Doppelgewinde einer Schraube; ~ threaded screw die Doppelschraube; ~ T iron das doppelte T Eisen; ~ track, ~ way (Eisenb.) das Doppelgeleise.

dowel s der Diebel, Dübel, Holzdübel.

down line s (T) diejenige Telegraphenleitung, welche zur Beförderung von Telegrammen in der Richtung von der Hauptstation der Linie nach den anderen Stationen dient; (Eisenb.) das Geleise für Züge, welche in der Richtung von der Hauptstation der Linie nach den anderen Stationen gehen; ~ motion (Masch.) die niedergehende Bewegung; ~ office (T) die an der down line (s. dies.) gelegene Telegr.-Station; ~ signal (Eisenb.) das Abfahrtssignal; ~ train (Eisenb.) der Zug von der Hauptstation.

to draft, to draught v Zeichnungen, Risse entwerfen; to ~ a document eine Urkunde entwerfen; to ~ instructions eine Dienstanweisung entwerfen.

draft s die Tratte, der Wechsel.

draftsman s dass. wie draughtsman.

to drag v schleppen, auf dem Boden hinziehen; (T) von den Morsezeichen gesagt, wenn sie in einer zusammenhängenden Linie auf den Streifen erscheinen.

drag s die Baggerschaufel.

drain s der Abzugskanal, der Entwässerungsgraben; die Gosse; side ~ (Eisenb.) der Bahngraben; ~ pipe die Drainröhre; die Abzugsröhre.

draining ditch s der Entwässerungsgraben.

draught s (Techn.) der Riß, die Zeichnung; rough ~ der erste Entwurf.

draughtsman, draftsman s der Zeichner.

to draw v zeichnen, entwerfen; to ~ bills Wechsel trassiren, ziehen;

to ~ to a scale nach einem Maßstabe zeichnen.

draw box *s* (*T*) der (Kabel=)Einziehkasten, der Untersuchungsbrunnen; ~ bridge die Zugbrücke; ~ pipe die Einziehröhre, in welche die Kabel mit Drähten oder Tauen eingezogen werden; ~ plate das Drahtzieheisen; ~ pole, stretching pole (*T*) die Stange für Spann=Isolatoren; ~ tongs *pl* die Froschklemme.

drawback *s* of duty der Rückzoll, die Vergütung der entrichteten Zollgebühr.

drawee *s* der Trassat, derjenige, auf den ein Wechsel gezogen, aufgestellt ist.

drawer *s* der Trassant, Aussteller eines Wechsels; die Schublade, das Schubfach; (*MA*) der Untersetzkasten.

drawing *s* das Zeichnen, die Zeichenkunst, die Zeichnung, der Entwurf; ~ in outline der Umriß; plan ~ das Planzeichnen; profile ~, sectional ~ die Durchschnittszeichnung; wire ~ das Drahtziehen, die Drahtzieherei.

drawing bench *s* die (Draht=)Ziehbank; ~ down das Ausrecken des Eisens; ~ hole das Drahtziehloch; ~ mill das Drahtziehwerk, die Drahtmühle; ~ rod (*HA*) die Zugstange.

dray *s* der niedrige Wagen, Karren; ~ cart der Karren, Rollwagen; ~ man der Rollfuhrmann; ~ man's ladder die Schrotleiter.

to dredge *v* baggern, ausbaggern.

dredge *s* der Bagger.

dredging boat *s* das Baggerboot; ~ bucket der Baggereimer; ~ engine die Baggermaschine.

dried *a* getrocknet, trocken; air-~ lufttrocken.

drift *s* der Abtrieb (beim Legen des Kabels in fließendem Wasser); ~ bolt (Mech.) der Treibbolzen; ~ ice das Treibeis; ~ sand der Triebsand, Treibsand.

to drill *v* bohren, drillen.

drill *s* (Techn.) der Bohrer, Drillbohrer, Drehbohrer; centrifugal ~ der Zentrifugalbohrer; ~ with lever, lever ~, ratchet ~ die Bohrknarre, der Ratschbohrer; rock ~ der Steinbohrer; twist ~, spiral ~ der Schraubenbohrer.

to drive *v* treiben, in Bewegung, in Gang setzen; rammen; to ~ a coach kutschiren; to ~ in eintreiben, einschlagen.

driver *s* der Rammblock; (Masch.) der Treiber, das Treibrad; der Mitnehmer, die Knagge; der Kutscher; mail ~ der Postkutscher, Postillon.

driving axle *s* die Treibachse, Triebwelle; ~ gear das Getriebe; ~ shaft die Treibachse, Triebwelle; ~ wheel das Treibrad.

drop handle *s* der Handgriff des Nadeltelegraphen; ~ letter (*P*) der an eine im Bestellbezirk der Aufgabepostanstalt wohnende Person adressirte Brief; ~ newspaper (*P*) die an eine im Bestellbezirk der Aufgabepostanstalt wohnende Person adressirte Zeitungssendung; ~ rate (*P*) die Ortsbestellgebühr.

drum *s* die Trommel, Rolle, Welle, Walze; ~, wire ~, coiling ~ die Drahtleier.

Drummond's light, hydro-oxygen-light *s* das Drummond'sche Signallicht (durch Einwirkung einer Knallgasflamme auf Kreide oder Aetzkalk hervorgebracht).

dry-rot *s* der Hausschwamm, die trockene Fäulniß des Holzes.

ductile *a* dehnbar, geschmeidig, streckbar; schmiedbar, hämmerbar.

ductility *s* die Geschmeidigkeit; test for ~ die Prüfung (des Drahtes) auf Torsionsfestigkeit.

due *a* fällig; to become ~ fällig werden; to fall ~ fällig sein; when ~ bei Verfallzeit.

due *s* die Gebühr.

duplex telegraphy *s* die Doppel=Telegraphie, das Gegensprechen (die Versendung zweier Telegramme zu gleicher Zeit in entgegengesetzten Richtungen).

dutch tongs *s/pl* die Froschklemme.

dutiable article *s* der zollpflichtige Gegenstand.

duty *s* 1) die Steuer; der Zoll, die Abgabe; ~ upon exportation der Ausfuhrzoll; ~ upon importation der Einfuhrzoll; ~ free zollfrei; ~ off unverzollt; ~ on verzollt; ~ paid zollbezahlt; 2) der Dienst; ~ pay die Zahlung für besonders geleistete Dienste, die Funktionszulage.

dyke, dike *s* der Uferdamm.

dynamic, dynamical *a* dynamisch; ~ unity die Arbeitseinheit.

dynamometer *s* der Dynamometer, Kraftmesser.

E.

ear *s* das Ohr, das Oehr, der Henkel, die Oese; ~ of a rope die Oese; ~ tube das Hörinstrument.

earth *s* die Erde, der Boden, das Erdreich; (T) der Erdfehler; to make ~ (T) mit der Erde in Verbindung stehen; to put to ~ (T) mit Erde verbinden; dead ~ (T) die tödtende Erde (ein Ende des gerissenen Drahtes liegt auf der feuchten Erde oder im Wasser); intermittent ~ (T) zeitweiser Erdkontakt (der Wind bringt die Drähte bisweilen in Berührung mit einem leitenden Körper, der Erdverbindung hat); partial ~ (T) zeitweiser Erdkontakt (in Folge mangelhafter Isolatoren, oder der Draht liegt auf mangelhaften Leitern [Mauern, Holz] auf, die mit der Erde in Verbindung stehen); total ~ dasf. wie dead ~; ~ borer der Erdbohrer; ~ current der Erdstrom; ~ magnetic *a* erdmagnetisch; ~ pipe die Thonröhre; ~ plate die Erdplatte; ~ terminal die Erdklemme; ~ wire der Erddraht.

earthen *a* irden, thönern; ~ ware die Töpferwaare, das Steingut, die gebrannte Thonwaare.

earthy *a* erdig, erdartig; ~ wire (T) der mit der Erde in Berührung gekommene Draht.

to ease *v* entlasten (ein Bauwerk u. dergl.); to ~ a working part (Masch.) den Spielraum zwischen zwei arbeitenden Theilen etwas vergrößern.

ebonite *s* das Ebonit, der Hartkautschuk.

eccentric *s* der exzentrische Kreis, das Exzentrik; ~ governor der Exzentrikregulator; ~ rod die Exzentrikstange; ~ shaft die Exzentrikwelle.

eccentric, eccentrical *a* exzentrisch, verschiedene Mittelpunkte habend.

eccentricity *s* die Exzentrizität (beim Kabellöthen: Draht- u. Löthstelle liegen außerhalb des Mittelpunktes der Kabelseele).

eddy current *s* der durch Selbstinduktion in einem Drahte hervorgerufene elektrische Strom.

to edge *v* schärfen, schleifen, wetzen; abkanten.

edge *s* der Rand, der Saum; die Kante; ~ tools *pl* die Schneidwaaren, schneidenden Werkzeuge.

educt *s* (Chem.) das Edukt (der aus einer Substanz abgeschiedene Körper, welcher in derselben als solcher, als fertig gebildeter Bestandtheil, vorhanden war).

eduction *s* (Dampfmasch.) die Ausströmung, Entweichung; ~ pipe die Abzugsröhre, Entweichungsröhre; ~ valve das Auslaßventil.

effect *s* (Mech.) die Kraftwirkung, mechanische Leistung einer Maschine; heating ~ die Wärmewirkung; lost ~, impeding ~ die Nebenleistung, verlorene Kraft; useful ~ die Nutzleistung, Effektivkraft; whole ~ die Totalleistung, der Totaleffekt.

effective *a* (Mech.) wirksam; ~ power die Effektivkraft.

efficiency *s* (Masch.) die Leistung, der Nutzungswerth.

efflorescence *s* (Chem.) die Auswitterung, Effloreszenz.

efflorescent *a* (Chem.) effloreszirend.

efflux s (Phys.) das Ausströmen, der Ausfluß.

elastic a elastisch, federkräftig; ~ force die Spannkraft, Elastizität; ~ gum der Kautschuk; ~ limit die Elastizitätsgrenze; ~ tubing der Gummischlauch.

elasticity s die Elastizität, Schnellkraft, Spannkraft; ~ of compression die Druckfestigkeit; ~ of extension die Zugelastizität, absolute Festigkeit; ~ of flexion die Biegungselastizität, relative Festigkeit; limit of ~ die Elastizitätsgrenze; modulus of ~ der Elastizitätsmodulus; ~ of torsion die Drehungselastizität.

elbow s (Masch., Techn.) das Knie, Knierohr, der Winkel, die Krümmung; ~ joint die Knieverbindung; ~ lever (MA) der Winkelhebel; ~ pipe (Masch.) das Knierohr.

elder s das Hollunderholz, Fliederholz; ~ pith das Hollundermark; ~ tree der Hollunderbaum.

electric s (Phys.) der elektrische Körper, Nichtleiter; non-~ der nichtelektrische Körper, Leiter.

electric, electrical a elektrisch; ~ alarm das elektrische Schlagwerk; ~ annunciator der elektrische Haustelegraph, (F) die Klappe; ~ battery die elektrische Batterie; ~ storm das elektrische Gewitter (mächtiger Erdstrom).

electrician s der Elektriker, Elektro-Techniker.

electricity s die Elektrizität; atmospherical ~ die atmosphärische Elektrizität; contact ~ die Berührungs-Elektrizität; dynamical ~ die dynamische (in Bewegung befindliche) Elektrizität; frictional ~ die Reibungs-Elektrizität; induced ~ die Induktions-Elektrizität; magneto-~ die Magnetoelektrizität; negative ~, resinous ~ die negative, Harz-Elektrizität; positive ~, vitreous ~ die positive, Glas-Elektrizität; static ~ die statische Elektrizität (Elektrizität im Zustande der Ruhe); thermo-~ die Thermo-Elektrizität; Voltaic ~, galvanic ~ die Berührungs-Elektrizität, der Galvanismus.

electrification, electrisation s die Elektrifizirung; ~ by induction die elektrische Induktion.

to electrify v elektrisiren, elektrisch machen.

electro - chemical a elektrochemisch; ~ deposit s der galvanische Niederschlag; ~ dynamics s/pl die Elektrodynamik; ~ magnet s der Elektromagnet; bifurcate ~ magnet s das Hufeisen, der zweischenkelige Elektromagnet; ~ magnetic a elektromagnetisch; ~ magnetic ringing apparatus s (Eisenb.) das elektrische Geläute, Läutewerk; ~ magnetism der Elektromagnetismus; ~ negative a negativ elektrisch; ~ positive a positiv elektrisch.

electrode s die Elektrode.

electrolysis s die Elektrolyse.

electrolyte s der Elektrolyt.

electromagnet s der Elektromagnet.

electrometer s der Elektrometer, Elektrizitätsmesser; quadrant ~ der Quadranten-Elektrometer.

electromotive force s die elektromotorische Kraft.

electron, electrum s der Bernstein.

electroscope s das Elektroskop; gold-leaf ~ das Goldblatt-Elektroskop.

electrostatic a elektrostatisch.

electrostatics s/pl die Elektrostatik, elektrische Gleichgewichtslehre.

element s das Element, der Grundbestandtheil; galvanic ~, Bunsen's ~, Daniell's ~ etc. das galvanische, Bunsen'sche, Daniell'sche u. s. w. Element.

elevation s der Aufriß, Standriß; ~ of temperature die Temperatur-Erhöhung; back ~ die Hinteransicht; front ~ die Vorderansicht; sectional ~ der Längen- oder Querschnitt; side ~ die Seitenansicht.

elevator s (Mech.) der Aufzug; der Fahrstuhl (in Hotels).

elongation s die Längenausdehnung des Drahtes; ~ test die Prüfung des Drahtes auf absolute Elastizität.

embanking s (Eisenb.) die Aufschüttung.

embankment *s* die Eindämmung, der Damm, der Erddamm.

embezzlement *s* die Unterschlagung, der Unterschleif.

to emboss *v* austreiben, ausbuckeln; erhaben arbeiten (Metalle); ~ing register, ~ing instrument (*T*) der Reliefschreiber.

embosser *s* (*T*) der Reliefschreiber.

emery, emeril *s* der Schmirgel; ~ paper das Schmirgelpapier, Rostpapier.

emission *s* (Techn.) die Ausströmung, das Ausfließen; ~ of postage stamps die Ausgabe von Postfreimarken, die Serie.

to emit *v* (Techn.) ausströmen lassen; (Phys.) ausstrahlen; in Umlauf setzen, ausgeben.

employee, employé *s* der Arbeitnehmer; der Beamte, Angestellte.

employer *s* der Arbeitgeber.

empty *a* (*P*) der leere Behälter für Pacete bei der Pacetpost (zu ergänzen: basket, bag, wrapper); empties *pl* leere Packungen.

enclosure *s* die Einfriedigung, der Zaun; die Einlage in einem Briefe; die Anlage; ~ bag (*P*) der Versteckbeutel.

end *s* das Ende; ~ play of a screw der leere, todte Schraubengang; ~ shake (Masch.) die unregelmäßige Bewegung der Enden einer Welle; ~ way of the grain die Hirnfläche des Holzes.

endless *a* ohne Ende, geschlossen; ~ chain die Kette ohne Ende; ~ screw die Schraube ohne Ende.

to endorse *v* indossiren, giriren.

endorsement *s* das Indossament, das Giro.

energy *s* (Phys., Mech.) die Energie, Spannkraft, lebendige Kraft; available ~ die verwendbare, verwandelbare (in Arbeit zu verwandelnde) Energie; conservation of ~ die Erhaltung der Kraft; intrinsic ~ die latente, gebundene, innere Energie.

to engage *v* (Masch., Techn.) in Gang, in Bewegung setzen, eingreifen lassen; engaging wheel das Treibrad.

engine *s* die Maschine, das Triebwerk; (Eisenb.) die Locomotive; steam ~ die Dampfmaschine; portable steam ~ die Locomobile.

engineer *s* der Ingenieur, der Techniker; der Maschinenbauer; assistant ~ der Unter-Ingenieur; chief ~ der Ober-Ingenieur.

engineering *s* die Ingenieurkunst, das Ingenieurwesen.

to enlarge *v* vergrößern, erweitern; ~d scale der vergrößerte Maaßstab.

entry *s* der Eingang; die Einfuhr (von Waaren); der Posten, Rechnungsartikel; der Eingang (von Geldern); die Eintragung (in die Bücher); duty of ~ der Einfuhrzoll; to make an ~ einen Posten in die Bücher eintragen; single, double ~ die einfache, doppelte Buchführung.

envelop, envelope *s* das Couvert, der Briefumschlag; penalty ~ (*P*) der für die amtliche Korrespondenz gebrauchte Briefumschlag, auf dem die Geldstrafe (penalty) vermerkt ist, welche derjenige verwirkt, der diesen Umschlag zu anderer als dienstlicher Korrespondenz verwendet; registered letter ~ (*P*) der Briefumschlag für eingeschriebene Briefe; special request ~ (*P Am*) der auf besondere Bestellung von der Postbehörde hergestellte mit Firma u. s. w. bedruckte Briefumschlag; stamped ~ (*P*) der Briefumschlag mit eingedrucktem Postwerthzeichen; taped ~ der mit grüner Schnur (tape) kreuzweis umwundene Briefumschlag, um die Einschreibung zu kennzeichnen.

Epsom salt *s* das Bittersalz, die schwefelsaure Magnesia.

equation *s* (Math.) die Gleichung.

equidistant *a* gleich weit entfernt.

equilibrium *s* das Gleichgewicht; stable ~ das sichere, stabile Gleichgewicht; unstable ~ das unsichere, labile Gleichgewicht.

to equip *v* ausrüsten, ausstatten.

equipment *s* die Ausrüstung; ~ of an office die Amtseinrichtung; mail ~ Kursausstattungs-Gegenstände.

erase v ausschaben, auslöschen, wegstreichen; the „erase" signal (T) das Irrungszeichen.

to erect v erbauen; to ~ poles Telegr.-Stangen aufstellen; to ~ the wire den Draht auf die Stangen aufbringen.

erection s die Errichtung, Erbauung, der Bau; die Aufstellung der Telegr.-Stangen.

erg s (Phys., Mech.) das Erg, die Arbeitseinheit.

escape s die Entweichung; (T) die Ableitung des Stroms in Folge ungenügender Isolirung, der Nebenschluß, der Stromverlust; ~ wheel s. escapement wheel.

escapement s (Masch.) die Hemmung; dead-beat ~ die schleifende Hemmung; repose ~ die ruhende Hemmung; step-by-step ~ die ruckweise Hemmung; ~ wheel, balance wheel das Hemmungsrad, das Steigrad.

to establish v a post office ein Postamt einrichten; an ~ed post office clerk ein fest angestellter Postbeamter.

establishment s das Etablissement, die Fabrik; (P) die Rangordnung der (englischen) Postbeamten; major ~ die Postbeamten vom Sortirer an aufwärts; minor ~ die Postbeamten vom Sortirer abwärts; ~ book (P) die Personalliste einer Verkehrsanstalt.

to estimate v schätzen, veranschlagen, taxiren.

estimate s der Anschlag, Kostenanschlag; die Aufstellung des Etats.

estimation s der Anschlag, Voranschlag; die Schätzung; ~ of demand die Bedarfsabschätzung; ~ by the sight das Augenmaaß.

ether s (Chem., Phys.) der Aether.

ethereal oils s/pl (Chem.) ätherische Oele.

to evaporate v verdunsten.

evaporation s die Verdunstung.

evasion s of payment of postage die Umgehung der Portozahlung.

even a gerade (von einer Zahl gesagt); eben, flach, eine gleichmäßige Oberfläche habend.

to examine v accounts Rechnungen prüfen.

to excavate v ausgraben, aushöhlen.

excavation s die Ausgrabung; (T) das Stangenloch.

excentric a s. eccentric.

excess s der Ueberschuß; ~ of work (Techn.) die übermäßige Arbeit.

exchange s die Börse; der Kurs; account of ~ das Wechselkonto; bill of ~ der Wechsel, die Tratte; course of ~ der Wechselkurs; in ~ in Tausch; under the quoted ~ unter dem Kurse; ~ advice der Kursbericht; ~ circulation der Wechselumlauf; ~ list der Kurszettel; ~ post office die Auswechselungs-Postanstalt, Uebergabe-Postanstalt, Grenzpostanstalt.

to excite v erregen; exciting solutions, excitants pl erregende Flüssigkeiten.

exempt a frei, befreit; ~ from duty zollfrei; ~ from postage portofrei.

exemption s from office & service die Befreiung der Postbeamten von bürgerlichen Verpflichtungen (z. B. Uebernahme von Vormundschaften, kommunalen Aemtern, Theilnahme an Schwurgerichten als Geschworene u. s. w.); ~ from postage die Portofreiheit.

to exfoliate v sich abblättern, abschiefern.

to exhaust v erschöpfen, entleeren, aussaugen; to ~ of air luftleer pumpen.

exhauster s der Exhaustor.

exhibit s der Ausstellungsgegenstand; die Eingabe, schriftliche Vorstellung.

exhibition s die Ausstellung; international ~, universal ~ die Weltausstellung.

to expand v (Phys.) ausdehnen; sich ausdehnen.

expansibility s (Phys.) die Ausdehnbarkeit.

expansion s (Phys.) die Ausdehnung, die Expansion; die Expansionsvorrichtung.

expansive force s die Spann=
kraft, Expansivkraft.
expenditure s die Ausgaben.
expenses s/pl die Ausgaben,
Kosten; additional ~, extraordinary
~, incidental ~ die Nebenausgaben,
unvorhergesehene Ausgaben; main=
taining ~, ~ of maintenance die
Unterhaltungskosten, Reparaturkosten;
office ~ die Verwaltungskosten, Bü=
reaukosten; petty ~ die kleinen Un=
kosten; sundry ~, sundries verschie=
dene, gemischte Ausgaben, Varia;
travelling ~ die Reisekosten; work=
ing ~, ~ of working die Betriebs=
kosten.
expiration s der Ablauf, die
Verfallzeit.
to expire v zu Ende gehen;
(von Zahlungen) verfallen, fällig
werden.
**explosives, explosive sub-
stances** s/pl Sprengstoffe, Spreng=
mittel.
express boat s das Schnell=
schiff; ~ carriage (Eisenb.) der Eil=
zugswagen; ~ cost of porterage, ~
porterage charge die Expreßbestell=
gebühr; ~ messenger der Eilbote;
~ train der Eil=, Kurier=, Expreß=
zug.
to extend v (Techn.) dehnen,
ausdehnen, recken, strecken.

extra attendance s (P) die
Ueberstunden; ~ bag (P) der außer=
gewöhnlich zur Absendung kommende
Briefbeutel (derselbe darf ohne
Briefkarte versendet werden); ~ best
best (zu ergänzen: wire) der beste
zum Telegr.=Bau verwendete Lei=
tungsdraht; ~ clerk (P) der Aus=
hülfsbeamte; ~ current (T, Phys.)
der Extrastrom; ~ duty (P, T) der
besondere Dienst, die Ueberstunden;
~ stamp box (P) die Abtheilung
im Briefkasten für Spätlingsbriefe;
~ stamp letter (P) der Spätlings=
brief (für welchen eine Zuschlagsge=
bühr von 1 Penny entrichtet werden
muß).
eye s das Auge, das Oehr, die
Oese, die Schleife; ~ of an axe or
hatchet die Haube, das Helmloch;
~ of a bolt das Auge eines Bol=
zens, die Schließenritze; ~ of a
borer der Ring des Bohrers; ~ of
a hammer das Ohr, Helmloch,
Stielloch; ~ of a needle das Na=
delöhr; ~ bolt der Augbolzen; ~
bolt & key der Bolzen mit Splint
oder Vorstecker; ~ screw der Schraub=
benring, die Ringschraube; ~ splice
die Augspliffung (ein Ring oder eine
Oeffnung am Ende eines Taues).
eyelet s das Oehr; ~ hole die
Hakenöse der alten Isolatoren.

F.

to face v bekleiden, belegen; to
~ the letters (P) die Briefe so sor=
tiren, daß die Adressen nach oben
liegen; to ~ a wall eine Mauer
verblenden, verkleiden.
face s die Fläche, Seite, Außen=
seite, Vorderseite; die Bekleidung;
~ of a hill der Abhang eines Hü=
gels; lateral ~ die Seitenfläche; ~
wheel (Masch.) das Kronrad.
facing s die Bekleidung mit
Blendsteinen, Verblendung; ~ of a
door die Bekleidung einer Thür; ~
of a wall die Mauerverblendung; ~
slip (P) der Klebezettel.
factor s of safety (Techn.,
Mech.) der Sicherheitsmodulus.

factory s die Fabrik; die Fak=
torei, Handelsniederlassung in frem=
den Welttheilen; ~ price der Fa=
brikpreis.
fagot, faggot s das Holzbün=
del; ~ of wire das Drahtbündel; ~
iron das Packeteisen, Schroteisen;
~ steel der Bundstahl.
to fail v falliren, Bankerott
machen; (T) ausbleiben (von den
Morsezeichen gesagt).
failure s die Fallissement, der
Bankerott; ~ of junction (Eisenb.)
das Verfehlen des Anschlusses; ~ of
junction form (P) das Formular
für die Meldungen der Bahnpost=
ämter über verfehlten Zuganschluß.

to fall *v* fallen; to ~ due (von Wechseln) fällig, zahlbar werden; to ~ out well gut rentiren.

fan *s* (Masch.) der Ventilator, Windfang; ~ wheel das Windrad.

fannion *s* das Absteckfähnchen (beim Vermessen).

faq (fair average quality) schöne Mittelsorte.

farad *s* (T, Elektriz.) das Farad, die Einheit der elektrischen Kapazität.

fashioned bar iron *s* das Formeisen, façonnirte Stangeneisen.

fast *a* fest, solide; schnell; (Eisenb.) durchgehend; ~ pulley, ~ sheave (Mech.) die feste Riemenscheibe, Festscheibe; ~ speed apparatus (T) der automatische Apparat, Schnellschreiber; ~ train (Eisenb.) der Schnellzug.

to fasten *v* befestigen, fest machen; to ~ with bolts verbolzen; to ~ with pegs pflöcken; to ~ with plaster eingipsen.

fastening *s* der Verschluß, das Band; die Verschlußvorrichtung.

to fathom *v* (auf See) lothen, den Grund auspeilen; (zu Lande) abklaftern, nach Faden messen.

fathom *s* die Klafter, der Faden (1,829 Meter); ~ line die Lothleine, Senkleine; ~ wood das Klafterholz, Brennholz.

faucet *s* der Zapfen, der Hahn (an einem Fasse); ~ of a pipe der Muff; ~ joint (Masch.) die Randverbindung, der eingelassene Rand; die Verzapfung; ~ pipe die Röhre mit Muff.

fault *s* der Fehler; (T) der Fehler in der Leitung; to localise a ~ einen Fehler eingrenzen; to remove a ~ einen Fehler beseitigen; ~ finder (T) das Untersuchungs-Galvanoskop mit astatischem Nadelpaar zur Auffuchung von Fehlern in Kabelleitungen.

faulty *a* (Techn.) fehlerhaft; a ~ wire ein Draht mit Fehlerstellen.

feather-edged file *s* die Einstreichfeile, Schraubenkopffeile.

fee *s* die Gebühr; die Sporteln (auch *pl* fees); late ~, late letter ~ (P) die Gebühr für Spätlingsbriefe.

to feed *v* a lamp eine Lampe speisen; to ~ the paper (T) den Papierstreifen weiter bewegen.

feed *s* (Masch., Mech.) die Zuführung, der Vorschub; ~ action, ~ motion die Vorschubbewegung; ~ cock der Speisehahn, Füllhahn; ~ pipe die Speiseröhre, das Speiserohr; ~ wheel das Zuführungsrad.

felling *s* das Holzfällen; season for ~ of trees die Fällzeit; ~ place der Schlagplatz.

female screw *s* die Schraubenmutter, Mutterschraube; ~ thread of a screw das Muttergewinde.

fender *s* das Schutzgitter (am Kamin); (Eisenb.) der Bahnräumer (an der Lokomotive); der Abweiser, Prellpfahl; ~ post (Eisenb.) der Haltepfahl.

ferric *a* acid (Chem.) die Eisensäure; ~ chloride das Eisenchlorid, Ferrichlorid.

ferricyanide, ferridcyanide *s* (Chem.) das Ferricyanid; ~ of potassium, potassium ~ das Ferricyankalium, Kaliumeisencyanid, rothe Blutlaugensalz.

ferrocyanide *s* (Chem.) das Ferrocyanid, eisenblausaure Salz; ~ of potassium das Ferrocyankalium, gelbe Blutlaugensalz.

ferrous *a* chloride (Chem.) das Eisenchlorür, Ferrochlorid; ~ oxide, protoxide of iron das Eisenoxydul; ~ sulphate das Ferrosulphat, schwefelsaure Eisenoxydul, der Eisenvitriol.

ferrule, ferrel *s* das Eisenband, die Zwinge, der lose Muff.

ferry, ferry boat *s* die Fähre, das Fährboot, der Prahm; flying ~ die fliegende Brücke; railway ~ das Trajektboot; ~ bridge die fliegende Brücke.

fibre *s* die Faser, die Fiber; hemp ~ die Hanffaser; ligneous ~, ~ of wood, woody ~ die Holzfaser, Längenfaser im Holz.

fibrous *a* faserig, sehnig (von Metallen); ~ fracture der sehnige Bruch (von Metallen); ~ iron das sehnige, zähe Eisen; ~ texture die sehnige Textur (von Metallen).

field *s* das Feld; magnetic ~

das magnetische Feld; ~ book das Notizbuch beim Telegr.-Bau, das Abpfählbuch; ~ line (*MT*) die Feld-Telegr.-Linie; ~ telegraph der Feld-Telegraph; ~ telegraph detachment, ~ telegraph train die Feldtelegraphen-Abtheilung.

figure *s* (Geom.) die Figur; (Math.) die Ziffer, das Zahlzeichen; ~ of merit (*T*) der kleinste Strom in Milliwebers, bei dem ein Telegr.-Apparat noch gut arbeitet; ~ blank (*HA*) das Zahlenblank; ~ changing lever (*HA*) der Wechselhebel.

filament *s* die Faser, das Fäserchen; carbon ~ die Kohlenfaser.

to file *v* (Mech.) feilen, abfeilen; to ~ a document ein Aktenstück (bei Gericht) einreichen; to ~ paper Papier aufziehen, in die Akten einheften.

file *s* die Schnur oder eine andere Vorrichtung zum Aufreihen von Papier; papers on ~ in den Akten befindliche, zu den Akten gebrachte Sachen; (Mech.) die Feile; arm ~ die Armfeile, große viereckige Feile; bit ~ die Grundfeile; flat ~, hand ~ die flache Feile, Ansatzfeile, Handfeile; half-round ~ die halbrunde Feile; pointed ~ die Spitzfeile; rat tail ~ der Rattenschwanz, eine kleine runde Feile; rifle ~ die Raspelfeile; rough ~, coarse ~ die Grobfeile; square ~ die vierkantige Feile; three-square ~, triangular ~ die dreikantige Feile.

filing *s* das Feilen; ~s *pl* die Feilspäne; iron ~s *pl* die Eisenfeilspäne.

to fill *v* füllen, anfüllen; to ~ the wire (*T*) eine Telegr.-Leitung voll in Anspruch nehmen.

fillet *s* of a screw das Schraubengewinde.

film *s* (Techn.) das Häutchen, die dünne Schicht; ~ of moisture die Feuchtigkeitsschicht; ~ of oxide die Oxydhaut.

fine *s* die Geldstrafe; to inflict a ~ upon *smb* J—n in eine Geldstrafe nehmen.

to fine *v* mit einer Geldstrafe belegen; (von Metallen) frischen.

fined *a* (von Metallen) gefrischt; ~ iron das Frischeisen.

finery *s* das Frischfeuer, der Frischofen, ~ furnace, ~ hearth der Frischheerd; ~ process die Feineisenbereitung.

fining forge, fining hearth *s* das Frischfeuer.

fir *s* die Föhre, Kiefer; die Tanne, Fichte; common pitch ~ die gemeine Fichte, Pechtanne, Harztanne; Scotch ~ die gemeine Kiefer; spruce ~ die Sprucetanne.

first cost *s* der Einkaufspreis, Selbstkostenpreis.

fiscal year *s* das Etatsjahr, Rechnungsjahr.

to fish *v* (Masch., Techn.) verlaschen; to ~ a beam einen Balken mit Seitenverstärkung versehen; to ~ a piece of timber ein Zimmerholz durch Anblattung verstärken.

fish *s* (Masch.) die Lasche; ~ bolt der Laschenbolzen; ~ piece die Seitenverstärkung (an Gebäuden).

fished *a* (Masch., Techn.) verlascht, verbunden; ~ beam der durch zwei Blätter verstärkte Balken.

fissure *s* die Spalte, der Riß, die Ritze; ~s *pl* in steel die Härterisse im Stahl.

to fit *v* zurüsten, ausstatten; befestigen, anschlagen; passen; to ~ up an office (*P*, *T*) ein Amt einrichten; to ~ out a pole (*T*) eine Stange armiren.

fitting *s* die Ausrüstung; (Mech.) das Einpassen, die Verbindung; ~s *pl* die vollständige Einrichtung, Ausrüstung; (Mech.) das Triebwerk; ~-in die Einfügung; ~-out die Ausrüstung; ~ piece das Bindestück; ~-up die Einrichtung; die Aufstellung.

to fix *v* befestigen, anheften, festmachen; to ~ an engine eine Maschine aufstellen, montiren; to ~ in einpassen.

fixed *a* fest; feuerbeständig; gebunden; ~ air (Chem.) die fixe Luft, Kohlensäure; ~ body (Chem.) der fixe, feste Körper; ~ bridge die stehende, feste Brücke; ~ light das feste Leuchtfeuer; ~ point der Festpunkt (beim Vermessen); ~ pulley (Mech.) die feste Rolle, Festscheibe.

fixing screw s die Klemm-schraube.

fixture s die Befestigung; ~s pl die niet- und nagelfesten Gegenstände in einem Gebäude.

flake s eine in Schichten lagernde Masse; die Schicht, Lage; die Flocke.

flaky a flockig, schuppig, schieferig.

flange s (Masch.) der Flantsch, Rand, die Flantsche; ~s pl die vorstehenden Enden bei doppelt T Eisen (s. web); ~ joint die Flantschenverbindung in Röhren; ~ pipe die Röhre mit Flantschen, Flantschenröhre.

to flare, to flare over v (Techn.) sich schief neigen, sich erweitern, ausfallen.

flat a flach, platt; ~ bar iron s das Flacheisen; ~ file die Flachfeile; ~ headed bolt der Scheibenbolzen; ~ iron das Flacheisen; ~ plyer, ~ nosed plyers pl die Flachzange.

to flaw v Risse bekommen, brüchig werden.

flaw s der Riß, Sprung, Spalt; der Schiefer (im Draht).

flawed a brüchig (vom Eisen gesagt), unganz.

flawy a rissig, brüchig.

flexibility s die Biegsamkeit.

flexible a biegsam.

flexure s (Phys., Mech.) die Biegung.

to flicker v flackern (vom elektrischen Lichte gesagt).

flooring s die Balkenlage; der Fußboden.

fluid s (Phys.) der flüssige Körper, die Flüssigkeit; non-elastic ~ die tropfbare Flüssigkeit; electric, magnetic ~ das elektrische, magnetische Fluidum.

fluke s of an anchor der Ankerflügel.

flush a gleich, in derselben Ebene liegend; to make ~ a surface with another abgleichen, zwei Ebenen in eine bringen; ~ box der Brunnen oder Kasten für das Einziehen von Kabeln (so genannt, weil die Brunnen mit der Erd- oder Straßenoberfläche gleich gelegt sind); ~ rivet die versenkte Niete.

to flute v (Techn.) kanneliren, auskehlen, ausriefeln.

flute s (Techn.) die Rinne, Kannelirung, Auskehlung.

flux s der Fluß, das Flußmittel (Mittel um das Flüssigwerden eines Körpers zu befördern); das Einströmen der Elektrizität in das Kabel.

fly s (Mech.) das Schwungrad; der Windfang; expanding ~ der Windfang; ~ nut die Flügelmutter; ~ wheel das Schwungrad; das Windfangrad.

flyer s (Mech.) die Unruhe; (HA) der Flügel am Kreissektor.

flying level s die flüchtige Aufnahme, das Croquis; ~ pinion der Windfang, das Flügelrad.

fob (free on board) frei an Bord zu liefern.

foil s die Folie (Metall); tin ~ die Zinnfolie; ~ tin das Blattzinn, Stanniol, die Folie.

to fold v a letter einen Brief zusammenlegen; ~ing ladder die Klappleiter.

foliated a blätterig, geblättert; schieferig.

foolscap s das Schreibpapier in Folioformat, Herrenpapier.

foot s der Fuß (als Maaß = 0,3048 Meter); der Fuß, die Basis; der Zollstock; cubic ~ der Kubikfuß; square ~ der Quadratfuß; ~ board das Trittbrett; ~ messenger der Fußbote; ~ path, ~ way das Trottoir.

footing s der Sockel; das Fundament; ~ beam der Spannriegel.

force s (Mech.) die Kraft, Gewalt; accelerating ~, accelerative ~ die beschleunigende Kraft; centrifugal ~ die Zentrifugalkraft; centripetal ~, central ~ die Zentripetalkraft; component ~ die Seitenkraft, Komponente; ~ of continuance, ~ of inertia das Beharrungsvermögen; ~ of gravity die Schwerkraft; laboring ~ die mechanische Leistung einer Kraft; moving ~ die bewegende Kraft, Triebkraft; resisting ~ die Widerstandskraft.

forecast s of the weather die Wettervorausbestimmung.

15

foreign *a* ausländisch, fremd; ~ telegram das internationale Telegramm; ~ trade der Handel mit dem Auslande.

foreman *s* der Vorarbeiter.

to forfeit *v* verwirken, verlustig gehen; verfallen.

forfeiture *s* der Verlust, die Verwirkung; die Strafe, Buße.

to forge *v* schmieden, hämmern; fälschen; ~d telegram das von einem Unberufenen aufgegebene Telegramm.

to fork *v* sich gabeln (vom elektrischen Strome gesagt).

forked *a* gabelförmig gespalten.

forking *s* die Gabelung, das Theilen in Zweige.

form *s* die Form; das Formular; A ~ f. A; B ~ f. B; C ~ f. C; ~ of authority (*T*) das Formular für ein bezahltes Antworttelegramm; message ~, telegram ~ das Telegrammformular; forwarded telegram ~ das Formular für Ursprungs-Telegramme; received telegram ~ das Formular für Ortstelegramme; transmitted telegram ~ das Formular für Durchgangstelegramme.

formation *s* die geologische Bodenbildung; (Eisenb.) das Planum; ~ level (Eisenb.) das Planum, die Kronlinie.

forward *a* vorwärts; brought ~, carried ~ (in Rechnungen) Transport der vorhergehenden Seite; ~ letters *pl* (*P*) die Briefe, die von einer Postanstalt einer zweiten zugeführt und von letzterer sortirt und einer dritten Postanstalt übersandt werden; die umzuarbeitende Korrespondenz; ~ office (*P*) das Zwischenamt, welches die *forward letters* behandelt, die Umleitungs-(Umspeditions-)Postanstalt.

to forward *v* senden, absenden, befördern; to ~ the mail die Post umarbeiten; ~ed telegram das zur Beförderung aufgegebene Telegramm, das Ursprungstelegramm; ~ed telegram form das Formular für Ursprungstelegramme.

fox *s* (Mech.) der Keil, Splint; ~ key (Mech.) der Keil mit Gegenkeil; ~ tail (Mech.) die Fuge in Gestalt eines Fuchsschwanzes; ~ tail saw die Blattsäge, der Fuchsschwanz; ~ wedge (Mech.) der Gegenkeil.

fracture *s* der Bruch, die Bruchstelle, der Riß (in Metallen, in der Telegr.-Leitung); crystalline ~ der körnige Bruch; fibrous ~ der sehnige Bruch; foliated ~ der blätterige, schieferige Bruch; granulated ~ der körnige Bruch; splintry ~ der splitterige Bruch.

frame *s* der Rahmen, das Gestell, Gerüst, die Einfassung; ~ work das Fachwerk, Bindewerk, die Fachwand.

to frank *v* (*P*) unter portofreiem Rubrum versenden; ~ing privilege *s* das Recht zu portofreier Korrespondenz.

frank *s* (*P*) der die Portofreiheit bestätigende Vermerk auf dem Briefumschlage (muß von der Hand desjenigen herrühren, der zur portofreien Versendung von Briefen 2c. befugt ist).

freight *s* die Fracht; die Befrachtung (eines Schiffes); das Frachtgeld; ~s *pl* die Frachtgüter; ~ by parcels die Stückfracht; ~ account die Frachtrechnung; ~ depot (Eisenb.) der Güterschuppen; ~ service (Eisenb.) die Güterabfertigung; ~ train (Eisenb.) der Güterzug.

friction *s* die Reibung; ~ brake der Bremsdynamometer; ~ gear, ~ gearing das Friktionsgetriebe, Getriebe von Friktionsrädern; ~ roller die Reibungsrolle, Leitrolle, Friktionswalze; ~ wheel das Friktionsrad.

frictional *a* electricity die Reibungselektrizität.

front *s* die Front, die Vorderseite (von Postkarten, Briefen 2c.); ~ elevation, ~ view die Voderansicht; ~ side die Straßenseite der Telegr.-Stangen.

frustum, frustrum *s* (Geom.) das Bruchstück, der abgestumpfte Körper; der abgestumpfte Kegel; ~ of a pyramid die abgestumpfte Pyramide.

fulcrum, fulcre *s* of a balance der Drehpunkt, Stützpunkt einer Waage; ~ of a lever der Drehpunkt, Stützpunkt, die Unterlage eines Hebels, die Wucht; die Tragfläche, Druckfläche.

to fumigate *v* räuchern, ausräuchern (zu Desinfektionszwecken).

fumigation *s* die Durchräucherung (in der Quarantaine einer aus einem cholera-behafteten Lande kommenden Post).

fund *s* der Fond, das Grundkapital; ~s *pl* der Geldvorrath, die Baarschaft; public ~s *pl* Staatsgelder, Staatspapiere; sinking ~ der Schuldentilgungsfond, das Amortisationskapital.

furlong *s* ein Längenmaaß ($1/8$ engl. Meile = 220 yards = 201,164 Meter).

furnace *s* der Ofen, Schachtofen, Schmelzofen; air ~, draught ~, wind ~ der Windofen; blast ~ der Hochofen; ~ steel der Schmelzstahl, Rohstahl, Frischstahl.

furniture *s* das Hausgeräth; die Ausrüstung; ~ of a magnet die Magnetarmatur.

furrow *s* (Techn., Mech.) die Rinne, die Nuth; der Schraubengang, das Schraubengewinde; ~ of a roller die Nuth einer Walze.

to fuse *v* schmelzen, zerfließen.

fusible *a* schmelzbar; ~ metal das Schnellloth; ~ wire (*T*) der Abschmelzdraht (an Blitzableitervorrichtungen).

fusing *a* schmelzend; ~ agents *pl* Schmelzmittel; ~ point der Schmelzpunkt.

fusion *s* das Schmelzen, die Schmelzung (von Metallen); die Verschmelzung (zweier Verwaltungen).

G.

to gage *v* s. to gauge.

to gain *v* 1) gewinnen; (*T*) voreilen (vom A-B-C-Apparat); 2) schräg ausladen, sich (nach irgend einer Richtung) erweitern.

gain *s* (Techn.) die schräge Ausladung, die Erweiterung; der Gewinn, Nutzen, Profit; (Mech.) der Einschnitt, die Kerbe, Fuge; account of ~ & loss das Gewinn- u. Verlustkonto; ~ of exchange der Gewinn aus dem Kurs.

gait *s* of a wheel die Länge der Umdrehung eines Rades.

gallon *s* (auch imperial ~, imperial standard ~) die Gallone, ein englisches Flüssigkeitsmaaß, 4,54 Liter enthaltend. Die amerikanische Gallone (standard ~) hält 321 Kubikzoll = 8,3389 Pfund Wasser bei 41° Fahrenheit.

galvanic *a* galvanisch; ~ battery die galvanische Batterie; ~ cell, ~ element das galvanische Element; ~ pile die galvanische, Volta'sche Säule; ~ trough der galvanische Trogapparat.

galvanism *s* der Galvanismus, die Berührungselektrizität.

to galvanize *v* galvanisiren; to ~ iron Eisen verzinken.

galvanometer *s* das Galvanometer; astatic ~ das astatische Galvanometer; coil ~ das Spulen-Galvanometer; dead-beat ~ das aperiodische Galvanometer; marine ~ das Marine-Galvanometer; mirror ~ das Spiegel-Galvanometer; shunted ~ das Galvanometer mit Nebenschluß; sine ~ das Sinus-Galvanometer, die Sinusbussole; sine-tangent ~ die Sinus-Tangenten-Bussole; tangent ~ das Tangenten-Galvanometer, die Tangenten-Bussole.

galvanoscope *s* das Galvanoskop; horizontal ~ das Horizontal-Galvanoskop; vertical ~ das Vertikal-Galvanoskop.

gang *s* of workmen die Bande, der Trupp Arbeiter.

gaped *a* rissig, gespalten.

garnish bolt *s* der Bolzen mit verschnittenem Kopf.

gas *s* (Phys., Chem.) das Gas,

der luftförmige Körper; das Leuchtgas; combustible ~es pl die brennbaren Gase; hydrogen ~ das Wasserstoffgas; illuminating ~ das Leuchtgas; noxious ~es pl die schädlichen Gase, der Hüttenrauch; oxygen ~ das Sauerstoffgas; ~ coke der Gaskoke; ~ conduit die Gasleitung; ~ flue das Gasrohr; ~ pipe die Hauptgasleitungsröhre; ~ retort carbon die Gasretortenkohle; ~ tar der Gastheer, Steinkohlentheer.

gaseous, gasiform a (Phys.) gasartig, gasförmig.

to gauge, to gage, to guage v aichen, ausmessen, kalibriren.

gauge, gage, guage s die Leere, das Maaß, Urmaaß; (Eisenb.) die Spurweite des Schienengeleises, Schienenweite; broad ~ die Spurweite von mehr als 4' 8½''; narrow ~ die Spurweite von weniger als 4' 8½''; rain ~ (Phys.) der Regenmesser; slide ~, sliding ~ die Schiebleere; standard ~ (Eisenb.) die Normalspurweite (von 4' 8½''); steam ~ das Manometer, der Dampfdruckmesser; ~ of way (Eisenb.) die Spurweite des Schienengeleises; wire ~ die Drahtleere, Drahtklinke.

to gear v (Masch.) verzahnen, eingreifen lassen; to ~ together (Masch.) in einander greifen.

gear, gearing s (Masch.) das Triebwerk, Gezeug; die Transmission, Fortpflanzung der Bewegung; ~, wheel ~ (Masch.) das Ineinandergreifen von Rädern, Zahnräderwerk, die Verzahnung; bevel ~, bevelled ~ (Masch.) die schiefe Verzahnung; das konische Zahnrad; conical ~, angular ~ die konische Verzahnung, Winkelverzahnung, das konische Triebwerk; connecting ~, intermediate ~ das Vorgelege, Zwischengetriebe; cylindrical ~, right ~, spur ~ die Kronverzahnung, zylindrische Verzahnung, Leitung durch Stirnräder; disengaging ~ die Vorrichtung zum Ausrücken, die Ausrückung; driving ~ das Getriebe; in ~ in Gang, in Bewegung; in Verbindung; out of ~ außer Verbindung mit den Bewegungstheilen, ausgerückt; running ~ das gehende Getriebe, Triebwerk in Bewegung; spur ~ die zylindrische Verzahnung; der Zahnkranz eines Kammrades; to take into ~ with one another eingreifen (von Zahnrädern gesagt); to throw into ~ in Eingriff bringen; to throw out of ~ ausrücken; toothed ~ das Zahnrad=, Kammradgetriebe, Zahnräderwerk.

general a account die Hauptrechnung, das Hauptkonto; ~ drawing die Uebersichtszeichnung, Hauptzeichnung; ~ post office das Generalpostamt.

generator s der Erzeuger; steam ~ (Dampfmasch.) der Kessel, in welchem Dampf erzeugt wird.

geometric, geometrical a geometrisch; ~ drawing die geometrische, nach Maaßen aufgetragene Zeichnung; ~ pen die Reißfeder.

geometry s die Geometrie, Meßkunst; subterraneous ~ die Markscheidekunst.

German a silver das Neusilber, Weißsilber, Argentan; ~ steel der Rohstahl, Schmelzstahl.

gib, jib s (Masch.) der Krahnbalken, Ausliger, Rollenholm.

gill s ein englisches Flüssigkeitsmaaß (je nach der Gegend 0,288 oder 0,144 Liter).

gimblet, gimlet s der Bohrer, Holzbohrer, Nagelbohrer; shell ~ der Hohlbohrer mit Zahn.

gin s (Mech.) der Bock, Hebebock, das Hebezeug; temporary ~ das Nothhebezeug; ~ block der an einem Hebebock oder Krahn angebrachte Flaschenzug; ~ cheeks die Schenkel, Beine eines Hebebocks.

gip s (Masch.) das Achslager.

giration, gyration s (Mech.) die Drehung.

girder s der Tragbalken, Bindebalken, Träger; box ~ der viereckige hohle Eisenträger; double ~ der in zwei Längentheile zersägte und in anderer Lage wieder zusammengebolzte Träger; iron ~ der eiserne Balken, Eisenträger; joggled ~ der Zahnbalken, verzahnte Träger; lattice ~, lattice work ~ der Gitterträger; truss ~, trussed ~ der verstärkte

Träger; tubular ~ der Röhrenträger.

giver s of a bill der Wechselaussteller, Wechselzieher, Trassant.

glide contact s (T) der Gleitkontakt, Schleifkontakt.

to glow v glühen.

glow lamp, glowing lamp s die Glühlampe.

to glue v kleben, aufkleben; leimen.

glue s der Leim; marine ~ der Marineleim (eine Lösung von Kautschuk in Steinkohlenöl).

gluey a klebrig, leimig.

gold-leaf electroscope s das Goldblatt-Elektroskop.

gong s der Gong (eine Art Lärmglocke); (T) die Weckerglocke.

goods s/pl die Güter, das Gut, die Fracht, Ladung; die Waaren; bulky ~ die Sperrgüter, sperriges Gut; ~ on freight die Frachtgüter; ~' depot (Eisenb.) die Güterhalle, der Güterschuppen; ~' service (Eisenb.) die Güterabfertigung; ~' station (Eisenb.) der Güterbahnhof.

gorge s of a pulley (Masch.) die Rinne, Rille, der Einschnitt.

gouge s das Hohleisen, der Hohlmeißel; ~ bit der Bohrlöffel.

governor s (Masch.) der Regulator; pendulum ~ (HA) der Pendel-Regulator.

gradient s die Neigung, Steigung, Neigungsfläche, das Gefälle; ascending ~, rising ~ die Steigung; descending ~ die Neigung, der Fall, das Gefälle; ~ of the face of a cutting, ~ of a sloped wall der Böschungswinkel; steep ~ die starke Neigung, der starke Fall; ~ post, indicator of ~ (Eisenb.) der Neigungsanzeiger.

to graduate v (Phys.) in Grade abtheilen, graduiren; mit einer Skala versehen.

graduated a arc (Phys.) der Gradbogen; ~ circle der Theilkreis; ~ gauge glass die graduirte Glasröhre.

graduation s die Abtheilung in Grade, Gradtheilung.

grain s die englische Gewichtseinheit = 0,0648 Gramm (7000 grains = 1 Pfund avoir-du-poids oder Handelsgewicht, 5760 grains = 1 Pfund troy oder Apothekergewicht); ~ of wood die Längenrichtung der Holzfasern, der Wuchs'; die Längenfaser im Holze, Holzfaser; ~ wood das Aberholz, Langholz.

grained a (von Metallen) gekörnt, körnig; coarse ~ grobkörnig; fine ~ feinkörnig.

gram, gramme s das Gramm (das Gewicht eines Kubikzentimeters Wasser bei 4° C. = 15,43 grains).

grant s die Erlaubniß, Konzession; ~ of a railway die Konzession zum Bau einer Eisenbahn.

grantee s der Konzessionar.

grantor s der Ertheiler einer Konzession.

granular a (von Mineralien) körnig; ~-crystalline körnig-krystallinisch.

to granulate v (von Metallen) granuliren, körnen; mit einer rauhen Oberfläche versehen; ~d gekörnt.

graphite, black lead, plumbago s (T) der Graphit; ~ resistance (T) der Graphit-Widerstand.

grapnel, grapling s der Bootsanker; die Sonde (um Kabel vom Meeresgrund aufzuholen), der Enterhaken; ~, grapple der Enterhaken.

grater s die Raspel.

gratuity s die außergewöhnliche Vergütung; die Gnadenpension für solche dienstunfähig gewordene Beamte, die eine noch nicht zehnjährige Dienstzeit zurückgelegt haben.

gravitation s (Phys.) die Schwerkraft; law of ~ das Gravitationsgesetz; ~ battery (Elektriz.) die Gravitationsbatterie.

gravity s (Phys.) die Schwere, das Gewicht; die Schwerkraft; the centre of ~ der Schwerpunkt; specific ~ das spezifische Gewicht; ~ battery die Daniell'sche Batterie (so genannt, weil die Flüssigkeiten vermöge ihres spezifischen Gewichts sich übereinander lagern).

grey a minium die Diamantfarbe.

to groove v aushöhlen, furchen; einschneiden, nuthen, auskehlen.

groove s die Aushöhlung, Furche, Rinne, Kerbe; das Drahtlager des Isolators; top ~ das obere Drahtlager des Isolators; side ~ das seitliche Drahtlager des Isolators.

grooved a genuthet, ausgehöhlt; ~ wheel (Masch.) das Schnurrad, das gezahnte Rad.

gross a amount der Bruttobetrag; ~ effect of an engine (Masch.) die dynamische Leistung, Totalleistung einer Maschine; ~ receipt die Bruttoeinnahme; ~ weight das Bruttogewicht; ~ work (Masch.) die Totalarbeit.

to ground v a wire (T) eine Leitung mit Erde verbinden.

ground plate s die Grundplatte, Erdplatte; ~ switch (T) der Umschalter für Erdverbindung.

grown a soil der gewachsene, natürliche Boden.

guage s f. gauge.

guard s (Eisenb.) der Schaffner, Konducteur; (T) die Schutzvorrichtung an englischen Isolatoren, um losgelöste Leitungsdrähte aufzufangen; „by ~" (P) „Sofort" (Vermerk auf Dienstbriefen hoher Staatswürdenträger); hook ~ (T) die hakenförmige Schutzvorrichtung an Isolatoren; hoop ~ (T) die reifenförmige Schutzvorrichtung an Isolatoren; mail ~ (P) der Eisenbahn-Postschaffner; ~ pole der Prellpfahl; ~ stone der Prellstein.

guide s (Masch.) die Führung, Leitung; der Führer; official postal ~, post office ~ das Posthandbuch für das Publikum; ~ pulley, ~ roller (Masch., T) die Leitrolle zur Führung des Papierstreifens an den Telegr.-Apparaten; paper ~ (T) die Papierführung.

guild s die Gilde, Zunft, Innung; ~ hall das Innungshaus; das Rathhaus, Stadthaus (besonders das von London).

gullet s die Wasserrinne, der Wasserlauf; der Graben; (Techn.) die zylindrische Feile.

gulley, gully s die Wasserrinne, der Wasserlauf.

gum s das Gummi, der Pflanzenschleim; ~ Arabic, Arabic ~ das Gummi arabicum; elastic ~, elastic das Gummi elasticum, der Kautschuk, das Federharz; ~ lac der Gummilack, Harzlack; ~ resin das Gummiharz.

Gunter's chain s die Gunter'sche Meßkette (66 Fuß).

guttapercha s die Guttapercha; vulcanized ~ die vulkanisirte Guttapercha; ~ sheath, ~ envelope der Guttapercha-Ueberzug; ~ sheet die Guttaperchatafel; ~ tubing die Röhren aus Guttapercha.

guy wire s (T) der Anker.

gyn s f. gin.

gypsum s der Gips, schwefelsaure Kalk; ~ of, plaster of Paris der Gips; paste of ~ der Gipsbrei.

gyration s die kreisende Bewegung, Kreisbewegung, Drehung.

H.

hackly a fracture of metals der hakige Bruch.

to hand v in a telegram, to ~ in a letter (P & T) ein Telegramm, einen Brief aufgeben.

hand s die Hand; der Arbeiter; der Zeiger; ~ cart der Handkarren; ~ shears pl (T) scheerenartig verbundene Stangen (beim Stangensetzen zum Nachschieben der Telegr.-Stangen gebraucht); ~ vice der Feilkloben, Handkloben.

handle s der Griff, das Heft, die Handhabe.

to harden v 1) härten; to case-~ härten, hartgießen, die Oberfläche des Eisens durch Einsatz in Stahl verwandeln; to ~ caoutchouc den Kautschuk härten, hornifiren; 2) binden, erhärten (von Zement, Gips u. s. w.).

to hatch v schraffiren (in Zeichnungen).

to haul v aufziehen, aufwinden;

to ~ ashore an's Land ziehen; to ~ in einholen; ~ing-in box der Einziehkasten (bei der Kabellegung).

hawk bill, hawk bill plyer s die Löthzange.

hawse s die Schiffsklüse, das Loch für das Ankertau; ~ box die Klüsenfütterung.

hawser s das Kabeltau.

head s der Kopf, die vordere Seite, Stirn; der Titel, Kopf, die Ueberschrift (auf Rechnungen u. s. w.); bolt ~ der Schraubenkopf; ~ bolt der Kopfbolzen; ~ clerk der erste Kommis, der Geschäftsführer; ~ letter carrier (P) der Oberbriefträger; ~ office (P) das Hauptpostamt; ~ postmaster der Vorsteher eines Hauptpostamtes; ~ way die Höhe im Lichten; to make ~ way (Techn.) vorangehen.

heart s das Herz, der Kern, das Innerste einer Sache, Mittelstück; ~ wood das Kernholz.

heat s die Hitze, Wärme; available ~ die nutzbare Wärme; evolution of ~ die Wärmeentwicklung; welding ~ die Schweißglühhitze, Schweißhitze; ~ lightening das Wetterleuchten.

heating s die Heizung, Erhitzung, Erwärmung; (Masch.) das Warmlaufen; ~ by hot air die Luftheizung; ~ by steam die Dampfheizung; ~ apparatus der Heizapparat; ~ effect die Wärmewirkung; ~ tube, ~ pipe, ~ flue das Wärmrohr; das Heizungsrohr.

to heave v up (Mech.) aufziehen, aufwinden.

hedge s die Hecke, der Zaun, die Einfriedigung; quickset ~ die lebendige Hecke.

heel s die Ferse, die schräge Abdachung; (Masch.) die Nase, der Ansatz; ~ of a mast, of an iron shod pole der Fuß eines Mastes, einer mit eisernem Fuß versehenen Stange.

height s die Höhe; ~, elevation die Höhe, Erhebung; barometric ~ die Barometerhöhe; ~ of fall (Mech.) die Fallhöhe; ~ of projection (Phys.) die Wurfhöhe, Kulminationshöhe; ~ in projection der Aufriß; ~ of water die Höhe des Wassers, der Wasserstand.

"Held for postage" matter (P Am) Postsachen, die unfrankirt oder ungenügend frankirt bei einer Postanstalt eingegangen sind und gegen Entrichtung des darauf haftenden Portos ausgegeben werden.

helical, helicoïdal a schraubenförmig, schneckenförmig, spiral; ~ line, helix (Geom.) die Spirallinie, Schneckenlinie, Schraubenlinie.

heliograph s der Heliograph; Sonnenlicht-Telegraph (Apparat, um durch reflektirtes Sonnenlicht Signale zu geben).

heliostate s der Heliostat, Lichtwerfer.

heliotrope s der Heliotrop (Instrument, um entfernte Bewegungen sichtbar zu machen).

helix s (s. auch helical) schraubenförmiges, spiralförmiges Gewinde; ~ of wire (T) die spiralförmige Drahtumwindung des Elektromagneten.

to helve v (Techn.) behelmen, bestielen, anschäften.

helve s (Techn.) der Stiel, Helm, Griff.

hematite iron s das aus Rotheisenstein gewonnene Eisen.

hemp s der Hanf; packing of ~ die Hanfumwickelung; ~ packing (Masch.) die Hanfliderung.

hempen a hanfen, von Hanf; ~ strand die Hanftrense; ~ thread der Hanffaden, Hanfzwirn.

to hew v behauen, beschlagen; glatt behauen.

hexode working s (T) die gleichzeitige Versendung von 6 Telegrammen mittels des Delany'schen Apparats.

hickory wood s das Hickoryholz, das nordamerikanische Nußbaumholz.

high furnace, ~ blast furnace s der Hochofen, Schachtofen; ~ pressure (Dampfmasch.) der Hochdruck, Ueberdruck; ~ road, ~ way die Landstraße, Kunststraße, Heerstraße; ~ water das Hochwasser, die

Fluth; ~ water mark das Zeichen des hohen Wasserstandes, Fluthzeichen.

hight *s* die Druckhöhe, das Gefälle.

to hinge *v* mit Angeln, mit Scharnieren versehen, einhängen.

hinge *s* (Mech.) der Bandhaken; ~, ~ joint das Scharnier, Gelenk, Gewinde; ~ band das Gelenkband, Scharnierband.

to hire *v* dingen, anwerben, miethen.

hire *s* der Lohn.

hitch *s* der Knoten (im Draht).

hoar frost *s* der Rauhreif.

hoe *s* die Hacke, Haue.

to hoist *v* (Techn.) aufziehen, aufwinden.

hoist *s* das Hebezeug, die Hebemaschine, der Aufzug; das Aufziehen.

to hold *v* enthalten, fassen; (Mech.) angreifen, packen.

holder *s* (Techn.) der Halter (des Fernsprechers); der Inhaber (eines Wechsels); stock ~ der Aktieninhaber, Aktionär.

holding bolt *s* (Masch., Techn.) der Riegelbolzen, Zugbolzen, Verbindungsbolzen.

to hollow *v* aushöhlen, auskehlen, abkehlen, kanneliren; to ~ with a mortise chisel ausstemmen.

hollow *s* die Aushöhlung; die Hohlleiste.

holophote, search light *s* das Rekognoszirungslicht.

home instrument *s* (*T*) der eigene Apparat; ~ freight die Rückfracht, Herfracht; ~ trade der inländische Handel, Binnenhandel.

homogeneous *a* (Techn., Phys.) homogen; ~ steel der Homogenstahl (direkt aus Schmiedeeisen hergestellte Gußstahl).

to honor *v* a bill, a draft einen Wechsel, eine Tratte honoriren.

hood *s* die Haube, die Kappe (einer Telegr.-Stange).

to hook *v* mit einem Haken befestigen, anhaspen, anhaken; to ~ in (Techn.) einhaken, einhängen.

hook *s* der Haken; die Nocke (der Britannia-Löthstelle); ~s & eyes *pl* Haken und Oesen; ~ bolt der Hakenbolzen; wire ~ der (in der Glocke mit Zement befestigte) Drahtträger des Pendel-Isolators.

hooked *a* hakenförmig; ~ on angeheftet.

horizontal *a* horizontal, wagerecht; ~ engine die liegende Maschine; ~ escapement die Zylinderhemmung.

Horse Power *s* die Pferdekraft; ~ shoe magnet der Hufeisenmagnet.

hose *s* (Techn.) der Schlauch; guttapercha ~ der Guttaperchaschlauch; India rubber ~ der Gummischlauch.

hundred-weight *s* (abgekürzt cwt.) der Zentner (112 engl. Pfd. = 50,8 Kilogramm).

hydrate *s* (Chem.) das Hydrat; ~ of potash das Kalihydrat, Aetzkali; ~ of soda das Natronhydrat, Aetznatron.

hydraulics *s/pl* die Hydraulik, Lehre von der Bewegung des Wassers und der flüssigen Körper, Mechanik der flüssigen Körper.

hydric *a* (Chem.) mit Wasserstoff verbunden; ~ chloride der Chlorwasserstoff.

hydrochloric *a* acid, chlorhydric acid (Chem.) die Chlorwasserstoffsäure, Salzsäure, der Chlorwasserstoff.

hydrodynamics *s/pl* die Hydrodynamik, Wasserkraftlehre, Wasserbewegungslehre.

hydroelectric *a* battery die galvanische Batterie.

hydrogen *s* (Chem.) der Wasserstoff, das Wasserstoffgas; ~ gas das Wasserstoffgas.

hydrogenated, hydrogenized *a* mit Wasserstoff verbunden.

hydrogenous *a* Wasserstoff enthaltend.

hydrographic *a* map die Seekarte.

hydrostatic, hydrostatical *a* (Phys.) hydrostatisch; ~ balance (Phys.) die Waage zur Bestimmung des spezifischen Gewichts fester Kör-

per; ~ level die Senkwaage, Wasserwaage; ~ pressure der Wasserdruck.
hydrostatics *s/pl* (Phys.) die Hydrostatik, Lehre vom Gleichgewicht des Wassers und der flüssigen Körper.

hypomochlium, fulcrum *s* (Masch.) das Hypomochlium, der Drehpunkt, Stützpunkt eines Hebels.
hypsometry *s* (Phys.) die Höhenmessung.

I.

illuminating *a* gas das Leuchtgas; ~ power die Leuchtkraft (z. B. des Gases).
to imbed *v* a cable by means of a dredging engine ein Kabel einbaggern.
immalleable *a* nicht hämmerbar (von Metallen gesagt).
to impart *v* mittheilen.
imperial *s* die Imperiale (eines Wagens), das mit Sitzen versehene Verdeck mancher Postkutschen; der Gepäckbehälter auf dem Verdeck der Postkutschen; eine Sorte Papier großen Formats, das Groß-Regalpapier.
impermeable, impervious *a* (Phys.) undurchdringlich, undurchlässig; ~ to water wasserdicht.
impetus *s* (Mech., Phys.) der Antrieb, Stoß; die Größe der Bewegung.
implement *s* (Techn.) das Geräth, Werkzeug; hand ~s *pl* das Handwerkszeug.
to impregnate *v* zubereiten, tränken, imprägniren.
impregnation, impregnating *s* of woods for their conservation die Tränkung, Zubereitung von Hölzern zur Konservirung derselben.
impression *s* der Druck, Abdruck, Abzug; die Auflage; das Exemplar; ~ of a stamp der Stempelabdruck; foul ~ der Fehldruck; ~ roller (M & HA) die Druckwalze.
incandescence *s* das Weißglühen, die Weißgluth; ~ lamp, incandescent lamp die Glühlampe.
inch *s* (abgek. in.) der Zoll (2,54 cm); square ~ der Quadratzoll; cubic ~ der Kubikzoll; ~ rod der Zollstock.
incidental *a* charges, ~ expenses *pl* die Nebenkosten.

inclination *s* die Neigung, die Inklination; das Gefälle; magnetic ~, dip die magnetische Inklination; ~ compass die Inklinationsbussole.
to incline *v* geneigt sein, schief stehen, auf eine Seite hängen.
incline *s* das Gefälle, die Neigung, Steigung; die geneigte, schiefe Ebene.
inclined *a* plane die schiefe Ebene; (Eisenb.) die Rampe; ~ plane road die Zahnradbahn.
to inclose *v* s. to enclose.
inclosure *s* s. enclosure.
to increase *v* the battery power (T) die Batteriekraft vermehren, verstärken.
increase *s* die Verstärkung, Vermehrung; ~ of salary die Gehaltszulage, s. auch increment.
increment *s* die Vermehrung; ~ of the current das Anwachsen des Stromes; ~ of salary die Gehaltszulage; ~ key (T) die Taste bei der Quadruplex-Telegraphie.
indecent *a* telegram ein Telegramm mit anstößigem Inhalt.
indemnification *s* die Entschädigung.
to indemnify *v* entschädigen.
indemnity *s* die Schadloshaltung.
to indent *v* (Techn.) verzahnen, mit Zahneinschnitten versehen; to ~ a beam einen Balken verzapfen, einzapfen.
indent *s* der Einschnitt, Zahneinschnitt, der Zahn.
indentation *s* die Einzähnung, der Einschnitt.
indented *a* ausgezackt.
index *s* der Zeiger; ~ of a balance die Zunge, der Zeiger einer Waage; ~ of a book das Inhaltsverzeichniß; ~ of correspond-

ence das Journal der Registratur; ~ letter (P) der Merkbuchstabe im Tagesstempel (zeigt die Stunde der Aufgabe oder Beförderung einer Postsendung an).

India rubber *s* der Kautschuk, das Gummi elasticum, das Federharz; vulcanized ~ rubber der vulkanisirte Kautschuk; ~ rubber pipe die Kautschukröhre, der Kautschukschlauch; ~ rubber tubing die Kautschukröhren.

indicator *s* (*T*) der Empfänger des A.B.C.-Apparats; (*F*) die Klappe; (Dampfmasch.) der Spannungsmesser; speed ~ (Masch.) der Geschwindigkeitsmesser; ~ telegraph der Schreib-, Druck- oder Zeigertelegraph; ~ of gradient (Eisenb.) der Neigungszeiger.

to indorse, to endorse *v* indossiren, zediren, giriren.

indorsee, endorsee *s* der Indossat.

indorsement, endorsement *s* das Indossament, Giro.

indorser, endorser *s* der Indossant; antecedent ~ der Vormann; immediate ~ der nächste Vormann; subsequent ~ der Nachmann.

to induce *v* (Phys., Elektriz.) induziren.

induction *s* (Phys., Elektriz.) die Induktion; ~ coil die Induktionsrolle; ~ current der Induktionsstrom; ~ key die Induktionstaste.

inductive *a* capacity die Induktionskapazität.

inductor *s* der Induktions-, Erregungsapparat.

inertia *s* (Phys.) die Trägheit, das Beharrungsvermögen; moment of ~ das Trägheitsmoment.

influence *s* (Phys.) die Influenz, elektrostatische Induktion.

infinity plugs *s/pl* (*T*) der erste und der letzte Stöpsel in Rheostaten (wenn beide stecken, ist kein Widerstand eingeschaltet).

to infringe *v* postal laws die Postgesetze verletzen.

infusible *a* (Phys.) unschmelzbar.

ingot *s* der Klumpen gegossenen Metalls, der Barren; ~ iron das Flußeisen; ~ steel der Flußstahl.

ink *s* die Tinte; (*T*) die Farbe; copying ~ die Kopirtinte; printing ~ die Druckerschwärze; stamping ~ die Stempelfarbe; ~ reservoir, ~ well (*MA*) der Farbebehälter; ~ roller (*MA*) die Farbwalze; ~ writer (*T*) der Farbschreiber.

inker, Morse inker *s* (*T*) der Morse-Farbschreiber.

inland letter *s* der an einen im Inlande Wohnenden gerichtete Brief; ~ parcel das für einen im Inlande Wohnenden bestimmte Packet; ~ telegram das interne Telegramm; ~ trade der Binnenhandel.

inner sack *s* (*P Am*) der Versteckbeutel.

inquiry *s* s. enquiry.

to insert *v* (Techn.) einsetzen, hineinstecken; to ~ a peg (*T*) einen Stöpsel einstecken; to ~ an instrument, an intermediate station (*T*) einen Apparat, eine Zwischenstation einschalten.

insertion *s* (*T*) die Einschaltung; die in ein Blatt gerückte Anzeige, das Inserat.

inside *s* die Innenseite; ~ screw die Mutterschraube, Schraubenmutter, inwendige Schraube; ~ width die Weite im Lichten.

inspector *s* der Aufseher, Inspektor; ~ of the road der Wegeaufseher; (Eisenb.) der Bahnaufseher; ~ of works der Bauaufseher; post office ~ (*Am*) der Postinspektor.

installation *s* (Techn.) die Einrichtung (z. B. einer Fernsprech-, einer elektrischen Lichtanlage u. s. w.).

instalments *s/pl* die Zahlungstermine; to pay by ~ terminweise, in Raten zahlen; payment by ~ die Ratenzahlung.

instant *s* (abgek. inst.) der laufende Monat.

instantaneous *a* current (Elektriz.) der Strom von augenblicklicher (sehr kurzer) Dauer.

instrument *s* (*T*) der Telegraphen-Apparat; (Techn.) das Werkzeug, Instrument; ~ counter (*T*) der Apparattisch; ~ room (*T*) der Apparatensaal.

insufficiently prepaid (*P, T*) ungenügend frankirt.
to insulate *v* isoliren, mit Nichtleitern umgeben.
insulating spring *s* (*HA*) die Isolirfeder; ~ stool (Elektriz.) der Isolirschemel; ~ stud (*MA*) der Isolirkontakt.
insulation *s* (Elektriz.) die Isolirung, Isolation; ~ resistance der Isolationswiderstand.
insulator *s* der Isolator; bell ~, bell shaped ~ der glockenförmige Isolator; double bell ~ der Doppelglockenisolator; lightning rod ~ der Blitzableiter-Isolator, Stangenblitzableiter; mushroom ~ der Isolator in Gestalt eines Pilzes; saddle ~ der die Hauptleitung (in England) tragende auf den Stangenende angebrachte Isolator; shackle ~ der Abspann-Isolator; spare ~ der Reserve-Isolator; suspended ~, swinging ~ der Baumisolator, Pendelisolator.
insurance *s* die Versicherung; ~ fee (*P*) die Versicherungsgebühr.
intelligence *s* die Nachrichten; ~ office das Nachrichtenbüreau; shipping ~ die Schifffahrtsnachrichten.
intensity *s* (Physi., Techn.) die Intensität, Stärke; ~ battery die Intensitätsbatterie (die Elemente sind hintereinander geschaltet); ~ galvanometer das Intensitäts-Galvanometer (mit langer Spule); ~ magnet der Elektromagnet mit vielen Windungen feinen Drahtes.
to intercept *v* a letter, a telegram einen Brief, ein Telegramm abfangen.
intermediary service *s* (*P, T*) der Vermittlungsdienst.
intermediate *a* poles *pl* (Magn.) die Folgepunkte eines Magneten; ~ station (*T*, Eisenb.) die Zwischenstation; ~ trade der Zwischenhandel, Tauschhandel.
to intermit *v* (Mech., Techn.) aussetzen, unterbrechen.
intermittent, intermitting *a* aussetzend; ~ current (*T*) der intermittirende Strom.

internal *a* connections *pl* (*T*) die Verbindungen im Innern des Amtes.
to interrupt *v* unterbrechen.
interruption *s* die Unterbrechung; (*T*) die Stromunterbrechung.
interruptor *s* (*T*) der Stromunterbrecher.
interval *s* between the poles der Zwischenraum, Abstand zwischen den Telegr.-Stangen.
inventory *s* das Ausstattungsverzeichniß, das Inventar; ~ pieces die Inventarienstücke.
inverse *a* umgekehrt, entgegengesetzt; in the ~ ratio, inversely *adv* im umgekehrten Verhältniß, umgekehrt.
inversion *s* die Umkehrung; (Chem.) die Umwandlung.
to invert *v* umkehren, umwandeln; ~ed commas *pl* die Anführungszeichen (im Deutschen „ ", im Englischen " ").
investigation *s* die Untersuchung.
invoice *s* die Rechnung, Faktura; per ~ laut Faktur; ~ amount der Fakturbetrag.
inward account *s* (*P*) die Nachweisung der angekommenen Briefe, die mit Zuschlagsporto belegt sind; letters ~s die angekommenen Briefe.
Ions *s/pl* (Elektriz.) die Jonen (Produkte der Elektrolyse).
iron *s* das Eisen; das Gußeisen; das Schmiedeeisen, das Stabeisen; angle ~, angular ~ das Winkeleisen; band ~ das Bandeisen; bar ~ das Stabeisen; black ~ das Schwarzblech; black short ~ das schwarzbrüchige Eisen; cast ~ das Gußeisen, der Eisenguß; charcoal blast ~ das Holzkohlenroheisen; cold short ~ das kaltbrüchige Eisen; drawn ~, drawn out ~ das Walzeisen; fined ~ das Frischeisen; flat ~, flat bar ~ das Flacheisen, Flachstabeisen; hot short ~, red short ~ das heißbrüchige Eisen; pig ~ das Roheisen; plate ~ das Eisenblech; raw ~ das Roheisen; rolled ~ das Walzeisen; round ~, round bar ~ das Rundeisen; sheet ~ das dünne

Eisenblech; short ~ das brüchige Eisen; soft ~ das weiche, hämmerbare Eisen; specular cast ~, spiegeleisen das Spiegeleisen; square ~ das Quadrateisen; steely ~ das Feinkorneisen; unwrought ~ das Roheisen; very short ~ das faulbrüchige Eisen; weak ~ das unganze Eisen; welded ~ das geschweißte Eisen; wrought ~ das Schmiedeeisen.

iron band *s* das Eisenband, Winkelband; ~ bar, ~ rod die Eisenstange; ~ coated mit Eisen bekleidet, bedeckt; ~ construction der Eisenbau; ~ cramp der Klammerhaken; ~ crow, ~ crow bar die Brechstange, das Brecheisen; ~ filings die Eisenspäne; ~ frame das eiserne Gestell, Eisengerippe; ~ hoop der eiserne Reif; ~ jack das eiserne Zahnrad; ~ keeper (*HA*) der Schwächungsanker; ~ ore das Eisenerz; ~ pin der eiserne Stift, Dorn; ~ plate die Eisenplatte, das Eisenblech; black ~ plate das Schwarzblech; corrugated ~ plate das gewellte Eisenblech; rolled ~ plate das Walzblech, gewalzte Eisenblech; ~ rod die Eisenstange; fine ~ rollers *pl* das Feineisenwalzwerk; ~ scale, ~ scales *pl* der Hammerschlag, Glühspan; ~ sheath die Eisenumhüllung (eines Kabels); ~ strap das Eisenband; ~ works *pl* die Eisenhütte, das Eisenwerk.

iron *s* (Techn.) das eiserne Werkzeug; soldering ~ der Löthkolben; tooling ~ das Glätteisen.

isinglass *s* die Hausenblase, der Fischleim; ~ stone der Talkstein, das Marienglas.

isochronal, isochronous *a* gleichzeitig.

isochronism *s* die Gleichzeitigkeit, gleichzeitige Bewegung.

isosceles *a* (Math.) gleichschenkelig.

to issue *v* ausgeben, emittiren; to ~ bills Wechsel ausstellen; to ~ new postage stamps neue Postwerthzeichen ausgeben.

issue *s* die Emittirung, Ausgabe; die Lieferung; bank of ~ die Zettelbank, Notenbank.

item *s* of an account der Posten in einer Rechnung, ~s *pl* in a newspaper Notizen, Vermischtes.

to itemize *v* Rechnungsposten spezifizirt eintragen, spezialisiren.

itinerants *s/pl* (*P*) Beamte der Postverwaltung, welche keinem bestimmten Amte zugetheilt sind und vermöge ihres Dienstes viel unterwegs sind (z. B. Leitungsrevisoren, Ingenieure, Leitungsaufseher u. s. w.; auch die englischen Postvertrauensärzte sind *itinerants*).

itinerary *s* (Eisenb.) der Fahrplan; das Handbuch für Reisende.

ivory *s* das Elfenbein; ~ black das Elfenbeinschwarz, Beinschwarz.

J.

jack *s* der Bock, das Gerüst; (Mech.) die Winde; (*F*) der Stöpsel für die Klappen; ~ cord (*F*) die Stöpselschnur; ~ knife das Klappmesser; (*T*, *F*) der Stöpsel; (*Am*) eine Art Umschalter.

jag *s* die Kerbe, Zacke; ~ bolt der Bolzen mit Widerhaken.

to jam *v* einzwängen, hineinklemmen.

jam nut *s* (Masch.) die Stellmutter, Gegenmutter, Doppelmutter.

jar *s* das Gefäß; Leyden ~ die Leydener Flasche.

jet *s* der Strahl (Wasser-, Lichtstrahl); ~ burner der Strahlbrenner.

jew's pitch *s* der Asphalt, das Judenpech, Erdpech.

jib *s* (Masch.) der Krahnbalken, Krahnbaum, Krahnarm.

to job *v* Arbeit in Akkord nehmen; mit Staatspapieren handeln.

job *s* die Arbeit in Akkord, Stückarbeit; ~ work die Akkordarbeit.

to join *v* verbinden; to ~ by casting anschmelzen; to ~ by forging anschmieden; to ~ the battery to the instrument (*T*) die Batterie

an den Apparat legen; to ~, to ~ up in circuit (T) einschalten.
joining up in circuit (T) die Einschaltung.
to joint v aneinanderpassen, zusammenfügen; (T) löthen, eine Löthstelle anfertigen; ~ed ladder die mit Scharnieren versehene, zusammenlegbare Leiter.
joint s (Masch.) die Verbindung, Fuge; (T) die Löthstelle; bellhanger's ~ dasf. wie twisted ~; Britannia ~ die Britannialöthstelle, Wickellöthstelle; cable ~ die Kabellöthstelle; guttapercha ~ die Guttaperchalöthstelle; twisted ~, bellhanger's ~ die Würgelöthstelle.
joint account s die gemeinschaftliche Rechnung; ~ box der Brunnen oder Kasten, in dem die Kabellöthstellen liegen; ~ hook der Löthhaken; ~ stock das Aktienkapital; ~ stock Company die Aktiengesellschaft; ~ testing die Löthstellenprüfung.
jointer s der Löther (Anfertiger von Kabellöthstellen).
journal s das Journal, Tagebuch; (Masch.) der Zapfen, Wellzapfen; ~ bearing das Achslager; ~ box die Achsbüchse, Lagerbüchse.
journey s (P) die Fahrt der Bahnpostbeamten.

judicature fee stamp s der Gebührenstempel der (englischen) Gerichtsverwaltung.
to jump v (Techn.) bohren, ein Bohrloch schlagen.
jumper s der Bergbohrer, Steinbohrer; ~ hole das Bohrloch.
junction s die Verbindung, Vereinigung; (Eisenb.) der Anschluß von Nebenlinien; die Verbindungsstation; failure of ~ (Eisenb.) das Verfehlen des Anschlusses; ~ of the overhead wire with the cable (T) Verbindung der oberirdischen Linie mit dem Kabel; ~ line, ~ railway die Verbindungsbahn; ~ post (T) die Ueberführungssäule; ~ station (Eisenb.) die Anschlußstation.
juncture s die Verbindung; der Verbindungspunkt; die Fuge; (Mech.) das Gelenk; articulated ~ (Mech.) die Gliederverbindung.
jurisdiction s der, das Geschäftsbereich.
to jut, to jut out v hervorragen, hervorspringen, einen Vorsprung bilden.
jut s (Techn.) der Vorsprung.
jute, Indian grass s die Jute, der Jutehanf, ostindischer Hanf; ~ twine der Jutezwirn; ~ yarn das Jutegarn.

K.

kaolin, China clay, porcelain earth s der Kaolin, die Porzellanerde.
kathode s (Elektriz.) die Kathode.
to keep v halten; to ~ the books buchhalten, die Bücher führen; to ~ book by double entry italienisch Buch halten; to ~ book by single entry einfach Buch halten; to ~ up fortsetzen, beibehalten.
keeper s der Aufseher, Wärter; (Mech.) der Halter; (Eisenb.) der Weichensteller; ~ of a magnet der Anker eines Magneten; ~ & registrator of official papers der Registrator.
to key v (T) den Schlüssel drücken; (Mech.) keilen, festkeilen; to ~ on (Mech.) auftreiben, auffeilen.

key s (T) die Taste, der Schlüssel; der Schlüssel zu einer Geheimschrift; (Mech.) der Keil, der Bolzen; ~ of a bolt (Mech.) der Keil, Splint, Vorstecker; screw ~ der Schraubenschlüssel; ~ board (HA) die Klaviatur; ~ lever (MA) der Tastenhebel; ~ screw (Mech.) der Schraubendreher, Schraubenschlüssel.
to kill v the wire den Draht strecken (durch Aufwinden auf Trommeln).
kilogram, kilogramme s das Kilogramm (Gewicht von 1000 Gramm = 2 Zollpfund).
kilogrammeter s das Kilogrammmeter, Meterkilogramm (die Leistung, welche nöthig ist, um 1

Kilogramm in 1 Sekunde 1 Meter hochzuheben).
kilometer, kilometre s das Kilometer (1000 Meter).
kinematics s/pl (Phys.) die Kinematik, Lehre von der Bewegung.
to kink v Kinke oder Schleifen bekommen (von Tauen, Kabeln u. s. w. gesagt).
kink s der Kink, die Schleife (im Draht).
kish s der Garschaum, Eisenschaum (beim Schmelzen von Erzen sich ausscheidender Graphit).
kishy a graphitisch, graphitartig; ~ pig, ~ pig iron das schwarze, übergare Roheisen.
knag s der Knorren, Knoten, Ast im Holz.
knaggy a knorrig, knotig, astig.
knee s (Masch.) das Knie, Kniestück, der Winkel; ~ joint das Kniegelenk; ~ pipe die Knieröhre; ~ timber das Knieholz.
knob s der Knopf, der Handgriff; ~ of an arbor (Masch.) der Wellbaumen, Hebedaumen; screw ~ der Schraubenkopf; ~ in wood der Knorren, Knoten im Holze.
knobby a knotig, knorrig.
knot s der Knoten, Knopf; der Knoten, die Seemeile (der 60ste Theil eines Grades, die allgemein angenommene Einheit bei der Kabelfabrikation); ~ in wood der Knorren im Holz.
knotty a knorrig, knotig (vom Holz).
to kyanize v timber Holz kyanisiren (in einer Sublimatauflösung tränken, um es in der Erde oder in Wasser vor Fäulniß zu bewahren; nach dem Erfinder Kyan benannter Prozeß).

L.

to label v mit Etikett versehen; (P) einen Zettel mit Angabe des Bestimmungsorts aufkleben.
label s das Etikett; (P) der Packetaufgabezettel, der Leitzettel; address ~ (P) der Packetaufgabezettel; number ~, number paper ~ (P) das Titelschild enthaltend die Angabe der Anzahl der Postpackete, die in einem Korbe, Sacke oder dgl. enthalten sind; penalty ~ (P Am) der Aufklebezettel (enthält die Androhung der Geldstrafe [penalty], welche denjenigen trifft, welcher den zur dienstlichen Korrespondenz dienendes Couvert für Privatkorrespondenz verwendet).
ladder s die Leiter; folding ~, jointed ~ die Klappleiter, die zusammenlegbare Leiter; ~ beam, cheek der Leiterbaum; ~ step die Leitersprosse; ~ wheel (Masch.) das Sprossenrad.
laden a in bulk mit Sturzgütern geladen; ~ in parcels mit Stückgütern geladen.
lading s die Ladung, Schiffsladung, Fracht; bill of ~ der Verladungsschein, das Konnossement.
lag s der Ueberstand, übergreifende Theil (von Brettern, Deckeln u. dgl.); ~ screw die Schraube mit übergreifendem Kopf.
lamellar, lamelliform, lamellated a blätterig (von Mineralien gesagt).
laminable a plättbar, streckbar, sich in dünne Plättchen bearbeiten lassend (von Metallen gesagt).
to laminate v strecken, plätten; to ~ wire Draht plätten.
laminating rollers s das Walzwerk.
land carriage s der Landtransport, die Landfracht; der Lohn dafür; ~ mark der Markstein, Feldstein, Grenzstein; ~ slip der Erdrutsch; ~ surveying die Landmessung, Landaufnahme, topographische Aufnahme.
landing s der Treppenabsatz, das Landen, die Landung; das Ausladen (eines Schiffes) (Eisenb.) der Perron; carriage ~ (Eisenb.) die Laderampe; ~ place der Landungsplatz.
to lap v boards (Techn.) Bretter

übereinander schlagen; to ~ over (Techn.) übergreifen, überstehen.

lap s (Techn.) die Ueberlappung; ~ joint der Stoß durch Uebereinandergreifen; ~ welded übereinandergeschweißt.

lapping s (Techn.) das Uebereinandergreifen.

lapsed a money order (P) die verfallene Postanweisung (die ein volles Jahr vom Tage der erfolgten Einzahlung ab nicht zur Auszahlung vorgezeigt worden ist).

larch, larch fir, larch tree s die Lärche, Lärchentanne.

to lash v anbinden, festbinden; abschnüren, mit der Schnur abmessen, laschen.

lash s die Lasche (bei Holzarbeiten).

lashing s (Techn.) die Befestigung.

last s die Last, Schiffslast (2 Tonnen = 2000 Kilogramm).

late a spät; ~ letter s (P) der Spätlingsbrief; ~ fee, ~ letter fee (P) die Einlieferungsgebühr für Spätlingsbriefe; ~ letter mail (P) die Spätlingsbriefkarte, der Spätlingsbriefkartenschluß; ~ parcel (P) das Spätlingspostpacket; ~ posting (P) die Aufgabe von Postsachen nach Schluß der Dienststunden; too ~ box (P) der Briefkasten, dessen Inhalt für die abgehende Post nicht mehr geleert wird.

latent a (Phys.) latent, gebunden; to become ~ (Phys.) gebunden werden.

lateral a seitlich; ~ fastening (Techn.) die Seitenbefestigung.

lath s die Latte; ~ nail der Lattennagel; ~ wood das Splittholz; ~ work die Lattung, Belattung, der Lattenverschlag, das Lattenwerk.

lathe, turning lathe s die Drehbank, Drechselbank.

latten s das Blech; das Messingblech; (auch brass ~, ~ brass); shaven ~ das dünnste Messingblech; ~ iron das Eisenblech; ~ wire der Draht aus Messingblech.

lattice s das Gitter, Gatter; open ~, clear ~ das durchbrochene Gitterwerk; ~ bridge die Gitterbrücke; ~ truss der Gitterbalken, hölzerne Gitterträger; ~ work das Gitterwerk; der Lattenzaun, Gitterzaun.

law s on bills of exchange das Wechselrecht; commercial ~ das Handelsrecht; ~ of customs das Zollgesetz; ~ of gravitation (Phys.) das Gravitationsgesetz; ~ stationer der Händler mit Schreibmaterialien, welche zur Abfassung gerichtlicher Urkunden nötig sind; ~ suit der Prozeß.

to lay v legen; to ~ attachment Beschlag legen; to ~ out a cable ein Kabel verlegen; to ~ out a line (T) eine Linie traciren; to ~ out the wire den Draht auslegen.

lay days, laying days pl die Liegetage, die Liegezeit (von Handelsgütern).

layer s die Lage, Schicht; die Haut, der Ueberzug; ~ of guttapercha die Guttaperchaschicht; concentric ~ of trees der Jahresring.

laying s of a cable die Kabellegung; ~ out die Abmessung nach der Schnur, das Abschnüren, Abstecken; ~ out of a line (T) die Tracirung einer Linie; ~ out of roads der Straßenbau.

to lead v verbleien; mit Blei überziehen; mit Blei ausgießen.

to lead v führen, leiten; ~ing-in wires pl (T) die Einführungsdrähte.

lead s das Blei; das Senkblei, Bleiloth; black ~ der Graphit, das Reißblei; deep ~ das Tiefloth; ~ covered wire das Bleikabel.

lead s (T) die Leitungsführung.

leaden a bleiern, von Blei.

leaf s (Techn.) das Blatt; der Bogen Papier; ~ of a door der Thürflügel; gold ~ das Blattgold; ~ of gold das Goldblättchen; ~ of a table das Tischblatt; ~ of a window der Fensterflügel.

league s die Meile (die englische *league* mißt 6076 yards; auf See ist *league* = 3 Seemeilen).

leakage s das Lecken, Rinnen, Auslaufen; (T) das Entweichen von Elektrizität, der Elektrizitätsverlust, Stromverlust, der Nebenschluß.

leaky *a* line (*T*) eine Linie mit Nebenschluß.

learner *s* (*T*) der Telegraphen-Anwärter; ~'s room (*T*) das Uebungszimmer; ~'s telegraph instrument der Uebungsapparat.

lease *s* der Pachtkontrakt, Miethkontrakt.

to leather *v* (Masch.) mit Liderung versehen, libern.

leathering *s* (Masch.) die Liderung.

ledge *s* (Techn.) die Leiste, der Stab; die Tragleiste.

lee side *s* die dem Winde abgekehrte Seite.

left-handed *a* (Techn.) links drehend; ~~ screw, left-hand screw die linke Schraube.

leg *s* (Techn.) die Stütze, Strebe; ~ of a magnet der Schenkel eines Magneten; ~ of a triangle (Geom.) der Schenkel eines Dreiecks.

legal *a* cap s. cap paper; ~ tender das gesetzliche Zahlungsmittel.

length *s* die Länge; (Eisenb.) der Bahnkörper; a short ~ of cable ein kurzes Ende Kabel; a cable's ~ eine Kabellänge (120 fathoms).

lengthening *s* die Holzverlängerung, Anstückung; ~ piece das Verlängerungsstück.

lens *s* (Opt.) die Linse, das Linsenglas; converging ~, condensing ~ die Sammellinse, Konverlinse; divergent ~, dispersing ~ die Zerstreuungslinse, Konkavlinse.

lessee *s* der Pächter, Miethsmann.

to lessen *v* (Techn.) dünner machen, verdünnen; to ~ in value an Werth abnehmen.

lessor *s* der Verpachter, Vermiether.

to let *v* vermiethen; to ~ in (Techn.) einfalzen, einlassen; to ~ out a contract einen Kontrakt vergeben; to ~ out on hire vermiethen (bewegliche Dinge); to ~ out a piece of work eine Arbeit vergeben; to ~ a ship to freight, to charter ein Schiff verfrachten.

letter *s* der Brief, das Schreiben; ~ of advice der Avisbrief; ~ of attorney die Vollmacht; ~ of conveyance der Frachtbrief; ~ exempt from postage der portofreie Brief; ~s patent *pl* das Patent, die Patenturkunde; dead ~ (*P*) der unbestellbare Brief; franked ~ der unter portofreiem Rubrum versandte Brief; late ~ s. late; ordinary ~ der gewöhnliche Brief; prepaid ~ der frankirte Brief; insufficiently prepaid ~ der ungenügend frankirte Brief; registered ~ der eingeschriebene Brief; ship ~ der Schiffsbrief, s. auch *ship letter*; single ~ der einfache Brief; steam-boat ~ der auf einem Dampfer aufgegebene Brief; unpaid ~ der unfrankirte Brief; way ~ der unterwegs (auf einer Unterwegsstation) aufgegebene Brief; by ~ *adv* brieflich; ~ bag der Briefbeutel; ~ balance die Briefwaage; ~ basket der Briefkorb; ~ bill die Briefkarte; ~ blank (*HA*) die Buchstabenblanktaste; ~ box der Briefkasten; ~ box clearer der Briefkastenleerer; ~ carrier der Briefträger; head ~ carrier der Oberbriefträger; ~ cover der Briefumschlag; ~ drop, slip of the ~ box der Briefkasteneinwurf; ~ mail der Briefkartenschluß; ~ mark das Postzeichen; ~ packet, ~ package, bundle of ~s das Briefpacket; das Briefbund; ~ paper das Briefpapier; ~ post die Briefpost; articles of the ~ post Briefpostgegenstände; ~ slit die Oeffnung am Briefkasten; ~ stamp der Briefstempel.

letting down *s* of steel das Anlassen, Nachlassen des Stahls; ~ out die Vergebung einer Arbeit; ~ out in contract die Verdingung einer Arbeit im Ganzen, in Akkord.

to level *v* abwägen, nivelliren; planiren, einebnen, abgleichen; (Techn.) zurichten, behauen, beschneiden; to ~ with the ground dem Erdboden gleich machen; wagerecht sein mit ...

level *a* wagerecht, horizontal.

level *s* die ebene, wagerechte Fläche oder Linie, das Niveau, das Nivellirinstrument, die Waage, Li-

belle; ~ with bulb of air, spirit ~ die Wasserwaage mit Libelle, Spirituswaage; mercurial ~ die Quecksilberwaage; ~ of the sea der Meeresspiegel; ~ crossing der Niveauübergang, Wegübergang im Niveau; ~ line der Wasserstand, die Wasserlinie; ~ plane das Eisenbahnplanum; ~ ruler das Richtscheit.
levelling s das Abgleichen, Nivelliren; die Planirung, Einebnung.
levelling instrument s das Nivellirinstrument; ~ plummet die Bleiwaage, Setzwaage; ~ rule der Nivellirmaaßstab; ~ screw die Stellschraube.
lever s (Mech.) der Hebel; der Hebebaum; ~ of a balance der Waagebalken, Hebel; ~ of the key (MA) der Tastenhebel; bent ~, angle ~ der gebrochene Winkelhaken, Kniehebel, Winkelhebel; ~ of the first kind der Hebel der ersten Art, zweiarmige Hebel (wo der Stützpunkt zwischen den beiden Endpunkten liegt); ~ of the second kind der einarmige Hebel, Traghebel; straight ~ der gerade Hebel; two-armed ~ der zweiarmige Hebel; ~ arm der Hebelarm; ~ detent der Sperrkegel.
leverage s (Mech.) das Verhältniß der Hebellänge, Hebelverhältniß; ~, ~ power die Hebekraft, Hebelkraft.
Leyden jar, Leyden phial s die Leydener Flasche; ~ battery die Batterie von Leydener Flaschen.
liability s die Verbindlichkeit, die Haftpflicht; ~ law das Haftpflichtgesetz.
to liberate v an alarm (T) einen Wecker auslösen.
libration s (Mech.) die Schwingung, Pendelschwingung.
lid, cover s der Deckel; hinged ~ der Klappdeckel.
life annuity s die Leibrente, Lebensrente, f. annuity; ~ boat das Rettungsboot; ~ insurance, ~ assurance die Lebensversicherung.
lift s das Heben, Aufheben; der Aufzug, Fahrstuhl; ~ of an arbor (Masch.) der Welldaumen, Hebedaumen; column ~ das Druckrohr,

Steigrohr einer Druckpumpe; ~ & descent die auf= und niedergehende Bewegung; steam ~ der Dampfaufzug.
lifter s (Masch.) der Welldaumen, Hebedaumen.
lifting cog s (Masch.) der Hebedaumen, Welldaumen; ~ gear (Mech.) das Hebezeug; ~ jack die Hebewinde, Bauwinde; ~ shaft (Mech.) die Hebewelle.
light s das Licht; ray of ~ der Lichtstrahl; electric ~ das elektrische Licht; ~ boat, ~ ship das Feuerschiff, Leuchtschiff.
light a leicht; ~ wire (T) die leichte Leitung; ~ weight falsches, betrügerisches Gewicht.
to lighten v erleuchten, erhellen; leichter machen, entlasten; to ~ down a piece of wood ein Stück Holz verjüngen, schwächen.
lighting s das Anzünden, Anstecken; die Beleuchtung, Erleuchtung; (von Metallen) das Härten und Anlassen; (Techn.) das Blankmachen.
lightning s der Blitz; flash of ~ der Blitzstrahl; ~ bridge protector f. protector; ~ conductor der Blitzableiter; ~ discharge der Blitzschlag; ~ discharger (T) der Blitzableiter; ~ guard f. guard; ~ protector (T) der Blitzableiter f. auch protector; ~ point die Blitzableiterspitze; ~ rod die Blitzableiterstange, der Blitzableiter.
limature s der Feilstaub, die Feilspäne.
limb s das Glied, der Arm (eines Hebels u. s. w.); ~ of a magnet der Schenkel eines Magneten.
lime s (Mineral) der Kalk; quick ~ der ungelöschte, gebrannte Kalk; wet ~ der abgelöschte Kalk; ~ light, Drummond light das Drummond'sche Licht, Kalklicht.
limestone s der Kalkstein.
limit s die Grenze; elastic ~ (Phys.) die Elastizitätsgrenze; ~ of weight (P) die Gewichtsgrenze.
to line v füttern, ausfüttern, bekleiden, beschlagen; (Mech.) mit

16

einer Kreideschnur eine Linie schlagen, schnüren; to ~ with masonry ausmauern; to ~ out abschnüren, den Schnurschlag machen; (bei der Vermessung) avisiren, abfluchten.

line s die Linie; (Eisenb.) die Trace; aërial ~ (T) die Luftlinie, die an Stangen entlang geführte Telegr.-Linie; ~ of no dip, magnetic equator (Phys.) der magnetische Gleicher; dotted ~ die punktirte Linie; double ~ (Eisenb.) die doppelgleisige Bahnlinie; down ~ (T) die Telegr.-Leitung, welche zur Beförderung von Telegrammen in der Richtung von der Hauptstation der Linie nach den anderen Stationen dient; (Eisenb.) das Geleise für Züge, welche in der Richtung von der Hauptstation der Linie nach den anderen Stationen gehen; land ~ (T) die Landlinie; oberirdische Linie; to lay out the ~ traciren; main ~ (T) die Hauptlinie; (Eisenb.) das Hauptgeleise; medial ~, neutral ~ die Mittellinie, neutrale, indifferente Linie; overhead ~ (T) die oberirdische Leitung, overhouse ~ (T) die über Häuser geführte Leitung; railway ~ die Eisenbahnlinie, der Schienenstrang, das Geleise; secondary ~ (Eisenb.) die Nebenlinie; submarine ~ (T) die unterseeische Linie; telegraphic ~, telegraph ~ die Telegraphenlinie; trunk ~ (T, Eisenb.) die Hauptlinie, von welcher sich Nebenlinien abzweigen; underground ~ (T) die unterirdische Leitung; up ~ (T) die Telegraphenleitung, welche zur Beförderung von Telegrammen in der Richtung nach der Hauptstation der Linie dient; (Eisenb.) das Geleise für Züge, welche in der Richtung nach der Hauptstation der Linie gehen; ~ current (T) der Linienstrom; ~ current working (T) das Arbeiten mit Arbeitsstrom ohne Relais; ~ man, ~ repairer (T) der Linienarbeiter, Leitungsaufseher; ~ sounder (T) der im Linienstrom eingeschaltete Klopfer ohne Relais; ~ wire (T) der Liniendraht; die Telegraphenleitung.

lineal, linear a (Geom.) linear; der Länge nach; ~ construction die Konstruktion mit Hilfe gerader Linien; ~ drawing das geometrische Zeichnen; ~ measure das Baumaaß.

lining s (Techn.) das Futter, die Fütterung, Bekleidung, Ausfütterung.

lining out s die Abfluchtung, Einfluchtung; ~ wall die Verkleidungsmauer, Futtermauer (von Gebäuden).

to link v (Mech.) verketten, durch Ketten mit einander verbinden.

link s die Fackel; (Masch., Mech.) das Gelenk, Glied; ~ of a chain das Kettenglied, Gelenk, der Kettenring.

linseed s der Leinsamen; ~ oil das Leinöl; boiled ~ oil, ~ oil varnish der Leinölfirniß.

to liquefy v schmelzen, flüssig machen; zur Flüssigkeit verdichten, kondensiren.

liquid a (Phys.) (tropfbar) flüssig.

liquid s (Phys.) die Flüssigkeit, der tropfbarflüssige Körper.

to liquidate v accounts Rechnungen bezahlen; to ~ debts Schulden liquidiren, saldiren.

liquidation s die Liquidation, Liquidirung, die Ausgleichung einer Schuld oder Forderung; die Auflösung einer Aktien-Gesellschaft.

list s die Liste, das Verzeichniß; der Kurszettel, das Kursbuch.

litharge s die Glätte, Bleiglätte.

litre s das Liter.

live a axle (Masch.) die bewegende, Bewegung mittheilende Achse; ~ load f. load.

to lixiviate v (Chem.) auslaugen.

lixiviating cask, ~ tank, ~ tub, ~ vat, ~ vessel s der Auslaugebottich.

lixiviation s (Chem.) die Auslaugung.

Lloyd's s ein Versammlungsplatz in der Börse zu London für alle Interessenten des Assekuranzgeschäfts, wo alle auf Schifffahrt und überseeischen Handel bezüglichen Nachrichten zuerst bekannt gemacht und in der *Lloyd's list* veröffentlicht werden.

to load v laden; to ~ a battery (Phys.) eine Batterie laden; to ~ in bulk Stürzgüter laden; to ~ in parcels Stückgüter laden.
load s die Last, Belastung, Ladung; (Masch.) die höchstmögliche Leistungsfähigkeit; (T) die Belastung, die äußeren Kräfte, die auf das Telegraphen-Gestänge wirken; ~ on the structure (T) die Belastung des Gestänges; dead ~ das Gewicht, das bei Prüfung von Materialien noch hinzugefügt wird, nachdem das proof load (s. dasf.) erreicht war; live ~ die Belastung von Materialien, wenn das Gewicht auf einmal, nicht nach und nach angebracht wird; proof ~ die Grenze der Belastung; working ~ die wirkliche Belastung; die Tragkraft.
loadstone, lodestone s der Magneteisenstein, natürliche Magnet.
loam s der Lehm, magere Thon, die Ziegelerde.
loan s die Anleihe; government ~ die Staatsanleihe; public ~ die öffentliche Anleihe.
lobby s (P Am) der Schaltervorflur; das Vorzimmer, der Vorsaal, die Vorhalle.
localization s of a fault (T) die Eingrenzung eines Fehlers.
to locate, to localize v a fault (T) einen Fehler eingrenzen.
location s der Beschäftigungsort, Stationsort eines Beamten; (Am) das vermessene und abgesteckte Stück Land.
local a örtlich, lokal; ~ action in a battery (Phys.) die Wirkung in der Batterie; ~ battery (T) die Lokal-(Stations-)Batterie; ~ current working das Arbeiten mit Arbeitsstrom unter Anwendung von Relais; ~ inker (T) der Stations-Farbschreiber; ~ line (T) die Ortslinie, Stadtlinie; ~ traffic (Eisenb.) der Ortsverkehr, Lokalverkehr.
to localize v s. to locate.
to lock v schließen, zuschließen, verschließen; (Mech.) festhalten, zum Stehen bringen; eingreifen; to ~ a wheel ein Rad hemmen; to ~ together by a detent (HA) verkuppeln.

lock s (Mech.) der Haken, die Spannkette; das Schloß; ~ nut (Masch.) die Gegenmutter; ~ saw die Lochsäge, Stichsäge.
locking s das Befestigen; ~ of toothed wheels das Ineinandergreifen gezahnter Räder; ~ wheel das Sperrrad.
locomotive, locomotive engine s die Lokomotive, der Dampfwagen; arranging ~ die Rangirmaschine.
to lodge v (Mech.) einsetzen, anbringen; to ~ a document ein Dokument deponiren.
log s der Klotz, Block; ~ of wood der Holzklotz, Holzblock.
long hundred weight s 120 Pfund; ~ measure das Längenmaaß.
longitudinal section s der Längendurchschnitt.
to loop v (T) zwei Leitungen zu einer Schleife verbinden.
loop s die Schlinge, Schleife; (T) der zu seinem Ausgangspunkt zurückkehrende Draht, die Schleife; ~ test (T) die Bestimmung eines Fehlers in der Schleife; die Schleifenprobe.
loose a lose; ~ articles Gegenstände ohne Verpackung, (P) bloßgehende Stücke; to work ~ (Mech., Masch.) Spielraum haben.
to lop v the branches of trees ausästen.
to lose v letters (vom Zeiger des Empfängers des A.B.C. Apparats gesagt) zurückbleiben.
to lower v the rates die Gebühren herabsetzen.
lowest bidder s (bei Arbeitsvergebungen) der Mindestbietende; ~ contractor der Mindestfordernde.
lowry, lory s (Eisenb.) der Blockwagen, die Lori, der Rollwagen zur Abfuhr von Gütern von der Bahnstation.
lozenge s (Geom.) die Raute, der Rhombus, das schiefwinklige Parallelogramm.
to lubricate v (Masch.) schmieren.
lubricating oil s (Masch.) das Schmieröl.
lubrication s (Masch.) das Schmieren, Oelen.

16*

lug *s* die Ruthe (ein Längenmaaß); der Haken, Henkel, die Zinke, der Ansatz.

luggage *s* (Eisenb.) das Gepäck, Passagiergut; ~ office das Gepäckbüreau; ~ train der Güterzug; ~ van, ~ waggon der Güterwagen.

lumber *s* das Bauholz; das Rundholz; ~ room die Geräthekammer, Rumpelkammer.

lump *s* der Klumpen (Metall); in the ~, by the ~ im Ganzen, in Bausch und Bogen; ~ work die Arbeit in Bausch und Bogen.

lutation *s* (Chem.) die Verkittung.

to lute *v* (Chem.) kitten, verkitten, verkleben.

lute *s* (Chem.) der Kitt; der Dichtungsring (von Gummi u. s. w.); fire ~ der feuerbeständige Kitt; hermetic ~ der luftdicht schließende Kitt.

M.

machine *s* (Masch.) die Maschine, das Kunstgetriebe, Triebwerk; ~ band der Treibriemen; ~ pulley die Riemenscheibe; covering ~ die Guttaperchapresse; paying-out ~ die Kabelauslegemaschine.

machinery *s* die Maschinerie, der Mechanismus, das Triebwerk; by ~ durch Maschinenkraft.

magazine *s* das Magazin, Waarenlager, die Niederlage, der Speicher.

magnesia *s* die Magnesia, Talkerde, Bittererde; carbonate of ~ die kohlensaure Magnesia; sulphate of ~ die schwefelsaure Magnesia, das Epsomsalz, Bittersalz.

magnesite *s* der Magnesit, Talkspath, Bitterspath (neutrale kohlensaure Magnesia).

magnesium *s* das Magnesium; ~ light das Magnesiumlicht; ~ oxide das Magnesiumoxyd, die Magnesia.

magnet *s* der Magnet; artificial ~ der künstliche Magnet; bar ~ der Magnetstab, Stabmagnet; bell-shaped ~ der Glockenmagnet; electro ~ der Elektromagnet; fagot ~ der zusammengesetzte Magnet; horse shoe ~ der Hufeisenmagnet; lamellar ~, laminated ~ der Blättermagnet; molecular ~ der Molekular-, Elementarmagnet; native ~, natural ~ der natürliche Magnet, Magneteisenstein; plate ~ der Bandmagnet; writing ~ der Schreibmagnet.

magnetic *a* magnetisch; earth-~ erdmagnetisch; ~ attraction die magnetische Anziehung; ~ battery, ~ magazine die magnetische Batterie, das magnetische Magazin; ~ induction, ~ influence die magnetische Induktion; ~ needle die Magnetnadel; ~ telegraph der magneto-elektrische Telegraph.

magnetization *s* die Magnetisirung.

magnetism *s* der Magnetismus; electro-~ der Elektromagnetismus; residual ~ das magnetische Residuum; terrestrial ~, earth ~ der Erdmagnetismus.

to magnetize *v* magnetisiren, magnetisch machen.

magneto-electric, magneto-electrical *a* magneto-elektrisch; ~-electricity der Elektromagnetismus; ~ electric sender, ~-electric transmitter der elektro-magnetische Transmissions-Apparat.

to mail *v* (P) mit der Briefpost versenden, einen Brief oder überhaupt einen Briefpostgegenstand aufgeben, in den Briefkasten stecken.

mail *s* (P) die Briefpost, die Briefschaften; das Briefpacket; alteration of ~s die Aenderung im Gange der Postzüge; conveyance of ~s by road die Postbeförderung auf Landwegen; despatch of ~ die Briefabfertigung; closed ~ das geschlossene Briefpacket; direct ~ das direkte Briefpacket; letter ~ die Briefpost; der Briefkartenschluß; open ~ der Einzeltransit; overland ~ die Ueberlandpost; ~ articles *pl* Gegenstände der Briefpost; ~ bag der Briefbeutel; extra ~ bag der Extra-Briefbeutel (der einzige Beutel, der ohne Briefkarte versandt werden darf; er enthält Postsendungen, die in Folge verspäteter Auflieferung nicht mehr dem regelmäßigen Kartenschluß haben beigefügt werden können); ~ bag apparatus die Vorrichtung zum Auffangen der Briefbeutel, die Fangvorrichtung; ~ boat das Postschiff; ~ box (Am) der Briefkasten; ~ car (Am), (in England stets) ~ carriage der Eisenbahn-Postwagen; ~ carrier der Postbote, Postillion; ~ cart der Postwagen; ~ cart contractor der Postfuhrunternehmer; ~ cart driver der Postkutscher, Postillion; ~ catcher (Am) der Fangapparat, s. auch catcher; ~ clerk der Beamte der Briefpost; railway ~ clerk der Postbeamte im Fahrdienst; ~ coach die Postkutsche; ~ coach service der Postfuhrdienst; ~ coach service contractor der Postfuhrunternehmer; ~ depredation der Postdiebstahl; ~ guard der Postbegleiter, Postschaffner; ~ guard employed on indoor duties der Postschaffner im innern Dienst; ~ guard employed on outdoor duties der Postschaffner im Bestellungsdienst; railway ~ guard, guard in charge of the ~ der Postschaffner im Begleitungsdienst; der Eisenbahnpostschaffner; ~ matter die Postsendung; ~ messenger der Postbote; ~ packet das Postdampfschiff, das Postpacketboot; ~ porter der Postunterbeamte; ~ pouch die Posttasche; ~ route der Postkurs, die Poststraße; ~ sack (Am) der Briefbeutel; ~ service der Postbeförderungsdienst; railway ~ service der Eisenbahnpostdienst; travelling ~ service der Postbegleitdienst; der Eisenbahnpostdienst; ~ train der Postzug, (Eisenbahnzug, der die Post mitführt); ~ transit das geschlossene Briefpacket; open ~ transit der Einzeltransit.

mailability s die Versendbarkeit, die Möglichkeit der Versendung durch die Post.

main s das Hauptrohr einer Gas- oder Wasserleitung; der Kanal; das Festland; die hohe See; water ~ die Wasserleitungsröhre; ~ line (T, Eisenb.) die Hauptlinie; ~ pipe die Hauptleitungsröhre; ~ road die Hauptstraße, Landstraße; ~ spring (MA) die Triebfeder.

maintenance s die Instandhaltung; Unterhaltung; die Instandhaltungsarbeiten; cost of ~ die Instandhaltungskosten.

make and break (T) die Selbstunterbrechung; ~ ~ ~ current der intermittirende, pulsatorische Strom.

male s of a screw, ~ screw die Schraube, Schraubenspindel.

malleability, malleableness s die Hämmerbarkeit, Dehnbarkeit, Streckbarkeit (von Metallen).

malleable a hämmerbar, streckbar (von Metallen); ~ cast iron der hämmerbare, schmiedbare Eisenguß; ~ iron das hämmerbare, streckbare Eisen, Schmiedeeisen, Stabeisen.

mallet s (Techn.) der hölzerne Hammer, Schlägel.

management s die Geschäftsführung, der Betrieb, die Verwaltung.

manager s der Geschäftsführer, Dirigent, Direktor; (Eisenb.) der Betriebsdirektor.

managing a clerk der Prokurist; ~ director der geschäftsführende Verwaltungsrath einer Aktiengesellschaft.

mandril, manderil, mandrel s (Techn.) der Dorn; der Schlagbohrer.

manganate s das mangansaure Salz; ~ of potassium das mangansaure Kali, Kaliummanganat.

manganese s das Mangan; der Braunstein; dioxide, bioxide, peroxide, black oxide of ~ das Mangansuperoxyd; der Braunstein, Pyrolusit.

manganic a acid die Mangansäure; ~ oxide das Manganoxyd.

manifold a mannigfaltig, vielfältig; ~ book (P) das Durchdruckbuch (die englische Post übt das

Verfahren, bei Annahme nachzuweisender Sendungen die zu Kontrollzwecken dienenden Notizen auf dem Wege des Durchdrucks in mehrfacher Ausfertigung herzustellen); ~ form of entry (P) das Formular für mehrfache Eintragungen auf dem Wege des Durchdrucks; ~ writer der Hektograph.

to manipulate v (Techn.) behandeln; (T) telegraphiren.

manipulator s (T) die Absendevorrichtung am Zeigertelegraphen, der Schlüssel, die Taste.

manometer s (Phys.) der Luftdichtigkeits-, Luftdruckmesser, das Manometer; (Dampfmasch.) der Dampfdruckmesser.

to map v out graphisch darstellen.

map s die Karte, Landkarte; hydrographical ~ die Seekarte; ~ of the mail routes die Postkurskarte.

margin s (Techn.) der Rand; die Grenze, der Spielraum.

marine glue s der Marineleim, Schiffsleim (eine Mischung von India-Rubber, Naphtha und Schellak).

to mark v (Techn.) bezeichnen; (T) Zeichen auf dem Streifen hervorbringen; to ~ out a line (T) eine Linie abstecken; to ~ out by stakes abpflöcken; ~ing iron der Brennstempel, das Brenneisen; ~ing pole der Absteckpflock, Markirpfahl (beim Vermessen).

mark s (Techn.) die Marke, das Merkzeichen, Kennzeichen; das Signal; trade ~ die Schutzmarke; ~ pile, ~ pole der Markirpfahl (beim Vermessen).

masonry s die Steinmetzarbeit; das Mauerwerk, Gemäuer; cemented ~ die Mauerung in Mörtel.

mast s der Mast (des Schiffes); (T) die große Telegraphenstange; pole ~ der Mast aus einem Stück.

mastic s der Mastix, das Mastixharz; der Kitt; asphalt ~ der Asphaltmastix, der Asphaltkitt; ~ cement der Mastixcement, Steinkitt; ~ varnish der Mastixfirniß.

material s das Material; ~s pl die Materialien; die Baumaterialien; raw ~s die Rohmaterialien, Rohstoffe.

mathematics s/pl die Mathematik; mixed ~ die angewandte Mathematik; pure ~, speculative ~ die reine Mathematik.

mattock s der Karst, die Haue; die Breithacke.

maturity s of a bill die Verfallzeit, Zahlungszeit eines Wechsels.

meadow ore s das Wiesenerz, der Raseneisenstein.

mean s die Mitte, das Mittel; geometrical ~ das geometrische Mittel.

mean a number die Durchschnittszahl; ~ proportion das Durchschnittsverhältniß.

to measure v (Techn.) messen, ausmessen.

measure s (Techn.) das Maaß; cubic ~, solid ~ das Körpermaaß; dry ~ das Maaß für trockene Gegenstände; fluid ~ das Flüssigkeitsmaaß; long ~, lineal ~ das Längenmaaß; square ~, superficial ~ das Flächenmaaß; tape ~ das Bandmaaß; unit of ~ die Maaßeinheit.

measurement s das Maaß, die Messung; ~ of a ship der Tonnengehalt eines Schiffes.

measuring chain s die Meßkette; ~ staff die Meßlatte; ~ tape die Meßschnur; ~ unit die Maaßeinheit.

mechanic s der Handwerker.

mechanic, mechanical a mechanisch; ~ powers pl die einfachen Maschinen, mechanischen Potenzen; ~ science die Mechanik.

mechanician, mechanist s der Mechaniker, Maschinenbauer.

mechanics s/pl die Mechanik, die Lehre von der Bewegung.

mechanism s (Masch.) der Mechanismus, das Getriebe, Triebwerk einer Maschine.

megafarad s das Megafarad (eine Million Farads).

megaveber s das Megaveber (eine Million Webers).

megavolt s das Megavolt (eine Million Volts).

megohm s das Megohm (eine Million Ohms).
membrane s die Membran, das häutige Zellgewebe.
to mend v ausbessern; to ~ cast iron Gußeisen schweißen.
mercuric a chloride das Quecksilberchlorid, das Sublimat; ~ nitrate das salpetersaure Quecksilberoxyd.
mercury, quicksilver s das Quecksilber; alloy of ~ die Quecksilberlegierung, Amalgam; compound of ~ die Quecksilberverbindung; perchloride of ~ das Quecksilberchlorid, Quecksilbersublimat; thread of ~ der Quecksilberfaden, die dünne Quecksilbersäule.
message s, telegraphic ~ das Telegramm; ~ form das Telegrammformular.
messenger s der Bote; mail ~ der Postbote; mounted ~ der berittene Bote; special foot ~ der Expreß-Fußbote.
metal s das Metall; die Bronze; das Gußeisen; (Eisenb.) die Kiesfüllung, Beschotterung; base ~ das unedle Metall; fine ~ das Feineisen; noble ~ das edle Metall; road ~ der Straßenbewurf.
metallic a metallisch; ~ circuit (T) der aus Drähten gebildete Stromkreis (mit Ausschluß der Erde).
meteor steel s der Meteorstahl.
meteorological a meteorologisch.
meteorology s die Meteorologie, Witterungskunde.
meter, metre s das Meter (der zehnmillionste Theil des Erdquadranten); cubic ~ das Kubikmeter; square ~ das Quadratmeter.
mica s der Glimmer, das Katzensilber, Katzengold.
microfarad s das Mikrofarad (ein Millionstel Farad).
microhm s das Mikrohm (ein Millionstel Ohm).
microveber s das Mikroveber (ein Millionstel Weber).
microphone s das Mikrophon; ~ transmitter der Mikrophon-Geber.
microvolt s das Mikrovolt (ein Millionstel Volt).

mil . . . der tausendste Theil eines Zolls.
mile, statute mile s die englische Meile (1760 yards = 1609,315 Meter); English geographical ~, nautical ~ die englische geographische Meile, Seemeile (1855 Meter oder der 60ste Theil eines Grades des Gleichers); German ~ die deutsche geographische Meile (7420,438 Meter = dem 15ten Theile eines Grades des Gleichers oder 4 Seemeilen); German metrical ~ die metrische Meile, deutsche Reichsmeile (7500 Meter = 4,66 englische Meilen); ~ mark, ~ stone, ~ post der Meilenstein, Meilenzeiger.
to mill v (Techn.) rändeln; walzen (Metalle).
mill s die Mühle, das Hüttenwerk, Hammerwerk; das Puddelwerk; rolling ~ das Walzwerk; ~ board der starke Pappendeckel (in den die Telegramm-Formulare eingeheftet sind).
milled a edge (Techn.) die Rändelung, der gekräuselte Rand; ~ head of a screw der geränderte Schraubenkopf.
mineral a mineralisch; ~ coal die Steinkohle; ~ oil die Naphtha, das Erdöl; ~ pitch der Asphalt, das Erdpech, Bitumen; ~ tar der Bergtheer, Erdtheer.
minimum s das Minimum; ~ limit die Minimalgrenze.
minium, red lead s der Mennig, die Mennige, das rothe Bleioxyd; grey ~ die Diamantfarbe.
minute book s das Protokoll (einer Behörde, Aktiengesellschaft u. s. w.).
minutes s/pl das Protokoll.
mirror s der Spiegel; ~ galvanometer das Spiegelgalvanometer.
Miscellaneous a „Insgemein" (der Titel eines Kostenanschlags).
misdemeanor s in office das Amtsvergehen.
misdirected matter s (P) die zur Post eingelieferte, aber so ungenügend adressirte Sendung, daß ihre Beförderung nicht erfolgen kann.

missing letter office (*P*) das Büreau, welchem die Nachforschungen nach verlorenen Brief- u. s. w. Sendungen obliegen.

mitre, miter *s* der halbe rechte Winkel, Winkel von 45°.

module, modulus *s* der Modul, Modulus; ~ of elasticity der Elastizitätsmodul; ~ of the line wire (*T*) der auf eine bestimmte Länge reduzirte Widerstand einer Linie aus Normaldraht, s. auch chain curve; ~ of tenacity der Zähigkeitsmodul.

moisture *s* (Phys.) die Feuchtigkeit; der Feuchtigkeitsgehalt; deposit of ~ der Feuchtigkeitsniederschlag; film of ~ die Feuchtigkeitsschicht.

molecular *a* (Phys., Chem.) molekular; ~ weight das Molekulargewicht.

molecule *s* (Phys.) das Molekül.

moment (*pl* moments), **momentum** (*pl* momenta) *s* das Moment; ~ of a force, of a power das statische Moment einer Kraft; ~ of flexure, ~ of flexion das Biegungsmoment; ~ of inertia das Trägheits-, Beharrungsmoment; ~ of resistance das Widerstandsmoment.

money *s* das Geld; circulation of ~ der Geldumlauf; base ~ das schlechte, geringwerthige Geld; coined ~ das falsche Geld; current ~ das gangbare Geld; ready ~, cash das baare Geld; ~ bag der Geldbeutel, Geldsack; ~ balance die Geldwaage; ~ bill die Geldkarte; ~ box die Geldkasse; ~ chest die Geldkiste, der Geldschrank; ~ letter der Brief, in welchem bei Eröffnung im „Dead Letter Office" Geld gefunden wird (nicht Geldbrief im deutschen Sinne); ~ order die Postanweisung; service ~ order die gebührenfreie Postanweisung, mittels welcher Gehälter an abwesende Beamte gesandt werden; ~ order advice die Benachrichtigung, welche von der Aufgabepostanstalt an die Bestimmungspostanstalt gerichtet wird; ~ order office das Postamt mit Postanweisungsverkehr; das Postanweisungsamt; ~ order service der Postanweisungsdienst; ~ packet, roll of ~ das Geldpacket; ~ scales *pl* die Geldwaage.

monkey *s* (Mech.) die Aushebungsvorrichtung; der Rammblock; ~ ram der Rammblock, die Pfahlramme; ~ spanner, ~ wrench der englische Schraubenschlüssel.

moor *s* das Moor, der Morast, das Sumpfland; ~ stone der Raseneisenstein, das Sumpferz.

morass *s* der Sumpf, Morast; ~ ore der Raseneisenstein, das Sumpferz.

Morse apparatus, ~ instrument, ~ telegraph *s* der Morse'sche Telegraph, der Morse; ~ code, ~ alphabet das Morse-Alphabet; ~ ink writer, ~ inker der Morse'sche Farbschreiber; ~ key der Morseschlüssel; ~ sounder der Klopfer nach dem Morsesystem, bei dem die Signale nach dem Gehör aufgenommen werden.

to mortise *v* (Techn.) verzapfen; to ~ one piece of timber into another einzapfen, einfugen; to ~ two pieces of timber durch Zapfenloch und Zapfen zusammenfügen, verzapfen.

mortise, mortice *s* (Techn.) das Zapfenloch; die Nuth; indented ~ das versetzte Zapfenloch; ~ bolt der hölzerne Zapfennagel, Vorstecker; ~ chisel der Lochbeitel, das Stemmeisen.

motion *s* (Mech.) die Bewegung; das Triebwerk; circular ~, ~ in a circle die Kreisbewegung; eccentric ~ die exzentrische Bewegung; reciprocating ~, alternate ~, alternating ~ die hin- und hergehende Bewegung; rotary ~, rotatory ~ die Kreisbewegung, Achsendrehung; undulatory ~ die Wellenbewegung.

motive power, motive force, motor *s* (Masch., Mech.) die bewegende Kraft, Bewegkraft, Triebkraft.

mould *s* (Techn.) die Form; die Schablone; ~, casting mould die Gießform.

to mount *v* aufstellen, montiren; to ~ by parbuckle (Mech.) schroten, aufschroten.

mouth *s* die Oeffnung, Mün-

to move v bewegen; sich bewegen.

moveable a (Mech.) beweglich, lose; ~ pulley die lose, bewegliche Rolle.

moveables s/pl die Mobilien.

movement s (Mech.) die Bewegung; gyratory ~ die Rotationsbewegung, Drehung; undulatory ~ die Schwingung.

moving a bewegend, beweglich; ~ force (Masch.) die bewegende Kraft, Triebkraft; paper ~ lever (HA) der Papierführungshebel; ~ on rollers auf Rollen laufend.

mucilage s der Pflanzenschleim; das Gummi arabicum.

mucilaginous a schleimig.

mud s der Schlamm, Modder; black ~ der Batterierückstand (der Niederschlag auf der Zinkplatte des Daniell'schen Elements); ~ box der Kasten für den Batterierückstand; ~ drag der Baggerapparat.

muff s der Muff, die Muffe (übergeschobene Hülse zur Verbindung zweier Röhren).

multiple a vielfach, vielfältig; ~ cable die Vereinigung mehrerer Kabel zu einem; ~ copies pl (T) die Vervielfältigung eines Telegramms; ~ received form (T) das Formular für ein Telegramm mit mehreren Adressen.

multiplex thread s die mehrfache Schraube; ~ thread screw das mehrfache Schraubengewinde; ~ telegraphy die mehrfache Telegraphie.

multiplier s der Multiplikator; bobbin of a ~ die Multiplikatorrolle.

municipal a telegraph station (T) die Kommunalstation.

muriate s (Chem.) die salzsaure Verbindung, das salzsaure Salz; ~ of potash das salzsaure Kali, Chlorkalium.

muriatic acid s die Salzsäure, Chlorwasserstoffsäure.

mushroom s der Pilz; ~ insulator der Isolator in Form eines Pilzes.

myriare s das Quadratkilometer (10 000 Ar oder eine Million Quadratmeter).

N.

to nail v nageln, annageln, vernageln; to ~ down, to ~ up zunageln.

nail s (Techn.) der (metallene) Nagel; barbed ~ der Nagel mit Widerhaken; the ~ bites der Nagel zieht an; clasp ~ der Nagel mit flachem Kopf; dog ~ der Schloßnagel; scupper ~ der Nagel ohne Kopf; plate ~ der Nagel mit flachem Kopfe; tree ~ der hölzerne Nagel; ~ head der Nagelkopf; ~ nippers pl die Nagelzange; ~ passer der Nagelbohrer.

naphtha s die (das) Naphtha, das Erdöl, Steinöl, Petroleum; coal ~ das leichte Steinkohlentheeröl.

narrow a eng, schmal; ~ gage (gauge) s die Schmalspur (Spurweite von weniger als 4′ 8½″); ~ gage railway die Schmalspurbahn.

narrows s/pl die enge Einsegelung.

native a gediegen; natürlich; roh (von Metallen gesagt); ~ copper das gediegene Kupfer.

navvy s der Erdarbeiter, Eisenbahnarbeiter.

neat, net a netto; ~ weight das Nettogewicht.

neck s of the insulator der Hals, das seitliche Drahtlager des Isolators; ~ of a crane (Masch.) der Krahnbalken.

needle s die Nadel, der Zeiger; (Mech.) der Windebalken; (Phys.) die Magnetnadel; dipping ~, inclinatory ~ die Inklinations-, Neigungsnadel; double ~ telegraph (T) der Doppelnadeltelegraph; magnetic ~, compass ~ die Magnetnadel; single ~ telegraph (T) der Einnadeltelegraph; ~ alphabet (T) das Alphabet des Zeigertelegraphen; ~ clerk (T) der den Nadelapparat bedienende Beamte; ~ instrument, ~ telegraph der Nadeltelegraph.

negative a (Phys.) negativ; ~ electricity die negative, Harz-Elektrizität; electro-~ elektronegativ.

to negotiate v a bill einen Wechsel begeben, verkaufen.

net a netto; ~ amount der Nettobetrag; ~ balance die Nettobilanz, der reine Saldo; ~ gain, ~ profit der Reingewinn; ~ proceeds, ~ produce der Reinertrag; ~ receipts pl die Nettoeinnahme; ~ weight das Nettogewicht.

netfob (netto free on board) netto frei an Bord.

neutral a (Chem., Phys.) neutral; ~ line die neutrale Mittellinie eines Magneten, Indifferenzlinie.

neutralization s (Chem.) die Neutralisation.

to neutralize v (Chem.) neutralisiren.

news s die Nachricht; das Zeitungspapier; ~ agent der Verkäufer von Zeitungen und Zeitschriften; ~ paper, newspaper die Zeitung; ~ paper wrapper das Kreuzband, Streifband; privileged ~ papers pl (P) diejenigen (amtlichen) Zeitungen, die gebührenfrei versendet werden: *London, Edinburgh & Dublin Government Gazettes & Police Gazette.*

to nick v kerben, einschneiden; abstecken (bei Vermessungen); to ~ out auszacken, mit Einschnitten am Rande versehen.

nick s in the head of a screw der Einschnitt, Einstrich am Schraubenkopfe.

nickel s das Nickel; ~ plating das Vernickeln, die Vernickelung.

nickeliferous a nickelhaltig; ~ iron das nickelhaltige gediegene Eisen, Meteoreisen.

night service s der Nachtdienst.

„nil" (nihil) Ausdruck für „leer", „vacat" in Leernachweisungen.

to nip v abkneifen, abzwicken, abkneipen.

nip s der Knick (im Drahte).

nipper plyers s/pl die Biege- und Kneipzange.

nippers s/pl die Kneipzange, Beißzange, Zwickzange; cutting ~ die Beißzange.

nitrate s (Chem.) das salpetersaure Salz, die salpetersaure Verbindung; ~ of silver, argentic ~ das salpetersaure Silberoxyd.

nitre s (Chem.) der Salpeter, der Natronsalpeter.

nitric a acid (Chem.) die Salpetersäure; ~ peroxide, hyponitric acid die Untersalpetersäure.

nitrogen, azote s (Chem.) der Stickstoff; nitromuriatic acid, nitrohydrochloric acid s (Chem.) das Königswasser, die Salpetersalzsäure.

nitrosulphuric acid s (Chem.) die salpetrige Schwefelsäure, Salpeterschwefelsäure.

nitrous acid s (Chem.) die salpetrige Säure.

nixes pl (Am) umfaßt alle Postsendungen, welche ohne Adresse aufgegeben werden, oder in der Adresse wohl den Bestimmungsort aber nicht die Angabe des betreffenden Staates enthalten, kurz alle Postsendungen, welche wegen ungenügender oder unrichtiger Adresse nicht befördert werden können.

nixey s (Am) der nach einem Orte ohne Postanstalt gerichtete Brief.

nob s (Techn.) der Knopf.

nock s die Kerbe, der Einschnitt.

nomination s (P) die erste Anstellung eines Postbeamten.

non- acceptance s of a bill die Nichtannahme eines Wechsels; ~ attendance (P) das Nichterscheinen zum Dienst; ~ conductor (Phys.) der Nichtleiter; ~ delivery (P & T) die Nichtbestellung, Unbestellbarkeit; ~ transmission of a telegram die Nichtbeförderung eines Telegramms.

Nonius, vernier s der Zehntelzeiger, Nonius, Vernier.

noose s die Schleife, Schlinge.

normal a senkrecht, perpendikulär, ~ line, normal die Normale, das Einfallsloth.

northern light, aurora borealis s das Nordlicht.

Norway fir *s* die Harztanne; ~ spruce fir die gemeine Tanne.

nose *s* die Nase, der vorspringende Vordertheil eines Gefäßes; ~ bit der Hohlbohrer mit einem Zahne; ~ key der Gegenkeil.

not-accounted for charges *s/pl* unvorhergesehene Ausgaben.

to notch *v* (Techn.) kerben, auskerben.

notch *s* (Techn.) die Kerbe, der Einschnitt, die Nuthe; (HA) die Falle der Stahlscheibe; ~ of a screw head der Einstrich am Schraubenkopfe.

to note, to note down *v* buchen, in die Bücher eintragen.

note *s* die Nota, Rechnung; der Schein, Schuldschein; ~ of the course of exchange der Geldkurszettel; ~ of exclamation das Ausrufungszeichen; ~ book das Notizbuch; ~ paper englisches dickes Briefpapier.

notice *s* (P & T) der Benachrichtigungszettel; ~ to the public, regulation ~ (P) der Aushang, der Postbericht; ~ board (T, Eisenb.) die Warnungstafel; ~ plate (P) die Stundenplatte am Briefkasten.

noxious *a* fumes, ~ gases *pl* die schädlichen Dämpfe, Gase.

number *s* die Zahl; ~ label, ~ paper label (P) das Titelschild mit Angabe der Anzahl der in einem Korbe, Sacke oder dergl. enthaltenen Postpackete; ~ peg (Eisenb.) der Nummerpfahl; ~ stone (Eisenb.) der Nummerstein.

nut *s* (Mech.) die Nuß, das Nußgewinde, Kugelgewinde; ~, screw ~ die Mutter, Schraubenmutter, inwendige Schraube; check ~, jam ~ die Gegenmutter; square ~ die vierkantige Mutter; winged ~, finger ~ die Flügelschraubenmutter; ~ key (Mech.) der Schraubenschlüssel; ~ screw die Nußschraube, die Schraubenmutter; ~ wrench der Schraubenschlüssel.

O.

oak, ~ tree *s* die Eiche, der Eichbaum; ~ wood, ~ timber das Eichenholz.

oaken, made of oak *a* eichen, aus Eichenholz.

oakum *s* das Werg, die Hede; black ~ getheertes Werg; white ~ ungetheertes Werg.

to obliterate *v* (P) entwerthen; obliterating stamp (P) der Entwerthungsstempel.

observation *s* die Beobachtung; ~ of an angle (Geom.) die Messung eines Winkels.

observatory *s* die Sternwarte.

to observe *v* visiren (bei Vermessungen); peilen; to ~ an angle einen Winkel messen.

obtuse *a* (Geom.) stumpf; ~ angled, ~ angular stumpfeckig, stumpfwinklig.

occupied *a* line (T) die besetzte Linie.

odd *a* (von Zahlen) ungerade; ~s & ends *s/pl* Abfälle.

off-charges *s/pl* die abzurechnenden Kosten.

office *s* das Büreau; das Komtor; (Techn.) die Werkstatt; ~ connections *pl* (T) die Zimmerverbindungen; ~ establishment (P, T) die Amtseinrichtung; ~ hours *pl* die Büreaustunden; ~ window (P, T) der Schalter; ~ wires *pl* (T) die Zimmerdrähte.

officer *s* der Beamte; ~ of the post office der Postbeamte.

official *a* dienstlich; vorgeschrieben; ~ business die Dienstsache; ~ hours *pl* die Dienststunden.

offset *s* die Gegenforderung, Gegenrechnung.

Ohm *s* (Elektriz.) das Ohm, die Widerstandseinheit, der Normalwiderstand.

Ohmad *s* das Ohmad (Bezeichnung der Widerstandseinheit zu Ehren Ohms an Stelle der B.A.U.)

okonite *s* das Okonit (ein Isolationsmaterial bestehend aus 38 %

reinen Kautschuks und im übrigen aus Kohlenwasserstoff und kieselsauren Salzen).

to open *v* an account ein Konto eröffnen; to ~ a line for traffic (*T*) eine Linie dem Verkehr übergeben.

open *a* offen; ~ mail, ~ mail transit (*P*) der Einzeltransit; ~ wire (*T*) die Luftleitung; ~ing table (*P*) der Entkartungstisch.

operation *s* der Betrieb, die Wirksamkeit; (Mech.) die Bewegung eines Getriebes; mail routes in ~ Postkurse im Betriebe; a telegraph line in ~ eine im Betriebe befindliche Telegraphenlinie.

operator *s* (*T*) der Apparatbeamte; (Masch.) die Arbeitsmaschine, Kraftmaschine.

order *s* die Bestellung; die Verfügung; (Techn.) die bestellte Arbeit; post office ~, money ~ (*P*) die Postanweisung.

ordinance *s* der Erlaß.

ore *s* das Erz; das rohe Metall.

origin *s* der Ursprung; land of ~ das Ursprungsland; office of ~ das Ursprungsamt.

original *a* message (*T*) das Ursprungstelegramm.

orthograph, orthographic projection *s* der Aufriß, Standriß, die Vertikalprojektion; (Geogr.) die senkrechte Projektion der Erde auf die Ebene des Aequators; external ~ der Aufriß, die Fassade; internal ~ der Durchschnitt.

to oscillate *v* (Mech.) oszilliren, hin- und herschwanken.

oscillation, oscillatory motion *s* (Mech.) die Schwingung, schwingende Bewegung, Pendelbewegung; axis of ~ die Schwingungsachse.

ounce *s* (abgekürzt oz.) die Unze (31,103 Gramm des Troygewichts; 28,350 Gramm des avoir du poids).

out *adv* of gear (Masch.) ausgerückt; ~ of joint aus den Fugen gewichen; ~ of line nicht in gerader Linie.

outfit *s* die Ausrüstung, Ausstattung.

outlaid expenses, outlays *s/pl* gehabte Auslagen.

outlet *s* der Abzug, die Abzugsrinne, der Abfluß; ~ of a pipe die Mündung einer Röhre; ~ pipe das Abführungsrohr; das Dampfauslaßrohr der Dampfmaschine; ~ valve der Abzug, das Ablaßventil.

to outline *v* im Umriß zeichnen, skizziren.

outline *s* der Umriß, die Skizze.

outward *a* freight die Ausfracht, Hinfracht; ~ account (*P*) die Nachweisung der beförderten Briefe, auf denen Zuschlagporto haftet; ~ office (Eisenb.) das Abgangsbüreau für Güter; ~ passage die Hinreise; ~ trade der Ausfuhrhandel.

outwards *adv* nach außen; letters ~ beförderte Briefe.

overground, overhead, overland line *s* (*T*) die oberirdische Linie, Luftlinie.

overland mail *s* (*P*) die Ueberlandpost.

to overlap *v* (Techn.) übergreifen.

overseer *s* (Techn.) der Aufseher; ~ of the customs der Zollaufseher; ~ of a port der Hafenmeister; ~ of a railway line der Bahnaufseher.

to overtax *v* a letter einen Brief zu hoch austariren.

oxidation *s* (Chem.) die Oxydation; degrees of ~ die Oxydationsstufen.

oxide *s* (Chem.) das Oxyd (die Verbindung eines Körpers mit Sauerstoff).

to oxidize *v* (Chem.) oxydiren; ~d compound die Sauerstoffverbindung.

oxy-acids, oxacids *s/pl* (Chem.) die Sauerstoffsäuren.

oxygen *s* (Chem.) das Oxygen, der Sauerstoff; ~ gas das Sauerstoffgas; ~ salts *pl* Sauerstoffsalze.

ozokerite, earth wax *s* der Ozokerit, das Mineralwachs, Erdwachs; refined ~ das raffinirte Erdwachs, Ceresin.

P.

pace s der Schritt (als Maaß = 2½ Fuß).

to pack v packen, einpacken, verpacken; (Masch.) lidern.

pack s der Pack, das Packet, der Güterballen; ~ cloth die Packleinwand; ~ thread der Bindfaden, Packzwirn; ~ wax der Packlack.

package s der Güterballen, das Kollo; die Verpackung; der Packerlohn.

packed a (Masch.) mit Liderung versehen.

packet s das Packet (irgend welcher Art, aber nicht Postpacket); das Packetboot, Postpacketschiff; ~ boat, ~ ship das Packetboot, Postpacketboot; ~ line die Postpacketbootfahrt; ~ note eine Art Schreibpapier (im Format von 9 auf 11 Zoll); ~ service der Postschiffsdienst.

packfong s das Packfong (ein Metall), Argentan, Neusilber.

packing s das Packen, die Verpackung; (Masch.) die Liderung, Packung; ~ of hemp die Umpackung mit Hanf; India rubber ~ die Kautschukliderung; ~ case die Packkiste; ~ charges pl die Verpackungskosten; ~ good die Emballage; ~ paper das Packpapier.

pad s das Kissen, Polster; (T) das Heft Telegramm - Formulare; (Eisenb.) das Stoßkissen, der Buffer; stamping ~ (P) der Färbeapparat zum Stempeln; ~ saw der Fuchsschwanz, die Blattsäge; ~ screw die Endschraube; ~ way die Landstraße.

„paf" (packing, assurance, freight) Verpackung, Versicherung und Fracht.

painter s (Techn.) die Fangleine, das Fangtau.

pakfong s s. packfong.

to pale v umpfählen, einpfählen.

pale s der Pfahl.

paling s die Einpfählung; der Pfahlzaun.

pall, pawl s (Masch.) der Sperrkegel, die Sperrklinke.

pallet s das Brettchen; die Schaufel; ~ of the escapement (HA) der Hemmungslappen.

pamphlet s die Flugschrift, Broschüre.

pane s die Tafel, dünne Scheibe; das Fach, Feld, die Füllung (an Gebäuden); ~ of a wall das Mauerfeld, das Fach einer Mauer.

panel, pannel s das Feld, die Füllung, das viereckige Stück zum Einsetzen; ~ of a window die Fensterfüllung; ~ work das Fachwerk.

pantelegraph s (T) der Pantelegraph, Caselli'sche Drucktelegraph.

paper s das Papier; die Zeitung; das Papiergeld; ~s pl Papiere, Briefschaften; to read a ~ einen Vortrag halten; blotting ~ das Löschpapier; bull ~ das Konzeptpapier; carbonic ~ das Durchdruckpapier; commercial ~s pl (P) Geschäftspapiere; copy ~, copying ~ das Kopirpapier; drawing ~ das Zeichenpapier; letter ~ das Briefpapier; packing ~ das Packpapier; punched ~ (T) der gelochte Papierstreifen; stamped ~ das Stempelpapier; wrapping ~ das Packpapier; writing ~ das Schreibpapier; ~ feeder (M & HA) die Papierzuführungs - Vorrichtung; ~ pulp das Papier-Maché, der Papierteig, die zerstampfte Papiermasse; ~ roll (T) die Papierrolle; ~ slide (M & HA) die Papierführung; ~ sorter, female ~ sorter der Hülfsarbeiter, die Hülfsarbeiterin in der englischen Post, deren Dienst im Sortiren solcher aktenreifen Sachen besteht, die auf das Postsparkassen- und Postanweisungswesen Bezug haben.

paraffin, paraffine s das Paraffin; ~ candle die Paraffinkerze.

parallax s (T) der Fehler beim Ablesen z. B. der Winkelgrade eines Meßinstruments.

parallel s (Geom.) die Parallellinie; der Parallelkreis; ~s pl (Eisenb.) die Parallelschienen, Doppel T Schienen.

parallel a parallel, gleichlaufend; ~ circle der Parallelzirkel; ~ motion die Parallelbewegung.

parallelogram *s* of forces (Mech.) das Parallelogramm der Kräfte.

paramagnetic *a* bodies *pl* die paramagnetischen Körper (welche durch ein magnetisches Feld in der Richtung des Feldes selbst magnetisirt werden).

parameter *s* die Konstante einer Kurve; s. auch chain curve.

paratonnere *s* der Blitzableiter.

to parbuckle *v* aufschroten.

parbuckle *s* das Schrottau; to lower by means of a ~ hinabschroten; to mount (to raise) by means of a ~ aufschroten.

parcel *s* (P) das Postpacket, die Packetsendung; numbered ~ das numerirte (eingetragene) Packet (mit angegebenem Werthe); postal ~ das Postpacket; delivery of ~s die Packetbestellung; exchange of ~s der Päckereiverkehr; transmission of ~s der Päckereidienst, die Packetbeförderung; ~s bill der Packetabgangs-, -eingangszettel; ~s delivery die Packetbestellung; ~ post die Packetpost; ~ room die Packkammer; ~ vehicle der Päckereiwagen; supplementary ~ vehicle der Päckereibeiwagen.

to pare *v* (Techn.) das Rauhe abarbeiten, abschaben, abraspeln; schneiden, beschneiden; zurichten, bearbeiten.

paring *s* das Zurichten; ~s *pl* die Abfälle; ~ chisel der dünne Stechbeitel; ~ knife das Schälmesser; ~ machine die Stanzmaschine.

parish *s* das Kirchspiel, die Landgemeinde; ~ road der Vizinalweg, Gemeindeweg.

parliamentary *a* train (Eisenb.) der (vom englischen Parlament allen Bahnen vorgeschriebene) Zug mit Wagen dritter Klasse, der an allen Stationen hält.

partial *a* earth (T) s. earth.

party *s* die Parthie; competent ~ der Sachverständige; contracting ~ der Kontrahent; the ~ failing die Partei, welche den Kontrakt nicht beobachtet; the ~ observing die Partei, welche den Kontrakt beobachtet; ordering ~ der Auftraggeber, Kommittent.

pass *s* der Erlaubnißschein; (T) die Legitimation der Telegramm-Aufgeber, welche ein Stundungskonto haben; ~, form of ~ (T) das bezahlte Telegramm - Antwortsformular.

passing place *s* (Eisenb.) die Weiche, Weichstelle.

to paste *v* kleistern, kleben; to ~ on ankleben; to ~ up aufkleben, aufziehen.

paste *s* der Teig, Brei, die zähe Masse; die Pappmasse; ~ of gypsum der Gipsbrei; paper ~ das Papierzeug; ~ work, work in ~ die Papparbeit.

pasteboard *s* der starke Pappendeckel.

patent *s* (Techn.) das Patent, Privilegium; to apply for a ~ ein Patent anmelden; the ~ is expired, lapsed das Patent ist erloschen, abgelaufen; to grant a ~ ein Patent ertheilen; to take out a ~ ein Patent nehmen; ~ agent der Patentagent; ~ office das Patentamt, Patentbüreau; ~ rolls *pl* das Patentregister; ~ specification die Patentbeschreibung.

patentee *s* der Patentinhaber.

path, foot ~ *s* der Pfad, Fußweg; ~ of the current (T) der Stromweg.

pattern *s* (Techn.) das Muster, Modell, Musterstück.

pavement *s* das Pflaster; asphalt ~ das Asphaltpflaster; concrete ~ das Betonpflaster; ~ of flags, flag ~ das Fliesenpflaster.

paver, pavier *s* der Pflasterer, Steinsetzer.

paving *s* das Pflastern, die Pflasterung; boulder ~ das Kieselpflaster, Rundsteinpflaster; ~ stone der Pflasterstein.

pawl, paul, pall *s* (Mech.) die Sperrklinke, der Sperrhaken, Sperrkegel.

to pay *v* zahlen, bezahlen; to ~ down, to ~ in cash baar bezahlen; to ~ in einzahlen; to ~ by instalments terminweise, ratenweise bezahlen; to ~ out a cable ein Kabel abrollen, auslegen; ~ing-out apparatus, ~ing-out machine die Auslegevorrichtung bei der Kabellegung.

pay *s* die Zahlung; der Lohn, das Gehalt; ~ day der Zahltag, Lohntag.

payable *a* zahlbar, fällig, abgelaufen (von Wechseln); to make a bill ~ einen Wechsel zahlbar machen, domiziliren.

payee *s* (P) der Empfänger einer Postanweisung; ~ of a bill der Inhaber, Vorzeiger eines Wechsels.

payer *s* of a bill der Bezogene, Trassat.

paying-out apparatus *s* f. to pay.

payment *s* die Zahlung, Bezahlung; die ausgezahlte Summe; ~ on account die à Contozahlung; anticipated ~ die Vorausbezahlung; ~ in part die theilweise Zahlung; abschlägliche Zahlung; ~ received Zahlung erhalten, quittirt (Quittungsformular).

peak *s* die Spitze; ~ of a mountain die Bergspitze.

peat *s* der Torf; das Torfmoor; ~ bog das Torfmoor; ~ litter die Torfstreu.

peck *s* ein Hohlmaaß, 2 Gallonen (9,1 Liter) enthaltend.

pecker *s* (T) das Relais.

pedal *s* das Fußbrett, Trittbrett; (T) die Doppeltaste des Einnadel-Telegraphen.

to peg *v* festpflöcken, mit einem Pflock befestigen; dübeln, zusammendübeln; to ~ out a line (T) eine Linie abpfählen.

peg *s* der Zapfen, der Holzdübel; die Knagge; der Pflock; das Markirpfählchen; (T) der Stöpsel; split ~ der geschlitzte Stöpsel; spring ~ der federnde Stöpsel.

penal law, penal statute *s* das Strafgesetz.

penalty *s* die Strafe, Strafbestimmung, Strafbarkeit; die Geldstrafe; ~ envelope (P Am) ein Briefumschlag, der nur für dienstliche Schreiben gebraucht werden darf (so genannt, weil die Strafandrohung für mißbräuchliche Benutzung auf der Vorderseite aufgedruckt ist); ~ label (P Am) der Zettel, auf dem die Strafe aufgedruckt ist, die denjenigen trifft, der zur Privatkorrespondenz dienstliche, die gebührenfreie Beförderung sichernde Couverts u. dgl. benutzt.

pencil *s* der Stift, Bleistift, Zeichenstift; (T) der Stift; der Pinsel; ~ of carbon der Kohlenstift; black lead ~, lead ~ der Graphitbleistift; red ~, red chalk ~ der Rothstift.

pendulum *s* (Mech.) das Pendel; der Perpendikel; der Regulator (der Dampfmaschine); electric ~ das elektrische Pendel; ~ ball (HA) die Pendelkugel; ~ regulator der Pendelregulator; ~ rod die Pendelstange, der Pendelarm.

penthode working *s* (T) die gleichzeitige Versendung von fünf Telegrammen auf dem Delany'schen Apparat.

per ... (Chem.) über ... (die höchste Verbindungsstufe andeutend).

per advance *adv* im Voraus; ~ balance per Saldo; as ~ account laut Rechnung; ~ waggon per Achse.

percentage *s* die Prozente; die Tantième; (Chem.) die Prozentigkeit, der Gehalt an ...

perch *s* die Ruthe, Meßruthe, Meßstange; ein Längenmaaß von 5,03 Meter.

perchlorate *s* (Chem.) das überchlorsaure Salz.

perchloric *a* (Chem.) überchlorsauer; ~ acid die Ueberchlorsäure.

perchloride *s* (Chem.) das Chlorid; ~ of iron das Eisenchlorid; ~ of mercury das Quecksilberchlorid.

to perforate *v* durchlöchern, durchbohren; to ~ the paper slip (T) den Streifen lochen.

perforator *s* (T) die Lochmaschine, der Schriftlocher; hand ~ der Handschriftlocher; keyboard ~ der Tastenschriftlocher.

to perform *v* leisten; work ~ed (Mech.) die mechanische Arbeit, geleistete Arbeit.

performance *s* (Mech.) die mechanische Leistung (z. B. einer Dampfmaschine).

permanence, permanency *s* (Phys.) der Beharrungszustand.

permanent *a* way (Eisenb.) der Oberbau, Bahnoberbau.

permanganate *s* (Chem.) das übermangansaure Salz.

permanganic *a* acid die Uebermangansäure.

permeability *s* (Phys.) die Durchdringbarkeit, Durchlässigkeit.

permeable, permable *a* (Phys.) durchlässig.

peroxide *s* (Chem.) das Superoxyd; ~ of hydrogen, hydric ~ das Wasserstoffsuperoxyd; ~ of manganese das Mangansuperoxyd, der Pyrolusit, Braunstein.

perpetual *a* screw, endless screw (Mech.) die Schraube ohne Ende.

perquisites *s/pl* die Nebenbezüge eines Angestellten.

pervious *a* (Phys.) durchlässig.

petty charges, petty expenses *pl* kleine Ausgaben, Verschiedenes.

pewter *s* das verarbeitete, gemischte Zinn.

pewterer *s* der Zinngießer; ~'s solder das Zinnloth; ~'s temper eine Legirung von Kupfer und Zinn, welche dem Zinn zugefügt wird.

philosophy, natural philosophy *s* die Naturwissenschaft.

phonic wheel *s* das phonische Rad.

phonograph, phonautograph *s* der Phonograph.

phosphate *s* (Chem.) das phosphorsaure Salz, Phosphat.

phosphatic *a* (Chem.) phosphorhaltig, phosphorreich; ~ pig iron phosphorhaltiges Roheisen.

phosphide *s* (Chem.) of copper die Verbindung des Phosphors mit Kupfer, die Phosphorbronze.

phosphor-bronze wire *s* der Phosphorbronzedraht.

phosphuret *s* (Chem.) die Phosphorverbindung, das phosphorhaltige Metall.

phosphuretted, phosphoretted, phosphorized *a* (Chem.) mit Phosphor verbunden, phosphorhaltig.

photophone *s* das Photophon.

physical *a* agents *pl* die Naturkräfte; ~ philosophy die Physik.

physicist *s* der Physiker.

physics *s/pl* die Physik, Naturwissenschaft, Naturlehre.

to pick *v* picken, hauen; to ~ up the mail die Post einnehmen (von Schiffen gesagt).

pick *s* (Techn.) die Picke, Haue, der Spitzhammer, das Spitzeisen; broad ~ die Breithaue; ~ axe die Pickart, Spitzhacke.

pick-up circuit *s* (T) die Leitung, in die an Stelle einer gestörten Theilstrecke das entsprechende Stück einer anderen Parallelleitung eingeschaltet ist.

picked *a* (Techn.) ausgesucht (von Rohmaterialien).

picker *s* jedes spitze Werkzeug; der Spitzhammer, die Spitzhaue.

picket *s* (Techn.) der Pfahl, Pflock; der Knebel; ~, tracing ~ der Absteckpflock, das Markirpfählchen.

piece *s* das Stück; ~ work die Arbeit auf's Stück, im Geding, Akkordarbeit; to do ~ work auf's Stück arbeiten.

pier *s* der Pfeiler, Strebepfeiler; der Damm, Hafendamm, Molo; ~ of a bridge der Brückenpfeiler.

pierage *s* das Dammgeld, der Hafendammzoll.

to pierce *v* anbohren; to ~ through durchstechen, durchbohren.

piercer *s* (Techn.) der Bohrer, Nagelbohrer.

pig, pig iron *s* das Roheisen; ~ of iron die Gans, Eisengans.

pigeon hole *s* das Fach in einem Schreibtisch.

pike *s* (Techn.) die Spitze eines Dinges; die Pieke; die Haue, Keilhaue; der Dorn.

to pile *v* pfählen, verpfählen; in Haufen setzen, aufstapeln; to ~ up aufschichten.

pile *s* (Techn.) der Haufen; der Pfahl; die (galvanische) Säule; to drive, to ram a ~ einen Pfahl ein-

piling — plank

rammen; ~ of wood der Holzstoß; ~ block der Rammknecht; ~ bridge die Pfahlbrücke, Jochbrücke; ~ driver die Ramme.

piling s das Pfahlwerk; die Rammarbeit.

pillar s der Pfeiler; ~ letter box der Säulenbriefkasten.

pillow s (Masch.) die Pfanne, das Lager, Zapfenlager; ~ block das Zapfenlager.

to pin v (Techn.) mit Stecknadeln anheften; verbolzen; einzapfen.

pin s (Techn.) die Stecknadel; der Stift, Vorsteckstift, Pflock, Bolzen; crank ~ (Masch.) der Kurbelzapfen; ~ of a joint der Pflock, Bolzen im Scharniere; ~ with screw head der Schraubenbolzen; set ~ die Stellschraube; wood ~, wooden ~ der Holznagel, Dübel; ~ bit die Bohrspitze, das Bohreisen; ~ drill der Zapfenbohrer, Zentrumbohrer; ~ switch (T) der Stöpselumschalter.

pincers, pinchers s/pl die Zange, Beiß, Kneipzange; crooked ~ die Krummzange; flat ~ die Plattzange.

to pinch off v abkneipen.

pincher, pinching bar s das Brecheisen, Hebeeisen, die Hebestange.

pinching nut s die Stellmutter, Gegenmutter, Klemmmutter.

pine s die Fichte, die Kiefer; pitch ~ die Pechtanne.

pinion s (Masch.) das Getriebe; das Zahnrad; das kleinere von zwei ineinander greifenden Zahnrädern; ~ & rack, rack & ~ Zahnstange und Rad, Zahnstange mit Getriebe; endless screw & ~ Schraube ohne Ende mit Schraubenrad; spur ~ das Stirnrad, Stirngetriebe.

pint v die Pinte (ein englisches Flüssigkeitsmaaß = $1/8$ Gallone = 0,577 Liter).

pipe s (Techn.) das Rohr, die Röhre; der Schlauch; die Röhrenleitung; joint ~ die Verbindungsröhre; main ~ das Hauptrohr, die Hauptleitungsröhre.

piston s (Mech.) der Kolben, Stempel; ~ rod die Kolbenstange.

to pitch v pechen, verpichen; to ~ piles Pfähle (im Wasser) einschlagen; to ~ a tent ein Zelt aufschlagen.

pitch s (Techn.) die Neigung, Steigung; das Pech; (Masch.) die Zahntheilung eines Zahnrades; die Tonhöhe; ~ of a screw (Mech.) die Steigung, Ganghöhe des Gewindes, Höhe des Schraubenganges; ~ pine die Pechtanne, Harztanne; ~ wheels pl (Masch.) ineinander greifende Räder.

pitcher s (Techn.) die Brechstange, das Brecheisen; die Haue, Hacke.

pitching s das Verpechen, Verpichen; die Senkung, Neigung.

pith s das (Pflanzen-)Mark; elder ~ das Hollundermark; ~ ball das Hollundermarkkügelchen; ~ ball electroscope das Hollundermarkkugel-Elektroskop.

pivot s (Mech.) der Zapfen, die Angel, der Stift, auf dem sich etwas dreht; ball ~ der Kugelzapfen; ~ of a horizontal shaft der liegende Zapfen, Wellzapfen; sole, socket of a ~ die Zapfenmutter; ~ of a vertical shaft, vertical ~ der stehende Zapfen, Spurzapfen; ~ screw die Zapfenschraube.

plain s die Ebene, Fläche.

plain a eben, glatt, flach.

to plan v (Techn.) einen Plan zu etwas machen, entwerfen.

plan s der Plan, Entwurf; der Grundriß; principal ~ der Hauptplan, Ueberfichtsplan; projected ~ der Entwurf; raised ~, elevation ~ der Aufriß, das Profil; sectional ~ der Horizontalschnitt; site ~ der Situationsplan.

to plane v (Techn.) ebnen; hobeln, abhobeln; to ~ the soil den Boden ebnen, planiren, abgleichen.

plane s die Ebene, Fläche; die Oberfläche; inclined ~ die schiefe Ebene; inclined ~ road die Zahnradbahn; level ~ das Planum, die Planie; oblique ~ die schiefe Ebene.

plane s der Hobel.

plane a eben, flach; ~ surveying die Feldmeßkunst.

plank s die Bohle, Planke; s. auch timber; ~ nail der Brettnagel.

17

planking s die Bekleidung mit Brettern, Verschalung.

plant s (Techn.) die Betriebsanlage; electric light ~ die elektrische Lichtanlage; telephone ~ die Fernsprechanlage.

plaster s der Putz, die Tünche, der Bewurf (an Gebäuden); to fasten with ~ eingipsen; ~ of Paris der gebrannte Gips; ~ cast der Gipsabguß.

to plate v plattiren, bekleiden, überziehen, versilbern.

plate s die Tafel, die Platte; die Metallplatte; das Blech; bottom ~ die Bodenplatte; collecting ~ die untere Kondensatorplatte; condensing ~ die obere Kondensatorplatte; copper ~ das Kupferblech; draw ~ das Drahtzieheisen; drawing ~ die Scheibe; die Scheibenbank (beim Drahtziehen); earth ~ (T) die Erdplatte; front ~ die vordere Wange, Gestellplatte des *HA*; hind ~ die hintere Wange, Gestellplatte des *HA*; ~ of iron die Eisenplatte; iron ~, sheet iron das Eisenblech; junction ~ die Verbindungsplatte; rolled ~ das gewalzte Blech, Walzblech; steel ~ das Stahlblech; (*HA*) die Lippe des Schlittens; ~ discharger, ~ lightning discharger (T) der Plattenblitzableiter; double ~ sounder f. double; ~ spring die Blattfeder.

to platinize, **to platinate** v platiniren, mit Platin belegen.

platinum, **platina** s das Platin; native ~ das gediegene Platin; spongy ~ der Platinschwamm; ~ foil, ~ sheet das Platinblech; ~ point die Platinspitze.

to play v (Mech.) spielen, sich frei bewegen, im Gange sein.

play s (Mech.) das Spiel; der Spielraum; downward ~ der Spielraum nach unten; end ~ der leere Gang; upward ~ der Spielraum nach oben.

pliers, **plyers** s/pl die Zange, Drahtzange; cutting ~ die Schneidzange, Drahtzange; flat ~, flat-nose ~ die Flachzange, Plattzange; round ~, round-nosed ~ die Rundzange.

plinth s die Plinthe, untere Platte an Säulenfüßen, die Säulenplatte, Sockelplatte.

to plot v entwerfen; to ~ a line (*T*, Eisenb.) eine Linie traciren.

plot s das Stück Land; der Plan, Riß, Entwurf; grass ~ der Rasenplatz; ground ~ der Grundriß; ~ of situation der Situationsplan.

to plug v (Techn.) zupflöcken; (*T*) stöpseln.

plug s (Techn.) der Pflock, Zapfen; (*T*) der Stöpsel; infinity ~s f. infinity; taper ~ der konische Pfropfen.

to plumb v nach der Bleiwaage richten, abloten.

plumb s das Bleiloth, Senkblei, Richtblei; ~ level die Bleiwaage, Grundwaage; ~ line die lothrechte, vertikale Linie, die Lothleine; ~ rule das Richtscheit.

plumb a lothrecht, bleirecht, senkrecht, im Loth.

plumbago s der Graphit, das Reißblei.

plumber s der Bleiarbeiter, Bleilöther; ~'s solder das Weichloth, Schnellloth.

plumbic a chloride das Bleichlorid, Chlorblei.

plummet s das Senkblei, Lothblei, Loth; ~ level, levelling ~ die Bleiwaage, Senkwaage.

plunger s (Masch.) der Taucherkolben; (T) die Taste des Schriftlochers.

plunging battery s (*T*, Elektriz.) die Tauchbatterie.

to ply v regelmäßig fahren, segeln (von Fahrzeugen und Wagen gesagt, welche regelmäßige Fahrten zwischen gewissen Orten ausführen); to ~ down abfahren.

pneumatic, **pneumatical** a pneumatisch; ~ bell die pneumatische Schelle; ~ dispatch tube, ~ tube das Rohr einer Rohrpost; ~ railroad die atmosphärische Eisenbahn.

pneumatics s/pl (Phys.) die Pneumatik (Lehre von der Bewegung der elastisch-flüssigen Körper).

to point v (Techn.) spitzen, zuspitzen.

point s der Punkt, die Spitze;

~ lightning discharger (*T*) der Spitzenblitzableiter.
pointed *a* spitz; ~ file die Spitzfeile; ~ screw die Spitzschraube.
pointer *s* der Zeiger; ~ instrument (*T*) der Zeigertelegraph.
points *s/pl* (Eisenb.) die Weiche.
pointsman *s* (Eisenb.) der Weichensteller.
poise *s* die Schwere, das Gewicht; die Waage; water ~ die Wasserwaage, Libelle.
polar *a* aurora, ~ light das Nordlicht; ~ wire (*T*, Elektriz.) der Poldraht.
polarisation *s* die Polarisation.
to polarise, to polarize *v* polarisiren; polarised relay (*T*) das polarisirte Relais.
polarity *s* die Polarität; die Eigenschaft, sich in verschiedenen Richtungen verschieden zu verhalten; das Besitzen zweier Pole.
pole *s* die Stange; (Phys.) der Pol; ein englisches Maaß = 5½ yards; consequent ~s *pl* die magnetischen Folgepunkte; like ~s *pl* (Phys.) gleiche Pole; ~s *pl* of contrary names ungleichnamige Pole; ~s *pl* of the same name gleichnamige Pole; telegraph ~ die Telegraphenstange; unlike ~s *pl* ungleiche Pole; ~ & way-leave rent die Miethe für Aufstellung von Stangen und Benützung von Wegen für die Telegr.-Linien; ~ bracket die Isolatorstütze (die an der Stange befestigt wird); ~ cap die Stangenkappe; ~ changing key die Wechselstromtaste; ~ clip das Stangenband, Ziehband für eiserne Stangen; ~ lifter der Stangenaufheber; ~ pieces *pl* die Polschuhe; ~ roof die Stangenkappe; ~ saddle die eiserne Stangenkappe, in welcher die Hauptleitung tragende Isolator befestigt ist, s. auch saddle; ~ stay rod der Anker aus einem Stück; ~ steps *pl* die Klettervorrichtung an der Telegr.-Stange, bestehend in eisernen Stufen.
to polish *v* (Techn.) poliren, glänzend machen.
polish *s* (Techn.) die Politur, der Glanz.

ponderability *s* (Phys.) die Wägbarkeit.
ponderable *a* (Phys.) wägbar.
ponderance *s* (Phys.) das Gewicht, die Schwere, Schwerkraft.
ponderous *a* (Phys.) schwer, gewichtig.
pony sounder *s* (*T*) der Klopfer, Klopfsignalapparat kleiner Form.
porcelain *s* das Porzellan; ~ earth die Porzellanerde, der Kaolin; ~ insulator der Porzellanisolator.
porosity *s* die Porosität.
porous *a* (Phys.) porös; (von Metallen) blasig; ~ cup, ~ cell der poröse Becher (im Batterie-Element).
port rule *s* (*T*) das Lineal, in welches die Typen des ersten Morse-Apparats eingesetzt wurden.
porterage *s* die Bestellgebühr (für Telegramme, Expreßbriefe u. s. w.).
position *s* of a pole (*T*) der Stangenstandpunkt.
to post *v* letters, parcels Briefe, Packete aufgeben; to ~ at the counter am Schalter aufgeben; to ~ in the box in den Briefkasten werfen.
post *s* die Post; die Poststation; der Pfosten, Ständer; der Rechnungsposten; die starke Telegraphenstange; der Prellstein; binding ~ (*T*) die Klemmschraube.
post card, postcard *s* die Postkarte; ~ chaise der Postwagen; ~ day der Posttag; ~ man, postman of the walk der Revierbriefträger; ~ mark das Postzeichen, der Poststempel; ~master der Postmeister; ~master general der General-Postmeister; ~mistress die Vorsteherin einer Postanstalt; ~ office das Postamt; ~ branch ~ office das Zweigpostamt; crown ~ office das reichseigene Postgebäude; exchange ~ office die Auswechselungspostanstalt, Uebergabepostanstalt; general ~ office das General-Postamt; head ~ office das Hauptpostamt; local ~ office die Ortspostanstalt; presidential ~ office (*Am*) das Postamt, dessen Besetzung seitens des Präsidenten der Vereinigten Staaten von

Amerika erfolgt; railway ~ office, travelling ~ office das Bahnpostamt; das fahrende (Eisenbahn=Postamt; sub ~ office (gewöhnlich nur sub office) die Postagentur, f. sub...; ~ office bill, ~ office law das Postgesetz; ~ office building das Postgebäude; ~ office circular das Postamtsblatt; ~ office counter der Postschalter; ~ office of despatch die Absendungs=, Abfertigungs=, Aufgabe=Postanstalt; ~ office of destination die Bestimmungs=Postanstalt; ~ office guide das Posthandbuch; ~ office inspector (Am) der Postinspektor; ~ office of origin die Aufgabe=Postanstalt; ~ office savings bank die Postsparkasse; ~ office telegraphs pl die staatliche Telegraphie; ~ office window der Postschalter; ~ paid franko, frei; ~ road die Poststraße; ~ town Stadt mit einem Hauptpostamt; ~ waggon (Eisenb.) der Postwagen.

postage s das Porto; exempt from ~ portofrei; ~ due stamp das Taxwerthzeichen; ~ stamp die Postmarke, Briefmarke, das Postwerthzeichen.

postal a building das Posthaus, Postgebäude; ~ car (Eisenb. Am) der Eisenbahnpostwagen; ~ card die Postkarte; ~ clerk der Postbeamten; ~ collection order der Postauftrag; ~ congress der Postkongreß; ~ convention der Postvertrag; ~ employee der Postbeamte; ~ flag die Postflagge; ~ guide das Postbuch, Posthandbuch; ~ law das Postgesetz; ~ officer der Postbeamte; ~ order die Postanweisung für Beträge bis zu 20 Schillingen; ~ privilege der Postzwang; ~ service der Postdienst; technical ~ service der Postbetriebsdienst; ~ traffic der Postverkehr.

poster s (Am) der öffentliche Anschlagzettel.

postman s der Briefträger; ~ of the walk der Revierbriefträger; rural ~ on foot der Landbriefträger zu Fuß; mounted rural ~ der fahrende Landbriefträger; town ~ der Stadtbriefträger.

to postmark v a letter (P Am) einen Brief mit dem Tagesstempel bedrucken; ~ing stamp der Tagesstempel, Entwerthungsstempel.

potash, potassa s die Pottasche, das Kali; bicarbonate of ~, acid carbonate of ~ das doppelt kohlensaure Kali; bichromate of ~ das doppelt chromsaure Kali; yellow prussiate of ~, prussiate of ~ das gelbe Blutlaugensalz, Ferrocyankalium; ~ alum der Kalialaun.

potassic salts s/pl die Kalisalze, Kaliumsalze.

potassium s das Kalium, Kalimetall; bicarbonate of ~ das doppelt kohlensaure Kalium (Kali); bichromate of ~ das doppelt chromsaure, rothe chromsaure Kali; chromate of ~ das gelbe chromsaure Kali; ferri-cyanide of ~ das Ferricyankalium, rothe Blutlaugensalz; ferrocyanide of ~ das Ferrocyankalium, gelbe Blutlaugensalz.

potential a potentiell; ~ energy die potentielle, mögliche Energie.

potential s das Potential.

potter s der Töpfer; ~'s clay, ~'s earth der Töpferthon, die Töpfererde.

pottery s das Töpferhandwerk; die Töpferwaaren, das Thongeschirr; das Steingut.

pouch s der Beutel; (P) die Briefträgertasche; Her Majesty's ~ die Mappe mit Briefschaften an die Königin von England oder von derselben.

pound s (abgek. lb.) das Pfund; das Pfund Sterling; ~ avoir du poids, avoir du poids ~ das englische Handelspfund (7000 grains = 0,4534 Kilogramm); ~ troy, troy ~ das Troypfund, Gewichtseinheit für Apotheker, Edelsteine und Edelmetalle (5760 grains = 0,3731 Kilogramm); ~ rate (P) die nach dem Gewicht (Pfd.) berechnete Gebühr.

poundage s (P) der Prozentsatz, der den Postmastern in England für den Verkauf von Post= & Telegr.=Freimarken für je ein Pfund Sterling des Erlöses zugestanden ist.

power s (Math.) die Potenz; (Mech.) die Kraft; die bewegende

Kraft; die Nutzleistung (auch effective ~, mechanical ~); ~ of attorney, full ~ die Vollmacht; conducting ~ (Elektriz.) das Leitungsvermögen; horse ~ (abgekürzt H. P.) die Pferdekraft; maintaining ~ die Federkraft, Elastizität; resisting ~ die Widerstandskraft; supporting ~ die Tragkraft; tractive ~ die Zugkraft; unit of ~ die Krafteinheit.

pram, praam s der Prahm (ein großes flaches und offenes Fahrzeug zum Fortschaffen schwerer Lasten); die Fähre.

preamble s (of a message) der Kopf eines Telegramms.

precedence s (T) der Vorrang in der Beförderung.

precipitant, precipitating substance s (Chem.) das Fällungsmittel.

to precipitate v (Chem.) fällen, niederschlagen; sich setzen, sich abscheiden, niederfallen.

preconcerted a language (T) die verabredete Sprache.

„pref" (preference) mit Vorzug (in Handels- und Geschäftstelegrammen oft gebrauchter Ausdruck).

prefix s (T) das Zeichen, das bei der Uebermittelung eines Telegramms zuerst gegeben wird und die Art des Telegramms anzeigt, so daß der aufnehmende Beamte weiß, welches Formular er für die Niederschrift des Telegramms zu benutzen hat.

premises s/pl das Haus nebst Zubehör, der Grund und Boden, das Grundstück.

to prepay v a letter einen Brief frankiren; prepaid reply (T) die vorausbezahlte Antwort.

prepayment s of letters die Frankirung; obligatory ~, compulsory ~ der Frankirungszwang.

preservation s die Erhaltung; ~ of timber die Imprägnirung des Holzes.

to press v the key (T) den Schlüssel drücken.

pressure, pression s (Mech.) das Drücken, Pressen; der Druck, die Druckkraft; ~ of the wind der Winddruck; atmospheric ~, pneumatic ~ der atmosphärische Druck, Luftdruck; back-~, counter-~ der Gegendruck; full ~ der Volldruck; high-~ der Hochdruck; low-~ der Niederdruck.

price s der Preis; cost ~ der Selbstkostenpreis; estimated ~ der Taxwerth; gross ~ der rohe Preis; ~ of labor der Arbeitslohn; selling ~ der Verkaufspreis; upset ~ das Angebot.

to prick v stechen, einstechen; to ~ the wire (of a cable [T]) eine Nadel in ein mehradriges Kabel stecken, um durch Einschaltung eines „detector" die fehlerhafte Leitung zu finden.

pricked a wurmstichig (vom Holz gesagt).

pricker s (Techn.) der Stachel, der Pfriem.

primary a coil (Elektriz.) die primäre Spule, Hauptspirale; ~ current der primäre Strom; ~ wire, ~ inducing wire der primäre Draht (der Draht, in dem der induzirende Strom zirkulirt).

prime a condition vortrefflicher Zustand; ~ conductor (Elektriz.) der Hauptleiter; ~ cost der Einkaufspreis, Selbstkostenpreis.

printed a matter (P) Drucksachen.

printing axle, ~ axis s (HA) die Druckachse; ~ cam (HA) der Druckdaumen; ~ instrument (T) der Druckapparat; ~ lever (HA) der Druckhebel; ~ mechanism das Druckwerk; ~ roller (HA) die Druckrolle, Druckwalze.

priority s (T) der Vorrang in der Beförderung.

private a bag (P) der Postbeutel, die Posttasche von Privatpersonen; ~ box (P) das Abholungsfach.

probation s of a candidate (P & T) die Probezeit eines Anwärters (dauert in England sechs Monate).

probationary a appointee der auf Probe angenommene Beamte; ~ employment die probeweise Verwendung; ~ term die Probedienstzeit.

probe *s* (T) die Sonde, Kabelsonde.

proceeds *s/pl* der Ertrag, Gewinn; the gross ~ der Brutto-Ertrag; the net ~ der reine Ertrag, Netto-Ertrag.

process, proceeding *s* (Techn.) das Verfahren, die Methode, die Arbeit; (Chem.) der Prozeß.

to profile *v* im Durchschnitt zeichnen, profiliren.

profile *s* der Querdurchschnitt, Durchschnittsriß, das Profil.

profit *s* der Nutzen, Gewinn; account of ~ & loss das Gewinn- und Verlustkonto; clear ~ der Reingewinn.

prohibited *a* verboten; ~ articles (P) Gegenstände, deren Versendung durch die Post verboten ist; ~ district (P) derjenige Bezirk in Irland, in welchem nach Anordnung des *Lord Lieutenant of Ireland* jeweilig Waffen u. s. w. nicht eingeführt werden dürfen.

to project *v* (Techn.) ausladen, vorstehen; entwerfen, projektiren.

projection *s* (Techn.) der Vorsprung, Ansatz, die Nase, die Ausladung; orthographical ~ of a telegraph office die Skizze der Einrichtung eines größeren Telegr.-Amts; isometrical ~ dasf. wie vorstehend für ein kleineres Telegr.-Amt; upright ~ der Aufriß, Standriß.

prolongation *s* (Techn.) die Verlängerung, der Ansatz; ~ of days for payment die Prolongation, Verlängerung des Zahlungstermins.

promissory note *s* der Handschein, Handwechsel, eigene, trockene Wechsel (Wechsel auf den Aussteller).

promotion *s* die Beförderung (in ein höheres Amt).

prompt *s* die Zahlungsfrist; at the ~ zur bestimmten Zahlungszeit; ~ note der Mahnzettel.

prong *s* (Techn.) der Zacken, die Zinke; das spitzige Werkzeug.

to prop *v* stützen, unterstützen, absteifen; (T) verstreben.

prop *s* die Stütze; der Pfahl; (T) die Strebe; (Mech.) der Stützpunkt, Drehpunkt.

to propagate *v* the electricity, the sound etc. die Elektrizität, den Schall u. s. w. fortpflanzen.

propagation *s* (Physf.) die Fortpflanzung.

to propel *v* (Masch.) treiben, forttreiben.

propeller *s* (Masch.) der Treiber, Propeller, Treibapparat.

property *s* das Eigenthum; die Eigenschaft; landed ~ Ländereien, liegende Gründe; personal ~ das bewegliche Gut, die bewegliche Habe; real ~ das unbewegliche Vermögen, die Immobilien; state ~ das Staatseigenthum; found ~ report (P) die Nachweisung aufgefundener Gegenstände.

proportion *s* (Math., Chem.) das Verhältniß, die Proportion; continual ~ (Math.) die stetige Proportion; due ~ das richtige Verhältniß, die Symmetrie; ~ of ingredients das Mengungsverhältniß; rule of ~ (Arith.) die Proportionsrechnung, Regel de Tri; ~ of volume das Volumenverhältniß; ~ of weight das Gewichtsverhältniß.

proportional *s* (Math.) das Glied, die Proportionale; the mean ~ die mittlere Proportionale.

propulsion *s* (Mech.) das Treiben, Forttreiben.

pro rata *a* u. *adv* pro Rata, d. h. nach dem Verhältniß eines Jeden, der etwas zu erhalten oder zu zahlen hat; ~ ~ proportion der pro Rata Theil.

protection *s* of a bill die Bezahlung, Annahme eines Wechsels; to find due ~, to meet due ~ akzeptirt werden; to show due ~ to a draft eine Tratte akzeptiren, annehmen, honoriren, bezahlen.

protector *s* (Techn.) die Schutzvorrichtung; carbon ~ (T) der Blitzableiter mit Kohlenspitzen; comb ~ (T) der Spitzenblitzableiter; lightning ~ (T) der Blitzableiter; lightning ~ to poles (T) der Stangenblitzableiter; lightning ~ with fusible wire (T, F) der Spindelblitzableiter; lightning ~ with opposing points (T) der Spitzenblitzableiter; lightning ~ with serrated plates (T) der Schneidenblitzableiter; lightning bridge ~, Varley's ~ (T) der Varley'sche Blitzableiter; plate lightning ~ (T) der Plattenblitzableiter; reel ~ (T) der Varley'sche Blitzableiter zum Schutz der Nadelapparate; vacuum ~ (T) der Varley'sche Blitzableiter (für versenkte Telegr.-Linien; Erdleitung und Liniendraht sind in eine theilweise luftleere Glasröhre eingeschmolzen).

to protest v a bill of exchange einen Wechsel protestiren, protestiren lassen.

protest s der Protest, Wechselprotest; instrument of ~ das Protestinstrument; ~ for non-acceptance der Protest wegen Mangels an Annahme; ~ for non-payment der Protest wegen Mangels an Zahlung; ~ charges pl Protestkosten, Protestspesen.

proto..., prot... (Chem.) (als Vorsilbe) die erste, unterste Stufe (der Oxydation, Schwefelung u. s. w.).

protosulphate s of mercury das schwefelsaure Quecksilberoxydul.

protoxide s die erste, unterste Oxydationsstufe eines Elements, das Oxydul; ~ of iron das Eisenoxydul.

proviso s die Bedingung, der Vorbehalt in einem Dokumente, die Klausel.

proxy s der Prokurator, Bevollmächtigte, Geschäftsträger; der bevollmächtigte Stellvertreter (bei Generalversammlungen von Aktiengesellschaften u. s. w.); die Vollmacht.

prussiate s das eisenblausaure Salz, die Ferrocyanverbindung; s. potash.

pry pole s of a gin (Masch.) der Schenkel, die Stütze eines Hebezeugs.

to puddle v puddeln, im Flammofen frischen; umrühren.

puddle rolling mill s das Rohschienenwalzwerk, Puddelwalzwerk; ~ steel, puddled steel der Puddelstahl.

puddling bar s die Rohschiene; ~ furnace der Puddelofen; ~ process die Frischarbeit im Flammofen; ~ rolling mill das Rohschienenwalzwerk.

pulley s (Masch.) die Rolle, der Kloben, Rollkloben, die Flasche, Winde; fast ~, fixed ~ die feste Rolle; guide ~ die Leitrolle; loose ~ die lose Rolle; moveable ~ die lose, bewegliche Rolle; ~ block, ~ case, ~ drum, ~ frame das Gehäuse einer Rolle, der Kloben, Rollkloben; ~ sheave, ~ wheel die Rolle einer Flasche.

pulp s der Brei; der Lumpenbrei (bei der Papierfabrikation); paper ~ das Papierzeug, die Papiermasse.

pulsation s (T) der Stromimpuls, kurze Strom.

pump s die Pumpe; air ~ die Luftpumpe; aspiring ~ die Saugpumpe; vacuum ~ die Vacuumpumpe, Luftpumpe.

to punch v bohren, durchbohren, durchschlagen; (T) Löcher in den Papierstreifen stanzen.

punch s (T) die Stanze des Schriftlochers; (Techn.) das Locheisen (Werkzeug zum Schlagen von Löchern).

punner bar s die Ramme des Marshall'schen Erdbohrers.

to purchase v kaufen, einkaufen; (Mech.) aufheben, aufwinden (mittels eines Hebels).

purchase s der Kauf, Einkauf; (Mech.) das Anwenden der Hebekraft; der Gang, die Bewegung einer Maschine; die Hebevorrichtung; double ~ der Flaschenzug mit zwei Rollen; simple ~ die einfache Rolle mit Seil, der Rollenzug, Kloben.

push, electric push, push piece s (F) der Druckknopf.

pusher s (HA) der Stößer des Schlittens.

to putty v (Techn.) kitten, verkitten.

putty, puttying s (Techn.) der Kitt.

pyrites, pyrite s der Pyrit, Eisenkies, Schwefelkies; copper ~

der Kupferkies; magnetic ~ der Magnetkies.

pyrolusite, peroxide of manganese *s* der Pyrolusit, das Manganſuperoxyd, der Braunſtein.

Q.

quadrant *s* (Geom.) der Quadrant, vierte Theil eines Kreiſes; ~ electrometer (*T*) das Thomſon'ſche Quadranten-Elektrometer.

quadratic *a* equation (Math.) die quadratiſche Gleichung; complete ~ equation, affected ~ equation die unreine quadratiſche Gleichung; incomplete ~ equation, pure ~ equation die reine quadratiſche Gleichung.

quadri..., quadro... (Chem.) vierfach, Tetra..., z. B. quadrichloride of tin, tetrachloride of tin das Zinntetrachlorid, Zinnchlorid.

quadrilateral *s* & *a* das Viereck; viereckig, vierſeitig.

quadruplex *a* telegraphy, ~ working (*T*) die vierfache Telegraphie, Beförderung von vier Telegrammen auf demſelben Drahte zu gleicher Zeit.

quality *s* die Beſchaffenheit.

quantity *s* (Math., Geom. u. ſ. w.) die Größe; to join up in ~ the cores of an electro magnet (*T*) die Kerne eines Elektromagneten nebeneinander ſchalten; to join up a battery in ~ (*T*) eine Batterie großplattig ſchalten; ~ of motion (Mech.) die Bewegungsgröße, das Moment; ~ detector, ~ galvanometer (*T*, Elektriz.) das Quantitäts-Galvanometer (Galvanoſkop mit wenigen Umwickelungen ſtarken Drahtes); ~ magnet (*T*, Elektriz.) der Quantitäts-Magnet (mit kurzer dicker Spule).

quart *s* ein engliſches Flüſſigkeitsmaaß = ¼ gallon.

quarter *s* der vierte Theil, das Viertel; das Vierteljahr; ein engliſches Hohlmaaß = 8 bushel = 6 Scheffel Berliner Maaß; der Viertelzentner (28 Pfund engliſch = 12,7 Kilogramm); the four ~s *pl* die vier Himmelsgegenden; ~ day der Quatember.

quarterly *a* accounts *pl* der vierteljährliche Rechnungsabſchluß.

Queen's head *s* (*P*) Ausdruck für die engliſche Briefmarke; ~ metal das Königinmetall, Weißmetall (Legirung von Zinn, Blei, Wismuth und Spießglanz); ~ ware eine Sorte gelben engliſchen Steinguts.

quick lime *s* der lebendige, ungelöſchte, gebrannte, ätzende Kalk, Aetzkalk; ~ sand der Triebſand, Flugſand.

quicksilver *s* das Queckſilber; native ~ das gediegene Queckſilber.

quintal *s* der (ausländiſche) Zentner (zum Unterſchied von dem engliſchen hundredweight); gewöhnlich der metriſche Zentner = 50 kg.

quinto... (Chem.) fünffach, Penta...

quire *s* das Buch Papier (24 Bogen).

to quoin *v* (Techn.) einkeilen, keilen, verkeilen, Keile eintreiben.

quoin *s* die ausſpringende Ecke; (Techn.) der Keil.

quota *s* die Quote, der Antheil, die Dividende.

quotation *s* der (Handels-)Preis, die Preisnotirung; ~s *pl* of specie, der Geldkurszettel; signs of ~, ~s *pl* die Anführungszeichen, Gänſefüßchen.

to quote *v* notiren (prices etc.), angeben, berechnen, verzeichnen; to be ~d at ... notirt ſein mit ..., koſten; at the price ~d zu dem verzeichneten Preiſe.

R.

to rabbet, to rebate *v* (Techn.) einfügen, aneinander ſchäften.

rabbet, rebate *s* (Techn.) der Anſchlag; die Fuge, Nuth; ~ joint

die Spündung; ~ ledge die Schlag=
leiste.
to rack *v* recken, dehnen.
rack *s* das Gerüst, Gestell; ~
for batteries, battery ~ (*T*) der Bat=
terieständer; (Mech.) die Zahnstange,
gezahnte Stange; ~ & pinion (Mech.)
Zahnstange und Getriebe; ~ file
die Zahnfeile; ~ ladder (Masch.)
der Leiterbalken, die Stangenleiter,
Krahnleiter; ~ railway die Zahnrad=
bahn; ~ wheel das Sperrrad; ~
work (Mech.) das Zahnwerk, Zahn=
getriebe.
radial *a* (Geom.) strahlig, einen
Radius bildend oder betreffend.
radiant *a* strahlend; ~ heat
(Phys.) die strahlende Wärme.
radiation *s* (Phys.) die Strah=
lung, die Verbreitung des Lichtes,
Schalles u. s. w. von einem Mittel=
punkte aus; ~ of heat die Wärme=
strahlung.
radius (*pl* radii) *s* (Geom.) der
Radius, Halbmesser; ~ of curvature
der Krümmungshalbmesser; ~ of
the curve (Eisenb.) der Kurven=
radius.
rafter *s* der Sparren; ~s *pl*
das Sparrenwerk; ~ nail der Spar=
rennagel.
rag bolt *s* (Techn.) der Bart=
bolzen (mit scharfer Spitze und nach
oben gerichtetem Zacken, so daß er,
einmal eingetrieben, nicht wieder
herausgezogen werden kann, ohne
das Holz zu zerreißen).
to rail *v* mit einem Gitter oder
Geländer, mit Pfosten umgeben, ein=
friedigen.
rail *s* das Gitter, Geländer,
die Einfriedigung aus Pfosten und
Querhölzern; das Querstück; der
Riegel; (Eisenb.) die Schiene; der
Schienenweg.
railing, hand railing *s* das
Geländer, die Einfriedigung aus
Pfosten und Querhölzern.
railroad *s* die Eisenbahn s. auch
railway; der Schienenweg; ~ with
double way, ~ with two lines, ~
with two sets of tracks die zwei=
spurige, doppelspurige, doppelgeleisige
Schienenbahn; ~ with single way,
~ with a single set of tracks die
einspurige, eingeleisige Schienenbahn.
railway, railroad *s* die Eisen=
bahn; branch ~ die Zweigbahn, Ne=
benbahn; circular ~, belt ~, en=
circling ~ die Gürtelbahn, Ring=
bahn; double ~ die zweigeleisige
Bahn; elevated ~ die Hochbahn;
broad gauge ~ die breitspurige
Bahn; narrow gauge ~ die schmal=
spurige Bahn; horse ~ die Pferde=
bahn; ~ in operation die im Be=
triebe befindliche Bahn; ~ in pro=
gress die im Bau befindliche Bahn;
return ~ die Kopfbahn; rope ~ die
Drahtseilbahn; single ~ die einge=
leisige Bahn; system, net, network
of ~s das Eisenbahnnetz; under=
ground ~, subterranean ~ die unter=
irdische Bahn; ~ board die Eisen=
bahnverwaltung, das Eisenbahndirek=
torium; ~ car (*Am*), (in England) ~
carriage, ~ waggon der Eisenbahn=
wagen; ~ committee der Eisenbahn=
verwaltungsrath; ~ crossing der
Eisenbahnübergang; ~ division s.
division; ~ gage, ~ gauge, ~ guage
die Spur, Spurweite; ~ guard der
Bahnwärter; ~ mail guard (*P*)
der Bahn=Postschaffner; ~ pass (*T*)
das Formular (in England), das
einem Eisenb.=Dienst=Telegramme
behufs dessen gebührenfreier Beför=
derung beigefügt sein muß, wenn der
Eisenb.=Gesellschaft eigene Telegr.=
Beförderungs=Mittel nicht zu Ge=
bote stehen; ~ plant die Absperrungs=
vorrichtung, Barrière; das Betriebs=
material einer Eisenb.; ~ post office
(*P*) das fahrende Postamt; Bahn=
postamt; ~ sleeper die Bahn=
schwelle; ~ station der Bahnhof,
die Eisenb.=Station; delivering ~
station (*T*) die zur Annahme und
Bestellung von Telegrammen befugte
Eisenb.=Station; ~ terminus der
Hauptbahnhof am Ende einer Bahn=
linie; ~ time table der Fahrplan;
~ traffic der Eisenbahnverkehr; ~
transportation der Eisenbahntrans=
port; ~ truck der offene Güter=
wagen.
rain cap *s* of the insulator die
(eiserne) Schutzkappe des Isolators.

to raise v aufziehen; (Mech.) aufwinden.

raised a screw die hochköpfige Schraube.

to rake v (T) (das Gestänge) in schiefer Richtung legen; überhängen (von den Telegr.-Stangen gesagt).

to ram v rammen, einrammen, feststampfen; to ~ in einrammen.

ram, rammer s (Mech.) die Ramme, Handramme; ~'s head (Mech.) die Brechstange.

ramification s die Verzweigung, Verästelung.

rammer s s. ram.

ramp s die Auffahrt, Anfahrt, Rampe, schief ansteigende Fläche.

rampant a schräge ansteigend.

to range v a telegraph line eine Telegr.-Linie in gerader Linie anlegen.

range s die Reihe; die gerade Linie, gerade Fläche, der Spielraum; ~ of pipes der Röhrenstrang, die Röhrenleitung; ~ finder der Distanzmesser.

rare a (Phys.) verdünnt, fein.

rarefaction s (Phys.) die Verdünnung.

to rarefy v (Phys.) verdünnen; sich verdünnen.

rarity, rareness s (Phys.) die Dünnheit, Dünnigkeit.

to rasp v (Mech.) reiben, raspeln, abraspeln.

rasp s (Mech.) die Raspel.

rasper s (Techn.) das Schabeisen.

ratan, rattan s der Rotang; das ostindische Rohr.

ratch s (Masch.) die gezahnte Sperrstange; die Auslösung.

ratchet s (Mech., Masch.) der Sperrkegel, Sperrhaken, die Sperrklinke; ~ with catch das Gesperr; ~ & pawl das Zahnrad mit Sperrklinke; ~ brace, ~ drill (Mech.) der Ratschbohrer, die Ratsche; ~ wheel, rack wheel, racket wheel das Sperrrad mit Sperrklinke; ~ wheel with fine teeth (HA) das Friktionsrad.

to rate v klassifiziren; veranschlagen, taxiren.

rate s die Rata, der verhältnismäßige Antheil; der Tarif; die Gebühr; at the ~ of ... zum Preise von ...; ~ of exchange, ~ of currency der Geldkurs; ~ of interest der Zinsfuß; ~ of postage (P) der Portosatz; ~ of wages der Lohnsatz.

ratio s (Math.) das Verhältniß; direct ~ das gerade, direkte Verhältniß; inverse ~ das umgekehrte, indirekte Verhältniß; to be in the inverse ~ sich umgekehrt verhalten.

rat tail, rat tail file s die Rattenschwanzfeile.

rattan s s. ratan.

ray s (Phys.) der Strahl; (Opt.) der Strahl, Lichtstrahl; emergent ~ der austretende Strahl; incident ~ der einfallende Strahl; luminous ~ der leuchtende Strahl; refracted ~, ~ of refraction der gebrochene Lichtstrahl.

reach s of a river etc. die Breite, Strombreite.

ream s of paper das Ries Papier (20 Buch = 480 bis 500 Bogen); ten ~s pl zehn Ries, der Ballen.

to ream, to ream out, to rime v a drilled hole ein Bohrloch ausräumen, nachbohren, nachschneiden.

reamer, rimer s die Räumahle, das Räumeisen.

to rebate v (Techn.) s. to rabbet; kanneliren, ausriefen, riefeln.

rebate s (Techn.) s. rabbet; die Riefung, Kannelirung.

rebatement s der Rabatt, Nachlaß (am Preise), Abzug.

to receipt v quittiren; to ~ in full eine Generalquittung geben, per Saldo quittiren.

receipt s das Rezept; der Empfang, der Empfangsschein, die Quittung; ~s pl die Einnahme; ~ in full die Generalquittung; to give a ~ eine Quittung ausstellen, quittiren; gross ~s pl die Nettoeinnahme; on ~ of this bei Empfang oder angesichts dieses; ~s & expenditures pl Einnahme und Ausgabe.

to receive v empfangen, erhalten; to ~ a telegram ein Telegramm aufnehmen.

received *a* in account in Gegenrechnung empfangen; ~ on account auf Abschlag empfangen; ~ telegram das angekommene zur Bestellung gelangende Telegramm, Orts-Telegramm.

receiver *s* der Sammelraum, das Sammelgefäß; der Behälter; (*T*) der Empfangs-Apparat; (*P*) der Inhaber eines *receiving house*, s. das.

receiving apparatus *s* (*T*) der Empfangsapparat; ~ house, ~ office, town ~ house (*P*) die Briefannahmestelle, Postanstalt mit beschränkten Annahme-Befugnissen (sie muß innerhalb des Bestellbezirks eines Hauptpostamts liegen); ~ station (*T*) die Empfangsstation.

to recharge *v* a battery (*T*) eine Batterie neu ansetzen.

recipient *s* der Empfänger (eines Betrages).

reciprocal *a* action (Mech.) die Wechselwirkung; ~ debt die Gegenschuld.

reciprocating *a* motion (Masch.) die hin- und hergehende, auf- und niedergehende Bewegung.

reclamation *s* die Reklamation, Zurückforderung; die Beschwerde.

reconnaissance, preliminary survey *s* die Auskundung der Telegr.-Linie, Rekognoszirung.

reconstruction *s* die Wiederherstellung; work of ~ die Wiederherstellungsarbeiten.

to record *v* registriren, aufzeichnen; to ~ a bill einen Wechsel notiren lassen; ~ing instrument (*T*) der registrirende, aufzeichnende Apparat, der Telegr.-Apparat, der bleibende Zeichen hinterläßt; mail matter to be ~ed (*P*) nachzuweisende Postgegenstände (Werthbriefe u. s. w.).

record *s* (*P*) der Nachweis (von Werthbriefen u. dgl.); ~s *pl* die Akten.

recorder *s* der registrirende, aufzeichnende Apparat; Thomson's syphon ~ (*T*) der Thomson'sche Heber-Apparat.

to recoup *v* one's self sich schadlos halten, sich entschädigen.

to re-direct, to redirect *v* (*P*) weiterbefördern.

re-direction, redirection *s* (*P*) die Weiterbeförderung.

to reduce *v* (Arithm.) reduziren, verkleinern; (Metalle) reduziren, schmelzen; to ~ a drawing eine Zeichnung in verjüngtem Maaßstabe darstellen; to ~ money fremde Münze auf einheimische berechnen; to ~ the tariff den Tarif herabsetzen; to ~ wood Holz bearbeiten.

reducing scale *s* der verjüngte Maaßstab (bei Zeichnungen).

reduction *s* (Techn.) die Umwandlung, Zurückführung; die Verdünnung, Verkleinerung; die Preiserniedrigung; die Reduktion (von Metallen); das Anfrischen.

redundant *a* messengers *pl* (*P*) überzählige Postboten, Posthülfsboten.

reel *s* der Haspel; die Spule; (*T*) die Walze, Rolle; ~, paper ~ (*M & HA*) der Papierhaspel; ~ protector (*T*) der Varley'sche Blitzableiter (zum Schutz der Nadel-Apparate).

to refine *v* verfeinern, reinigen; (Chem.) läutern, reinigen; to ~ copper das Kupfer hammergar machen; to ~ iron das Eisen frischen; to ~ steel den Stahl gerben, raffiniren.

refined *a* copper das Garkupfer; ~ iron das Gareisen; ~ steel der raffinirte, gegerbte Stahl, Gerbstahl.

refinement *s* (Techn.) das Reinigen, Raffiniren, Frischen.

to reflect *v* (Phys.) zurückwerfen, zurückstrahlen, reflektiren; ~ing galvanometer *s* Thomson's Reflexions- oder Spiegel-Galvanometer.

reflux *s* der Rückstrom; flux & ~ Ebbe und Fluth; (*T*) das Einströmen der Elektrizität in das Kabe. und das Ausströmen derselben aus dem Kabel.

to refresh *v* a battery (*T*) die Batterie beschicken.

to refund *v* the amount paid den gezahlten Betrag erstatten.

refuse *s* der Abfall, die Abfälle, Rückstände; der Ausschuß, die Ausschußwaare; woolen ~ die Putzwolle.

to register *v* (Masch.) anzeigen;

to ~ a letter (*P*) einen Brief einschreiben, rekommandiren.
register *s* das Register, Inhaltsverzeichniß; (*T*) der bleibende Zeichen gebende Telegr.-Apparat.
registration fee *s* (*P*) die Einschreibegebühr.
registry *s* die Registratur.
to regulate *v* the line wire (*T*) den Liniendraht reguliren.
regulating *a* apparatus (Mech.) der Regulirapparat; ~ screw, adjusting screw (*T*, Mech.) die Stellschraube; ~ wheel das Stellrad.
regulation *s* die Anordnung; ~ notice (*P*) der Aushang, aus dem die Angabe der Dienststunden, Zeit der Kastenleerung, die Schlußzeit für die Packetannahme u. s. w. zu ersehen ist; der Postbericht.
to reheat *v* wieder erhitzen; schweißen (von Metallen gesagt).
to re-imburse, to reimburse *v* wieder bezahlen, wieder erstatten, decken; to ~ one's self sich wieder bezahlt machen.
re-imbursement, reimbursement *s* die Deckung, vgl. to re-imburse.
rejector *s* (*HA*) der Stößer.
to relax *v* a spring (Mech.) eine Feder springen lassen, loslassen.
relay *s* (*T*) das Relais; ~ without an armature das ankerlose Relais; ~ with upright (with horizontally disposed) electro magnet das liegende (stehende) Relais; polarized ~ das Induktions- (polarisirte, Siemens'sche) Relais; repeating ~ das Translationsrelais; ~ circuit der durch das Relais geschlossene Lokalstromkreis; ~ magnet der Relais-Elektromagnet, Hülfs-Elektromagnet.
to release *v* (Techn.) losmachen, loslassen; (*HA*) abschnellen (vom Anker gesagt).
relief *s* (*P*, *T*) die (Dienst-)Ablösung.
to relieve *v* (im Dienst) ablösen.
remains *s/pl* (Techn.) die Rückstände.
remanent *a* magnetism der remanente Magnetismus.

to remit *v* übersenden, einsenden, übermachen (money, bills Geld, Wechsel).
remittance *s* die Uebermachung, die Wechsel- oder Geldsendung; die Rimesse, Tratte, der Wechsel; to make ~s remittiren; Geld einliefern.
remitter *s* (*P*) der Absender einer Postanweisung.
remnant *s* (Techn.) der Rest, Ueberrest; ~ of magnetism das magnetische Residuum.
removal *s* of an officer die Absetzung eines Beamten, Amtsentsetzung; die Versetzung eines Beamten; ~ expenses *pl* Umzugskosten.
to render *v* accounts Rechnung ablegen; to ~ damages Schadenersatz leisten; rentiren, Gewinn abwerfen.
rendering, rendition *s* of accounts die Rechnungsablegung.
to renew *v* (*T*, Techn.) auswechseln.
renewal *s* (Techn.) die Auswechslung eines abgenutzten Stücks, die Erneuerung; die Prolongation (of a bill eines Wechsels).
rent *s* (Techn.) der Riß, Sprung, die Spalte; die Rente, Miethe, der Miethszins; die Leibrente.
rental *s* der Miethsbetrag.
repair *s* (Techn.) die Ausbesserung, Instandsetzungsarbeiten; ~s *pl* die Instandsetzungsarbeiten; die Ausbesserungskosten (auch cost of ~).
to repeat *v* a telegram ein Telegramm wiederholen.
repeater *s* (*T*) der Uebertrager; high speed ~ der Uebertrager für automatische Apparate.
repeating points *s/pl* (*T*) Kontaktstifte für Uebertragungszwecke.
to repel *v* one another (Phys., Mech.) einander abstoßen.
repellency *s* (Phys.) die Abstoßungskraft, Repulsionskraft.
to replace *v* materials (*T*, Techn.) auswechseln.
to replenish *v* Leyden jars (Elektriz.) Leydener Flaschen laden.
replenisher *s* (Elektriz.) Thomson's Replenisher zum Laden von Leydener Flaschen.

reply *s* die Antwort; ~ paid (prepaid) (*T*) Antwort bezahlt.
report *s* der Bericht, die Nachweisung; die Zolldeklaration.
repose *s* (Mech.) die Ruhe; angle of ~ der Ruhewinkel; ~ escapement die ruhende Hemmung.
repulsion *s* (Phys.) die Abstoßung, Zurückstoßung; power of ~, repulsive power die Abstoßungskraft, Repulsionskraft.
requirements *s/pl* of the service das Dienstbedürfniß.
requisition *s* die Bedarfsnachweisung.
residual *a* electricity der elektrische Rückstand; ~ magnetism der magnetische Rückstand; ~ products *pl* (Techn.) Ueberprodukte; Nebenprodukte.
residuary *a* (Chem.) rückständig.
residue, residuum *s* (Techn.) der Rest, Rückstand, Ueberschuß; (Chem.) der Rückstand; (im Handel) der Rechnungsrest, Rechnungssaldo.
to resign *v* one's appointment aus seiner Stellung ausscheiden, seine Stellung aufgeben.
resignation *s* of an officer das freiwillige Ausscheiden aus einer Stellung; to tender one's ~ seinen Abschied einreichen.
resilience *s* (Techn., Mech.) das Zurückspringen eines elastischen Körpers in seine ursprüngliche Lage.
resin *s* das Harz; cake of ~ (Elektriz.) der Harzkuchen; gum ~ das Gummiharz, Schleimharz.
resino-electric *a* (Elektriz.) negativ elektrisch.
resinous *a* harzig, ~ bodies, ~ substances *pl* Harze; ~ electricity, resino-electricity die negative oder Harzelektrizität.
resistance, resisting force *s* (Phys.) der Widerstand, der Beharrungszustand; artificial ~ der künstliche Widerstand; electric ~ der elektrische Widerstand, Widerstand gegen elektrische Leitungsfähigkeit; exterior, external ~ der äußere Widerstand; conductivity ~ der Leitungswiderstand; insulation ~ der Isolationswiderstand, Widerstand der Nebenschließungen; internal ~ der innere Widerstand; reduced ~ der reduzirte Widerstand; specific ~ der spezifische Widerstand; bobbin of ~, ~ coil (*T*) die Widerstandsrolle; to insert ~ (*T*) Widerstand einschalten.
resonance *s* die Resonanz; ~ box der Resonanzkasten.
resounding board *s* der Resonanzboden.
rest *s* (Phys., Mech.) die Ruhe; to be at ~ (Mech.) in Ruhe sein; state of ~ der Ruhezustand (einer Telegr.-Linie); ~ stop (*T*) der Ruhekontakt.
resultant *s* (Mech.) die Mittelkraft, Resultante, Diagonalkraft.
to retard *v* verzögern.
retardation *s* die Verzögerung.
retarded motion, retarded velocity *s* die verzögerte Bewegung, verzögerte Geschwindigkeit.
re-transmission, retransmission *s* (*T*) die Weiterbeförderung.
to re-transmit, to retransmit *v* a telegram ein Telegramm weiterbefördern.
retrograde motion *s* die rückläufige Bewegung.
return *s* der amtliche statistische Bericht; die Produktionsübersicht; ~s *pl* die Einnahme; ~ ticket (Eisenb.) die Fahrkarte für Hin- und Rückfahrt; ~ wire (*T*) die Rückleitung.
Returned Letter Office *s* (*P*) das Büreau für unbestellbare Postsachen, Rückbrief-Amt.
reversal *s* of polarity das Umkehren der Polarität; ~s *pl* (*T*) die Reihenfolge von abwechselnd positiven und negativen Strömen.
to reverse *v* (Techn.) umstellen, umkehren.
reverse *a* current (*T*) der entgegengesetzte Strom, Gegenstrom.
reversible *a* umkehrbar, umstellbar.
reversibility *s* die Umkehrbarkeit.
reversing key *s* (*T*) die Umschaltungstaste, Taste für die Versendung von abwechselnd positiven und negativen Strömen.

revolution s (Mech., Masch.) die Umwälzung, Umdrehung, Rotation; die Tour, der Umgang einer Maschine.

reward s die Belohnung, die außergewöhnliche Vergütung.

rheochord s (Elektriz.) das Rheochord (zum Messen des elektromagnetischen Widerstandes).

rheomotor s (Elektriz.) der Rheomotor (Apparat zur Hervorbringung eines elektrischen Stromes).

rheophore s der Stromträger.

rheoscope s das Elektroskop.

rheostat s der Rheostat (Apparat zur Regulirung eines elektrischen Stromes).

riband (ribband) iron post s (T) die eiserne Telegr.-Stange aus Winkeleisen gitterförmig zusammengesetzt.

ribbon, riband s (Techn.) die dünne Leiste; der Metallstreifen; das Band; metallic ~ das Metallband.

rider s (HA) die Stahlleiste am Schlitten.

to rift v (Techn.) sich spalten, aufreißen, bersten.

rift s (Techn.) der Sprung, Riß, die Ritze, Spalte.

right-hand screw, right-handed screw s (Mech.) die rechte Schraube, das rechte Gewinde.

rigidity s (Techn.) die Festigkeit (z. B. des Telegr.-Gestänges); die Dauerhaftigkeit (Reaktion gegen Formveränderung); die Steifheit, Steifigkeit.

rim s der Rand, Reifen; die Zarge; ~ of a fly wheel (Masch.) der Schwungring.

rime s die Leitersprosse; (Phys.) der Reif, Rauhreif.

rimer s (Techn.) die Reibahle, Räumahle, der Spitzbohrer.

to ring v for a number (F) anklingeln; to ~ off (F) abklingeln; to ~ up a subscriber (F) einen Theilnehmer an der Fernsprech-Anlage anrufen.

ringing battery s (auch battery for ringing) (F) die Weckerbatterie.

ring-off system s (F) das Abklingungssystem (wenn zwei Stationen zu sprechen aufhören, ertönt ein Glocke zum Zeichen der beendete Unterhaltung).

to rive v (Techn.) spalten; sich spalten, aufreißen.

to rivet v (Techn.) nieten, vernieten.

rivet s (Techn.) das Niet, die Niete, der Nietnagel; flush ~ countersunk ~ das versenkte Niet Niet mit versenktem Kopfe; ~ join die Vernietung, Verbindung durch Stifte, ~ pin der Nietkeil, Nietstift.

road s die Straße, Landstraße high ~, main ~ die Heerstraße Chaussee; ~ authority die Wege Aufsichtsbehörde; ~ letter box (F der auf den öffentlichen Straßen aufgestellte Postbriefkasten (pilla letter box und wall letter bo [s. dies.] sind road letter boxes).

to rock v schwingen, schaukeln

rock s der Felsen; das Gestein Gebirge, die Felsart; ~ drill bi Bohrknarre, der Ratschbohrer; ~ oi das Erdöl, Bergöl, Steinöl; ~ shaft rocker shaft, rocking shaft (Masch. die oszillirende Welle.

rod s (Techn.) die Ruthe, Stange, der Stab; die englisch Ruthe (ein Längenmaaß von 5¹/ yards = 5,029 Meter); coupling ~ pl die Kuppelstangen, Verbindungs stangen; lightning ~ der Blitzab leiter; ~ coupling die Stangenkup pelung; ~ stay (T) der feste Anke (aus einem Stück).

to roll v (Techn.) rollen, um drehen, walzen; to ~ into plate, to ~ into sheets zu Blech walzen; t ~ out the wire (Telegr.-Bau) der Draht abrollen.

roll s (Techn.) die Rolle, Walze der Zylinder; paper ~ die Papier rolle.

rolled a bar iron das gewalzt Stabeisen; ~ iron das Walzeisen ~ timber das windbrüchige Holz der Windbruch; ~ wire der gewalzt Draht.

roller s (Masch., Mech.) di Rolle, Walze; die Drahtwalze; da Röllchen; (T) der Holzkern be Morserolle; ~s pl das Walzwerk friction ~ die Friktionswalze; impression ~ (M & HA) die Druckrolle; ink ~, inking ~ (MA) die Farbewalze.

rolling s (Techn.) das Walzen, Rollen, Strecken; ~ mill das Walzwerk, die Walzhütte; ~ stock (Eisenb.) das Rollmaterial, Betriebsmaterial, der Lokomotiv- und Wagenpark.

roof s of a pole (T) das Schutzdach einer Telegr.-Stange.

rope s das Seil, Tau, der Strick, die Leine; ~ ladder die Strickleiter; ~ railway die Seilbahn; ~ walk, ~ yard die Seilerbahn; ~ way die Seilbahn; endless wire ~ way die Drahtseilbahn.

rosace, rosette s die Rosette; (F) die Oeffnung, durch welche die Fernsprechkabel in das Vermittlungsamt eingeführt sind.

rosin s das entwässerte und von ätherischen Oelen befreite, geschmolzene Harz.

roster s die Liste, das Verzeichniß.

to rot v faulen, verfaulen, vermodern.

rot s die Fäule, Fäulniß; dry ~ die Trockenfäule, trockene Fäulniß, der Holzschwamm; wet ~ die nasse Fäulniß.

rotary, rotatory a (Masch.) sich drehend, rotirend; ~ fan (Masch.) das Zentrifugalgebläse, der Ventilator; ~ motion, rotation die drehende, rotirende Bewegung, Kreisbewegung, der Kreislauf; ~ power das Drehungsvermögen.

rotation s s. rotary motion.

rotative a dass. wie rotary.

rotatory a s. rotary.

rotten a faul, verfault; (vom Holze gesagt) brandig, stockig.

rough a roh, grob, unbearbeitet; (von Metallen) spröde; ~ balance die rohe Bilanz, der ungefähre Saldo; ~ calculation der Ueberschlag; ~ copy, ~ draft, ~ draught der erste Entwurf, die Skizze.

round a rund; ~ bar iron das Rundeisen; ~ file die Rundfeile; ~

nose pliers, ~ pliers die Rundzange, runde Drahtzange.

route s (Eisenb. u. f. w.) die Route; post ~ der Postkurs.

royal, royal paper s das Regalpapier (Zeichenpapier, Druckpapier und Packpapier).

royalty s die Abgabe an die Krone; die einem Patentinhaber für Benutzung des Patents bezahlten Gebühren; das Regal.

rubber s (Techn.) der Wetzstein; der Wischlappen; die Grobfeile, Raspel (auch rubber file); (Elektriz.) das Reibzeug, Reibkissen der (Elektrisir-Maschine); India ~, caoutchouc der Kautschuk, das Gummi elasticum.

rule s (Techn.) das Lineal; das Richtscheit; der Maaßstab; ~ of three, ~ of proportion (Arithm.) die Regel de Tri, Regel des Dreisatzes.

ruler s (Techn.) das Lineal, Richtscheit, Richtmaaß; clerk of the ~ der Kanzlist.

to run v laufen; (T) durchgleiten (vom Draht an den Isolatoren gesagt); to ~ ahead (HA) voreilen; to ~ too slow zu langsam gehen; the marks ~ together die (Morse-)Zeichen laufen zusammen; to ~ the wire (T) den Draht auf die Stangen aufbringen.

run s die Bahn, Laufbahn, Fahrbahn; die Eisenbahnfahrt; (P) die Fahrt des Bahnpostbeamten; bills at the long ~ lang laufende Wechsel; ~ in earth work die Anschwemmung; ~ steel der Flußstahl.

run a, meter ~ das laufende Meter.

rupture s der Bruch, das Brechen; die Bruchstelle (im Draht).

rural a postman, ~ letter carrier s. postman und letter carrier.

to rust v rostig machen, oxydiren; rosten, rostig werden.

rust s der Rost, Eisenrost, Metallrost; to gather ~ Rost ansetzen, rosten.

rymer, rimer s (Mech.) der Aufräumer, das Räumeisen.

S.

sack *s* der Sack; ein Maaß von 3 Scheffeln; (*P Am*) der Briefbeutel; inner ~ (*P Am*) der Versteckbeutel; ~ cloth die Sackleinwand, grobe Leinwand.

sacramental fast day *s* der sakramentalische Fasttag. (Diese Fasttage sind der schottischen Kirche eigenthümlich und werden als kirchliche Feiertage angesehen, an denen die Postanstalten Sonntagsdienst halten.)

saddle *s* der Sattel; ~ bracket (*T*) die gerade Isolatorstütze für den auf dem Zapfende der Stange befestigten Isolator, welcher die Hauptleitung trägt; ~ wire (*T*) die zu oberst auf der Stange angebrachte Telegraphen=Leitung, die Hauptleitung der Linie.

safety *s* die Sicherheit; factor of ~ (Mech.) der Sicherheitsmodul; ~ valve das Sicherheitsventil.

to sag *v* sich durchbiegen, niederhangen (vom gespannten Leitungsdrahte gesagt).

sag *s* (*T*) der Durchhang (des Drahtes).

sal *s* (Chem.) das Salz; ~ amarum, ~ catharticum das Bittersalz, die schwefelsaure Magnesia; ~ ammoniac das salzsaure Ammoniak, Chlorammonium, der Salmiak.

salary *s* das Gehalt; augmentation of ~, increase of ~ die Gehaltsaufbesserung, Gehaltszulage; scale of ~ die Gehaltsskala.

sale *s* der Verkauf; ~ by auction, public ~, auction ~ der Verkauf in Auktion, die öffentliche Versteigerung.

salient *a* hervorspringend, ausspringend; ~ angle der ausspringende Winkel; to be ~ hervorragen, einen Vorsprung bilden.

sally *s* der vorspringende Theil, Vorsprung; ~ port der Bootshafen.

salt *s* (Chem.) das Salz (Verbindung einer Säure mit einer Base); common ~, kitchen ~, culinary ~ das Salz im engeren Sinne, Kochsalz, Chlornatrium; Epsom ~ das Bittersalz.

saltpetre, potash-nitre *s* der Salpeter, Kalisalpeter.

sample *s* die Probe, das Muster; ~s *pl* of merchandise Waarenproben.

sand *s* der Sand; shifting ~, quick ~ der Flugsand, Treibsand, Triebsand; ~ battery (*T*) die Sandbatterie; ~ drift der Flugsand, Treibsand; ~ split der Riß (im Drahte).

sap *s* der Saft, Pflanzensaft; der Splint; ascent of ~ das Aufsteigen des Saftes; ~ wood das Splintholz.

to saturate *v* (Chem.) sättigen, tränken.

saturation *s* (Chem.) die Sättigung, Tränkung; to keep a solution at the point of ~ eine Lösung im Zustande der Sättigung halten.

saving bank *s* die Sparbank; ~ ~ deposit die Spareinlage.

to saw *v* sägen, schneiden.

saw *s* die Säge; ~ dust battery (Elektriz.) die Sägespänbatterie (eine Daniell'sche Batterie, in welcher die porösen Zylinder durch Sägespäne ersetzt sind).

scaffold *s* das Gerüst; Baugerüst; to erect a ~ ein Gerüst aufschlagen; flying ~ das Hängegerüst, schwebende Gerüst.

scaffolding *s* das Gerüst; das Aufschlagen eines Gerüstes; die Materialien zu einem Gerüst.

to scale *v* schuppen, abschuppen; (Metalle) glühen; to ~ off sich abschuppen.

scale *s* der Maaßstab; die Graduirung, Grabeintheilung, Skala; die Schale einer Waage; der Splitter (im Drahte); ~s *pl* of iron der Glühspan; ~s *pl*, pair of ~s die Waage; counter ~s *pl* die gewöhnliche Waage; decimal ~s *pl* die Dezimalwaage; enlarged ~ der vergrößerte Maaßstab; lever ~ die Hebelwaage; plain ~ die natürliche Größe; platform ~ die Brückenwaage; reduced ~, ~ of reduction der verjüngte Maaßstab, die Reduktionsskala; vernier ~ die Noniuseintheilung; ~ beam der Waagbalken.

scalene s (Geom.) das ungleichseitige Dreieck.
scalene, scalenous a (Geom.) ungleichseitig.
scaly a schuppig.
to scantle v timber das Bauholz zuschneiden, zurichten.
scantling s das Maaß, der Maaßstab, wonach die Dimensionen eines Stückes Bauholz in Beziehung auf seine Breite, Länge und Dicke bestimmt werden, das vorgeschriebene Maaß eines Bauholzes u. s. w., s. auch timber.
scapement, escapement s (Techn.) die Hemmung, der Abfall; dead beat ~ die ruhende, schleifende Hemmung; ~ wheel das Hemmrad.
to scarf v (Techn.) zusammenblatten, aufpfropfen; abschrägen.
scarf s (Techn.) das Blatt, die Blattung, Laschung; ~ with indents die Verhakung u. Verzahnung; ~ joint die Verbindung zweier Planken oder Hölzer.
scarp s die Böschung, Abdachung; die Böschungsmauer.
scenographical a perspektivisch (in Zeichnungen).
scenography, scenic drawing s die perspektivische Abbildung eines Körpers; Perspektivzeichnung.
schedule s das (Waaren-)Preisverzeichniß; ~ of arrivals & departures (Eisenb.) das Kursbuch; ~ of steamers appointed to convey the mails to foreign countries Postdampfschiffsverbindungen nach dem Auslande.
scheme s (Techn.) das Schema, Urbild, Vorbild; der Plan, Entwurf, das Projekt.
scoop s die Schippe, Schaufel; der Schöpfer; der Bohrlöffel.
to score v (Techn.) kerben, einkerben, einschneiden; auf die Rechnung setzen, anschreiben.
score s die Zahl von zwanzig; die Kerbe, der Einschnitt.
scoria s (pl scoriae) die (Metall-)Schlacke.
scoriaceous a schlackig, schlackenartig.

scorification s die Schlackenbildung, Verschlackung.
to scorify v zu Schlacke werden.
scotch s (Techn.) die Kerbe, der Einschnitt; der Unterlegkeil, Zwischenkeil.
scotch a fir die schottische Kiefer (pinus silvestris).
to scour v (Techn.) scheuern, putzen; sich abscheuern (vom Kabel auf sandigem Untergrund gesagt).
scrap s (Techn.) das kleine Stück, Schnitzchen; ~ iron das Abfalleisen, die Eisenabfälle; ~ steel der Bruchstahl.
to scrape v (Techn.) schaben, abschaben, abkratzen; to ~ out ausradiren.
scraper s der Schaber; (T) der Batteriesch aber.
to screw v schrauben; ein Schraubengewinde schneiden; to ~ down, fast, in, on zu-, fest-, ein-, anschrauben; to ~ off ab-, losschrauben; to ~ up die Schrauben anziehen, zuschrauben.
screw s die Schraube; adjusting ~, regulating ~ die Stellschraube; attachment ~ die Klemmschraube; binding ~, clamping ~ die Druckschraube, Klemmschraube; double thread ~, double threaded ~ die zweifache, doppelte Schraube; external ~ die Schraubenspindel; female ~, inside ~, nut ~ die Schraubenmutter; left ~, left-handed ~ die linke, linksgängige Schraube; regulating ~ die Stellschraube; right-hand ~, right-handed ~ die rechte, rechtsgängige Schraube; set ~, thumb ~ die Stellschraube, Druckschraube, Klemmschraube; ~ with split head die Schnittschraube; sunk ~, countersunk ~ die versenkte Schraube; turn ~ der Schraubenzieher, Schraubenschlüssel; winged ~, finger ~, thumb ~ die Flügelschraube, Flügelmutter; wood ~, ~ nail die Holzschraube (eiserne Schraube zum Schrauben in Holz); wooden ~ die Holzschraube, hölzerne Schraube.
screw auger s der Schraubenbohrer, Schneckenbohrer; ~ bolt der Schraubenbolzen; ~ nut, female ~

die Schraubenmutter; ~ clamp die Schraubenklemme; ~ drill der Drillbohrer; ~ driver, turn ~ der Schraubenzieher; ~ key, ~ spanner, ~ wrench der Schraubenschlüssel; ~ knob der Schraubenkopf; ~ nail, wood ~ der Schraubennagel, die Holzschraube; ~ thread das Schraubengewinde, der Schraubengang; ~ vice der Schraubstock; universal ~ wrench der englische Schraubenschlüssel, Universalschraubenschlüssel.

scroll s die Rolle, Spirale, Schnecke; der Streifen auf einem Dokument, der die Stelle des Siegels vertritt und die Buchstaben „L. S." trägt; a ~ of paper eine Rolle Papier.

sea s das Meer, Weltmeer, die See; ~ cable das Seekabel; deep ~ cable das Tiefseekabel; ~ chart die Seekarte; ~ conveyance (P) die Seebeförderung, der Seetransport; ~-going steamer das Seedampfschiff; ~-going vessel das Seeschiff; ~ lead das Tiefloth, schwere Loth; ~ mile die Seemeile (1855 Meter); ~ postage (P) das Seeporto; ~ rate (P) das Porto für Beförderung von Briefen über's Meer; ~ trading company die Seehandlungsgesellschaft; ~ transit (P) der Seetransit.

to seal v (Techn.) siegeln, versiegeln; (mit Kitt, Gips, Mörtel, Blei) befestigen, eingießen; zuschmelzen; to ~ hermetically hermetisch verschließen; to ~ up zusiegeln.

seal s das Siegel, Petschaft, der Stempel; (Techn.) die Verkittung; ~ pipe (Techn.) das Wasserverschlußrohr.

seam s der Saum, die Naht; rivet ~ die Vernietung; soldered ~ die Löthnaht.

searcher s der Güterbesichtiger, Visitator (beim Zollamt); (T) die Regulirvorrichtung am Klopfer-Relais.

to season v timber Bauholz trocken werden lassen.

second s die Sekunde; ~ of exchange die Sekunde, der Sekundawechsel.

secondary a battery (Elektriz.) die Polarisationsbatterie; ~ coil (Elektriz., T) die Nebenspirale, induzirte Spirale; ~ current, ~ circuit (Elektriz., T) der sekundäre, induzirte Strom; ~ wire (Elektriz., T) der Nebendraht, induzirte Draht.

secrecy s, official ~ (P & T) das Amtsgeheimniß.

secretary s der Secretair (ein hoher Staatsbeamter, dem die Leitung eines Zweiges der Verwaltung anvertraut ist, nicht Secretair [Subalternbeamter] im deutschen Sinne); ~ of the General Post Office der Abtheilungschef im General-Postamt; ~ of State der Staatssecretair; ~ of the Treasury der Staatssecretair des Schatzamts; ~ of war der Kriegsminister.

section s das Schneiden, Zerschneiden; der Theil, die Abtheilung; die Durchschnittsansicht, das Profil; cross ~, lateral ~, transverse ~ der Querschnitt, das Querprofil; horizontal ~ der Horizontalschnitt, Grundriß; longitudinal ~ der Längendurchschnitt; ~ of a railway das Bahnprofil.

sectional a area der Querschnitt; ~ elevation der Längsschnitt, Querschnitt; ~ plan der Horizontalschnitt.

sector s (Geom.) der Sektor, Kreisausschnitt; (HA) das Sektorrad.

secular a variations pl die säkularen Variationen der Magnetnadel.

to secure v (Techn.) fest machen, befestigen.

securing pieces pl of a telegraph plant Sicherungsmittel des Telegr.-Gestänges.

security s die Bürgschaft, Gewährleistung, Kaution; die Hypothek, das Unterpfand; ~ies pl Werthpapiere; to become ~ for the payment of a debt sich für die Bezahlung einer Schuld verbürgen; ~ for a bill die Wechselbürgschaft; to give ~, to stand ~ Bürgschaft leisten.

sediment s (Chem. u. s. w.) der Satz, Bodensatz, Niederschlag.

seismograph, **seismometer**, **seismoscope** s das Seismoskop (Instrument zur Beobachtung von Erdbeben).

to seize *v* mit Beschlag belegen.
seizure *s* die Beschlagnahme, der Arrest.
self-acting *a* (Masch.) selbstwirkend, sich selbst regulirend; ~-~ **insulator** (*T*) Isolator, der den durchgleitenden Draht von selbst festklemmt; ~-~ **make & break** (Elektr., *T*) der automatische Selbstunterbrecher; ~-~ **movement** die selbstthätige Bewegung.
semaphore *s* (*T*) der optische Telegraph, Flügeltelegraph, Semaphor; ~ **station** (*T*) die Semaphorstation.
semi- **circle** *s* der Halbkreis; ~ **steel** der Puddelstahl.
sender *s* (*P*) der Absender; (*T*) der Versendungsapparat.
seniority *s* das Dienstalter, die Ancienität.
sensitive *a* empfindlich (von Telegr.-Apparaten gesagt).
sensitiveness *s* die Empfindlichkeit (von Instrumenten, Apparaten u. s. w.).
separate *a* **touch** (Magn.) der getrennte Strich (beim Magnetisiren des Eisens).
separator *s* die Trennschicht (bei den Hooper'schen Kabeln).
series *s* (Math.) die Reihe, Zahlenreihe; **to connect up a battery in** ~ (*T*) die Elemente einer Batterie hintereinander schalten.
serrate, serrated *a* zackig, gezackt, gezahnt.
serrature *s* der sägenartige Einschnitt, die Auszackung.
to serve *v* **a cable** ein Kabel bekleiden; **serving of hemp** die Hanfumwickelung.
service *s* der Dienst; **day** ~ der Tagesdienst; **night** ~ der Nachtdienst; **permanent** ~ der ununterbrochene Dienst; ~ **instructions** *pl* (*T*) dienstliche Angaben im Kopfe des Telegramms; ~ **message** (*T*) das Diensttelegramm.
serving *s* **of a cable** die Bekleidung eines Kabels, ~ **of hemp, hemp** ~ die Hanfbekleidung.
to set *v* (Techn.) setzen, stellen, richten, einrichten; sich einstellen, sich richten (vom Magneten gesagt); **to** ~ **up a battery** (*T*) eine Batterie ansetzen.
set *s* der Satz, die Gruppe; (*T*) der Zustand, wenn Telegr.-Material durch äußere Einflüsse dauernde Veränderung erlitten hat; ~ **bolt** (Mech.) der Schraubenbolzen, Kopfbolzen; ~ **iron** das Streicheisen; ~ **off** der Vorsprung, Maueransatz; ~ **screw** die Stellschraube, Klemmschraube.
setting-out *s* **of a line** das Abstecken (einer Eisenb.- od. Telegr.-Linie).
to settle *v* sich sacken, sich setzen; einen Niederschlag bilden; **to** ~ **an account** eine Rechnung bezahlen; **to** ~ **a balance** saldiren, einen Saldo ausgleichen.
settlement *s* die Niederlassung; der Vertrag; ~ **of an account** der Abschluß, die Saldirung einer Rechnung; **account of** ~ die Schlußrechnung, Schlußbilanz.
sewer, sink, drain *s* die Kloake, der Abzugskanal.
sewerage *s* das Anlegen von Abzugskanälen; das abgeleitete, schmutzige Wasser; die Gesammtheit der Abzugskanäle, das Kloakensystem einer Stadt.
to shackle *v* **a wire** (*T*) eine Leitung durchschneiden und an der Schnittstelle einen Isolator einfügen; eine Leitung abspannen.
shackle *s* das Kettenglied; das Oehr; **screw** ~ der Schraubenbügel; ~ **bolt** der Schäckelbolzen; ~ **insulator** (*T*) der Isolator in Winkelpunkten; der Isolator für sehr weite Spannungen (der Draht ist in der Mitte eines Eisenbolzens befestigt, der durch den aus zwei Porzellanglocken [eine oben, eine unten] bestehenden Isolator hindurchgeht).
shaft *s* der Schaft; (Masch.) die Welle, Achse, der Wellbaum; **counter** ~ (Masch.) die Transmissionswelle; ~ **bearing** das Wellenlager.
shaken, shaky *a* rissig, kernrissig, voller Sprünge (vom Holz gesagt).
shank *s* (Techn.) der Stiel,

Schaft (eines Werkzeugs u. f. w.); ~ of a screw die Spindel einer Schraube.

share s die Aktie, der Antheilschein; das Einlagekapital; die Dividende; die Einzahlung, das Eingezahlte; ~ broker der Aktienmakler; ~ holder der Aktieninhaber, Aktionär.

to shave v (Techn.) schaben, abschaben; abhobeln.

shave, drawing knife s das Schnittmesser, Schnitzmesser.

shavings s/pl (Techn.) die Späne, Abfälle; die Hobelspäne; Bohrspäne; die Feilspäne.

shearing s (Mech., Phys.) die Scheerung, Abscheerung; ~ strain, ~ stress (Mech., Phys.) die Scheerspannung; ~ strength (Mech.) die Scheerfestigkeit.

shear-legs s/pl (T) scheerenartig verbundene Hölzer (beim Stangensetzen zum Nachschieben der Stangen gebraucht).

shears s/pl (Techn.) die (große) Scheere; (T) die scheerenartige Stange (zum Aufrichten der Telegr.-Stangen beim Einsetzen derselben in die Löcher.

sheath s (Techn.) die Scheide, die Schutzhülle; iron ~ of a cable die eiserne Schutzhülle eines Kabels.

to sheathe v (Techn.) überziehen; (ein Kabel) mit Schutzdrähten bekleiden.

sheathing armor s die Schutzhülle.

sheave s (Mech.) die Rolle, runde Scheibe; belt ~ die Riemenscheibe; fast ~ die feste Rolle; loose ~ die lose Rolle; rope ~ die Seilscheibe; running ~ die Leitrolle; ~ drum die Riemenscheibe.

shed s das Wetterdach, Schirmdach; der Schuppen; der Wagenschuppen; ~s pl (Eisenb.) die Nebengebäude; cable ~ die Kabelhütte.

sheer s (Mech.) der Krahn; ~ legs pl (Mech.) der Krahn; der Mastenkrahn.

sheet s (Techn.) die Platte, Tafel; das Blech; die Metallplatte; ~ of caoutchouc die Kautschukplatte; ~ of iron die Eisenblechtafel; tinfol

~s pl Stanniolblätter; ~ copper das Kupferblech; ~ iron das dünne Eisenblech.

shell s die Schale, Hülse; das Gehäuse; ~ of a pulley (Mech.) das Gehäuse, Klobengehäuse, die Flasche, Hülse; ~ auger der Löffelbohrer; ~ bit der Bohrlöffel; ~ lac, shellac der Schellack.

shellac s f. shell.

shelving a schräg, abschüssig.

shelving s das Regal, Batteriegestell.

shifting sand s der Treibsand, Flugsand; ~ spanner der Universal-Schraubenschlüssel.

ship s das Schiff; light ~ das Leuchtschiff; ~ letters pl (P) Schiffsbriefe (Briefe, welche mit Privatschiffen in einem Seehafen ankommen); ~ letter stamp (P) der Schiffsbriefstempel (mit dem *ship letters* gestempelt werden müssen).

to shoe v (Techn.) mit Eisen beschlagen; anschuhen.

shoe s (Techn.) der Schuh; der eiserne Beschlag; der Stangenschuh.

shore s das Ufer, Gestade, der Strand, die Küste; ~, strut (Techn.) die Steife, Spreize; ~ end (of a cable) das Küstenkabel; ~ test (T) die Kabelprüfung vom Lande aus (während der Kabellegung).

short a kurz; (von Metallen) faulbrüchig; ~ circuit (T) der kurze Schluß; cold ~ kaltbrüchig; hot ~, red ~ rothbrüchig; ~ price der Nettopreis.

to short-circuit v (T) eine Leitung kurz schließen.

to shunt v (T) eine Nebenschließung anbringen; (Eisenb.) den Eisenbahnzug auf ein Seitengeleis bringen.

shunt s (T) der Shunt, die (künstliche) Nebenschließung eines elektrischen Stromkreises (auch ~ of a circuit, ~ of a galvanometer); (Eisenb.) das Seitengeleis, Nebengeleis, die Weiche.

shunted a (T) mit künstlicher Nebenschließung versehen.

side s die Seite; ~ elevation der Seitenaufriß; ~ face die Seiten-

Ansicht; ~ pace das Bankett; ~ projection das Profil; ~ view die Seitenansicht.
sidewalk *s* das Trottoir.
to sight *v* (beim Vermessen) mit den Augen abmessen; to ~ out aboisiren, abfluchten.
sight *s* die Sicht (von einem Wechsel u. s. w. gesagt); das Absehen (beim Vermessen), das Visir; a bill payable at ~ ein nach Sicht zahlbarer Wechsel, Sichtwechsel; line of ~ die Visirlinie; long ~ die lange Sicht (eines Wechsels); point of ~ der Gesichtspunkt (bei Vermessungen); short ~ die kurze Sicht (eines Wechsels).
sign *s* das Zeichen; das Schild, Aushängeschild; conventional ~s *pl* (T) verabredete Zeichen; ~ board das Aushängeschild; ~ post der Wegweiser.
signal *s* das Signal; block ~ (Eisenb.) das Blocksignal; starting ~ (Eisenb.) das Abfahrtszeichen, station ~ (Eisenb.) das Einfahrtssignal; ~ code die Signalordnung, das Signalreglement; United States ~ corps das Signal-Corps der Vereinigten Staaten von Amerika (eine Art Feld-Telegr.-Truppe, im Frieden zu meteorologischen Beobachtungen verwendet); ~ room (T) der Apparatsaal.
silent *s* (T) der Ausschalter.
silicate *s* (Chem.) das kieselsaure Salz, die kieselsaure Verbindung, das Silikat; ~ of potassium das kieselsaure Kalium; ~ cotton die Schlackenwolle.
silicious *a* kieselartig, kieselhaltig.
sine, sinus *s* (Geom., Math.) der Sinus, die Winkelstütze; ~ galvanometer die Sinusbussole; ~ tangent galvanometer die Sinus-Tangentenbussole.
single *a* bill der Solawechsel; ~ block (Mech.) der einscheibige Block; ~ touch (Magn.) der einfache Strich; ~ acting (Mech.) einfach wirkend; ~ instrument (T) der Einnadeltelegraph; ~ needle code, ~ needle alphabet (T) das Alphabet des Zeigerapparats.

to sink *v* versenken; to ~ a shaft einen Schacht abteufen; ~ing fund der Tilgungsfond.
sinter *s* der Sinter (Nebenprodukt beim Erzschmelzen).
sinus *s* s. sine.
siphon, syphon *s* (Phys.) der Heber, Saugheber; ~ pipe das Heberrohr; ~ recorder (T) Thomson's Heberapparat.
site *s* die Lage, Situation; plan of ~ der Situationsplan.
to size *v* (Techn.) einer Sache die rechte Größe oder das gehörige Maaß geben; nach der Größe sortiren.
size *s* (Techn.) die Größe, Gestalt, das Format; der Maaßstab; der Leim; full ~, real ~ die natürliche Größe; ~ color die Leimfarbe; ~ roll das an eine Urkunde angehängte Stück Pergament.
skeleton *s* (Techn.) das Gerippe, Zimmerwerk; das Rahmholz, Rahmwerk; (beim Vermessen) das Netz; ~ bills, ~ bonds, ~ notes *pl* unausgefüllte Formulare von Wechseln, Schuldscheinen, Rechnungen u. s. w.; ~ drawing die Unrißzeichnung; ~ key der Dietrich; ~ telegram das Skelett-Telegramm (per Kabel mit wenigen Worten ankommt und von den Zeitungs-Agenturen oder Zeitungs-Redaktionen ausgearbeitet als Original-Telegramm abgedruckt wird).
skelp *s* (Techn.) die Eisenschiene, Rohrschiene; der Metallstreifen.
to sketch *v* skizziren, flüchtig aufnehmen; vorzeichnen.
sketch *s* die Skizze; die Aufnahme; eye ~, hasty ~, rough ~ die Skizze nach dem Augenmaaß; ~ book das Konzeptbuch des Kaufmanns.
skew *a* schief, schiefwinklig; windschief; ~ scarf das schräge Blatt, die schräge Anblattung.
slab *s* die Platte, Steinplatte; die Schwarte, das Schwartenbrett, s. auch timber; ~ of a table die Tischplatte, das Tischblatt; ~ board das Schalbrett; ~ iron das Blechmaterialeisen.

slabbing s das Behauen, Bearbeiten der Stämme; das Zersägen der Stämme zur Gewinnung der Schalbretter; ~ saw die Brettsäge, Längsfäge.

to slack, to slacken v locker machen, nachlassen; to ~ lime Kalk löschen; to ~ the motion (Mech.) die Bewegung schwächen.

slack a (Techn.) schlaff, locker; (Mech.) vielen Spielraum habend.

slag s die Schlacke; ~ wool, ~ hair die (durch Eingießen von geschmolzener Schlacke in Wasser gewonnene) Schlackenwolle.

slant, slanting a (Techn.) schief, schräg, quer.

sledge hammer s der große Schmiedehammer, Zuschlaghammer.

sleeper s der Querbalken; das Grundholz, die Schwelle, Bodenschwelle, Eisenbahnschwelle.

to slide v (Mech.) gleiten.

slide s die Gleitbahn; die Gleitfläche; der Schieber; ~ gauge die Schublere, Schieblere; ~ label (P) das Vorsteckschild; ~ valve das Schieberventil.

sliding a gleitend; ~ weight das Laufgewicht, verschiebbare Gewicht.

slip s der Streifen, das Stückchen; der Papierstreifen; (P) der Zettel; advice ~ der Benachrichtigungszettel; face ~ (P) der Klebezettel; forwarded messages ~ (T) der Streifen, der die Ursprungs-Telegramme enthält; registered letter ~ (P) der Zettel mit dem Verzeichniß der eingeschriebenen Briefe.

to slit v (Techn.) spalten, schlitzen; sich spalten.

slit s (Techn.) die Spalte, der Einschnitt; (P) der Briefkasteneinwurf; ~ in the head of a screw der Einstrich; die Kerbe.

to slope v abschrägen, schräge, schief machen; abböschen.

slope s die Schräge, die schräge Richtung oder schräge Fläche; die Böschung, der Abhang; ascending ~ die Steigung; falling ~ der Fall, das Gefälle.

sloping talus s die Böschung.

slot s (Mech.) die Kerbe, Furche, Nuth; der Einschnitt, Schlitz; (HA) die Oeffnung im Stiftgehäuse.

to smelt v (Metalle) schmelzen; to ~ down einschmelzen.

to smooth v (Techn.) ebnen, glätten.

smooth a (Techn.) glatt, geglättet.

smoothing iron s das Glätteisen.

snag s in wood der Knoten, Knorren im Holze.

to soak v (Techn.) einweichen, einwässern, durchweichen; to ~ in sich einsaugen.

socket s (Mech., Masch.) die Hülse, Dille, Röhre, Muffe; der Schuh; granite ~ der Steinsockel; ~ pipe die Röhre mit Muff.

socle, sockle s der Sockel, das Fußgestell (für eiserne Telegr.-Stangen).

soft a iron das weiche Eisen; ~ solder das Weichloth, Zinnloth, Schnellloth; ~ steel der ungehärtete Stahl.

to soften v (Techn.) erweichen, weich machen; to ~ iron Eisen geschmeidig machen; to ~ steel den Stahl weich machen, enthärten, anlassen.

to solder v löthen, zusammenlöthen, verlöthen.

solder s das Loth, das Löthzeug, Löthmittel; brass ~ das Messingschlagloth; hard ~, strong ~ das Hartloth, Schlagloth; soft ~, tin ~ das Schnellloth, Weichloth, spelter ~ das (zinkreiche) Messingschlagloth.

soldering s das Löthen, die Löthung; hard ~ das Hartlöthen; soft ~ das Weichlöthen; ~ bit, bolt, ~ iron, ~ tool der Löthkolben; ~ box die Löthbüchse; ~ furnace der Löthofen; ~ groove, ~ spoon der Löthlöffel; ~ lamp die Löthlampe; ~ pan die Löthpfanne; ~ stove der Löthofen; ~ support der Löthbock.

sole s (Techn.) die untere Fläche, Grundfläche; (Masch.) die Lagerplatte; ~ plate die Fundamentalplatte.

solicitor s of the Post Office der Rechtsbeistand.
solid s der Körper, der feste Körper; die Masse.
solid a (Techn.) fest, gediegen, massiv, solid.
soluble a löslich; readily ~ leicht löslich.
solution s (Chem.) die Auflösung.
to solve v (Chem. u. s. w.) lösen, sich lösen.
to sort v (P) sortiren.
sortation s of letters (P) die Briefsortirung.
sorter s (P) der Sortirer (Unterbeamte).
sorting carrier s (P) der Sortirbriefträger; ~ clerk der Sortirer (Beamte); ~ list (P) die Leitübersicht, Speditionsliste; division ~ list (P) die Vertheilungsliste für den betreffenden Kartirungsbezirk nach den von London ausgehenden Eisenbahnpostkursen; ~ tender (P) der Bahnpostwagen, in welchem die Postsachen sortirt werden.
to sound v sondiren, peilen, mit dem Bleiloth untersuchen.
sound s der Ton, Schall; to read a despatch by the ~ (T) ein Telegramm nach dem Gehör aufnehmen, abhören; ~ board, sounding ~ der Resonanzboden; ~ reading (T) die Aufnahme nach dem Gehör; ~ telegraphy die Telegraphie nach dem Gehör.
sounder s (T) der Klopfer, Klopfsignalapparat; Morse ~ (T) der Morse'sche Klopfapparat; double plate ~ (T) der Doppelklopfer (ein Telegr.-Apparat, welcher mit zwei verschiedenen Klängen erzeugenden Klopfvorrichtungen versehen ist. Der Strom geht vom Relais zu der einen Klopfvorrichtung, deren Schläge die Punkte, oder zur andern, deren Schläge die Striche der Morsezeichen darstellen); pony ~ der Klopfer kleiner Form.
sounding board s f. sound board; ~ box der Resonanzkasten; ~ lead das Loth, Senkblei.
south-magnetism s der Südmagnetismus.

to space v (T) den Zwischenraum zwischen den Morsezeichen lassen.
spacing current s f. current.
spade s der Spaten, die Schaufel, das Grabscheit.
span s die Spanne; der Bogen, das Gewölbe; die Spannung, Spannweite.
Spanish spoon s der Erdbohrer (zum Graben von Löchern für die Telegr.-Stangen).
spanner s der Spanner; ~, screw ~ der Schraubenzieher, Schraubenschlüssel; shifting ~ der Universalschraubenschlüssel.
spare a (Techn.) übrig, überflüssig, Reserve ...
spark s der Funken; electric ~ der elektrische Funken; ~ coil (T) der Widerstand, der den Funken an den Kontaktstücken des absendenden Apparats mildern soll, wenn die ganze Batterie wirkt; ~ drawer (Elektriz.) der Funkenzieher.
specific a spezifisch; ~ gravity, ~ weight (Phys.) das spezifische Gewicht; ~ resistance (Elektriz., T) der spezifische Widerstand.
specification s die Spezifikation, ausführliche Beschreibung, das Verzeichniß; (T) die Veranschlagung der Materialien-Transporte; ~ of the wire die Lieferungsbedingungen für den Telegr.-Draht.
specimen s die Probe, das Schema; das Probeexemplar.
specular a spiegelartig; ~ cast iron das Spiegeleisen.
speed s (Mech.) die Geschwindigkeit, der Gang einer Maschine; ~ governor der Geschwindigkeitsregulator; fast ~ instrument (T) der automatische Telegr.-Apparat; high ~ telegraphy die automatische Telegraphie.
spelter s das Handelszink; ~ solder das (zinkreiche) Messingschlagloth.
sperm, spermaceti s der Wallrath; ~ candle die Wallrathkerze (englische Normalkerze mit Flammenhöhe von 43—45 Millimeter; Verbrauch 7,77 Gramm [120 grains] in der Stunde).

spigot *s* der Zapfen, Hahn eines Fasses; ~ end of a pipe (Techn.) der Röhrenhals, das gerade Ende der Röhre; & faucet joint, ~ joint die Verbindung der Röhren mittels Muffen, Muffenverbindung.

to spike *v* (Techn.) nageln, vernageln, zusammennageln.

spike *s* (Techn.) der Bolzen, der eiserne Stift, Dorn, Niet.

spindle *s* die Spindel; (Mech., Masch.) der Zapfen einer Welle; die Achse.

spiral *s* (Geom., Techn.) die Spirale, Schneckenlinie, Schraubenlinie; ~ coil die Spirale, Spule.

spiral *a* spiralförmig, schneckenförmig, schraubenförmig; ~ drill der Schraubenbohrer; ~ spring die Spiralfeder.

to splice *v* spleißen, splißen; (T) eine Kabelspleißstelle anfertigen.

splice *s* die Spleißstelle; ~ joint die Kabellöthstelle.

splint *s* (Mech.) der Keil, Splint; (Techn.) der Splitter, Span.

to split *v* (Techn.) spalten, zertrennen; sich spalten, rissig werden.

split *s* der Spalt (im Draht); ~ burner der Schlitzbrenner.

spongy *a* schwammig; ~ iron schwammiges, bei niederer Temperatur aus Oxyd reduzirtes Schmiedeeisen; ~ platinum der Platinschwamm.

spool *s* die Spule; ~ wire der Spuldraht.

spoon bit *s* der Löffelbohrer.

sprig *s* (Techn.) der Nagel ohne Kopf.

spring *s* (Mech.) die Feder, Triebfeder; main ~ die Hauptfeder, Schlagfeder; plate ~ die Blattfeder; spiral ~ die Spiralfeder; ~ arbor der Federstift, Wellbaum; ~ balance, ~ yard die Federwaage; ~ barrel, ~ box die Federtrommel; ~ break piece (T) die Federunterbrechung (am Selbstunterbrecher); ~ jack (T Am) eine Art Umschalter für Untersuchungszwecke auf Zwischenstationen; ~ jack switch (T Am) der federnde Umschalter; ~ tight *a* federfest (ein Telegr.-Apparat ist federfest, wenn eine unverhältnißmäßige Federspannung das Arbeiten desselben hindert); ~ vice (Mech.) die Federschraube, der Federspanner.

springy *a* elastisch, spannkräftig, federnd.

spruce, spruce fir *s* die Pechtanne, Sprossenfichte.

spur *s* (Masch., Mech.) der Dorn, Stachel; ~ gear, ~ gearing, ~ pinion (Masch.) das Stirnradgetriebe; ~ wheel (Masch.) das Stirnrad, Kammrad, Spurrad.

to square *v* (Techn.) viereckig machen; vierkantig schmieden; to ~ an account eine Rechnung ausgleichen; to ~ timber das Holz beschlagen, vierkantig zuschneiden, behauen.

square *s* (Geom., Techn.) das Quadrat; das Winkelmaaß; in the ~ im Geviert; ~ file die vierkantige Feile; ~ headed bolt der Bolzen mit viereckigem Kopf; ~ iron das viereckige Stabeisen, Quadrateisen; ~ timber das Kantholz.

stability *s* (Mech.) die Standfestigkeit (einer Telegr.-Linie).

to staff *v* an office ein Amt mit Beamten besetzen.

staff *s* der Stab; das Personal (einer Behörde, eines Amtes).

stage *s* das Gerüst, Gestell; hanging ~ das schwebende Gerüst; heavy ~, ~ coach die Postkutsche; ~ waggon die Frachtpost, der Packwagen.

to stake *v* mit Pfählen besetzen, bepfählen; to ~ out abpflöcken, abstecken.

stake *s* die Stange, der kleine Pfahl, Pflock; das Markirpfählchen.

stalk *s* der Stengel, Stiel; die Stütze des Isolators.

to stamp *v* (Techn.) stampfen; (P) stempeln.

stamp *s* der Stempel; der Aufdruck, das Gepräge; date ~, dated ~ (P) der Datumstempel, Aufgabestempel; defacing ~ (P) der Entwerthungsstempel; double ~ (P) der Stempel, welcher zugleich Aufgabe- und Entwerthungsstempel ist; internal revenue ~s *pl* Stempelzeichen für fiskalische Zwecke, Quittungsstempel u. s. w.; obliterating ~ der

Stempel zur Entwerthung der Freimarken, Entwerthungsstempel; post ~, postage ~ die Briefmarke; postmarking ~ (*P Am*) der Aufgabestempel, Entwerthungsstempel; receipt ~ der Quittungsstempel; ship letter ~ (*P*) der Stempel, mit dem Schiffsbriefe gestempelt werden, vgl. *ship letters*; special delivery ~ das Postwerthzeichen für Eilbestellung; ~ duty, ~ fee die Stempelgebühr.

stamping ink *s* die Stempelfarbe; ~ iron das Stempeleisen; ~ machine die Stempelmaschine; ~ pad das Stempelkissen, die Stempelunterlage.

stanchion *s* die Stütze, der Ständer; die Steife.

stand *s* (Techn.) das Gestell, der Ständer; ~ box (*MA*) der Untersetzkasten.

standard *s* die feste, beständige Valuta; der Normalpreis; der Pfosten; der Ständer, Pfeiler.

standard *a* (Techn.) maßgebend, Normal…; ~ register (*T*) der Normal-(Morse-)Farbschreiber.

staple *s* der Stapelplatz, die Niederlage; (Techn.) die eiserne Krampe; ~ of a bolt die Schließklappe.

star route *s* (*P Am*) die Postbeförderung auf Landwegen durch Wagen und Pferde (so genannt, weil die bedienten Orte in den amtlichen Verzeichnissen mit einem Stern versehen waren, um sie von Dampfschiff- und Eisenb.-Post-Linien zu unterscheiden); ~ wheel (Masch.) das Sternrad.

to start *v* (Mech., Masch.) in Bewegung setzen, in Gang bringen.

statement *s* of account current der Rechnungsauszug, die Abschlußrechnung; ~ of duties der Zolltarif; ~ of exchanges der Stand der Kurse, Kursbericht; ~ of specie der Geldkurszettel.

statical *a* electricity die statische Elektrizität; ~ induction, influence die elektro-statische Induktion, Influenz.

statics *s/pl* die Statik, Lehre vom Gleichgewicht der Körper.

station *s* die Station; railway ~ die Eisenb.-Station, Haltestelle, der Bahnhof; ~ of departure die Abgangsstation; goods ~, freight ~ der Güterbahnhof; intermediate ~ (*T*, Eisenb.) die Zwischenstation; ~ master (Eisenb.) der Stationsvorsteher; ~ messenger (*P*) der Postschaffner.

stationary *a* (Mech., Masch.) stationär, feststehend.

stationer *s* der Schreibmaterialienhändler, Papierhändler; ~ der Händler mit den zur Ausfertigung gerichtlicher Dokumente erforderlichen Materialien.

stationery *s* die Schreibmaterialien (*pl*); die Schreibmaterialienhandlung.

statute mile *s* die englische Meile (der 60ste Theil eines Grades des Aequators = 1760 yards = 1609,315 Meter).

to stay, to stay up *v* (Techn.) stützen, absteifen.

stay *s* die Stütze, Steife; (*T*) der Anker; ~ block der Ankerpfahl, Ankerkloß; ~ bolt der Ankerbolzen; ~ hook der Ankerhaken.

to steel *v* (Techn.) stählen, verstählen, anstählen.

steel *s* der Stahl; annealed ~ der angelassene Stahl; blister ~, blistered ~ der Blasenstahl (das erste Stadium des Zementstahls, der durch Gerben in den Gerbstahl übergeht); case hardened ~, chilled ~ der Einsatzstahl; cast ~ der Gußstahl; cement ~, cemented ~, converted ~ der Zementstahl, Brennstahl, gebrannte Stahl; crude ~, raw ~ der Rohstahl; fined ~, charcoal ~ der gefrischte Stahl; flowing ~, ingot ~ der Flußstahl; puddling ~, semi ~ der Puddelstahl; refined ~, shear ~ der raffinirte Stahl, Gerbstahl; spring ~ der Federstahl; ~ pointed screw die Spitzenschraube; ~ yard die Schnellwaage, Balkenwaage.

steely iron *s* das stahlartige Eisen, Feinkorneisen (mit 0,25 bis 0,50 % Kohle).

step *s* die Stufe; ~ of an embankment (Eisenb.) das Bankett; ~-by-step motion telegraph der Zeigerapparat.

to stick *v* (Techn.) anstecken, anheften, ankleben; kleben, hängen bleiben.

stick *s* (Techn.) der Stock; iron ~ (T) der Knebel (zum Anspannen des Ankers an der Telegr.=Stange).

stiffness *s* die Dauerhaftigkeit, Widerstandsfähigkeit (Reaktion gegen Formveränderung).

stock *s* (Techn.) der Stock, Schaft, der hölzerne Theil mancher Werkzeuge; der Vorrath, das Quantum; das Inventar; ~s *pl* die Kluppe, Schraubenschneidekluppe; circulating ~, floating ~ das zirkulirende Kapital; ~ on hand der Vorrath; joint ~ das Aktienkapital; rolling ~ (Eisenb.) das Betriebsmaterial; ~ broker der Fondsmakler, Makler in Staatspapieren; ~ exchange die Fondsbörse.

to stop *v* (Techn.) stopfen, zustopfen, zumachen; halten, anhalten; to ~ the wire (T) die Leitung abspannen; to ~ the work den Betrieb einstellen.

stop *s* (Techn.) der Halt, Einhalt; die Hemmung; (Mech.) die Sperrung; (Eisenb.) die Haltestelle; (P) das Ueberlager, Stillager der Bahnpostbeamten; (T) das Kontaktstück am Kurbelumschalter, (back ~ das dem Beamten entferntere, front ~ das dem Beamten zunächst befindliche Kontaktstück); ~ cock (Mech.) der Sperrhahn; ~ motion (Mech.) die Abstellvorrichtung, Selbstauslösung; ~ pin der Begrenzungsstift; ~ screw die Hemmschraube.

storage *s* das Ansammeln; der Ansammlungsraum; die Lagermiethe, das Lagergeld; electrical ~ battery (Elektriz.) der Akkumulator.

store *s* der Vorrath; das Waarenlager; ~s *pl* (P & T) Materialien, Ausrüstungsgegenstände; small ~s *pl* die Nebenmaterialien; ~ keeper (P & T) der Materialienverwalter.

straight *a* gerade; vollkommen eben; to put wires ~ (T) die normale Verbindung wiederherstellen.

to straighten *v* (Techn.) richten, gerade machen.

to strain *v* (Masch.) überarbeiten; (Mech.) deformiren, eine Formveränderung hervorbringen.

strain *s* (Mech.) die Spannung, der Zug, Druck, die Beanspruchung, die Belastung; (T) die Formänderung, welche durch die Belastung der Telegr.=Linie bewirkt wird; breaking ~, transverse ~ die Zerbrechungskraft; resistance to breaking ~ die Zerbrechungs= (relative) Festigkeit; compressing ~, compressive ~, crushing ~ die Druckspannung; resistance to compressive ~ die Zerdrückungs= (rückwirkende) Festigkeit; shearing ~ die Scheerspannung; resistance to shearing ~ die Abscheerfestigkeit; tensile ~, tensive ~ die Zugspannung, Zugkraft; resistance to tensile ~, tensile strength die Zug= (absolute) Festigkeit; torsional ~ die Drehspannung; resistance to torsional ~ die Torsionsfestigkeit.

straining winch *s* die Drahtwinde.

strand *s* der Strang, die Litze eines Taues; ~ of seven copper wires ein siebenlitziger Kupferdraht; ~ wire die Drahtlitze.

stranded *a* wire der verseilte Drahtstrang.

stranding machine *s* die Verseilungsmaschine.

strap *s* (Techn.) das Band; iron ~ das eiserne Band; ~ of iron wire das Drahtband; ~ iron das Bandeisen; ~ wheel die Riemenscheibe.

stratum *s* (*pl* strata) das Lager, die Schicht; ~ of air die Luftschicht.

strength *s* (Mech., Phys.) die Festigkeit, die Tragkraft; ~ of compression, compressive ~ die Druckfestigkeit, rückwirkende Festigkeit; ~ of current (T) die Stromstärke; ~ of flexure, transverse ~ die Bruchfestigkeit, relative Festigkeit; ~ of gungsfestigkeit; shearing ~ die Scheerfestigkeit; tensile ~, ~ of extension die Zugfestigkeit, absolute Festigkeit; torsional ~ die Torsions=(Drehungs=) Festigkeit.

strengthening pieces *pl* of the telegraph structure Verstärkungsmittel des Telegraphen-Gestänges.
to stress *v* (Mech.) deformiren; ~ed body der Körper, der sich nicht in seinem natürlichen Zustande befindet.
stress *s* der Zwang, Druck, der auf einen Körper ausgeübt werden muß, um ihm eine bestimmte Formänderung zu geben (s. auch load & strain); die elastische Reaktion gegen eine Formänderung.
to stretch *v* (Techn.) spannen, ausspannen; strecken, recken, ausrecken.
stretching insulator *s* (T) der Spannisolator; ~ pole, ~ post (T) die Stange mit Spannisolator (an der der Draht festgebunden wurde, während er sonst nur lose aufliegt).
to strike *v* across (Elektriz.) überspringen (von elektrischen Funken gesagt).
strip *s* der schmale Streifen; ~ of copper der Kupferstreifen.
stroke *s* der Schlag, Streich, Stoß; (Techn.) die Vorzeichnung, der Riß; die Hubhöhe; back ~ (Elektriz.) der Rückschlag; ~ of a bell (T) der Schlag des Weckers.
structure *s* der Bau, das Bauwerk, Gebäude; die Bauart; Struktur; das Gefüge (von Mineralien); fibrous ~ die faserige Struktur; foliated ~ die blättrige Struktur.
to strut *v* abspreizen, absteifen, verstreben.
strut *s* die Strebe, Steife, Spreize.
strutting beam, strutting piece *s* der Spannriegel.
stub *s* (Am) der Theil eines Blattes im *check book*, der zurückbleibt, nachdem der *check* ausgerissen ist.
stud *s* (Techn.) der Stift; ~ bolt der Schraubenbolzen mit Gewinden an beiden Enden.
sub... (Chem.) basisch; ~ head *s* der Titel eines Kostenanschlags, des Etats u. s. w.; ~ office (P) ein Postamt mit beschränktem Dienst und beschränkten Befugnissen, eine Postagentur; ~ postmaster der Vorsteher einer *sub office*, Postagent.

subaquatic, subaqueous *a* line (T) die Unterwasser-Linie.
submarine *a* unterseeisch.
to submerge *v* eintauchen, versenken; to ~ a cable ein Kabel legen.
to subpoena *v* bei Strafe vor Gericht laden.
subpoena *v* die Vorladung vor Gericht; to serve *smb* with a ~ J—n vor Gericht laden.
subscriber *s* der Abonnent, Theilhaber.
subsistence allowance *s* das Zehrgeld (für Arbeiter, Telegr.-Bauaufseher u. s. w.).
substitute *s* der Stellvertreter; (P) der Hülfsarbeiter, der für ausgetretene, erkrankte u. s. w. Beamte eingestellt wird.
subterranean, subterraneous *a* unterirdisch.
sulphate *s* das schwefelsaure Salz; ~ of copper, cupric ~ das schwefelsaure Kupferoxyd, der Kupfervitriol; ferrous ~, of iron das schwefelsaure Eisenoxydul, der Eisenvitriol; ~ of magnesia, magnesic ~ die schwefelsaure Magnesia, das Bittersalz; ~ of zinc der Zinkvitriol.
sulphuric *a* acid (Chem.) die Schwefelsäure.
sulphurous *a* acid (Chem.) die schweflige Säure.
Sunday posting *s* (P) die Abwickelung des Postschalterverkehrs an Sonntagen.
sunk *a* (Techn.) versenkt; ~ screw die versenkte Schraube.
to superannuate *v* durch Alter (oder langes Leben) abnutzen, untauglich werden, pensionirt werden; a superannuated officer ein invalider, dienstuntauglicher Beamter.
superannuation *s* der Zustand, da eine Person oder Sache durch Alter untauglich geworden ist; die Pensionirung; ~ act das Pensionirungsgesetz.
superintendence *s* (T & P) die Aufsicht (beim Bau, im Dienst u.s.w.).
superintendent *s* der Oberaufseher; ~ on duty (T & P) der Oberaufsichtsbeamte; district ~ (P) der Bezirks-Postinspektor.

supervising *a* officer (*T*) der Leitungsrevisor.
supervision *s* (*T*) die Beaufsichtigung der Telegr.-Linien.
to supply *v* versorgen; liefern; to ~ a battery (*T*) eine Batterie speisen.
supply *s* die Unterhaltung, die Versorgung; Post Office supplies *pl* Amtsbedürfnisse.
support *s* die Stütze, der Träger; (*T*) der Leitungsstützpunkt.
supporting pieces *pl* (*T*) Unterstützungsmittel.
to surcharge *v* (*P*) Porto zutaxiren, mit Zuschlagporto, Strafporto belegen.
surcharge *s* (*P*) das Zuschlagporto, die Zuschlaggebühr, das Strafporto; die Werthänderung von Postfreimarken (durch Ueberdruck oder Aufdruck); ~ book das Verzeichniß derjenigen Briefe, die mit Zuschlagporto belegt worden sind.
surface *s* (Geom., Techn.) die Oberfläche, die Fläche; die Außenfläche; ~ of the formation (Eisenb.) das Planum; plane ~ die ebene Fläche.
surplus *s* der Ueberschuß.
surtax *s* (*P* & *T*) die Zuschlagtaxe, das Zuschlagporto.
to survey *v* besichtigen; vermessen; to ~ a route for a telegraph line eine Telegr.-Linie auskunden.
survey *s* die Aufnahme, Vermessung; preliminary ~ (*T*) die Auskundung, Rekognozirung.
surveying officer *s* (*T*) der Telegraphen-Bauführer.

surveyor *s* (*P*) der Postinspektor; ~ of the District (*P*) der Distrikts-Post-Inspektor; (Techn.) der Aufseher; der Feldmesser, Landmesser; ~'s chain die Meßkette; ~'s level der Grabbogen, die Markscheiderwaage; ~'s table der Meßtisch.
to suspend *v* aufhängen; to ~ an officer, to ~ an officer from duty einen Beamten seines Amtes entheben; to ~ payment die Zahlung einstellen.
suspension *s* die Amtsenthebung; (Techn.) die Aufhängung; ~ of payment die Zahlungseinstellung.
swamp ore, bog ore *s* der Raseneisenstein, Torfeisenstein, das Sumpferz, Sumpfeisenerz.
swift *s* die leichte Haspel; das einfache lose Rad (bei der Prüfung des Drahtes auf Festigkeit).
swinging insulator *s* (*T*) der Pendelisolator.
Swiss commutator *s* (*T*) der Linien-Umschalter.
to switch *v* (*T*) umschalten; einschalten; to ~ in resistance Widerstand einschalten.
switch *s* (*T*) der Umschalter, Kommutator; ~ board das Umschalterbrett, (*F*) der Klappenschrank.
swivel *s* (Mech.) der Drehring; ~ joint das Drehgelenk, Universalgelenk.
synchronism *s* der Synchronismus, die Gleichzeitigkeit.
synchronous *a* synchron, gleichzeitig; ~ printer *s* (*T*) der Hughes'sche Typen-Druckapparat.
syphon *s* s. siphon.

T.

Tab *s* (*P*) die rechte Seite des Empfangsbescheinigungsformulars über eingeschriebene Briefe.
table *s* der Tisch; (Math., Phys. u. s. w.) die Tafel, Tabelle; ~ of contents das Inhaltsverzeichniß, Register (eines Buches); ~ of conversion die Umrechnungstabelle; ~ of despatch (*P*) die Leitübersicht; ~ of interest die Zinstabelle; despatch ~ (*P*) der Abfertigungstisch; opening ~ (*P*) der Entkartungstisch; time ~ (Eisenb.) der Fahrplan; turn ~ (Eisenb.) die Drehscheibe; ~ board die Tischplatte; ~ connection (*T*) die Tischverbindung; ~ drawer der Tischauszug; ~ slab das Tischblatt; ~ stamp (*P*) der Entkartungsstempel (durch dessen Beidrückung der entkartende Beamte bescheinigt, daß

die Einschreibsendungen sämmtlich vorhanden gewesen sind).
tablet *s* das Täfelchen; ~s *pl* die Schreibtafel; ~ check (*T*) das Apparattagebuch.
tabular statement *s* die tabellarische Uebersicht.
to tack *v* heften, anheften.
tack *s* der Stift, die Zwecke; wire ~ der Drahtstift.
tackle *s* (Mech.) der Flaschenzug, Rollenzug, Kloben; block & ~, fall & ~ der Flaschenzug; ~ blocks *pl* die Takelblöcke; ~ fall das Seil des Flaschenzugs; ~ hook der Windehaken.
tag *s* der Stift; das Etikett, der Zettel zum Anhängen (an Gepäck u. s. w.), die Gepäckmarke.
tain *s* das ganz dünne Weißblech; die Zinnfolie, das Stanniol.
to take *v* nehmen; to ~ down (Masch., Bau) abbrechen, abnehmen; to ~ the level nivelliren; to ~ to pieces auseinandernehmen, zerlegen; anziehen (von Nägeln); binden (vom Mörtel).
talking wire (*T*) diejenige Eisenb.-Telegr.-Leitung, welche vom Eisenbahndienstpersonal für amtliche Mittheilungen benutzt wird.
to tally *v* nachzählen, kontroliren; mit der Rechnung u. s. w. stimmen.
to tamper *v* with telegrams mit Telegrammen unlautere Manipulationen vornehmen.
tangent *s* (Geom.) die Tangente; ~ galvanometer die Tangentenbuffole.
tank *s* der Behälter, das Reservoir; der Wasserbehälter.
to tap *v* anzapfen; to ~ a telegraph line sich in eine Telegr.-Linie einschalten, um Telegramme abzufangen.
tap *s* (Techn.) der Zapfen; der Schraubenbohrer; ~ wrench das Windeeisen, Wendeeisen.
to tape *v* off mit einem Bandmaaß ausmessen; ~d envelope (*P*) der Umschlag für eingeschriebene Briefe, welcher kreuzweis mit einer farbigen Schnur (grün oder blau) umbunden ist; ~d wire (*T*) der mit Band umwundene Draht.
tape *s* die Schnur, das schmale Band; (*T*) der Morseftreifen; red ~ rothes Band (mit dem in England amtliche Schriftstücke gebunden werden); measuring ~, ~ measure die Meßschnur, das Bandmaaß; ~ wheel (*MA*) der im Untersetzkasten angebrachte Papierrollen-Träger.
to taper, to taper off *v* (Techn.) zuspitzen, konisch oder spitz zulaufend machen; spitz zulaufen, sich verjüngen.
taper *s* (Techn.) das Spitze, Spitzzulaufende, die Verjüngung.
taper, tapered, tapering *a* (Techn.) spitz zulaufend, verjüngt; ~ scale der verjüngte Maaßstab.
tapper form *s* (*T*) die Doppeltaste des Einnadel-Telegr.-Apparats.
tappet *s* (Mech., Masch.) die Knagge, der Daumen, Hebedaumen; ~ rod, ~ shaft die Daumenwelle; ~ wheel das Daumenrad.
to tar *v* (Techn.) theeren.
tar *s* der Theer; coal ~, gas ~ der Steinkohlentheer; vegetable ~, wood ~ der Holztheer, vegetabilische Theer.
to tare *v* tariren (die Tara vom Bruttogewicht in Abzug bringen).
tare *s* die Tara (das Gewicht der Verpackung, sowie die Vergütung dafür); real ~ die reine Tara.
tarif, tariff *s* der Tarif; der Zolltarif.
task work *s* (Techn.) die Akkordarbeit; ~ worker der Arbeiter auf's Stück, Akkordarbeiter.
to tax *v* taxiren, schätzen, veranschlagen; to ~ a letter (*P*) einen (unfrankirten) Brief austaxiren.
tax *s* der Zoll, die Steuer, Abgabe.
team *s* das Gespann, der Zug (Pferde oder Ochsen); ~ of workmen die Schaar, Gesellschaft Arbeiter.
technic, technical *a* technisch; ~ service (*P & T*) der praktische Dienst.
technician *s* der Techniker.
technics *s/pl* die Technik.

technology s die Technologie, Gewerbekunde.

telegram, telegraphic despatch, telegraphic message s das Telegramm; ~s *pl* & paper slip das Telegramm-Material; code ~ das Telegramm in Geheimschrift; cypher ~ das chiffrirte Telegramm; foreign ~ das internationale Telegramm; forwarded ~ das Ursprungs-Telegramm; ~ franked by a railway pass das gebührenfreie Eisenbahndiensttelegramm; inland ~ das interne Telegramm; ~ with multiple addresses das zu vervielfältigende Telegramm; news ~ das Zeitungs-Telegramm; ordinary ~ das gewöhnliche Telegramm; press ~ das Zeitungstelegramm; received ~ das Ankunfts-Telegramm; ~ re-directed to a second address, ~ to follow das nachzusendende Telegramm; ~ to be repeated das verglichene Telegramm; semaphoric ~ das semaphorische Telegramm; service ~ das Diensttelegramm; skeleton ~ das Skelett-Telegramm, s. skeleton; special ~ das besondere Telegramm; transmitted ~ das Durchgangstelegramm; urgent ~ das dringende Telegramm.

to telegraph, to transmit by telegraph v telegraphiren.

telegraph s der Telegraph; copying ~ der Kopirtelegraph; dial ~, disk ~, indicator ~, needle ~, pointer ~ der Nadeltelegraph; double needle ~ der Doppelnadeltelegraph; duplex ~ der Duplex-(Doppel-)Telegraph; field ~, military ~ der Feldtelegraph; printing ~ der Drucktelegraph; recording ~ der Telegraph, der das Alphabet durch konventionelle Zeichen übermittelt; submarine ~ der unterseeische Telegraph; type printing ~ der Typendruckapparat; writing ~ der Schreibtelegraph.

telegraph alphabet s das Telegraphen-Alphabet; ~ apparatus, ~ instrument, ~ register der Telegr.-Apparat; ~ cable das Telegr.-Kabel; ~ clerk der Telegr.-Beamte, Telegraphist; ~ construction der Telegr.-Bau; ~ Department die Telegr.-Abtheilung; ~ engineer der Telegr.-Bauführer; ~ engineering department die Telegr.-Bau-Abtheilung; ~ engineering staff das Telegr.-Bau-Personal; ~ key der Telegr.-Schlüssel; ~ learner (male ~ learner, female ~ learner) der Anwärter für den Telegr.-Dienst; ~ line die Telegr.-Linie; ~ master der Vorsteher einer kleineren Telegr.-Anstalt (Britisch-Indien); ~ office das Telegr.-Amt, die Telegr.-Anstalt; ~ plant die Telegr.-Anlage; ~ pole, ~ post die Telegr.-Stange; ~ service der Telegr.-Dienst; ~ stamp die Telegr.-Freimarke; ~ station die Telegr.-Station; ~ structure das Telegr.-Gestänge; ~ tender die Einladung zu einem Anbietungsverfahren auf Telegr.-Materialien; universal union der Allgemeine Telegr.-Verein; ~ wire der Telegr.-Draht, die Telegr.-Leitung.

telegrapher, telegraphist, operator, telegraph official, instrument clerk s der Telegraphist, Telegraphenbeamte; female telegraphist, female operator die Telegraphengehülfin.

telegraphic a telegraphisch; ~ apparatus, ~ instrument der telegraphische Apparat, Telegraph.

telegraphy s die Telegraphie; duplex ~ die Duplex-Telegraphie; multiplex ~ die Vielfach-Telegraphie; quadruplex ~ die Vierfach-Telegraphie.

telephone s der Fernsprecher, das Telephon; ~ call box die Fernsprechzelle; ~ call room, ~ call station die Fernsprechstelle; public ~ call station die öffentliche Fernsprechstelle; ~ connection der Fernsprechanschluß; ~ exchange, local ~ exchange das Stadtfernsprechvermittlungsamt; ~ line die Fernsprechlinie; ~ office das Fernsprechamt; ~ plant die Fernsprechanlage; ~ receiver der Fernsprech-Empfangsapparat; ~ sender der Absende-Fernsprecher; ~ service der Fernsprechdienst, Fernsprechbetrieb; ~ trunk line die Fernsprechverbindungslinie; ~ wire die Fernsprechleitung.

telephonic *a* circuit die Fern=
sprechleitung.
telephony *s* die Telephonie;
long distance ~ das Fernsprechen
auf weite Entfernungen.
tell-tale *s* (Techn.) die selbst=
thätige Anzeige= oder Zählvorrich=
tung; die Wächter=Kontrolluhr.
telpherage *s* die elektrische Lasten=
beförderung.
to temper *v* steel den Stahl
anlassen, nachlassen, ausglühen.
temper screw *s* (Mech.) die
Stellschraube.
tenacity *s* (Phys.) die Zähigkeit,
Festigkeit, die Kohäsion, das Zu=
sammenhalten der Körper.
to tender *v* anbieten; to ~ a
parcel for posting (P) ein Packet
aufgeben; to ~ one's resignation
seine Entlassung einreichen.
tender *s* die Eingabe, Offerte,
der Lieferungsantrag; das Angebot
auf eine Submission; (Eisenb.) der
Tender, Vorrathswagen; sorting ~
(P, Eisenb.) der Eisenbahnpostwagen,
in welchem die Postsachen sortirt
werden; legal ~ das gesetzliche Zah=
lungsmittel (im Handelsverkehr).
tenon *s* der Zapfen; ~ with key,
fox-tail wedged ~ der Zapfen mit
Keil, verkeilte Zapfen; ~ auger der
Zapfenbohrer.
tensile *a* strain (Mech., Phys.)
die Zugspannung, Zugkraft; resistance to ~ strain, ~ strength die
Zugfestigkeit, absolute Festigkeit.
tension *s* (Mech., Phys.) die
Spannung; die Spannkraft; ~ of
the steam (Dampfmasch.) die Dampf=
spannung; ~ roller die Spannrolle;
~ spring die Spannfeder.
term *s* der Termin; ~s *pl* die
Bedingungen; ~ of payment der
Zahlungstermin; ~s *pl* of sale die
Verkaufsbedingungen.
terminal *s* die Klemmschraube
am Ende einer Volta'schen Säule;
(T) die Abspannstange, ferner (in
England) jede Stange, an welcher
der Draht einen Winkel von nahezu
90° bildet.
terminal *a* carbon die letzte
Kohle (einer elektrischen Batterie); ~

pole (T) s. terminal; ~ station
(Eisenb.) der Hauptbahnhof am An=
fang oder Ende einer Bahnlinie, die
Kopfstation; ~ zink (T) das letzte
Zink in der Batterie.
terminus *s* dass. wie terminal
station.
terrestrial *a* current der Erd=
strom; ~ magnetism der Erdmagne=
tismus.
to test *v* probiren, prüfen; (T)
untersuchen, messen.
test *s* die Probe, Prüfung; (T)
Untersuchung, Messung; bending ~
die Prüfung des Drahtes auf Zähig=
keit; ~ for ductility, torsion ~
die Prüfung auf Drehungsfestigkeit;
elongation ~ die Prüfung auf abso=
lute Elastizität; loop ~ die Schlei=
fenprobe; ~ box s. testing box.
testing battery *s* (T) die Unter=
suchungsbatterie; ~ box, test box
der Untersuchungsbrunnen; ~ bracket
die Untersuchungskonsole; ~ insulator
der Untersuchungsisolator; ~ station
die Untersuchungsstation.
tetrode working *s* (T) die
gleichseitige Versendung von vier
Telegrammen auf dem Delany'schen
Apparat.
text book *s* das Lehrbuch.
texture *s* das Gefüge (von Me=
tallen), die Textur, Struktur, das
Aussehen auf dem Bruche; fibrous
~ die sehnige Struktur; grained ~,
granular ~ die körnige Struktur.
thermo-electric *a* (Phys.) ther=
mo-elektrisch; ~-~ pair, ~-~ couple
das thermo-elektrische Element.
thermo-electricity *s* (Phys.) die
Wärme=Elektrizität, die durch Wärme
hervorgebrachte Elektrizität.
thimble *s* der Ring, die Zwinge;
(T) die Doppelmutter (zur Befesti=
gung des Zugankers).
thorough-cut *s* (Eisenb. u. s. w.)
der Durchstich, Durchschnitt.
thread *s* der Faden; longitudinal
~ der Längsfaden; screw ~ (Mech.)
der Schraubengang, das Schrauben=
gewinde; double ~ das doppelte Ge=
winde; female ~ das Muttergewinde;
single ~ das einfache Gewinde.
threaded *a* mit einem Gewinde

versehen; double ~ screw die Schraube mit Doppelgewinde.

through *adv* durch; to be ~ (T) verbunden sein; ~ ticket *s* (Eisenb.) das durchgehende Billet; ~ traffic (T & P) der direkte Verkehr.

to throw *v* into gear (Mech.) in Gang setzen (Zahnräder u. s. w.), einrücken, in Eingriff bringen; to ~ out of gear außer Gang setzen, ausrücken.

throw *s* der Wurf, die Wurfweite; ~ of the needle of a galvanometer der Ausschlag der Nadel eines Galvanometers.

thumb *s* (Mech.) der Daumen, Zapfen; ~ screw (Mech.) die Flügelschraube.

to tick *v* off the mail (P) die Post abnehmen; to ~ off the bags in the arrival book die ankommenden Briefbeutel im Ankunftsbuche abstreichen.

tick list *s* (P) die Liste der transitirenden Kartenschlüsse.

ticket *s* (Eisenb.) die Fahrkarte, das Billet; to issue ~s Fahrkarten ausgeben; return ~ das Retourbillet; through ~ das durchgehende Billet; ~ porter der konzessionirte Packträger.

to tie *v* (Techn.) binden, mit Querbändern versehen; verankern.

tie *s* (Techn.) das Band, die Befestigung, das Verbindungsstück; (T) der Anker; ~ bolt der Ankerbolzen; ~ hook der Ankerhaken.

tight *a* dicht; air ~ luftdicht; spring ~ s. spring; water ~ wasserdicht.

to tighten *v* a spring eine Feder anziehen.

till *s* die Schublade, (besonders) die Geldschublade; (P) die Schalterkasse.

tilt *s* das Zelt (über einem Wagen oder Boote), die Wagendecke, Plane; ~ waggon (Eisenb.) der mit einer Plane bedeckte Frachtwagen.

tilted *a* iron das gehämmerte Eisen, Hammereisen; ~ steel der Gerbstahl.

timber *s* das Holz, Bauholz; standing ~ der lebende Baum; rough ~ der gefällte Baum; squared ~, sided ~ das bearbeitete Holz; die abgeschnittenen Stücke heißen slabs *pl*, das größte viereckige Stück, das von einem Baume geschnitten werden kann, heißt baulk; converted ~ gesägtes Holz; Holz in dicken Stücken scantling, in dünnen Stücken planks und boards (Bretter), ein dickeres Brett heißt plank, ein dünneres batten.

time *s* die Zeit; to be behind ~ Verspätung haben; loss of ~ of a screw der todte (leere) Gang einer Schraube; ~ ball der Zeitball; ~ bill (P) der Stundenzettel; ~ card (P) das Bestellbuch der Briefträger; ~ keeper der Chronometer, die Seeuhr; ~ signal das Zeitsignal; ~ table der Stundenplan, (Eisenb.) der Fahrplan.

timer *s* (Am) der Telegraphist.

to tin *v* verzinnen; mit Folie belegen.

tin *s* das Zinn; sheet ~ das Weißblech; ~ foil die Zinnfolie, das Stanniol.

to tip *v* (Techn.) beschlagen; to ~ with platinum das obere Ende eines Stiftes u. dgl. mit Platin versehen; to ~ with steel die Spitze (eines Werkzeugs) verstählen.

T iron *s* das T Eisen, T förmige Eisen; double T iron das Doppel T Eisen.

ton *s* die Tonne (als Gewicht = 20 cwt = 1016,048 Kilogramm); long ~ die Tonne von 21 Zentnern zu 112 Pfund.

tongs *s/pl* (Techn.) die Zange; dutch ~ die Froschklemme.

tongue *s* (Techn.) die Zunge (ein spitz zulaufender Theil); die Deichsel; ~ of a balance die Zunge am Waagbalken.

tool *s* das Werkzeug; cutting ~s, edge ~s *pl* Schneidwerkzeuge; ~ box, ~ chest der Werkzeugkasten.

tooling-iron *s* das Glätteisen, Streicheisen.

tooth *s* (Techn.) der Zahn; ~ wheel (Mech.) das Zahnrad.

toothed *a* (Techn.) gezahnt; ~ wheel das Zahnrad; ~ wheel work die Verzahnung.

toothing s die Verzahnung.
top s (Techn.) die Spitze, das oberste Ende; ~ of the form (T) der Kopf des Telegramm=Formulars; ~ of a pole das Zopfende einer Stange.
torsion s (Mech.) die Drehung, Torsion; die Verdrehung.
torsional a spring (Mech.) die Torsionsfeder; ~ strength die Drehungsfestigkeit.
to touch v berühren; to ~ a port einen Hafen anlaufen; (Techn.) greifen, angreifen (von einer Feile gesagt).
touch s die Berührung; der Strich; double ~ (Magn.) der doppelte Strich; separate ~ der getrennte Strich; single ~ der einfache Strich.
tow boat s das Schleppschiff; ~ path, towing path der Leinpfad.
towage s das Schleppen von Schiffen, bugsiren; der Lohn dafür; die Kettenschifffahrt.
town receiving house s (P) die Briefannahmestelle.
toy telephone s das Fadentelephon.
to trace v zeichnen, entwerfen, skizziren; to ~ a letter (P) einen Brief nachweisen.
trace s der Umriß, die Tracirung.
track s der Weg, die Bahn; (Eisenb.) der Schienenstrang, das Geleise; double ~ das Doppelgeleise; main ~ das Hauptgeleise; ~ path der Leinpfad.
tract s der Traktus, die Strecke, Straßenlinie.
traction s (Mech.) der Zug, das Ziehen, die Spannung; ~ wheel (Mech.) das Treibrad, Triebrad, Zugrad.
traffic s der Verkehr; carrying ~ der Speditionshandel; through ~ der Durchgangsverkehr; ~ manager (Eisenb.) der Betriebsinspektor, Verkehrschef; ~ returns pl die Verkehrsberichte.
train s (Eisenb.) der Bahnzug; down ~ der Zug von der Hauptstation nach der Provinz; up ~ der Zug von der Provinz nach der Hauptstation; ~ service (P) der Eisenbahn = Postdienst; ~ wire (Eisenb.) die zum Signalisiren der Züge benutzte Telegr.=Leitung.
transfer s (P) die Uebergabe der Post, des Dienstes; die Versetzung eines Beamten in ein höheres Amt; (Am) die Versetzung eines Angestellten; ~ sheet (P) der Uebergabebogen.
transformer s (Elektriz.) der Transformator.
transient a signals pl (T) nicht bleibende Zeichen.
transit s der Durchgang, der Transit; expenses of ~ die Transitkosten; rate of ~ die Transitgebühr, der Transitsatz; land ~, territorial ~ der Landtransit; maritime ~, sea ~ der Seetransit; exchange of correspondence (P) der Transitverkehr; ~ statistics pl die Durchgangsstatistik.
translating a station (T) die Uebertragungsstation.
translator s (T) der Uebertragungsapparat, Uebertrager.
transmission s (Physi.) die Leitung, Uebertragung; (Masch.) die Transmission, Leitung; ~ of telegrams die Abtelegraphirung, Beförderung von Telegrammen.
to transmit v übersenden, überschicken, überliefern; (T) befördern, abtelegraphiren; transmitted telegram das aufgenommene Telegramm, das per Draht weiterzubefördern ist, Durchgangstelegramm; (Masch.) übertragen.
transmitter s (T) der Geber, gebende Apparat.
transport s der Transport, die Uebertragung; charges of ~ die Transportkosten.
transportation s der Transport, die Beförderung.
transverse s (Geom.) die Transversale, Transversallinie, durchschneidende Linie; ~ beam der Querbalken; ~ section das Querprofil; ~ strength (Mech.) die Bruchfestigkeit, relative Festigkeit; die Biegungsfestigkeit.

19

travelling post office s das Eisenbahn-Postamt.

to traverse v durchfließen (vom elektr. Strome gesagt).

traverse s (Techn.) das Querstück, Querholz, der Querriegel; ~ section das Querprofil.

treasure trove s (P) das in Briefbeuteln, Briefkasten u. s. w. gefundene Geld oder irgend welche (einer Brief- oder Packetsendung) entfallene Werthsache.

tree s der Baum; ~, axle ~ (Masch.) die Welle, Achse.

to treenail, to trenail v (Techn.) mit hölzernen Nägeln verbinden, zusammenbolzen.

treenail, trenail, trennel s der Dübel, hölzerne Nagel, Pflock, Bolzen.

trembler s (Elektriz., T) ein elektrischer Glockenapparat; der Schnarrwecker.

trembling bell s (Elektriz., T) der Schnarrwecker.

trench s der Graben, Bahngraben; der Rinnstein.

trendle s (Mech.) die Rolle, Walze.

trennel s f. treenail.

trepan s (Mech.) der Bogenbohrer, die Bogendrille.

trestle, tressel, trussel s das Gerüst, Gestell, der Bock; ~ bridge, bridge on trestles die Bockbrücke; ~ work daſſ. wie ~ bridge.

trial s (Techn.) die Probe, Prüfung, das Experiment.

trig s der Unterlegkeil.

to trim v (Techn.) zurichten; beschneiden; behauen; to ~ in einstemmen, einsetzen, einlassen.

triode working s (T) die gleichzeitige Versendung von drei Telegrammen mit dem Delany'schen Apparat.

trip s (Eisenb.) die einmalige Fahrt zwischen zwei Orten; (P) die Fahrt des Bahnpostbeamten; ~ allowance, ~ money die Fahrtgelder.

trolly s (Eisenb.) der Streckenwagen, Bahnmeisterwagen, die Draisine.

trough s der Trog; (T) die Kabelrinne; galvanic ~ der galvanische Trogapparat; ~ battery (Elektriz.) die Trogbatterie.

troy, troy weight s das Troygewicht, Goldgewicht; ~ pound s. pound.

truck s der Tausch, Tauschhandel; (Eisenb.) der Blockwagen, der Rollwagen; der Streckenwagen; ~ system das Tauschwerthsystem (wonach die Arbeiter mit Waaren statt mit baarem Gelde abgelohnt werden).

truncated, truncate a (Geom.) abgestumpft; ~ cone der abgestumpfte Kegel.

trundle, trundle wheel s (Mech., Masch.) der Drehling, Drilling, das Getriebe.

trunk s (Techn.) der Stamm; (Eisenb.) die Hauptlinie einer Bahn; ~ line (T) die Hauptlinie, von welcher sich Nebenlinien abzweigen; (Eisenb.) die Hauptlinie; ~ wire (T) die Telegr.-Leitung, von welcher sich Nebenlinien abzweigen; ~ wires exchange (F) das Hauptvermittlungsamt für die Verbindungsleitungen zwischen verschiedenen Fernsprechämtern.

trunnion s (Mech.) der Zapfen, Kurbelzapfen, Drehzapfen.

to truss v (T) den Fußpunkt einer Stange mit Steinen oder Holz verstärken; eine Stange verankern; to ~ a piece of timber einen Balken armiren.

truss s (Techn.) der Bund; das Hängewerk, der Bock; ~es pl das Gerüst, der Bock; ~ beam der Eisenbalken; ~ bridge die Gitterbrücke, Brücke mit Hängewerk; ~ wire (T) der Anker.

trussed beam s der armirte, verstärkte Balken; ~ girder der armirte Träger; ~ poles pl (T) gekuppelte Stangen.

tube s (Techn.) das Rohr, die Röhre; die Löthmuffe; caoutchouc ~, elastic ~, India rubber ~ der Gummischlauch, Kautschukschlauch.

tubing s das Material zu Röhren; das Röhrenwerk.

tubular, tubulous a röhrenartig, röhrenförmig; tubular post die Rohrpost.

tug boat s das Schleppboot.
tun buoy s die Tonnenboje.
turf s der Rasen; der Torf.
to turn v (Techn.) drehen, wenden; sich drehen.
turn s (Techn.) die Drehung, Windung, Umwindung; (T) die Reihenfolge in der Telegrammbeförderung; ~ screw der Schraubenschlüssel.
turning bridge s die Drehbrücke; ~ joint das Scharnier, Gelenk.
twine s der Bindfaden; der Zwirn.

to twist v flechten, verseilen; seven wires ~ed into a single strand sieben Drähte zu einer Litze verseilt.
twist s das Drehen, die Drehung; die Raupe (einer Löthstelle); ~ drill der Schneckenbohrer; ~ test die Prüfung des Drahtes auf Drehungsfestigkeit.
type s die Type, der Druckbuchstabe; ~ printing apparatus (T) der Typendrucktelegraph; ~ shaft (HA) die Typenradachse; ~ wheel (HA) das Typenrad.

U.

to unbend v a spring (Mech.) eine Feder nachlassen.
unclaimed a mail matter (P) unbestellbare, nicht abgeholte Postsachen.
to uncouple v (Masch.) losmachen, auslösen, ausrücken.
uncostumed a unverzollt.
underground a unterirdisch; ~ line (T) die versenkte Leitung.
undulation s (Phys.) die Wellenbewegung; die Schallwelle.
undulatory a wellenförmig.
unentered a beim Zollamte nicht angegeben, unverzollt.
to ungear v (Masch.) aus dem Getriebe bringen, außer Gang setzen, auslösen.
uninsulated a (Elektriz.) nicht isolirt, mit der Erde leitend verbunden.
unipolar a einpolig.
unit s (Arithm., Phys.) die Einheit; ~ of capacity, farad die Einheit der elektrischen Kapazität, das Farad; ~ of current die Stromeinheit; dynamical ~ die Krafteinheit, Einheit der mechanischen Leistung; ~ of electromotive force, volt die Einheit der elektromotorischen Kraft, das Volt; ~ of force die Krafteinheit; ~ of resistance, ohm die Widerstandseinheit, der Normalwiderstand, das Ohm; ~ of work die Arbeitseinheit.
universal a joint (Mech.) das

Universalgelenk; ~ screw wrench der englische Schraubenschlüssel, Universalschraubenschlüssel; ~ switch (T) der Linien-Umschalter.
unmailable a matter (P) zur Beförderung durch die Post ungeeignete oder verbotene Gegenstände.
unoccupied a circuit (T) die unbesetzte Linie.
unpaid a unbezahlt; (P) unfrankirt.
to unscrew v abschrauben, aufschrauben, losschrauben.
unsettled a unerledigt (von einer Rechnung u. s. w.).
to unsolder v ablöthen, loslöthen, auflöthen.
unsquared a timber unbehauenes Bauholz.
unstable a equilibrium (Phys.) das unsichere, labile Gleichgewicht.
to unwind v (Techn.) aufwinden, abwickeln; sich aufwinden.
unwrought a (Techn.) unverarbeitet, roh; ~ iron das Roheisen.
up & down working s (T) abwechselndes Arbeiten in Reihen; ~ & down working proper (T) abwechselndes Arbeiten mit je einem Telegramm; ~ line (T) diejenige Telegr.-Linie, welche zur Beförderung von Telegrammen in der Richtung nach der Hauptstation der Linie dient; (Eisenb.) das Geleise für Züge, welche in der Richtung nach der Hauptstation der Linie gehen; ~ side

of a pole (*T*) die der *up station* (f. dief.) zugewendete Seite der Stange; ~ *signal* (Eisenb.) das Einfahrtssignal; ~ *station* (*T*) die Hauptstation einer Telegr.-Linie; ~ *train* (Eisenb.) der Zug von der Provinz nach der Hauptstation; ~ *wire* (*T*) eine Telegr.-Leitung der *up line* (f. dief.).

upkeep *s* *of fittings* die Unterhaltung technischer Einrichtungen.

upright *s* der Aufriß (eines Gebäudes); der Ständer, die Säule; (*T*) die große starke Stange.

usance *s* der Uso, die Wechselfrift, der Wechselgebrauch; *bill at* ~ der Usowechsel.

used *a* (Techn.) gebraucht, ausgenutzt, abgängig; ~ *up* aufgebraucht.

useful effect, useful work *s* (Masch.) der Nutzeffekt, die Nutzleistung.

utensils *s/pl* (Techn.) das Geräth, die Werkzeuge, Geräthschaften.

V.

vacancy *s* das Freiwerden einer Stelle, die Stellenerledigung.

to vacate *v* *an office* aus einem Amt ausscheiden.

vacuum *s* (Phys.) das Vacuum, der luftleere Raum; ~ *protector* (*T*) der Varley'sche Blitzableiter (für versenkte Linien; Erdleitung und Liniendraht sind in eine luftleere Glasröhre eingeschmolzen).

valid *a* giltig.

to validate *v* rechtsgiltig machen, bestätigen.

validity *s* die Giltigkeit.

valuation *s* die Schätzung, Taxation.

value *s* der Werth, die Valuta, der Münzwerth; ~ *in account* Werth in Rechnung; *current* ~ der gangbare Werth; *intrinsic* ~ der Gehalt.

valve *s* (Masch.) die Klappe, das Ventil; *air* ~ das Luftventil, die Windklappe; *delivery* ~ das Auslaßventil; *escape* ~, *safety* ~ das Sicherheitsventil.

van *s* der Transportwagen; (Eisenb.) der Gepäckwagen; *guard's* ~ (Eisenb.) der Wagen für das Begleitpersonal; *travelling* ~ (*P*) der Eisenbahnpostwagen.

vane *s* die Wetterfahne, Windfahne; (Masch.) das Register, die Klappe; *sight* ~ das Diopter; *electric* ~ das elektrische Flügelrad.

to vaporate, to vaporize *v* (Phys.) verdunsten, verdampfen.

vaporization *s* die Verdampfung, Verdunstung.

variable *a state* (*T*) der veränderliche Zustand (einer Telegr.-Leitung).

variation *s* die Veränderung; die Schwankung; ~ *of the compass* die Abweichung, Deklination der Magnetnadel; *line of no* ~, *agonic line* (Magn.) die agonische Linie.

variegated *a* (Techn.) bunt, gefleckt; ~ *copper ore* das Buntkupfererz.

to varnish *v* firnissen, überfirnissen.

varnish *s* der Firniß; *black* ~ ein Firniß zum Anstrich von Brettern u. f. w.; *coat of* ~ der Firnißüberzug; *lac* ~ der Lackfirniß.

vat *s* das große Faß; die Küpe, der Trog; *lixiviating* ~ der Auslaugebottich.

veber *s* (*T*) die B. A. Einheit der Quantität (der Elektrizität, die in einer Sekunde durch einen Stromkreis fließt, dessen elektromotorische Kraft = 1 Volt und dessen Widerstand = 1 Ohm ist).

vegetable *a coal* die Holzkohle; ~ *fibre* die Pflanzenfaser.

vehicle *s* das Gefährt; *postal* ~ der Postwagen (allgemein).

velocity *s* (Mech.) die Geschwindigkeit; *angular* ~ die Winkelgeschwindigkeit; *mean* ~ die mittlere Geschwindigkeit.

verification *s* die Bescheinigung, Beurkundung; ~ *of the receiving office* (*P*) das Empfangsanerkenntniß, die Bescheinigung über unbeanstandete Uebernahme der Post; ~ of a telegraph apparatus (T) die Untersuchung der verschiedenen Bestandtheile eines Telegraphenapparats; ~ certificate, bulletin of ~ (P) die Rückmeldung.

verifier s of accounts der Kalkulator.

to verify v beweisen; die Richtigkeit (einer Rechnung) bescheinigen.

vernier s der Vernier, Nonius, Sekundentheiler; ~ calliper die Schubleere mit Nonius.

vertex s (Geom.) der Scheitel, die Spitze; (T) der tiefste Punkt des Drahtes zwischen zwei Stangen, s. auch chain curve.

vertical a vertikal, stehend, lothrecht, senkrecht; ~ plan, ~ section der Aufriß, Standriß; ~ point der Scheitelpunkt, Zenith.

verticity s (Techn.) die Umschwingung, Rotation.

vessel s (Techn.) das Gefäß; das Fahrzeug, Schiff; communicating ~s pl (Phys.) kommunizirende Gefäße.

to vibrate v (Phys.) schwingen; (Masch.) schnarren, zittern.

vibrating a alarm der Schnarrwecker; ~ spring (HA) die Pendelstange.

vibration s (Phys.) die Schwingung; amplitude of ~ die Schwingungsweite.

vibrative motion, vibratory motion s (Phys.) die Schwingungsbewegung.

vice, vise s (Techn.) der Schraubstock; bench ~, table ~ der Bankschraubstock; hand ~, small ~, tail ~ der Feilkloben, Handkloben; ~ bench die Schraubstockbank.

virtual a (Mech., Phys.) virtuell, mit der Kraft zu wirken.

vis elastica s (Phys.) die Spannkraft; ~ inertiae das Beharrungsvermögen, die Trägheitskraft; ~ major die höhere Gewalt.

visual a angle der optische Winkel, Gesichtswinkel; ~ distance die Sehweite; ~ line die Sehlinie; ~ signalling die optische Telegraphie (im britischen Heere); ~ signalling troop (England) das optische Telegraphen-Korps.

vitreo-electric a glaselektrisch, positive Elektrizität enthaltend.

vitreous a glasig; ~ electricity die positive oder Glaselektrizität.

vitriol s der Vitriol; ~, oil of ~, sulphuric acid die Schwefelsäure; blue ~, copper ~ der blaue Vitriol, Kupfervitriol; green ~, iron ~, copperas der grüne Vitriol, Eisenvitriol; white ~, zinc ~ der weiße Vitriol, Zinkvitriol.

void a money order (P) dass. wie lapsed money order s. lapsed.

volatile a (Chem.) verfliegend; ~ oils pl flüchtige, ätherische Oele.

volatilization s (Chem.) die Verflüchtigung.

to volatilize v (Chem.) verflüchtigen, flüchtig machen.

volt s (Elektriz.) das Volt (Einheit der elektromotorischen Kraft).

voltaic a arc der Volta'sche Bogen, elektrische Lichtbogen; ~ electricity, voltaism der Galvanismus, die Berührungselektrizität.

voltameter s (Phys.) das Voltameter (Instrument zur Messung der durch einen elektrischen Strom entwickelten Gase).

volume s (Phys., Geom.) das Volumen, der körperliche Inhalt; atomic ~, specific ~ das Atomvolum.

volumetric a analysis (Chem.) die Maaßanalyse.

vote account s der Betrag, welcher vom Parlamente für die Gesammtausgaben (eines Verwaltungszweiges) bewilligt wird.

voted service account s der Betrag, welcher vom Parlamente für die Ausgaben (eines Verwaltungszweiges) im laufenden Jahre bewilligt wird.

voucher s der Beleg, Belag; der Empfangsschein.

vulcanite, ebonite s das Hartgummi, der Ebonit.

vulcanization, vulcanizing s of caoutchouc das Vulkanisiren, Schwefeln des Kautschuks.

to vulcanize v caoutchouc Kautschuk vulkanisiren, schwefeln.

W.

wafer, waffle *s* die Oblate.
wages *s/pl* (Techn.) der Lohn, die Löhnung, das Gehalt.
waggon, wagon *s* der Wagen, Lastwagen, Frachtwagen; (Eisenb.) der Güterwagen, Gepäckwagen; goods ~ der Güterwagen.
waif *s* das herrenlose Gut.
walk *s* (P) das Bestellrevier (des Briefträgers); der Bestellgang des Landbriefträgers; inward ~ der Bestellgang zurück zum Postamt; outward ~ der Bestellgang vom Postamte aus.
wall bracket *s* (T) die Mauerkonsole; ~ clamp der Maueranker; ~ hook die Rohrschelle; ~ letter box der im Mauerwerk angebrachte Postbriefkasten.
ware *s* die Waare; brown ~ gemeines Steinzeug, Steingut; china ~ das Porzellan; earthen ~ das irdene Geschirr, die Töpferwaare; hard ~ die Eisenwaaren.
to warp *v* (Techn.) zusammenziehen, krumm machen; sich zusammenziehen, sich verziehen, sich werfen (vom Holze gesagt).
to warrant *v* garantiren, gut sagen, sich verbürgen für . . .
warrant *s* die Anweisung auf Waaren, die in den Lagerhäusern der Regierung unter Zollverschluß aufbewahrt werden; ~ of attachment der richterliche Befehl für eine Beschlagnahme; ~ of attorney die Spezialvollmacht; ~ of caption der Steckbrief.
warranty *s* die Garantie, Sicherheit.
washer *s* (Mech.) die Unterlagsscheibe; das Bolzenblech, Mutterblech, die Zwischenlage; India rubber ~ (Mech., Masch.) der Kautschukdichtungsring, die Unterlagsscheibe von Gummi.
waste *s* (Techn.) der Abgang, Abfall; ~ paper die Makulatur, das Ausschußpapier, unbrauchbar gewordene Dienstpapiere, Drucksachen u. s. w.
watchman *s* (Eisenb.) der Bahnwärter; watchmen's inspector der Bahnmeister.
way bill *s* der Frachtbrief; (P) der Begleitzettel; Ladezettel; ~ leave (T) die Erlaubniß, einen Weg zum Setzen von Telegr.-Stangen zu benutzen; die Gebühr, die für solche Erlaubniß gezahlt wird; ~ letter (P) der einem Postboten unterwegs übergebene Brief.
to wear *v* away (Techn.) abgenutzt werden, abnehmen, schwinden; to ~ out sich abnutzen.
weather contact *s* (T) die Berührung der Drähte infolge schlechten Wetters.
web *s* (Techn.) die Tragrippe eines Eisenträgers, einer Stange aus doppeltem T Eisen.
to wedge *v* (Techn.) keilen, verkeilen; to ~ in einkeilen.
wedge *s* (Techn.) der Keil; ~ cut-out (T) der keilförmige Stöpsel zum *spring jack* (s. dens.) gehörig.
to weigh *v* wägen, wiegen; wiegen, schwer sein.
to weight *v* belasten, beschweren.
weight *s* das Gewicht; die Schwerkraft; das Gewichtsstück; die Belastung; bill of ~, certificate of ~ der Waageschein; dead ~ das Eigengewicht eines Gefäßes; even ~ das Gleichgewicht; hundred ~ s. hundredweight; sliding ~ (Mech.) das Laufgewicht; standard ~ das Normalgewicht.
to weld *v* schweißen, anschweißen; to ~ together zusammenschweißen.
welding heat *s* die Schweißhitze; ~ hot weißglühend, schweißwarm; ~ point die Schweißstelle.
wet-rot *s* die nasse Fäulniß.
wheel *s* das Rad; ~ & axle Rad und Welle; annular ~ das Zahnrad mit innerer Verzahnung; arbor ~ das Wellrad; balance ~ das Steigrad; bevel ~, conical ~, angular ~, mitre ~ das konische Rad, Winkelrad; crown ~ das Kronrad; cog ~, cogged ~ das Kammrad; cylindrical ~, cylindrical toothed ~, right ~, spur ~ das Stirnrad;

escape ~, escapement ~ das Steig=
rad, Hemmungsrad; fly ~, flying ~
das Schwungrad; friction ~ das
Friktionsrad; ratchet ~, rack ~ das
Sperrrad; type ~ (HA) das Typen=
rad; ~ barrow der Schubkarren,
Schiebkarren; ~ shaft, ~ spindle
die Radachse, Radwelle; ~ work das
Räderwerk, Laufwerk.

whim *s* der Göpel, Haspel, die
Aufziehmaschine; horse ~ der Pferde=
göpel; ~ engine der Göpel; ~ gin
der Pferdegöpel.

whimble *s* (Techn.) der Aus=
räumer.

whip *s* (T) ein Instrument zur
Messung von Kondensatoren (be=
stehend im Wesentlichen aus einer
Feder, welche zwischen zwei Kontakten
vibrirt, von denen einer mit der
Batterie, der andere mit einem Gal=
vanoskop verbunden ist).

„white" *s* das Merkzeichen beim
Abstecken u. s. w. einer Telegr.=Linie
(bestehend aus einer gespaltenen Latte,
in welche ein Stück weißen Papiers
gesteckt ist).

width *s* (Techn.) die Weite,
Breite; ~ of the bay die Jochweite,
lichte Brückenspannung; ~ in the
clear, inside ~ die Lichtenweite; ~
between the rails, ~ on the track
(Eisenb.) die Spurweite.

wimble *s* (Techn.) der Bohrer,
Fretbohrer; ~ scoop der Löffelbohrer.

winch *s* (Mech.) der Haspel, die
Winde, die Kurbel; (T) die Draht=
winde; hand ~ die Handwinde;
purchase ~ die Winde mit Vorge=
lege; ~ handle die Kurbel.

to wind *v* (Techn.) winden,
wickeln, aufwickeln; sich winden; to
~ off abwinden, abhaspeln; to ~ up
an account eine Rechnung abschlie=
ßen, saldiren.

wind & water line *s* die
Grundlinie (der Punkt, an dem die
Telegr.=Stange aus dem Erdboden
tritt).

windlass, windlace *s* (Masch.,
Mech.) die Winde, Hebewinde, der
Haspel; der Krahn; ~ of a gin die
Haspelwalze.

window *s* das Fenster; (T & P)
der Schalter; ~ delivery (P) der
Schalterdienst, die Schalterabholung;
~ tube (T) das Einführungsrohr.

wippe *s* (T) die Wippe.

to wire *v* mit Draht befestigen;
(T) Draht auf die Stangen auf=
bringen; telegraphiren.

wire *s* der Draht; (T) der Te=
legr.=Draht, die Telegr.=Leitung;
aërial ~ die oberirdische Leitung;
annealed ~ der ausgeglühte Draht;
binding ~ der Bindedraht; braided
~ der verseilte Draht; conducting
~ der Leitungsdraht; covered ~ der
besponnene Draht; drawing-in ~ der
Einziehdraht; drawn ~ der gezogene
Draht; earth ~ der Erddraht; fine
~ der Wickeldraht; main ~ die
Hauptleitung, (vgl. main line); open
~, overground ~, overland ~ die
oberirdische Leitung; overhouse ~
die über die Häuser weggeführte
Leitung; primary ~ der primäre
Draht (mit induzirtem Strome);
secondary ~ der sekundäre (indu=
zirte) Draht; special ~ die für be=
sondere Zwecke bestimmte Telegr.=
Leitung; spiral ~ die Drahtspirale;
strand ~ die Drahtlitze; subaquatic
~, subaqueous ~ die Unterwasser=
leitung; submarine ~ die unter=
seeische Leitung; subterranean ~ die
unterirdische Leitung; talking ~ die
Telegr.=Leitung, welche vom Eisen=
bahndienstpersonal für amtliche Mit=
theilungen benutzt wird; taped ~
der mit Band umwickelte isolirte
Draht; thin ~ der Wickeldraht; tie
~ der Bindedraht; train ~ die Te=
legr.=Leitung, welche zum Signali=
siren der Eisenbahnzüge benutzt wird;
trunk ~ die Telegr.=Leitung, von
welcher sich kleinere Verbindungen
abzweigen; underground ~ die unter=
irdische Leitung.

wire brush *s* die Drahtbürste;
~ carrier der Drahtträger des
Pendel=Isolators; ~ coil der Draht=
ring; ~ covering die Drahthülle;
~ drawing mill die Drahtziehe=
rei; ~ drawing plate das Draht=
zieheisen; ~ drum die Drahtleier
(Scheibe zum Drahtziehen); ~ finder
das Instrument, mit dem eine seh=

lerhafte Ader im Kabel gefunden wird; ~ gauge die Drahtleere, Drahtklinke; ~ hook dasſ. wie ~ carrier; ~ mill dasſ. wie ~ drawing mill; ~ rope das Drahtſeil; ~ straightening tool die Drahtwinde; ~ strainer der Drahtſpanner; ~ tack der Drahtſtift; ~ works die Drahtfabrik.

to withdraw v a letter, a telegram einen Brief, ein Telegramm zurückziehen.

wood s das Holz; der Wald; cross grained ~, end grained ~ das Hirnholz; grain of ~ die Holzfaſer; grain ~ das Aderholz, Langholz; heart ~ das Kernholz; sap ~ der Splint, das Splintholz; seasoned ~ das lufttrockene, ausgetrocknete Holz; ~ screw die Metallſchraube zum Einſchrauben in Holz.

wooden a screw die Holzſchraube, hölzerne Schraube.

wool s die Wolle; slag ~ die Schlackenwolle.

word rate, word tax s (T) die Worttaxe.

to work v (Techn.) arbeiten, bearbeiten, verarbeiten; im Betrieb, im Gange ſein; to ~ a line (T) eine Telegr.-Linie in Betrieb ſetzen.

work s (Techn.) die Arbeit, das Werk; der Betrieb; (Mech., Maſch.) das Getriebe; to set poles against their ~ (T) Stangen ſo ſetzen, daß ſie nach der Richtung des ſeitlichen Zuges etwas überſtehen; to set to ~ (at ~) in Betrieb ſetzen.

working s (Techn.) das Verarbeiten, Bearbeiten; die Hantirung; der Gang, Betrieb, das Spiel; in ~ order in betriebsfähiger Ordnung; ~ point der Angriffspunkt; ~ point of a telegraph line (T) der Zuſtand, in welchem eine Telegr.-Linie gerade noch betriebsfähig iſt; ~ power die Arbeitskraft.

works s/pl die Fabrik.

workshop s of the Telegraph Department (T) die Haupt-Apparat-Werkſtatt.

worm & wheel, screw & wheel s (Maſch.) das Schneckenradgetriebe; endless ~ die Schraube ohne Ende; ~ of a screw der Schraubengang, das Schraubengewinde; ~ gear dasſ. wie ~ & wheel; ~ screw der Schraubenzieher; ~ wheel das Schraubenrad, Schneckenrad.

to wrap v umwickeln; to ~ in, to ~ up einwickeln, einſchlagen.

wrapper s das Packtuch; (P) der Behälter für Packete bei der Packetbeförderung; das Streifband, Kreuzband.

wrapping paper s das Packpapier.

wrench, screw wrench s der Schraubenſchlüſſel, Schraubenzieher; monkey ~, shifting ~, universal screw ~ der engliſche (Univerſal-) Schraubenſchlüſſel.

wrongly-delivered letters pl (P) unrichtig beſtellte Briefe.

wrought a bearbeitet; ~ iron das Schmiedeiſen, Stabeiſen; das weiche Eiſen.

Y.

yacht s die Yacht, das Vergnügungsſchiff.

yard s die Yard, engliſche Elle (3 engliſche Fuß = 0,9144 Meter); der Hofraum; (in Zuſammenſetzungen: Hof, Anſtalt, z. B. wood ~ der Holzhof, impregnating ~ die Imprägnir-Anſtalt u. ſ. w.).

year s das Jahr; calendar ~ das Kalenderjahr; fiscal ~ das Rechnungsjahr; ~ under report das Berichtsjahr.

to yield v (Techn.) liefern; abgeben; einbringen, abwerfen, eintragen.

yield s der Ertrag; der Gehalt (von Metallen); (Techn.) die Ausbeute.

yoke s die Querverbindung der Elektromagnetſchenkel.

yute hemp, yute, jute s der Jutehanf.

Z.

zero *s* die Null, der Nullpunkt (des Thermometers u. s. w.); ~ adjusting lever (*HA*) der Einstellhebel; ~ stop das Kreuz auf dem Zifferblatte des A.B.C.-Apparats.

zinc *s* das Zink; chloride of ~ das Zinkchlorid; oxide of ~ das Zinkoxyd; sulphate of ~, vitriol of ~ der Zinkvitriol; ~ covered wire der galvanisirte Draht, Draht mit Zinküberzug; ~ rod der Zinkstab; ~ sender (*T*) der Zinksender (ein selbstthätiger Umschalter für den Kabelbetrieb).

zincing, zinking *s* das Verzinken, die Verzinkung; galvanic ~ die galvanische Verzinkung.

Anhang,

enthaltend die im

Englischen am häufigsten vorkommenden Abkürzungen.

1. Der Vermerk „(telegr. Zeichen)" bedeutet, daß der Buchstabe oder die Buchstaben-Zusammenstellung die dahinter angegebene Bedeutung im innern telegraphischen Verkehr Großbritanniens hat.

2. Diejenigen Buchstaben und Buchstaben-Gruppen, welche „Präfixe" sind, sind durch den Beisatz (*pref.*) gekennzeichnet. Vgl. „*prefix*" im englisch-deutschen Theile des Wörterbuchs.

A.

A. Das Formular für Ursprungstelegramme.
A (Reply). Das Formular für bezahlte Antworts-Telegramme.
A 1. Das Formular für Ursprungstelegramme mit eingedruckter Sixpenny Freimarke.
A $\frac{S}{E}$. Das Formular für Telegramme von der Fondsbörse.
A 1 $\frac{S}{E}$. Dasselbe wie vorstehend mit eingedruckter Freimarke.
A $\frac{S}{R}$. Das Formular für Ursprungstelegramme mit mehreren Adressen.
A. A. P. S. American Association for the Promotion of Science.
A. A. S. (Academiae Americanae Socius) Fellow of the American Academy.
A. B. (Artium Baccalaureus) Bachelor of Arts.
acc., acct. account.
agt. agent.
Al., Ala. Alabama.
A. M. (Artium Magister) Master of Arts; Ante Meridiem (Vormittags); Anno Mundi.
Amer. America.
amt. amount.
An. anno.
Ap., Apl. April.
A. R. Anno regni; Anna Regina (gemeint ist Queen Ann, Königin Anna 1702—14).
Ar., arr. Arrive, arrival.
Ark. Arkansas.
Art. Article.
A. S. Assistant Secretary; Assistant Surgeon.
Att., Atty. Attorney; **Atty. Gen.** Attorney General.
Av. Average; Avenue.
Avoir. Avoir du poids.

B.

B. Bachelor (ein akademischer Grad); battery; das Formular für Durchgangs-Telegramme; Base; Baron; Bay; Book.
b. born.
B. A. British America; Bachelor of Arts, s. **A. B.**
B. A. U. British Association Unit.
Bah. Bahamas.
Bal. Balance.
Bar. Barrel.
Barb. Barbadoes.
Bart. (auch **Bt.**) Baronet.
Bbl. Barrel, barrels.
B. C. Before Christ; Board of Control.
B. C. L. Bachelor of Civil Law.

B. D. Bachelor of Divinity.
Bd. Bond; bound.
Belg. Belgic.
Berks. Berkshire.
B. I. British India.
Bic. Bichromate.
Bk. Bank, book.
B. K. reversing battery key.
B. L. Bachelor of Laws.
B. LL. (Baccalaureus legum) Bachelor of Laws.
Bl. Barrel.
B. M. (Baccalaureus medicinae) Bachelor of Medicine.
B. O. Branch Office.
Bor. Borough.
BQ (*pref.*) „Reply to an urgent or ordinary service telegram."
Br. Bro. brother.
B. S. L. Botanical Society London.
Bush. Bushel.
B. W. G. Birmingham Wire Gauge.

C.

C. Das Formular für Ankunfts-Telegramme; carbon; conductor; Congress; Consul; (Centum) a hundred; Cent; centime; chapter.
C. A. Chief Accountant; Controller of Accounts.
Ca. California.
caf. cost, assurance, freight.
Cal. California; Calendar.
Cam., Camb. Cambridge.
Cap. Capital; Chapter.
Caps. Capitals.
Capt. Captain.
Carp. Carpentry.
Cash. Cashier.
Cat. Catalogue.
C. B. Companion of the Bath; Cape Breton.
C. C. County Commissioner; County Court; contra; credit; (compte courant) account current.
Cd. condenser.
C. E. Canada East; Civil Engineer.
Cent. (Centum) a hundred.
cf. confer.
C. G. Commissary General; Coast Guard.

c, g, s (centimeter, gramm, second, die Einheiten des absoluten Centimeter-Gramm-Sekundensystems).
C. H. Court House; Custom House.
Ch. Church; Chapter; Charles.
Chanc. Chancellor.
Chap. Chapter.
Chas. Charles.
Chem. Chemistry.
Civ. civil.
C. J. Chief Justice.
Cl. Clergyman; Clerk; Chlorine.
Cld. Cleared.
Clk. Clerk.
C. M. Common meter.
Cm. Commutator.
Co. Company; County.
c/o care of.
C. O. Crown Office; Colonial Office.
C. O. D. cash (or collect) on delivery.
C. of G. H. Cape of Good Hope.
Col. Colonel; Colonial; Column.
Coll. College; Collector; Collection; Colleague.
Com. Commissioner; Commodore; Committee; Commerce; Commentary; Common.
Comp. Compare; Compound.
Con. contra.
Cong. Congress.
Conn., Con., Ct. Connecticut.
Const. Constitution.
Cor. Mem. Corresponding Member.
Cor. Sec. Corresponding Secretary.
Coss. Consuls.
C. P. Candle Power.
C. P. C. Clerk of the Privy Council.
CQ (telegr. Zeichen) „All stations."
Cr. credit; creditor; chromium.
Ct. Connecticut; Count; Court; Cent; (centum) a hundred.
Cts., cts. cents.
cur. current (der laufende Monat).
C. W. Canada West.
Cwt., cwt. (lat. centum a hundred und engl. weight) a hundred weight.

D.

D., d. Duke; Duchess; day; died; dime; deputy; degree; (denarius, denarii) a penny, pence.
Dan. Danish.
D. C. District of Columbia.
D. C. L. Doctor of Civil Law.
D. D. (Divinitatis Doctor) Doctor of Divinity.
d. d. de dato.
De. Delaware.
Def., def. definition.
Deft., deft. Defendant.
Deg., deg. Degree, Degrees.
Del. Delaware; Delegate.
Den. Denmark.
Dep. Departed; Departure; Deputy; Department.
Dept. Department; Deponent.
DF (telegr. Zeichen) „Direct line free."
Dft., dft. Defendant.
D. H. Dead Head.
Dict. Dictionnary.
Dis., dis. Distance; distant.
Dis., Disct. Discount.
Dist. District.
Div. Dividend; Division.
D. K. Charge and discharge key (of a telegraph instrument).
D. L. O. Dead Letter Office.
D. O. District Office.
Dolls., dolls. Dollars.
Doz., doz. Dozen.
D. P. Doctor of Philosophy.
D. P. O. Distributing Post Office.
Dpt. Deponent.
Dr. Debtor; Doctor.
D. S. (Dal Segno) From the sign.
DS (*pref.*) Engineer's service telegram requiring immediate attention.
D. T. Dakotah Territory; (Doctor Theologiae) Doctor of Divinity.
Du. Dutch.
Dub. Dublin.
Dwt., dwt. (lat. denarius und engl. weight) pennyweight.
Dyn. Dynamics.

E.

E. East; Eastern (Postal District, London); Edinburgh, Eagle, eagles (Goldstück der Verein. Staaten v. Amerika im Werth von 10 Dollars).
ea. each.
Ebor. (Eboracum) York.
E. C. Eastern Central (Postal District, London).
Ed. Editor; edition.
Edin. Edinburgh.
Eds. Editors.
E. F. East Florida.
e. g. (exempli gratiâ) for example.
E. I. East India, East Indies.
E. I. C. S. East India Company's Service.
Elec. Electricity.
E. Lon. East Longitude.
E. M. F. Electromotive Force.
Ency., Encyc. Encyclopaedia.
Encyc. Brit. Encyclopaedia Britannica.
Eng. England, English.
Engin. Engineering.
Env. Ext. Envoy Extraordinary.
E. P. S. Electrical Power Storage.
Eq., eq. equivalent.
Esq., Esqr. Esquire.
et al. (et alibi) and elsewhere; (et alii or aliae) and others.
et seq. (et sequentes or et sequentia) and the following.
Ex. Example; Exception.
Exc. Excellency; Exception.
Exch. Exchequer; Exchange.
Exec. Executor.
Execx. Executrix.
Exon. (Exonia) Exeter.

F.

F. France; Fellow (ordentliches Mitglied einer literarischen oder wissenschaftlichen Gesellschaft); Folio; Friday.
F., f. Feminine; Franc, francs; Florin, florins; Farthing, farthings; Foot, feet; Fiat.
Fa. Florida.
Fahr. Fahrenheit.

faq. fair average quality.
Far. Farthing.
F. A. S. Fellow of the Society of Arts.
Fcp., fcp. Foolscap.
f. e. for example.
Feb. February.
FF (telegr. Zeichen) „Have finished signalling in figures."
f. i. for instance.
FI (telegr. Zeichen) „Going to signal in figures."
Fin. Finland.
Fl., Flor. Florida.
Fo., fo., Fol., fol. Folio.
For. Foreign.
fow. free on waggon.
Fr. France; Francis; French.
fr. from.
F. R. A. S. Fellow of the Royal Astronomical Society.
F. R. G. S. Fellow of the Royal Geographical Society.
Fri. Friday.
F. R. S. Fellow of the Royal Society.
Ft., ft. Foot, feet.
Fth Fathom.
Fur., fur. Furlong.
fwd. forward.

G.

G. Galvanometer; (telegr. Zeichen) „Go on" („bringen"); Guinea; Gulf.
Ga. Georgia.
G. A. General Assembly.
Gal., gal. gallon, gallons.
G. B. Great Britain.
G. B. & I. Great Britain and Ireland.
G. C. B. Grand Cross of the Bath.
G. D. Grand Duke; Grand Duchess.
Geo. George; Georgia.
Geog. Geography, Geographer.
Geol. Geology, geological, Geologist.
Geom. Geometry, Geometer.
Ger., Germ. German.
Gi., gi. Gill, gills.
G. O. General Order.
Gov. Governor.
Gov. Gen. Governor General.

G. P. Guttapercha.
G. P. O. General Post Office.
GQ (telegr. Zeichen) „Fresh line."
G. R. Georgius Rex.
Gr., gr. Grain, grains.
Gro., gro. Gross.
Gtt., gtt. (gutta, guttae) drops.

H.

H. Hydrogen.
H., h. High; Height; Harbor, Hour, hours.
Hants. Hampshire.
H. B. M. His (Her) Britannic Majesty.
H. C. House of Commons.
h. e. (hoc est, hic est) that is, this is.
H. E. I. C. Honorable East India Company.
H. G. Horse Guards.
Hhd., hhd. Hogshead, hogsheads.
H. I. H. His Imperial Highness.
Hind. Hindoo; Hindostan; Hindostanee.
H. L. House of Lords.
H. M. His (Her) Majesty.
H. M. S. His (Her) Majesty's steamer, ship, service.
Ho. House.
Hon. Honorable.
Hond. Honored.
H. P. Horse Power; Half Pay.
H. R. House of Representatives.
H. R. H. His (Her) Royal Highness.
H. S. H. His (Her) Serene Highness.
Hun., Hung. Hungary, Hungarian.
Hyd. Hydrostatics.
Hydraul. Hydraulics.

I.

Ia. Indiana.
Id. T. Idaho Territory.
i. e. (id est) that is.
II (telegr. Zeichen) Break signal between the address and the body of the telegram.
Ill. Illinois.
Imp. Imperial; Imperator.
In., in. Inch, inches.

Ind. India; Indian; Indiana.
Ind. T., Ind. Ter. Indian Territory.
In loc. (in loco) in its place.
Ins. Inspector; instant.
Ins. Gen. Inspector General.
inst. instant (the present month).
Int., int. Interest.
intens. intensive.
in trans. (in transitu) on the passage.
Io. Iowa.
I. O. U. I Owe You (eine Schuldverschreibung).
i. q. (idem quod) the same as.
Ir. Ireland, Irish.
I. R. O. Internal Revenue Office.
IQ (telegr. Zeichen) „not through".
I. S. Irish Society.
I. T. Indian Territory.
It., Ital. Italian; Italic, Italy.
Itin. Itinerary.

J.

J. Judge.
J. A. Judge Advocate.
J. A. G. Judge Advocate General.
Jam. Jamaica.
Jan. January.
J. C. D. (Juris Civilis Doctor) Doctor of Civil Law.
J. D. (Jurum Doctor) Doctor of Laws.
J. P. Justice of the Peace.
J. U. D. (Juris Utriusque Doctor) Doctor of both Laws (i. e. the Canon and the Civil Law).
Jul. July.
Jun. Juni.
Jun., jun., Junr., junr. Junior.
Jus. Justice.
Jus. P. Justice of the Peace.

K.

K. King; Knight; (Kalium) Potassium.
Kan. Kansas.
K. B. Knight of the Bath; King's Bench.
K. C. B. Knight Commander of the Bath.
Ken., Ky. Kentucky.
K. G. Knight of the Garter.
KK (telegr. Zeichen) „Words to be in parenthesis."
Km. Kingdom.
Knt. Knight.
KQ (telegr. Zeichen) „Say when you are ready."
Ks. Kansas.
Kt. Knight.
Ky. Kentucky.

L.

L. Lady; Latin; Lord; Low; London (nach Titeln); (liber) Book.
L., l. Lake; Lane; League, leagues; Line, lines; Link, links.
L., lb. (libra) a pound in weight.
L., l., £. A pound sterling.
La. Louisiana.
Ladp. Ladyship.
Lb., lb (libra) a pound in weight.
L. C. Lower Canada; Lord Chancellor.
L. C. J. Lord Chief Justice.
Ld. Lord.
L. D. Lady Day (Mariä Verkündigung); Light Dragoons.
Ldp., Lp. Lordship.
Lea., lea. League.
Lec. Leclanché element.
Leg., Legis. Legislature.
L. G. Life Guards.
L. H. A. Lord High Admiral.
L. H. C. Lord High Chancellor.
L. H. T. Lord High Treasurer.
L. I. Long Island; Light Infantry.
Lib. Librarian.
Lib., lib. (liber) book.
Lieut. Gov. Lieutenant Governor.
Lit. Literature, Literary.
Lit., lit. Literally.
LL (telegr. Zeichen) „Words to be underlined."
LL. B. (Legum Baccalaureus) Bachelor of Laws.
LL. D. (Legum Doctor) Doctor of Laws.
L. L. I. Lord Lieutenant of Ireland.
Lon., Lond. London.
Lon., lon., Long., long. Longitude.

Lou., La. Louisiana.
Lp., Ldp. Lordship.
L. P. S. Lord Privy Seal.
L. S. D. or **l. s. d.** (Libra, Solidi, Denarii) Pounds, Shillings, Pence.
Lv., lv. Livres.

M.

M. Monday; Monsieur; Morning; (Mille) Thousand; (Meridies) Meridian or Noon.
m. married.
M., m. Masculine; Moon; Month, months; Minute, minutes; Mill, mills; Mile, miles; (Manipulus) a handful; (Mensura) Measure; by measure.
M. A. Military Academy; Master of Arts ſ. **A. M.**
Ma., Minn. Minnesota.
Mach. Machinery.
Mad., Madm. Madam.
Mag. Magazine.
Manuf. Manufacturing.
Mar. March; Maritime.
Mass., Ms. Massachusetts.
Math. Mathematics; Mathematician.
M. B. (Medicinae Baccalaureus) Bachelor of Medicine ſ. **B. M.**
M. C. Member of Congress; Medical Certificate.
Mch. March.
M. D. (Medicinae Doctor) Doctor of Medicine.
M. D. Metropolitan District.
Md. Maryland.
Mdlle. Mademoiselle.
M. E. Military or Mechanical Engineer; Most Excellent.
ME (telegr. Zeichen) Greenwich mean time.
Me. Maine.
Mech. Mechanics.
Mem. Memorandum, memoranda.
Messrs. or **MM.** (Messieurs) Gentlemen; Sirs.
Metal. Metallurgy.
Meteor. Meteorology.
Met. Dis. Metropolitan District.
Mex., Mexic. Mexico, Mexican.

Mf., (*pl* mfs.) Microfarad.
Mfg., mfg. Manufacturing.
Mg. Magnesium.
M. G. Major General.
Mho von S. W. Thomson i. J. 1883 vorgeſchlagener Ausdruck zur Bezeichnung der Einheit der Leitungsfähigkeit.
M. Hon. Most Honorable.
Mi. Mississippi.
Mi., mi. Mill, mills.
Mich. Michigan; Michaelmas.
Mil. Military.
Min. Mineralogy.
Min., min. Minute, minutes.
Minn. Minnesota.
Min. Plen. Minister Plenipotentiary.
Miss., Mi. Mississippi.
Mlle. Mademoiselle.
MM. Their Majesties; (Messieurs) Gentlemen, Sirs.
Mme. Madame.
Mn. Manganese; Michigan.
M. O. Money Order.
Mo. Missouri.
Mo., mo. Month.
Mod. Modern.
Mon., Mond. Monday.
Mons. Monsieur, Sir.
Morn., morn. Morning.
Mos., mos. Months.
M. P. Member of Parliament; Member of Police.
M. P. C. Member of Parliament in Canada.
M. P. P. Member of the Provincial Parliament.
MQ (telegr. Zeichen) „Wait."
Mr. Master, Mister.
M. R. A. S. Member of the Royal Asiatic Society; Member of the Royal Academy of Science.
M. R. C. C. Member of the Royal College of Chemistry.
M. R. G. S. Member of the Royal Geographical Society.
M. R. I. Member of the Royal Institution.
M. R. I. S. Member of the Royal Irish Society.
Mrs. Mistress.
MS. Manuscript.
MSS. Manuscripts.

Mt. Mount, Mountain.
Mts. Mountains.
Mus. Museum; Music.

N.

N. Noon; North; Northern (Postal District, London); Note, Name; New; Number; Nitrogen.
N., n. Noun; Neuter; Nail, nails.
N. A. North America, North American.
Na. Nebraska.
Na., na. Nail, nails.
Nat. Natural; Natal; National.
Nat. Hist. Natural History.
Naut. Nautical.
N. B. North Britain, North British; New Brunswick; (Nota bene) Note well, take notice.
N. C. North Carolina.
N. E. North East; Northern Eastern (Postal District, London); New England.
N. E. No effects, „nicht akzeptirt, proteſtirt" (wird auf proteſtirte Wechſel geſchrieben, wenn der Traſſant die Valuta des Wechſels nicht bei dem Traſſaten vorher deponirt hat).
Neb. Nebraska.
Nem. con., nem. con. (Nemine contradicente) No one contradicting; unanimously.
Nem. diss. (Nemine dissentiente) No one dissenting; unanimously.
Neth. Netherlands.
netfob. netto free on board.
New M. New Mexico.
N. F. Newfoundland.
N. G. New Granada.
N. H. New Hampshire; New Haven.
Ni. Nickel.
N. J. New Jersey.
N. l., n. l. (Non liquet) The case is not clear.
N. L., N. Lat. North Latitude.
N. M. New Mexico.
NN. (telegr. Zeichen) „Nothing more coming".
No., no. Number.
N. O. New Orleans.

Non obst., non obst. (Non obstante) Notwithstanding.
Non seq., non seq. (Non sequitur) It does not follow.
Nor., norm. Norman.
Norw. Norway; Norwegian.
Nos., nos. Numbers.
Nov. November.
N. P. New Providence, Notary Public.
N. R. North River.
N. S. Nova Scotia; New Style (since 1752).
N. T. Nevada Territory.
N. u., n. u. Name, names unknown.
Num., Numb. Numbers.
Numis. Numismatics.
N. W. North West; Northern Western (Postal District, London).
N. W. T. North West Territory.
N. Z., N. Zeal. New Zewland.

O.

O. Ohio; Oxygen; Old; (beim Telegr.=Bau:) the pole of ordinary scantling.
Ob., ob. (Obiit) Died.
Obj., obj. Objective; Objection.
Obs. Observatory.
Obs., obs. Obsolete; Observation.
Obt., obdt. Obedient.
Oct. October.
OK (unter den Telegraphiſten allgemein übliches, aber dienſtlich nicht gebilligtes Zeichen für „all right" — angeblich dadurch entſtanden, daß ein unorthographiſcher Yankee den Ausdruck „all correct" abkürzen wollte und der Meinung war, derſelbe würde Oll Korrect geſchrieben). Der Ausdruck „OK" iſt auch vielfach in die gewöhnliche Sprache übergegangen)
Ol. (Oleum) Oil.
O. M. Old Measurement.
On. Oregon.
Op. Opposite.
Opt. Optics.
Opt., opt. Optative.
Or. Oregon.

Ord. Ordinary.
O. S. Old Style (previous to 1752).
Os. Osmium.
O. T. Oregon Territory.
O. V. Oil of Vitriol.
Oxf. Oxford.
Oxon. (Oxonia) Oxford.
Oz., oz. Ounce, ounces.

P.

P., p. Page; Part; Participle; Pole; Phosphorus; Pint.
Pa. Pennsylvania.
paf. packing, assurance, freight.
Par. Paragraph.
Parl. Parliament, Parliamentary.
Part., part. Participle.
Pay^{t.}, pay^{t.} Payment.
Pb. (Plumbum) Lead.
P. B. (Philosophiae Baccalaureus) Bachelor of Philosophy.
P. C. Privy Council, or Councilor.
P. D. (Philosophiae Doctor) Doctor of Philosophy.
Pd., pd. Paid.
P. E. I. Prince Edward's Island.
Penn. Pennsylvania.
Pent. Pentecost.
Per., Pers. Persia, Persian.
Per an., per an. (Per annum) By the year.
perh. perhaps.
Pers., pers. Person.
Pg. Portuguese.
Phar. Pharmacy.
Ph. B. (Philosophiae Baccalaureus) Bachelor of Philosophy.
Ph. D. (Philosophiae Doctor) Doctor of Philosophy.
Phil. Philip; Philosophy; Philosopher; Philosophical.
Phys. Physics; Physiology.
Pk., pk. Peck.
Pks., pks. Pecks.
P. L. Poet Laureate.
Pl., pl. Place; Plate; Plural.
P. L. C. Poor Law Commissioners.
Plff. Plaintiff.
P. M. Post Master; (Post meridiem) Afternoon.
P. M. G. Post Master General.
P. O. Post Office.
P. O. D. Post Office Department.
Pol. Polish.
P. O. O. Post Office Order.
Pop., pop. Population; Popularly.
Port. Portugal; Portuguese.
pp. Pages.
Pph., pph. Pamphlet.
Pr. Prince; Priest.
P. R. Porto Rico.
P. R. A. President of the Royal Academy.
Pref., pref. Prefix; Preference.
Pres. President.
Pres., pres. Present.
Prin. Principles.
prin. principally.
Print. Printing.
Pro tem., pro tem. (Pro tempore) for the time being.
Prox., prox. (Proximo) next or of the next month.
P. R. S. President of the Royal Society.
Prus. Prussia, Prussian.
P. S. Permanent Secretary; Privy Seal; Postscript.
Pt., pt. Pint; Part; Payment; Point; Port.
P. T., p. t. Post Town.
Pub. Public; Published; Publisher.
Pub. Doc. Public Documents.
P. v., p. v. Post village.
Pwt., pwt. Pennyweight.

Q.

Q. Question.
Q., q. (Quadrans) A farthing.
Q., Qu. Query; Question; Queen.
Q. B. Queen's Bench.
Q. C. Queen's Council or Counsel; Queen's College.
Q. l., q. l. (Quantum libet) As much as you please.
Qm., qm. (Quomodo) By what means.
Q. Mess. Queen's Messenger.
Q. P., q. pl. (Quantum placet) As much as you please.
Qr., qr. Quarter (28 pounds); Farthing; 25 cents American money; Quire.
Qrs., qrs. Quarters; Farthings; Quires.

Anhang.

Q. S. Quarter Sessions.
Qt., qt. Quart; Quantity.
Qts., qts. Quarts.
Qu. Queen; Question.
Qu., Quar. Quarterly.
Ques. Question.
Q. v., q. v. (Quod vide) which see.
Qy. Query.

R.

R. Railway; Rare; Resistance; (Rex) King; (Regina) Queen.
R., r. Rood, roods; Rod, rods; River; Red; Resides; Retired.
R. A. Royal Academy or Academician; Russian America.
R. & A. G. Receiver and Accountant General.
RD (telegr. Zeichen) Signal of acknowledgment.
Re, re (lat.) in Sachen.
R. E. Royal Engineers; Royal Exchange; Right Excellent.
Rec'd. Received.
Recpt. Receipt.
Rec. Sec. Recording Secretary.
Rect. Rector; Receipt.
Ref. Reference.
Reg., Regr. Register; Registrar; Regular.
Reg., Regt. Regent.
Rem. Remark, remarks.
Rep. Representative; Republic; Report; Reporter.
Repub. Republic.
Res. Resistance.
Retd. Returned.
Rev. Review; Revenue; Reverend.
Revd Reverend.
Revs. Reverends.
R. H. A. Royal Hibernian Academy.
Rhet. Rhetoric.
R. I. Rhode Island.
Riv., riv. River.
R. L. I. Registered Letter Irregularity.
R. M. Royal Marines; Royal Mail; Resident Magistrate.
R. M. S. Royal Mail Steamer; Railway Mail Service.
R. N. Royal Navy.
R. O. Receiving Office.

Ro (Recto) right-hand page.
RQ (telegr. Zeichen) „Repetition or correction required".
RR (telegr. Zeichen) „Words to be in inverted commas."
R. R. Railroad; Right Reverend.
R. S. Railway Station; Recording Secretary.
R. S. A. Royal Society of Antiquaries; Royal Scottish Academy.
R. S. L. Royal Society of London.
R. S. O. Railway Sub Office.
R. S. S. (Regiae Societatis Socius) Fellow of the Royal Society.
Rt. Right.
Rt. Hon. Right Honorable.
Rt. Rev. Right Reverend.
Russ. Russia, Russian.

S.

S. (Telegr. Bau) a stiff pole (besonders starke Stange); (pref.) An ordinary telegram, including a government telegram not entitled to priority; Sign; South; Southern (Postal District, London); Supplement; Sunday; Saturday.
S., s. Second; Shilling; See; Series; Son.
SA (pref.) A telegram franked by a railway pass.
S. A. South America; South Australia; South Africa.
Sat. Saturday.
SB (pref.) A government telegram entitled to priority.
S. B. South Britain (i. e. England and Wales).
S. C. South Carolina.
sc. (scilicet) to wit; namely.
Sch., sch., Schr., schr. Schooner.
Sci. Science.
scil. (scilicet) f. sc.
Scot. Scotland, Scotch, Scottish.
S. E. South East; South Eastern (Postal District, London).
Sec. Secretary.
Sec., sec. Second; Section.
Sec. Leg. Secretary of Legation.
Sen. Senate; Senator; Senior.

Sep., Sept. September.
Seq., seq. (Sequentes, sequentia) The following, the next.
Ser. Series.
Serb. Serbian.
Serv., Servt. Servant.
SG (*pref.*) An urgent service telegram.
Sh. Shunt.
Sh., sh., S., s. Shilling.
S. K. Short circuit key.
S. L. Solicitor at Law.
S. Lat., S. L. South Latitude.
Sld., sld. Sailed.
S. O. Sub Office.
Soc. Society.
Sp. Spain, Spanish.
SP (*pref.*) A press or news telegram.
Sq., sq. Square.
Sq. ft., sq. ft. Square feet.
Sq. in., sq. in. Square inches.
Sq. m., sq. m. Square miles.
Sq. r., sq. r. Square rods.
Sq. yds., sq. yds. Square yards.
Sr. Sir; Senior.
SR (*pref.*) A telegram to be repeated.
SRP (*pref.*) A telegram to which a reply has been prepaid.
SS., ss. (Scilicet) f. sc.; Steamship.
SSG (*pref.*) An urgent service telegram, of which the code time and number of words must be signalled.
St. Street; Stone; Strait.
S. T. Sorting Tender.
Stat. Statute, Statutes; Statuary.
Ster., Stg. Sterling.
Su. Sunday.
SU (*pref.*) An ordinary service telegram.
Sun., Sund. Sunday.
Sup. Superior; Supplement.
Sup. C. Superior Court.
Super. Superior.
Supp. Supplement.
Supt. Superintendent.
Surg. Gen. Surgeon General.
Surv. Surveying, Surveyor.
S. W. South West; South Western (Postal District, London).
Sw. Swedish, Sweden.
Switz. Switzerland.

T.

T. (Stempel auf unfrankirten oder unzureichend frankirten Briefen); Tuesday.
T., t. Town; Township; Territory; Ton.
Tal. qual., tal. qual. (Talis qualis) Just as they come, average quality.
Tan., tan. Tangent.
T. E. Topographical Engineers.
Ten., Tenn. Tennessee.
Term., term. Termination.
Tex. Texas.
Th., Thurs. Thursday.
TI (telegr. Zeichen) Daily time signal.
Tit. Title.
T. O. Turn over.
Tonn., tonn. Tonnage.
T. P. O. Travelling Post Office.
TQ (telegr. Zeichen) „Am I through?"
Tr. Translation; Translator; Transpose; Treasurer; Trustee.
Trans. Transactions; Translated, Translation, Translator.
Trav. Travels.
Ts. Texas.
Tu., Tues. Tuesday.
Turk. Turkey, Turkish.

U.

U. C. Upper Canada.
U. E. I. C. United East India Company.
U. G. R. R. Underground Railroad.
Uh. Utah.
U. J. D. (Utriusque Juris Doctor) Doctor of both Laws (i. e. the Canon and the Civil Law).
U. K. United Kingdom; A kind of the Daniell Battery.
um. Unmarried.
Univ. University.
Univ., univ. Universally.
Up. Upper.
UQ (telegr. Zeichen) „Attend to other instrument."

Anhang. 311

u. s. (ut supra) as above.
U. S. A. United States of America; United States Army.
U. S. M. United States Mail; United States Marine.
U. S. M. A. United States Military Academy.
U. S. N. United States Navy.
U. S. S. United States Senate; United States Ship (Steamer).
Usu., usu. Usually.
U. T. Utah Territory.

V.

V. Victoria; Viscount; Village; Volume; (Vide) see.
V., v., Vs., vs. (Versus) against (bei Prozessen: „N. N. vs. U. S." N. N. gegen die Vereinigten Staaten).
V. A. Vice Admiral.
Va. Virginia.
V. C. Vice Chancellor.
V. D. L. Van Diemen's Land.
VE (telegr. Zeichen) completion of message, the „understand" signal.
Ver. Vermont.
v. g. (verbi gratiâ) for example.
Vice Pres. Vice President.
Vis., Visc. Viscount.
Viz., viz. (videlicet) namely; to wit.
Vo (Verso) left-hand page.
V. P. Vice President.
V. R. Victoria Regina.
Vs., vs. (Versus) s. v.
V. S. (beim Telegr.=Bau) a very stiff pole.
Vt. Vermont.

W.

W. West; Western (Postal District, London); William; Wednesday.
W., w. Week.
W. A. West Africa; West Australia.
W. B. Wheatstone's Bridge.
W. C. Western Central (Postal District, London).
Wed. Wednesday.

W. F. West Florida.
Whf., whf. Wharf.
W. I. West India, West Indies.
Wis., Wisc. Wisconsin.
Wk., wk. Week.
W. Lon. West Longitude.
Wp. Worship.
Wpful. Worshipful.
W. R. William (Rex) King; West Riding.
W. T. Washington Territory.
Wt., wt. Weight.
W. Va. West Virginia.

X.

X (pref.) An ordinary telegram, including a government telegram not entitled to priority.
XA (pref.) A telegram franked by a railway pass.
XB (pref.) A government telegram entitled to priority.
XL (pref.) A telegram handed in at one office in London for transmission by wire through the Central Telegraph Office to another Office in London.
Xm., Xmas. Christmas.
XP (pref.) A press or news telegram.
XR (pref.) A telegram to be repeated.
XRP (pref.) A telegram to which a reply has been prepaid.

Y.

Y., yr. Year.
Y. B., Yr. B. Year Book.
Yd., yd. Yard.
Yds., yds. Yards.
YQ (telegr. Zeichen) „Two or more stations."
Yr., Yrs. Your, Yours.

Z.

Z. An insulator for telegraph lines made out of one piece.
ZM (telegr. Zeichen) „Weather."
ZQ (telegr. Zeichen) „Attend to switch."

MIX
Papier aus verantwortungsvollen Quellen
Paper from responsible sources
FSC® C105338

If you have any concerns about our products,
you can contact us on
ProductSafety@springernature.com

In case Publisher is established outside the EU,
the EU authorized representative is:
**Springer Nature Customer Service Center GmbH
Europaplatz 3, 69115 Heidelberg, Germany**

Printed by Libri Plureos GmbH
in Hamburg, Germany